The Systematics Association Special Volume Series 75

Unravelling the algae

the past, present, and future of algal systematics

The Systematics Association Special Volume Series

Series Editor

Alan Warren

Department of Zoology, The Natural History Museum

Cromwell Road, London SW7 5BD, UK

The Systematics Association promotes all aspects of systematic biology by organizing conferences and workshops on key themes in systematics, publishing books and awarding modest grants in support of systematics research. Membership of the Association is open to internationally based professionals and amateurs with an interest in any branch of biology including palaeobiology. Members are entitled to attend conferences at discounted rates, to apply for grants and to receive the newsletters and mailed information; they also receive a generous discount on the purchase of all volumes produced by the Association.

The first of the Systematics Association's publications *The New Systematics* (1940) was a classic work edited by its then-president Sir Julian Huxley, that set out the problems facing general biologists in deciding which kinds of data would most effectively progress systematics. Since then, more than 70 volumes have been published, often in rapidly expanding areas of science where a modern synthesis is required.

The *modus operandi* of the Association is to encourage leading researchers to organize symposia that result in a multi-authored volume. In 1997 the Association organized the first of its international Biennial Conferences. This and subsequent Biennial Conferences, which are designed to provide for systematists of all kinds, included themed symposia that resulted in further publications. The Association also publishes volumes that are not specifically linked to meetings and encourages new publications in a broad range of systematics topics.

Anyone wishing to learn more about the Systematics Association and its publications should refer to our website at http://www.systass.org

Other Systematics Association publications are listed after the index for this volume.

Frontispiece showing the diversity of the algae developed by Reen Pilkington from images contributed by authors

The Systematics Association Special Volume Series 75

Unravelling the algae

the past, present, and future of algal systematics

Edited by

Juliet Brodie

Natural History Museum
London, UK

Jane Lewis

University of Westminster
London, UK

 **CRC Press**
Taylor & Francis Group
Boca Raton London New York

CRC Press is an imprint of the
Taylor & Francis Group, an **informa** business

CRC Press
Taylor & Francis Group
6000 Broken Sound Parkway NW, Suite 300
Boca Raton, FL 33487-2742

First issued in paperback 2019

ISBN-13: 978-0-8493-7989-5 (hbk)
ISBN-13: 978-0-367-38819-5 (pbk)

Visit the Taylor & Francis Web site at
http://www.taylorandfrancis.com

and the CRC Press Web site at
http://www.crcpress.com

Table of Contents

Foreword

This is a fascinating and challenging book about a group of diverse living organisms that span the whole course of evolution. Numerous and extremely widely distributed across the planet, they are titans in the global control of the biosphere and have been main players and drivers in environmental change. Yet most of us don't even know of their existence, let alone their crucial importance to everyday life in the 21st century. So what more appropriate time for this publication than during the Tercentenary of Linnaeus, the scientist who first persuaded the world of natural historians that providing every distinct organism with a name was not only the foundation of scientific understanding but also of the progress of human economy. Linnaeus not only named names but was a pioneer in natural economy, and if alive today I am sure would play the stock market in the burgeoning enterprise of algal products. Every child lucky enough to visit the ocean will discover seaweeds, and almost certainly partake of their products. They will enjoy ice cream, snack on processed food, and use a growing number of everyday items, including antibiotics, all of which depend on organic chemicals extracted from these strange slippery plants. A rainbow of green, brown, and red living colour and a myriad of shapes and forms have sown the seeds of wonder, exploration, ecology, and taxonomy in many youngsters. Some such individuals at the start of their scientific careers and other well-established scientists have helped craft this wonderful book. Now in the laboratory and using new technologies like chromatography and genetic fingerprinting (both pioneered using alginates), their innermost secrets of ancestry and phylogeny are being revealed.

In addition to the macroscopic seaweeds, there are legions of minute algae whose intricate beauty is revealed only under the microscope, and with magnifications now probing the bounds of the atom, the information to hand is encyclopaedic. Information there to hone our understanding of life on earth and the staggering magnitude algae have played across evolutionary time. The happy band of phycologists already know that in their strange world, enslavement of once free-living organelles helped produce more complex cells and more complex organisms with greater capabilities. The minutiae of change have allowed the construction of the world's first stromatolites, solar-powered, self-repairing sea defences, and the shape of more massive things to come. There are reefs that are home to thousands of different fishes and other animals.

Extremophiles thrive where other life forms cannot exist, in heat and impossible concentrations of toxic chemicals. Single cells that gained a new individuality of purpose became eggs or sperm, and thus the carriers of diverse genes that shook up evolution, and stirred it with the lust of sex. Whiplash flagella or the sensual rhythm of beating cilia speed gametes to their chosen mates. The Oceans were not enough—the land was their final goal.

In our knowledge of the algae, the subject of this book, lie the answers to many important questions science has dreamed of answering for decades, so it should take pride of place in your library. A *vade mecum* for the future, which can only make us wonder what else this diverse group have hidden in and around their genes. I was very proud when one of my grandchildren was named Luca. The **L**ast **U**niversal **C**ommon **A**ncestor of all living things, whose descendants helped charge the biosphere with free oxygen trying and testing new armamentaria of antioxidants, enzymes, and substrates that are as important to us today as they were almost 4 billion years ago. Half bacteria, half alga, half plant, half animal, we do not know. But one day we will, and that is the excitement of this book.

Professor David Bellamy

Preface

The chapters in this book came about as a result of a two-day symposium in 2006 "Unravelling the algae: the past, present and future of algal molecular systematics," which was convened in order to review the state of the subject and assess the impact of molecular tools on the taxonomy of virtually all the different algal groups and to produce a volume setting out this material. The symposium included 16 talks by invited speakers, all of whom (sometimes with assistance of coauthors) contributed papers to the book. In reviewing the symposium and putting the material together for the book, we were keen to fill a gap we identified and were pleased that the chapter on the euglenophytes was so very willingly contributed by the chapter's authors during the production of the book. The symposium also attracted 27 posters, which also covered almost all the algal groups, and topics were diverse, ranging from the general to the specific, covering macroalgae to ultra-plankon and from phylogenies to the development of identification tools using molecular techniques. The meeting also provided the opportunity for some lively discussion that allowed participants to debate issues as they arose and to speculate on the shape of the future. A common thread throughout the talks and an overwhelming outcome from the resulting discussions was that a range of traditional and molecular approaches are required coupled with other techniques including the use of transmission electron microscopy in order to support interpretation.

We were able to run this symposium because of the generous support of the Systematics Association, Linnean Society, British Phycological Society, and the Natural History Museum and the support of the publisher, CRC. These organisations play a crucial role in the advancement of science via their sponsorship and resources, and this should never be underestimated. We were able to create this timely and unique book because of the intellect and generosity of time and materials from the authors, some of whom also peer reviewed other chapters, and many of whom supplied the illustrations for the book cover and the frontispiece. It was their tremendous spirit of endeavour that enabled us to achieve the goals of this project.

Acknowledgments

The Systematics Association, Natural History Museum, London (NHM), Linnean Society, British Phycological Society, CRC Press, and Abgene all provided sponsorship or support for the meeting. We acknowledge the assistance of Mandy Walton, Elinor and Sarah Rowing (University of Westminster), Julie Gray and Barbara Rinkel (NHM) for administrative, technical, and secretarial support and the students and staff of both institutions who assisted in the smooth running of the symposium. Artistic inspiration was provided by Reen Pilkington. Chapters were peer reviewed by Margaret Clayton, Katharine Evans, Paul Hayes, Lucien Hoffmann, Max Hommersand, David John, Marina Montresor, Gianfranco Novarino, Øjvind Moestrup, Barry Leadbeater, Fabio Rindi, and Max Taylor. We thank them for the efforts in improving the manuscripts. We also had excellent advice from members of the Systematics Association Council, in particular Alan Warren the Systematics Association Special Volumes Series Editor and Colin Hughes the Treasurer. We also had the support of two book proposal reviewers, Barry Leadbeater and Wytze Stam. We would also like to thank the production team at CRC, in particular John Sulzycki, Pat Roberson and Amy Rodriguez for their work. Finally, we are indebted to all the authors for their contributions and for enabling us to meet our manuscript submission deadline.

The Editors

Juliet Brodie is a senior researcher in marine algae at the Natural History Museum, London. She graduated from the University of Bristol with a degree in botany and zoology and has studied macroalgae for over 25 years. Her doctoral research was undertaken in Galway (the National University of Ireland) and focused on life histories, crossing studies, morphology, and photoperiodic responses of red algae. She continued as a postdoctoral researcher with the Smithsonian Institution, USA, undertaking taxonomic studies on tropical red algae. She embraced molecular techniques in the early days of their use in algal taxonomy and notably applied them to *Porphyra*, a large, economically important, taxonomically problematic genus of red algae. She is a leading authority in the world on the taxonomy of this genus and has co-authored a book on the subject. She specialises in molecular systematics, taxonomy, and ecology of red algae, and her experience covers many different parts of the world. Her other interests include the biodiversity and conservation of marine algae.

Jane Lewis is the Dean of the School of Biosciences at the University of Westminster. She graduated from the University of Wales, Bangor with a degree in marine biology and oceanography and since has studied microalgae for some 25 years. Her doctoral research was carried out in Scotland at the Scottish Marine Biological Association and Royal Holloway and Bedford New College (University of London), and was focused on the ecology of harmful algal blooms, particularly dinoflagellates. Studies of this nature crucially depend on accurate identification of potentially harmful species, and so her interest in microalgal taxonomy began. Her studies of dinoflagellates include a particular focus on the cyst stage of the life cycle, including their role in the ecology of harmful algal blooms and their use in the interpretation of the fossil record. Establishing cyst–theca relationships and working with *Alexandrium* species have, more recently, led to the application of molecular techniques in her research. However, she still favours "knowing" the organisms she is working on and encourages students to become proficient in microscopic as well as molecular techniques.

Contributors

Andrew E. Allen
Département de Biologie
Ecole Normale Supérieure
Paris, France

Robert A. Andersen
Provasoli-Guillard National Center for Culture
  of Marine Phytoplankton
Bigelow Laboratory for Ocean Sciences
West Boothbay Harbor, Maine

Chris Bowler
Cell Signalling Laboratory
Stazione Zoologica
Napoli, Italy
and
Département de Biologie
Ecole Normale Supérieure
Paris, France

Juliet Brodie
Department of Botany
Natural History Museum
London, United Kingdom

Thomas Cavalier-Smith
Department of Zoology
University of Oxford
Oxford, United Kingdom

Federica Cerino
Stazione Zoologica A. Dohrn
Naples, Italy
and
Department of Marine Sciences
Polytechnic University of Marche
Ancona, Italy

Niels Daugbjerg
Department of Phycology
Biological Institute
University of Copenhagen
Copenhagen, Denmark

Olivier de Clerck
Phycology Research Group
Ghent University
Ghent, Belgium

Bruno de Reviers
Département Systématique et Evolution
Muséum National d'Histoire Naturelle
Paris, France

Charles F. Delwiche
Cell Biology and Molecular Genetics
University of Maryland at College Park
College Park, Maryland

Stefano G.A. Draisma
Nationaal Herbarium Nederland
Universiteit Leiden Branch (NHN-L)
Phanerogams and Cryptogams of the
  Netherlands and Europe (PCNE)
Leiden, The Netherlands

Bente Edvardsen
Department of Biology
University of Oslo
Oslo, Norway

Nermin Adel El Semary
Department of Botany and Microbiology
University of Helwan
Cairo, Egypt

Katharine M. Evans
Royal Botanic Garden
Edinburgh, Scotland

Mark A. Farmer
Department of Cellular Biology
University of Georgia
Athens, Georgia

John D. Hall
Cell Biology and Molecular Genetics
University of Maryland at College Park
College Park, Maryland

Paul K. Hayes
School of Biological Sciences
University of Bristol
Bristol, United Kingdom

Ken-ichiro Ishida
Graduate School of Life and Environmental
  Sciences
University of Tsukuba
Tsukuba, Japan

Uwe John
Alfred Wegener Institute
Bremerhaven, Germany

Frederik Leliaert
Phycology, Research Group
Ghent University
Ghent, Belgium

Jane Lewis
School of Biosciences
University of Westminster
London, United Kingdom

Christine A. Maggs
School of Biological Sciences
Medical Biology Centre
Queen's University Belfast
Belfast, United Kingdom

David G. Mann
Royal Botanic Garden
Edinburgh, Scotland

Linda K. Medlin
Alfred Wegener Institute
Bremerhaven, Germany

Katja Metfies
Alfred Wegener Institute
Bremerhaven, Germany

Øjvind Moestrup
Department of Phycology
Biological Institute
University of Copenhagen
Copenhagen, Denmark

Jeanine L. Olsen
Department of Marine Benthic Ecology
  and Evolution
Centre for Ecological and Evolutionary Studies
University of Groningen
Haren, The Netherlands

Shuhei Ota
Graduate School of Natural Science
  and Technology
Kanazawa University
Kanazawa, Japan

Thomas Pröschold
Scottish Association for Marine Science
Culture Collection of Algae and Protozoa
Dunstaffnage Marine Laboratory
Dunbeg by Oban, Scotland

Florence Rousseau
Département Systématique et Evolution
Université Pierre et Marie Curie (Paris VI)
Paris, France

Patricia Sánchez-Baracaldo
School of Biological Sciences
University of Bristol
Bristol, United Kingdom

Richard E. Triemer
Department of Plant Biology
Michigan State University
East Lansing, Michigan

Heroen Verbruggen
Phycology Research Group
Ghent University
Ghent, Belgium

David M. Williams
Department of Botany
Natural History Museum
London, United Kingdom

Akinori Yabuki
Graduate School of Life and Environmental
  Sciences
University of Tsukuba
Tsukuba, Japan

Adriana Zingone
Stazione Zoologica A. Dohrn
Naples, Italy

Abbreviations

18S	Small subunit of the ribosomal operon
26S	Large subunit of the ribosomal operon
5.8S rDNA	Ribosomal DNA located between the SSU and LSU rDNA
ADP	Adenosine diphosphate
AFLP	Amplified fragment length polymorphism
ALL	Alariaceae, Lessoniaceae, Laminariaceae
ARB	Software for handling genetic data, named for "arbor," Latin for tree
ATP	Adenosine triphosphate
atpB	Gene coding for beta subunit of ATPase
BOLD	Barcode of Life Data System
bp	Base pair
CBC	Compensatory base changes
CCAP	Culture collection of algae and protozoa
CCM	Carbon concentrating mechanism
CCMP	Center for Culture of Marine Phytoplankton
CCW	Counterclockwise
CFB	Cytophaga-Flavobacterium-Bacteroides
CM	Cytoplasmic membrane
cox1	Mitochondrial cytochrome with oxidase subunit 1
CPS	Carbomyl phosphate synthase
CW	Clockwise
DNA	Deoxyribonucleic acid
DO	Directly opposite
DSOS	Dictyotales, Sphacelariales, Onslowiales, Syringodermatales
DSP	Diarrhoeic shellfish poisoning
ER	Endoplasmic reticulum
EST	Expressed sequence tag
F	Phenylalanine
FA	Fatty acid
FBA	Fructose 1,6-bisphosphate
GAPDH	Glyceraldehyde 3-phosphate dehydrogenase
GTPase	Guanine triphosphate
ICBN	International Code of Botanical Nomenclature
ICZN	International Code of Zoological Nomenclature
IM	Inner membrane
IPC	Internal periplast component
ITS	Internal transcribed spacers
ITS-1	Internal transcribed spacer region 1 located between SSU and 5.8S rDNA
ITS-2	Internal transcribed spacer region 2 located between 5.8S and LSU rDNA
KT boundary	Cretaceous–Tertiary boundary
LBA	Long-branch attraction
LGT	Lateral gene transfers
LHC	Light-harvesting complex
LINEs	Long interspersed nuclear elements

LPS	Lipopolysaccharide
LSU rDNA	Large subunit of the ribosomal operon
MB	Bayesian method
ML	Maximum likelihood
MLS	Multilayered structure of the flagellar apparatus in streptophyte green algae and higher plants
MP	Maximum parsimony
mRNA	Messenger ribonucleic acid
MY	Million years
NERC	Natural Environment Research Council
NJ	Neighbour joining
NSP	Neurotoxic shellfish poisoning
OM	Outer membrane
PCR	Polymerase chain reaction
PPM	Periplastid membrane
PPP	Periplastid processing peptidase
PS	Periplastid space
psbA	Chloroplast gene coding for core protein of photosystem II
PSP	Paralytic shellfish poisoning
RAPD	Random amplification of polymorphic DNA
***rbc*L**	Large subunit of ribulose bisphosphate carboxylase gene
rDNA	Ribosomal DNA
rDNA-ITS	Ribosomal DNA of internally transcribed spacer region
RFLP	Restriction fragment length polymorphism
RNA	Ribonucleic acid
rpL36	Gene encoding ribosomal protein
rRNA	Ribosomal RNA
SEM	Scanning electron microscope/microscopy
SINEs	Short interspersed nuclear elements
SNPs	Single nucleotide polymorphisms
SP	Signal peptidase
SPC	Superficial periplast component
SRP	Signal recognition particle
s.s.	*Sensu stricto*
SSU	Small subunit
SSU rDNA	Small subunit of the ribosomal operon
SZN	Stazione Zoologica Napoli
TAT	Twin arginine transferases
TEM	Transmission electron microscope/microscopy
TP	Transit peptidase
***tuf*A**	Gene-encoding protein synthesis elongation factor
UV	Ultraviolet
VLE	Vegetative lytic enzymes (sporangia lytic enzymes/autolysins)
Z	Zygote

1 Introduction

Juliet Brodie and Jane Lewis

CONTENTS

HOW THE BOOK CAME ABOUT

In April 2006, a unique occasion took place at the Natural History Museum, London, when phycologists working on the systematics of virtually all the different groups of algae from around the world came together for a symposium entitled "Unravelling the algae: the past, present and future of algal molecular systematics." The concept of the symposium was to review the state of algal systematics and was timely as it was the first such event to take place in the light of the impact of molecular studies on the subject over the last 20 years.

WHAT ARE THE ALGAE?

So what are the algae and why, in comparison to many other groups, have they often been neglected until recently? It is generally agreed that algae are photosynthetic organisms other than the land plants (although see Delwiche, chapter 2, this volume). It is often agreed that they are just the eukaryotic representatives, although the prokaryotic cyanobacteria have frequently been included with the algae. Indeed in this volume, we have included the cyanobacteria because they are so much a part of the "algal world" and they provide such fascinating insights into the genetics of organisms. In addition to their importance in endosymbiosis, the ingestion (enslavement) of a prokaryotic cyanobacterium is considered to be the basis for the evolution of the eukaryotic algae and consequently the land plants.

In 1753 the algae were placed in the Cryptogamia along with other "non-flowering plants" (Linnaeus, 1753) and since that time scientists have been intrigued by this heterogeneous group of photosynthetic organisms with an estimated 350,000 known species (Brodie & Zuccarella, 2007). They range in size from single cells to giant kelps over 60 m long, and some species have "animal" characteristics. The study of algae has been punctuated by a series of major breakthroughs, for example, the use of colour in the 19th century to split the algae into groups (Lamouroux, 1813; Harvey, 1836), the advent of the electron microscope which led to many new discoveries such as the structure of the flagellum, or scales in, e.g. the cryptomonads and chrysophytes (see Cerino and Zingone, chapter 11, and Andersen, chapter 15, this volume) and theories of endosymbiosis (see Delwiche, chapter 2, and Cavalier-Smith, chapter 3, this volume), but it has been since the discovery of the structure of DNA followed by the development of molecular tools which have not only extended our understanding of the immense diversity of algae but also enabled a revolution of ideas on their taxonomy and classification.

The reasons for our paucity of knowledge in respect of the algae is almost certainly related to the difficulty of defining what algae are and finding suitable tools for their identification. This is compounded by the sheer number of species, their ubiquity on the earth, and their often miniscule size that requires sophisticated techniques to even visualise let alone isolate species.

A major debate is over the question of monophyly versus polyphyly in the algae. Many scientists fall on the side of the algae being a heterogeneous group of organisms that are not monophyletic, but there are others (e.g., Delwiche, chapter 2, this volume) who argue for monophyly of all photosynthetic organisms, based on the origin of plastids. (We need to be able to think along both of these lines.) This does not necessarily leave us any the wiser as to what the algae are but we do now know many of the groups and increasingly we have a better idea of where they sit in the tree of life. This is eloquently demonstrated by Cavalier-Smith (chapter 3, this volume). The number of groups of algae has varied over the years, with as many as 16 phyla (van den Hoek et al., 1995). The latest classifications place "algae" into four of the five supergroups of eukaryotes (Keeling, 2004). It is their extraordinary range and diversity, both in numbers of species and of shape and form, which are increasingly leading scientists to pursue research into these organisms.

WHY THE INTEREST IN THE ALGAE?

So why is there so much current interest in the algae? They are crucial to the functioning of the planet. They are oxygen producers; they dominate the world's oceans and account for the production of a major fraction of the world's oxygen. They are a major source of food for plankton, fisheries, and, via the food chain, ultimately for humans. Some species are the basis of the structure of the ecosystem, for example, the giant kelp forests or maerl beds. Many algae are eaten directly as food in different parts of the world, e.g. nori (the wrapping of *Porphyra* around sushi) in Japan, dulse (*Palmaria palmata*) in Ireland, and cochayuyo (*Durvillea antarctica*) in Chile. A myriad of products are also derived from the algae, including alginates and carrageenans that are employed in a wide range of industries for the production of e.g. toothpastes, cosmetics, paper sizing, gels, emulsifiers, and bandages; in almost all walks of life it is likely that some algal product will be involved.

Evidence from the fossil record indicates that the origins of eukaryotic algae are extremely ancient (see, e.g., Maggs et al., chapter 6, and Moestrup and Daugbjerg, chapter 12, this volume). Although controversial, the earliest multicellular eukaryotic organism is a red alga, *Bangiomorpha pubescens* dating back to 1.2 billion years ago (Butterfield, 2000). The algae are therefore a fascinating repository of genetic information.

Unfortunately some algae can also be a nuisance. Through fish kills, the intoxication of shellfish, and unsightly water discolourations, "harmful algal blooms" (HABs) have received growing and worldwide attention. With man's increasing use of the coastal strip, their impact has become even more prominent over the last thirty years. There is also evidence that in some places there has been an increase in harmful algal events (Hallegraeff, 1993). The study of some algal groups has certainly benefited from this attention as it became apparent that very tiny morphological differences between toxic and non-toxic species were crucial to determining the likelihood of a harmful event (e.g., Hasle et al., 1996). Since the first conference on toxic dinoflagellate blooms (LoCicero, 1975) with 100 participants from 3 countries to the most recent one in Copenhagen with 550 participants from 57 countries, the interest in harmful algae has blossomed.

WHAT IS TAXONOMY AND HOW DOES IT WORK?

Taxonomy is a human construct. It is a means of distinguishing between organisms. The ultimate aim is to arrive at the perfect classification that reflects natural (monophyletic) relationships. It is, however, a consensus, and inevitably there are differing views and arguments, and this is reflected in the chapters of this book and is explored in detail by Williams (chapter 4, this volume). The book reflects current debate and this is another exciting part of science. When new ideas are introduced the debate is often fierce and can divide the scientific community. An example of this would be when the idea was originally introduced that e.g. mitochondria and plastids were derived from a symbiotic relationship between organisms (see Cavalier-Smith, chapter 3). Such controversial ideas spark fevered studies but eventually the concepts settle down and indeed, sometimes, become conventional wisdom—something we should never be afraid to challenge.

THE BOOK

This book is the product of the remarkable symposium and covers the state of algal systematics today as well as looking to the future. It is the first time that all the algal groups are considered in one volume from the perspective of the impact of molecular systematics and is a landmark review of the subject. It draws together under one umbrella the most up-to-date thinking about the subject and also points to the way forward, revealing gaps in knowledge and establishing future research directions in algal systematics.

Virtually all known algal groups are covered to a greater or lesser extent somewhere in this book. The one exception that does not have its own chapter is the Glaucophyceae, a tiny group of single-celled organisms with about four known species and an estimated 13 in total (Brodie & Zuccarello, 2007). It is nevertheless important in understanding the tree of life, and this little group falls with the green and red algae in Primoplantae, which arose from primary cyanobacterial symbiosis. However, we should never underestimate the hidden algal diversity that is certainly still to be discovered, particularly from the more extreme habitats on earth (e.g., extremophiles). There is no doubt that there are new discoveries to be made at all levels of classification; the development of new techniques using combinations of genomics, culturing, and knowledge of fine structure will reveal exciting new photosynthetic organisms.

The book is divided into three sections. The first after the introduction contains three chapters (Delwiche, Cavalier-Smith, and Williams) that are overviews of the subject from three very different perspectives. They contain opinions that are controversial and yet pave the way for more intellectual debate and discovery. Delwiche focuses on the plastid and how endosymbiosis events unite the algae and land plants as a monophyletic group. Cavalier-Smith, on the other hand, argues that the algae are polyphyletic because the eukaryotic algae arose as chimaeras of a bikont protozoa and a cyanobacterium and that there were at least four secondary symbioses that generated the diversity of the meta-algae. Williams' chapter in which he discusses classification, using the diatoms as the model for his arguments, is thought provoking, takes a long historical view, and argues philosophically about the interpretation of relationships. This is a valuable debate because all too often, the data can be interpreted to produce misunderstandings of relationships.

The second section of the book contains twelve chapters covering the major algal groups, including the cyanophytes, red algae, green algae, charophytes, chlorarachniophytes, haptophytes, cryptomonads, dinoflagellates, diatoms, brown algae, chrysophytes and euglenoids. Despite the diversity of organisms and range of approaches, there are underlying themes that recur throughout the chapters: molecular tools, although very powerful, do not answer all the questions about systematics and identification and need to be used hand in hand with other techniques. This is also where the value of bringing together phycologists working on such a wide range of organisms, yet united by the algae, can identify areas to be addressed in common to us all. Thus, it was demonstrated overwhelmingly in the symposium and clearly seen here, that it is inappropriate to focus on a single technique to further our understanding of the algae and their systematics. It is clear that to advance the subject, a range of traditional and molecular approaches are required. It became apparent during the symposium that it is also necessary to study fine structure of algae to support interpretation, so transmission electron microscope studies, in danger of becoming deeply unfashionable for funding agencies, will remain an important tool in this respect. To achieve all of these goals, it is also essential to maintain the skill base in such studies and this requires ensuring that there is a future for the next generation of phycologists.

The final two chapters in the volume (Bowler & Allen, chapter 17, and Medlin et al., chapter 18) look specifically to the future. They cover genomics and also review where we find ourselves and where we will go with algal systematics. The next great step is to develop a wider understanding of systematics at the species and sub-species level, and we need to be prepared to radically alter our thinking on species concepts. Already the data challenge traditional notions of species, and this is an exciting area for future study. There is no doubt that a much greater understanding of both diversity and relationships at the species level is needed to be able to refine and resolve higher-level classifications amongst the algae and ultimately to refine the tree of life.

Progress in algal systematics for the different algal groups covered in this book is variable and clearly displayed in the chapters following. From the relatively new recognition of the chlorarach-niophytes, where hardly any species are known, through the better known Chrysophyceae where work is in progress on the wealth of culture material that has to be better defined and understood, to the considerable understanding now of the rhodophyte algae. Nevertheless, this book reveals just how rapidly our understanding of algal systematics has moved on over the last 20 years. It has highlighted some of the crucial debates that are taking place and raised concerns that we do not lose sight of the goal of systematics. There is concern about the appropriate use and interpretation of molecular data, notably for such approaches as DNA barcoding (a short diagnostic sequence to distinguish between species) and its use in algal taxonomy. In Maggs et al. (chapter 6, this volume) this is discussed for the red algae (see also Robba et al., 2006). Without dismissing DNA barcoding, which is potentially a useful tool in, e.g., ecology, forensics, and conservation, we need to caution against confusing it with DNA taxonomy and phylogenetics, even though the data may be considered for those uses independently. One aspect of this that clearly emerged during the symposium is the need to establish a system to preserve labelled genomes for algal species. This is not an easy proposition as it will require long-term commitment and effort to achieve. However, there is clearly a need for a consensus on the proper long-term vouchering of analysed material to sit alongside any electronic record (e.g. GenBank) in order to ensure consistency of further investigation in the future.

Science never stands still, and the enquiry into the algae is now entering an extremely exciting phase in its history. We hope that this book will help to inspire the next generation of scientists to study the taxonomy/systematics of algae. We can only anticipate the next stage of this story. Perhaps it will be the establishment of a worldwide system of buoys automatically monitoring the microalgae of our oceans and feeding back the data via satellites to allow managers to predict harmful algal events or possibly a remote sensor to be fitted to an aeroplane that means we can establish the diversity of algae on the shore for conservation purposes. No doubt the development of handheld identification tools will be here within a decade or so. We hope that this book will provide a stepping stone along the way.

REFERENCES

Brodie, J. and Zuccarello, G.C. (2007) Systematics of the species-rich algae: red algal classification, phylogeny and speciation. In *The taxonomy and systematics of large and species-rich taxa: building and using the Tree of Life* (eds. T.R. Hodkinson and J. Parnell), Systematics Association Series, CRC Press, pp. 317–330.

Butterfield, N.J. (2000) *Bangiomorpha pubescens* n. gen., n. sp.: implications for the evolution of sex, multicellularity and the Mesoproterozoic-Neoproterozoic radiation of eukaryotes. *Paleobiology*, 26: 386-404.

Hallegraeff, G.M. (1993) A review of harmful algal blooms and their apparent global increase, *Phycologia* 32: 79–99.

Harvey, W.H. (1836) Algae. In *Flora Hibernica*, vol. 2 (ed. J.T. Mackay), Curry, Dublin, Ireland, pp. 157–256.

Hasle, G.R., Lange, C.B., and Syvertsen, E.E. (1996) A review of *Pseudo-nitzschia*, with special reference to the Skagerrak, North Atlantic and adjacent waters, *Helgoländer Meeresuntersuchungen*, 50: 131–175.

Keeling, P.J. (2004) Diversity and evolutionary history of plastids and their hosts, *American Journal of Botany*, 91: 1481–1493.

Lamouroux, J.V.F. (1813) Essai sur les genres de la famille de Thalassiophytes non articulées, *Annales du Muséum (National) d'Histoire Naturelle (Paris)*, 20, 21–47, 115–139, 267–293, seven plates.

Linnaeus, C. (1753) *Species plantarum, exhibentes plantas rite cognitas, ad genera relatas, cum differentiis specificis, nominibus trivialibus, synonymis selectis, locis natalibus digestas. Cryptogamia*, Salvi, Stockholm, Sweden, pp. 1061–1186.

LoCicero, V.R. (ed.) (1975) Proceedings of the First International Conference on Toxic Dinoflagellate Blooms. Massachusetts Science and Technology Foundation, Wakefield, Massachusetts.

Robba, L., Russell, S.J., Barker, G.L., and Brodie, J. (2006) Assessing the use of the mitochondrial *cox1* marker for use in DNA barcoding of red algae (Rhodophyta), *American Journal of Botany* 93: 1101–1108.

van den Hoek, C., Mann, D.G., and Jahns, H.M. (1995) *Algae. An introduction to phycology*, Cambridge University Press, Cambridge.

2 Algae in the warp and weave of life: bound by plastids

Charles F. Delwiche

CONTENTS

ABSTRACT

Oxygenic photosynthesis is responsible for the presence of an oxidizing atmosphere and the great majority of primary productivity on Earth. This vital metabolic process is carried out only by cyanobacteria, algae, and their descendants, the land plants. Photosynthesis in eukaryotes occurs in plastids, endosymbiotic organelles derived from cyanobacteria. Thus, oxygenic photosynthesis is unique to cyanobacteria and their descendants. However, because eukaryotes acquired photosynthesis via endosymbiosis either of a cyanobacterium or of another eukaryote that already had plastids, these organisms are chimaeras, and have major genomic contributions for two or more sources. There is reason to believe that some formerly photosynthetic eukaryotes have lost that ability, but the chimaeric nature of their genomes leaves genetic evidence of their photosynthetic past. This raises fundamental questions about the mechanisms that underlie the establishment of endosymbiotic associations, and the degree to which phototroph genes might be expected in the genomes of primitively non-photosynthetic predators.

INTRODUCTION

Algae are photosynthetic eukaryotes, although as normally used the term artificially excludes land plants. This also ignores cyanobacteria, which have at times been called "blue green algae." We will see below how cyanobacteria have a very special place among these organisms, although we cannot think of them as algae *per se*. The algae (Figure 2.1a through Figure 2.1f) include several of the most abundant eukaryotes on earth, and together with land plants are responsible for the bulk of global primary productivity. As a result, they form the base of marine and freshwater food chains, and their ancestors' remains became the world's great oil deposits. They also occur in terrestrial environments, and, although most species require at least a film of liquid water to be metabolically active, they can be found in a remarkable range of habitats, from snowfields to the edges of hotsprings, and from damp earth and the leaves of plants to sun-baked desert soils. Even the insides of rocks are open to colonization, with endolithic algae occupying tiny cracks in rocks

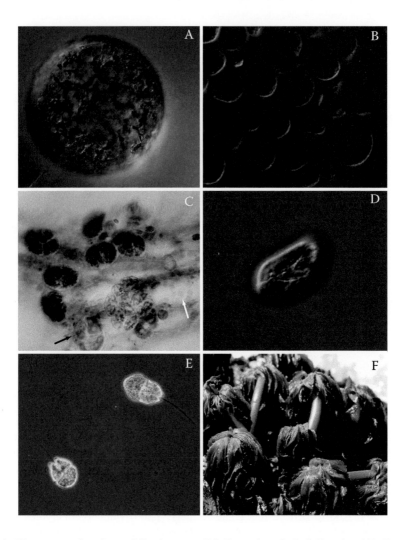

FIGURE 2.1 (Please see colour insert following page 76) Examples of algal diversity. (A) *Eremosphaera*, a unicellular green alga with cells close to 1 mm in diameter (from Jyme Bog, Oneida Co., WI). (B) *Porphyridium cruentum*, a unicellular red alga; note the striking difference in colour when compared to *Eremosphaera* (from cultured material, UTEX 161). (C) *Glaucocystis* sp., a glaucophyte, living on the surface of the angiosperm *Ceratophyllum*, also showing the green algae *Coleochaete* and *Cosmarium* (black arrows) and plastids of *Ceratophyllum* (white arrow; from an aquatic garden at a commercial nursery, Rockville, MD). (D) *Cryptomonas* sp., a cryptomonad, and one of the organisms with secondary plastids that retain degenerate eukaryotic nuclei termed "nucleomorphs" in association with the plastid (from a seasonal freshwater wetland in Caroline County, MD). (E) *Akashiwo sanguinea*, a dinoflagellate. The trailing flagellum is visible, and the nucleus is visible as a clear region near the center of the cell, while the single plastid occupies much of the remainder of the cell (from the Cheseapeake Bay, MD). (F) *Postelsia* sp., a macroscopic brown alga that lives on exposed rocky shore lines (at Bodega Head, CA).

in alpine and arctic environments that show no obvious signs of vegetation. Because they are key primary producers, algae play a vital role in the Earth's carbon cycle, and in fact, the Earth's atmosphere would not contain free oxygen at all if it were not for the activities of algae and cyanobacteria. Key to the health of many of the Earth's most important biomes, algae can also be detrimental. Although they produce oxygen during the day, they consume it at night, and large populations of algae can contribute to locally anoxic conditions, sometimes killing fish and other aquatic life. Even more lethal can be the production of algal toxins that can incapacitate or kill marine life and create discomfort and illness in humans in or near the water (Bold and Wynne, 1985; Falkowski et al., 2004; Fritsch, 1965; Graham and Wilcox, 2000; van den Hoek et al., 1995).

Despite their nearly universal distribution, great abundance, and environmental and economic importance, we often pass by them without a glance, and few would first think of algae when asked what was most important to them. The most conspicuous algae are so important that traditionally classifications arbitrarily cull them out and give them a kingdom of their own; the land plants, or embryophytes, are terrestrial organisms whose evolutionary origins lie deep within the green algae. Land plants, a group that includes everything from mosses and liverworts to ferns, pine trees, and flowering plants, are artificially excluded from the algae, and a natural classification would treat them as the "drier algae" rather than the "higher plants." However, because the evolutionary relationships among these organisms were obscure to early biologists, the structurally simple algae were considered to be a subset of land plants rather than *vice versa*.

Comparable in size and structural complexity to the land plants are the kelps, huge marine algae that form vast forests in temperate waters. These forests are every bit as spectacular as any terrestrial forest, although only a privileged few have really seen them because they are located underwater. But the vast majority of algae are relatively small and inconspicuous, if not microscopic. To appreciate their beauty and diversity requires careful attention, and generally the use of a magnifying glass or microscope. What then are these organisms that are all around us, and so vital, and yet are so easy to miss?

Ingenhousz (1779), in his seminal paper *Experiments Upon Vegetables*, recognized that algae were photosynthetic organisms. In the chapter "Notes Upon the Green Matter that Accumulates in the Bottom and Sides of Vessels in which Water is Left Standing," he notes that the matter "appears to be of the vegetable kind," and documents the production of oxygen. The key observation was that there are living things that resemble plants in their physiological processes, but that have a much simpler structure. With the clarity of hindsight we understand that land plants are large and complex but are the specialized descendants of a more diverse set of microorganisms. But to Ingenhousz these microscopic organisms seemed to be a special case of the much more familiar land plants. Consequently, the term "alga" came to refer to an organism that carried out oxygenic photosynthesis (i.e., producing oxygen, as do plants) but that lacked the characteristic plant structure. As it turns out, there is a vast diversity of organisms that fit this description. What they all share is the ability to perform oxygenic photosynthesis. When the physiology of photosynthesis was investigated in more detail, it led to a remarkable discovery: oxygenic photosynthesis occurs in specialized, subcellular organelles called "plastids" (they are most familiar as "chloroplasts," which is to say "green plastids," in green algae and plants) (Delwiche, 1999; Gray, 1989; Margulis, 1968). Plastids are semi-autonomous organelles and contain their own genomes and protein-expression apparatus. In fact, plastids are endosymbiotic organelles, derived from once free-living cyanobacteria. The realization that two key cellular organelles, mitochondria and plastids (including chloroplasts), are symbiotic in origin counts among the most striking biological insights of the 20th century.

From this discovery emerged what may be an even more extraordinary insight, albeit one that is more subtle; oxygenic photosynthesis is carried out only by cyanobacteria, or by their highly specialized descendants, plastids. Cyanobacteria are a natural (i.e., monophyletic) group, and with a very few exceptions all cyanobacteria are oxygenic phototrophs. Oxygenic photosynthesis occurs in algae and plants only because they have incorporated endosymbiotic cyanobacteria. Consequently, the air we breathe, the food we eat, the wood we use to build shelter, the fuel that powers our automobiles, airplanes, and steel mills, the raw material for both natural and synthetic fabrics, and a vast range of plastics and pharmaceuticals all derive from the unique metabolic capabilities of a single group of bacteria. No other organism is capable of this process. We are, truly, indebted to the cyanobacteria. And it is appropriate to think of algae (along with land plants) as a single type of organism; they are united by their possession of a plastid, even though their nuclear lineages are not monophyletic.

A second vital insight that came from the understanding that plastids are endosymbiotic organelles was that oxygenic photosynthesis is a property that can be acquired. A non-photosynthetic organism can become photosynthetic simply by eating and retaining key parts of another photosynthetic organism.

Thus, algae are all chimeras. (This is actually true of all eukaryotes because the mitochondrion is also an endosymbiotic organelle, but is a more profound insight for algae because there are many eukaryotes that lack plastids.) While red algae, green algae, and glaucocystophytes all have plastids that are directly derived from cyanobacteria, all other groups of algae have "secondary" plastids, which is to say that they acquired their plastids by eating another eukaryote that already had plastids (Delwiche, 1999). So eukaryotes can steal plastids, and by so doing become photosynthetic. This means that the nuclear genomes of algae need not be particularly closely related, and that lineages with similar plastids could in other ways be quite dissimilar. A major question, then, is how many times this process had occurred, and how many independent lineages of algae there are. Debate on this subject has been heated, with some authors inferring a very large number of events, and others insisting that incorporation of a foreign organelle is a very surprising event, and one would expect it to be rare.

I will return to that topic, but first I want to drive home the point that because photosynthetic eukaryotes are chimaeras, there is no single phylogeny that applies to all parts of the cell. The cytosol (i.e., nuclear) lineage and plastid lineage can potentially have very different histories. Within those organisms with primary plastids, there are at least three distinct genomes to consider: the nuclear, mitochondrial, and plastid genomes (Figure 2.2). It is possible for genes to be transferred

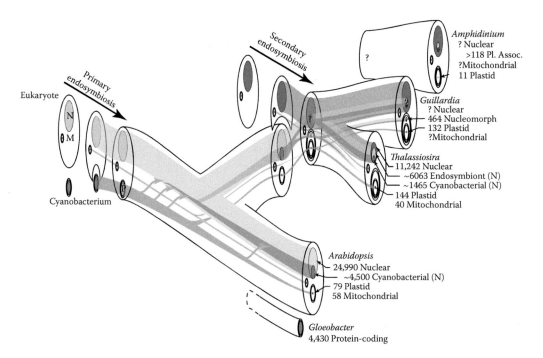

FIGURE 2.2 (Please see colour insert following page 76) Gene transfer in the evolution of chloroplast genomes. One of the consequences of endosymbiosis has been the modification of both the host and endosymbiont genomes. Primary endosymbiosis (left) refers to the acquisition of a cyanobacterium by a non-photosynthetic eukaryote (already equipped with a mitochondrion). Over time, many genes have been transferred to the nuclear genome (green arrows), while others have been lost outright. Some of the transferred genes are expressed in the plastid and assist in its function. Others are now expressed in the cytosol and are not necessarily involved in plastid function. Estimates of the number of genes in each genetic compartment and their ultimate evolutionary origins are shown to the right. (Arabidopsis Genome Initiative, *Nature*, 408: 796–815, 2000. With permission.) Secondary endosymbiosis (centre) refers to the acquisition of plastids from one eukaryote by a second eukaryote. As with primary endosymbiosis, substantial transfer of genes from the endosymbiont to the host has occurred, and the evolutionary history of the genomes of these organisms can be quite complex (right). (From Armbrust, E.V. et al., *Science*, 306: 79–86, 2004; Bachvaroff, T.R. et al., *Protist*, 155: 65–78, 2004; Douglas, S. et al., *Nature*, 410: 1091–1096, 2001. With permission.)

from one genome to the other. The most common direction of transfer is from the organelles into the nuclear genome, such that the nuclear genome is large and the organellar genomes are small, but transfers of other types are possible (Cho and Palmer, 1999). The genomes of organisms with secondary (or tertiary) plastids are even more complex, with at least five genomes to consider: the host nuclear and mitochondrial genomes, as well as those of the prey nucleus, mitochondrion, and plastid (Figure 2.2). What is more, organellar genes that have been transferred to the nuclear genome do not necessarily find expression in the organelle from which they were transferred; many such genes are expressed in the host cytosol and have become involved in processes that are not directly coupled to the mitochondrion or plastid (Armbrust et al., 2004; Keeling and Palmer, 2001). Thus, while all eukaryotes are chimeras, the algae are particularly complex mixtures of diverse ancestry, and understanding their biology requires understanding these distinct genetic contributions. Central to their identity as algae is the possession of plastids, so I will spend the remainder of this chapter addressing the origin and evolution of these remarkable organelles.

PLASTIDS

That plastids and mitochondria are derived from once free-living bacterial ancestors is not in serious question (Delwiche, 1999; Gray, 1999; Keeling, 2004b; Moreira and Philippe, 2001; Palmer, 2003). Mitochondria are widespread among eukaryotes and may have been present in the most recent common ancestor of all living eukaryotes, although in some cases they have been lost or reduced to barely recognizable "hydrogenosomes" (Roger and Silberman, 2002). Plastids, by contrast, are present only in a subset of eukaryotes, and although initially acquired by a limited group of organisms, they subsequently spread much more widely. Thus, the origin, diversification, and spread of plastids presents a complex problem in phylogenetic reconstruction, yet at the same time provides an opportunity to identify general principles underlying the acquisition and incorporation of symbiotic organelles, including extensive modification of both host and symbiont genome.

Plastids are most familiar as the chloroplasts of land plants, which are phylogenetically a highly specialized group of green algae adapted to life on land (Karol et al., 2001). The early recognition of a close relationship between land plants and green algae was in large part due to the similarity of pigmentation found in their plastids, although, not surprisingly, green algae as a whole show more diversity in their plastid ultrastructure and pigmentation than do plants. And green algae constitute only a small fraction of all plastid diversity, so to understand plastids requires a rather broad survey of eukaryotes, often delving into unfamiliar and obscure organisms (Delwiche et al., 2004; Falkowski et al., 2004). This can be a challenge, but it is a pleasure as well; these organisms are fascinating and often challenge our assumptions about the universal features of life. The taxonomic complexity of plastid evolution is further elaborated by the fact that plastids are symbiotic organelles that are capable of horizontal transmission (that is to say, secondary endosymbiosis can lead to the acquisition of the plastids of one eukaryote by a second eukaryote). Consequently, the characteristics of plastid and host cell follow a peculiar, mix-and-match pattern (Delwiche, 1999; Gibbs, 1981). The nomenclature used here follows that of Adl et al. (2005) whenever practical.

Acquisition of secondary plastids is likely an ongoing process (Marin et al., 2005; Okamoto and Inouye, 2005), but this chapter is primarily concerned with plastids that are obligate, heritable, and show the profound genome modification characteristic of long-term symbionts (Cavalier-Smith, 1999; Martin et al., 2002; Palenik, 2002).

A SINGLE ORIGIN OF PRIMARY PLASTIDS … PROBABLY

The plastids of green algae, red algae, and glaucophytes are surrounded by two unit membranes, which in glaucophytes are separated by a thin peptidioglycan cell wall (Bhattacharya and Schmidt, 1997). The glaucophytes are striking in that their plastids retain a bacterium-like cell wall, but as with all other true plastids, the glaucophyte plastid is able to import polypeptides from the cytosol, has undergone massive reduction of its genome (to roughly 5% the size of free-living cyanobacterial genome), and is clearly a fully obligate symbiont. Glaucophytes may be an outgroup to all other primary-plastid algae (reds and greens), or they may be a basal member of the red algal lineage, in which case they could be considered to be red algae that retain their ancestral flagella (Delwiche et al., 2004).

The presence in plant plastids of membrane-integral proteins derived from cyanobacteria in both the inner and outer membrane suggests that these correspond to the two membranes of the gram-negative cyanobacterial wall (Keegstra and Cline, 1999), which in turn implies that the primary plastid lies directly within the host cell cytoplasm (rather than within a compartment derived from a food vacuole). Note, however, that in at least some cases the outer membrane proteins may be inverted with respect to their orientation in cyanobacteria, perhaps as a result of the transfer of the gene encoding the protein to the nuclear genome (Reumann et al., 1999). This elegantly explains the apparent conversion of an export protein to an import protein, but at the same time reduces utility of this protein as evidence of homology of the outer plastid envelope to that of cyanobacteria.

Among the fundamental questions concerning the evolution of plastids is whether all of these organelles share a common origin, or if they are the result of convergent evolution, a topic that has been widely discussed (Keeling, 2004b; McFadden, 2001; Palmer, 2003; Stiller et al., 2003). Oxygenic photosynthesis is unique to cyanobacteria, and there is essentially no doubt that all plastids are derived from cyanobacteria (Delwiche et al., 1995). But whether they are all derived from a common plastid ancestor (i.e., all share an endosymbiotic origin) is a more difficult question.

There is considerable phylogenetic evidence suggesting that all three lineages of primary plastids (red, green, and glaucophyte) are monophyletic (McFadden, 2001; Moreira and Philippe, 2001). However, it is important to bear in mind that plastids could be monophyletic, and yet not derived from a common plastid ancestor. If closely related, free-living cyanobacteria were independently recruited to symbiosis, plastid phylogenies would show them as monophyletic. Relying on plastid phylogenies alone, it would be possible to demonstrate independent origins of plastids only if the free-living relatives of each of the independent lineages had been sampled. Given the relative paucity of phylogenetic data from cyanobacteria (*Gloeobacter*, the most structurally simple known cyanobacterium, has very rarely been isolated from nature), it would not be surprising if the key intervening taxa had simply not been identified, or have gone extinct.

A second line of evidence for monophyly of plastids comes from features of the cell that appear to be direct adaptations to, or consequences of, symbiosis. The most spectacular such feature is probably the protein-targeting mechanism that relies on distinctive target-peptide sequences to direct nuclear-encoded proteins to the plastid. Transit peptides of all three primary plastid lineages are very similar and seem to be largely interchangeable. This may be evidence of common origin, but too many questions remain about the origin and operation of protein-targeting mechanisms to make these data easy to interpret (McFadden, 2001). Another clearly post-symbiotic feature is the highly reduced genome, but while some similarities have been noted in gene content and genome organization among plastids (most strikingly the presence of a distinctive inverted repeat in most plastids), the processes governing genome reduction and organization are poorly understood, and consequently these observations are subject to diverse interpretations. The difficulty of testing the monophyly of plastids from plastid data alone has been emphasized by Stiller et al. (2003), who noted that the complement of genes present in the highly reduced genomes of plastids is plausibly the product of convergent evolution.

Phylogenies of the host cells (rather than the plastids) provide another informative line of evidence. If extant plastids were derived from a single, common plastid ancestor, then one would expect the host cells (and their mitochondria) to be monophyletic, with a phylogeny that matches that of the plastids. If they were acquired independently, then this need not be the case. (The remaining possibility, that plastids were acquired independently, but from closely related cyanobacteria and by host cells that were themselves closely related, is particularly difficult to test.) Phylogenetic analyses of both nuclear- and mitochondrial-encoded genes, with but a few exceptions, show the primary-plastid containing lineages to be monophyletic (Baldauf, 2003; Keeling, 2004b; Sánchez Puerta et al., 2004).

AND HOW MANY SECONDARY PLASTID LINEAGES?

Primary plastids are only the beginning of the story. Many eukaryotes rely on plastids that were acquired indirectly, by establishing a "secondary" endosymbiotic association with either a green or a red alga (Figure 2.2). In secondary endosymbiosis, one eukaryote acquires a eukaryotic endosymbiont, so that the plastid is contained within its original host cell, which is in turn inside of another host; even tertiary symbioses are known, with three eukaryotic cells nested like Russian dolls (Chesnick et al., 1996; Moreira and Philippe, 2001; Tengs et al., 2000). In some cases the eukaryotic endosymbiont may have been acquired intact, while in others it may be the case that only the organelle was acquired, by permanent retention of a "kleptoplastid."

Organisms with secondary plastids dominate the world's oceans and are responsible for a huge fraction of photosynthesis worldwide, but seem exotic because plants dominate the land, and their primary plastids are more familiar. There are six major groups of algae with secondary photosynthetic plastids (Table 2.1). Each of these groups is noncontroversially monophyletic, but their placement within the tree of life is much less certain, and some may cluster together into larger clades that would also include a substantial number of organisms that do not possess plastids (Delwiche et al., 2004).

Relationships among the chlorophyll c containing taxa in particular are (and always have been) confusing and controversial. There was an early recognition that taxa pigmented with chlorophyll b (particularly green algae and plants) were fundamentally different from those with other pigmentation patterns, which were referred to in aggregate as the brown lineage, or "Chromophyceae" (Chadefaud, 1950; Christensen, 1971, 1989; Doherty, 1955). Study of fine structure and pigmentation led to broad agreement that the brown algae, diatoms, chrysophytes, and several smaller groups are closely related; these "core" chromophytes have also been called "heterokonts," in reference to their two dissimilar flagella. Taylor (1976) accurately noted that many non-photosynthetic taxa,

---

TABLE 2.1
Major Groups of Algae with Secondary Photosynthetic Plastids

	Group	Membranes[1]	Pigments[2]	Nucleomorph[3]
1°	Green algae and land plants	2	b	n/a
2°	Chlorarachniophytes	4	b	Yes
2°	Euglenoids	3	b	No
1°	Glaucophytes	2	pbs	n/a
1°	Red algae	2	pbs	n/a
2°	Cryptomonads	4	c and pb	Yes
2°	Stramenopiles (or heterokonts)	4	c	No
2°	Haptophytes	4	c	No
2°	Dinoflagellates (those with peridinin)[4]	3	c	No
2°	Apicomplexa (non-photosynthetic)	4	—	No

[1] Number of membranes separating the plastid from the cytosol.

[2] Major accessory pigmentation syndromes (all have chlorophyll a at the reaction center as well as other accessory pigments): b = chlorophyll b; pbs = phycobilisomes; pb = phycobiliproteins, but not arranged in phycobilisomes; c = chlorophyll c.

[3] When plastid is secondary in origin, is there an associated nucleomorph?

[4] Several types of plastids are known from dinoflagellates (see text).

including oomycetes and thraustochytrids, were closely related to the heterokonts, and Patterson (1989) proposed a new name, "stramenopiles" (based on the tubular hairs on the flagella of many of these taxa) for a supergroup including the heterokonts and their non-photosynthetic relatives (Patterson, 1999). Following Adl et al. (2005), I have adopted the name stramenopiles, and understand it to include both photosynthetic and non-photosynthetic organisms. The heterokonts, if they prove to be monophyletic, would be a sub-group of stramenopiles. Two additional groups, the haptophytes and cryptomonads, have chlorophyll *c* pigmented plastids but have flagellar structures that are strikingly different from the stramenopiles. (Haptophytes have two similar flagella and in many cases a third, unique structure, the haptonema; cryptomonads have two dissimilar flagella, but with a distinctive arrangement and hairs.) Among the chlorophyll *c* taxa, nearly every possible grouping has been proposed at one time or another, often applying some variant of the term "chromophyte," such that it is very difficult to be sure how a given author is using that term without additional information

Operating on the *a priori* assumption that establishing novel symbiotic associations is rare, Cavalier-Smith (1999) proposed that all taxa with secondary plastids fell into two supergroups, the "cabozoa" comprising the two groups with secondary plastids pigmented with chlorophylls *a* and *b* (i.e., green algal in origin), and the "chromalveolates," uniting the four taxa with chlorophyll *c* (all of which have plastids thought to be of red algal origin). Although there seems to be little basis for the cabozoan hypothesis, the chromalveolate hypothesis has much to recommend it and has received considerable attention in the recent literature. The chromalveolate hypothesis has far-reaching implications, because if the chloroplasts of all the chlorophyll *c* containing taxa are the result of a single endosymbiotic event, then their host cells must necessarily be monophyletic, and an indeterminate number of members of that group would be inferred to have lost their plastids (Delwiche et al., 2004). The chromalveolates would constitute a substantial portion of eukaryotic diversity, and if they are derived from a plastid-containing ancestor, that could have important implications for the biology of the group, even the members that are now wholly non-photosynthetic.

The term "chromalveolate" is a fusion of "chromist"–Cavalier-Smith's term for the major lineage that includes brown-pigmented algae with complex secondary plastids (cryptomonads, stramenopiles, and haptophytes)–and "alveolate"–a second major group that includes dinoflagellates, many of which are brown pigmented. In addition to photosynthetic organisms, both of these major lineages include several non-photosynthetic groups. One of these, the Apicomplexa (which consists entirely of obligate parasites of animals), has been shown to contain relict, non-photosynthetic plastids (Gardner et al., 2002; Kohler et al., 1997; McFadden et al., 1996), and plastid genes have recently been reported from a non-photosynthetic dinoflagellate, *Crypthecodinium cohnii* (Sánchez-Puerta et al., 2006). There is at present no evidence of plastid genes reported from the single ciliate genome available (*Tetrahymena*), but analysis of the complete genome of the oomycete *Phytophthora*, an important plant pathogen, revealed many genes of probable plastid origin (Tyler et al., 2006). The data for *Phytophthora* are not strong but derive from several genes and seem to provide support for the chromalveolate hypothesis.

Among the fundamental evolutionary questions one would like to be able to answer for chlorophyll *c* containing taxa is how these organisms are related to each other; what is each group's closest relative ("sibling taxon"); and did they acquire their secondary plastids independently, or are they inherited from a common ancestor? These questions are subject to many of the same difficulties discussed for primary plastids. In particular, when relatively few organisms have been studied, the patterns observed may appear to be simpler than is really the case, and careful study of each group and its close relatives—which in many cases have not yet been identified with confidence—is essential.

It will be important to identify the sibling taxon of each of the host cell lineages. In a phylogenetic analysis of six nuclear, protein-coding genes, Harper et al. (2005) performed the most detailed analyses of nuclear-encoded genes to date, and found stramenopiles to be monophyletic with alveolates, but these were never observed to cluster with the other two chlorophyll *c* containing groups, haptophytes and cryptomonads. However, the relatively weak support for the placement of haptophytes and cryptomonads means that these analyses do not rule out the possibility of monophyletic chromalveolate host cells. Similarly, analyses of mitochondrial genes do not show monophyletic chromalveolates, but are too weakly supported to be interpreted as strong evidence against the hypothesis (Sánchez Puerta et al., 2004). None of the studies published to date has genuinely comprehensive taxon sampling; there remain a large number of described protists whose close relatives have not been identified (Adl et al., 2005), and study of these organisms will provide important context for the study of protist phylogeny.

Molecular phylogenetic analyses do indicate that chromalveolate plastids are monophyletic (Bachvaroff et al., 2005; Harper and Keeling, 2003; Patron et al., 2004; Yoon et al., 2002), but given that it is widely accepted that these plastids are derived from those of red algae, this may not be surprising.

More informative seem to be two nuclear-encoded, plastid-expressed genes (GAPDH and FBA), which are of a distinctly different type in stramenopiles, cryptomonads, haptophytes, and dinoflagellates than they are in red algae (Fast et al., 2001; Patron et al., 2004). These data, taken along with similarities of pigmentation and ultrastructure, suggest a fundamental "sameness" of chromalveolate plastids; the most obvious explanation for this would be that they are homologous, i.e., are descended from a common plastid ancestor. As with primary plastids, distinguishing independent acquisition of plastids from closely related organisms could be very difficult. There has been ample time for extinction to eliminate important lineages (in this case, of red algae); if chromalveolate plastids are indeed monophyletic, the lineage may be well over a billion years old (Yoon et al., 2004). Red algae are certainly old enough to allow this (Butterfield, 2000), although fossils unequivocally referable to photosynthetic lineages with secondary plastids are much younger (Tappan, 1980).

Although it is tempting to think of endosymbiosis as the result of a unitary event (a momentous case of indigestion), the many organisms that have specific and obligate, but not permanent, associations with algae or their plastids (Johnson et al., 2004; Okamoto and Inouye, 2005; Rumpho et al., 2000) suggest that extensive adaptation may have predated the permanent acquisition of plastids. Thus, it is entirely possible that several chromalveolate lineages have independently acquired similar plastids, perhaps even sequentially from one another. This is by no means inconsistent with known examples of plastid acquisition. At least two dinoflagellate lineages seem to have lost their peridinin-type plastid, only to acquire a tertiary endosymbiont from among the chromalveolates (Chesnick et al., 1997; Delwiche et al., 2004; Keeling, 2004a).

THE PROCESS OF PLASTID ACQUISITION

The acquisition of plastids by a host cell was certainly a complex process. First, the ancestral host cell must have made use of some form of predation that involved endocytosis (i.e., ingestion of the prey). Different mechanisms of endocytosis, such as phagocytosis and myzocytosis, may explain differences in the ultrastructure of secondary plastids (Delwiche, 1999). Given a predator/prey relationship among populations of cells, at least four additional fundamental steps seem necessary for the evolution of modern plastids, although the order in which these events occurred is a matter for discussion. These steps are effective transfer of metabolites from prey to host without destruction of the prey; transfer of genes to the host cell with functional targeting back to the prey; replication, maintenance, and retention of the prey through both mitosis and meiosis; and loss of the bulk of the prey genome.

At least two distinct models can be postulated for the sequence in which the events leading to endosymbiosis took place, with innumerable variants. In the "indigestion" model, the first event is assumed to have been a failure of the mechanism that leads to the destruction and degradation of the prey cell. Metabolites leaked from the intact prey are presumed to become available to the host cell following its retention. The availability of such metabolites would result in an increase in fitness for the newly chimeric cell, and this would lead to selection for retention of the prey (nascent symbiont) through multiple cell cycles. Gene transfer is assumed to have been a later event, facilitated by the intimate relationship between the two cells. In contrast, an "upkeep" model would suppose that some metabolites are generally available to the predatory cell prior to destruction of the prey, and that there might be selection for retention of intact prey long before there was any obligate association between the cells. One such form of adaptation would be acquisition of genes involved in plastid maintenance by the predator cell. Selection would favor evolution of progressively more complex adaptations to the retention of prey, and might well lead to specialization of one type of prey. Under this model, genetic mechanisms for maintenance of the prey would substantially precede retention of the prey through cytokinesis. The distinction between these models is subtle, but the implications for genome evolution are substantial. In the indigestion model, one would predict that all genes involved in plastid maintenance were derived from a single source, the prey cell. By contrast, the upkeep model would allow for gradual acquisition of maintenance genes by the predator, and these need not necessarily be derived from the same source, although they would need to be sufficiently similar to function in the retained prey. This distinction makes it possible to test between these models by examining the phylogenies of multiple nuclear-encoded, plastid-associated genes.

INSIGHTS FROM COMPLETE GENOMES

Because most complete eukaryotic genomes sequenced to date have been from a few derived clades (Metazoa, Fungi, Plantae, Apicomplexa), only a few are available that directly bear on issues of early plastid evolution. However, some interesting insights have already emerged. First, it is clear that both primary and secondary plastids are fully integrated components in chimeric cells (Figure 2.2). In the *Arabidopsis* genome, of the scale of 18% of all of the genes in the nuclear genome may be of plastid origin, many of them now functioning in the cytosol in roles unrelated to the organelle (Martin et al., 2002).

Remarkably, the contribution of the secondary (red algal) endosymbiont to its host may be even greater in the diatom *Thalassiosira*, where on the scale of a third of the nuclear genome might ultimately have been derived from the symbiont (Armbrust et al., 2004). Such inferences are tenuous at best because of the difficulties of accurately determining the evolutionary history of a gene and the small number of germane genomes available for analysis, but they hint at the complexity of the genomic interactions between endosymbiont and host. As noted above, the non-photosynthetic oomycete *Phytophthora*, which is relatively closely related to *Thalassiosira*, appears to have plastid genes within its genome (Tyler et al., 2006). Because both diatoms and oomycetes are heterokonts, this suggests that plastids may be ancestral for the heterokonts, and in turn provides support for the chromalveolate hypothesis.

Another dramatic example of the intimate association between the plastid and its host can be found in dinoflagellates, where the peridinin-type plastid genome seems to have been reduced to a handful of single-gene minicircles (Howe et al., 2003; but see Laatsch et al., 2004; Zhang et al., 1999), and where an unusually large complement of genes is now in the nuclear genome (Bachvaroff et al., 2004a; Hackett et al., 2004). This emphasizes the fact that the organelle and its host, although previously two independent organisms, are now one fully integrated whole.

Endosymbiotic organelles are among the most dramatic and remarkable phenomena of modern biology. They are, however, simply another form of interaction between two organisms, only qualitatively different from predation, parasitism, mutualism, and competition. The line that separates endosymbiosis from predation and parasitism is primarily the boundary at which the association becomes mutually obligate and heritable. Once the association is permanent, the nature of the interaction is fundamentally different because the fitness of both partners is linked, and their evolutionary fates are joined. However, the fate of predator and prey are also linked, and to draw too bright a line between predation, cultivation, and symbiosis would be to oversimplify the profound complexity of the natural world.

ACKNOWLEDGMENTS

Credit for the "drier algae" line goes to Brent Mishler, and John Hall helped with some details of the manuscript. I am particularly grateful to the organizers of the symposium for the invitation to speak and for their willingness to accept a greatly delayed manuscript. This work was supported in part by NSF grant #MCB-0523719.

REFERENCES

Adl, S.M., Simpson, A.G.B., Farmer, M.A., Andersen, R.A., Anderson, O.R., Barta, J.R., Bowser, S.S., Brugerolle, G.U.Y., Fensome, R.A., Fredericq, S., James, T.Y., Karpov, S., Kugrens, P., Krug, J., Lane, C.E., Lewis, L.A., Lodge, J., Lynn, D.H., Mann, D.G., McCourt, R.M., Mendoza, L., Moestrup, O., Mozley-Standridge, S.E., Nerad, T.A., Shearer, C.A., Smirnov, A.V., Spiegel, F.W., and Taylor, M.F.J.R. (2005) The new higher level classification of eukaryotes with emphasis on the taxonomy of protists. *The Journal of Eukaryotic Microbiology*, 52: 399–451.

Arabidopsis Genome Initiative (2000) Analysis of the genome sequence of the flowering plant *Arabidopsis thaliana*. *Nature*, 393: 796–815.

Armbrust, E.V., Berges, J.A., Bowler, C., Green, B.R., Martinez, D., Putnam, N.H., Zhou, S., Allen, A.E., Apt, K.E., Bechner, M., Brzezinski, M.A., Chaal, B.K., Chiovitti, A., Davis, A.K., Demarest, M.S., Detter, J.C., Glavina, T., Goodstein, D., Hadi, M.Z., Hellsten, U., Hildebrand, M., Jenkins, B.D., Jurka, J., Kapitonov, V.V., Kroger, N., Lau, W.W.Y., Lane, T.W., Larimer, F.W., Lippmeier, J.C., Lucas, S., Medina, M., Montsant, A., Obornik, M., Parker, M.S., Palenik, B., Pazour, G.J., Richardson, P.M., Rynearson, T.A., Saito, M.A., Schwartz, D.C., Thamatrakoln, K., Valentin, K., Vardi, A., Wilkerson, F.P., and Rokhsar, D.S. (2004) The genome of the diatom *Thalassiosira pseudonana*: ecology, evolution, and metabolism. *Science*, 306: 79–86.

Bachvaroff, T.R., Concepcion, G.T., Rogers, C.R., Herman, E.M., and Delwiche, C.F. (2004) Dinoflagellate expressed sequence tag data indicate massive transfer of genes to the nuclear genome. *Protist*, 155: 65–78.

Bachvaroff, T.R., Sánchez Puerta, M.V., and Delwiche, C.F. (2005) Chlorophyll *c* containing plastid relationships based on analyses of a multi-gene dataset with all four chromalveolates lineages. *Molecular Biology and Evolution*, 22: 1772–1782.

Baldauf, S.L. (2003) The deep roots of eukaryotes. *Science*, 300: 1703–1706.

Bhattacharya, D., and Schmidt, H.A. (1997) Division Glaucocystophyta. *Plant Systematics and Evolution*, 11: S139–S148.

Bold, H.C. and Wynne, M.J. (1985) *Introduction to the Algae*. Prentice Hall, Englewood Cliffs, NJ.

Butterfield, N.J. (2000) *Bangiomorpha pubescens* n. gen., n. sp.: implications for the evolution of sex, multicellularity, and the Mesoproterozoic/Neoproterozoic radiation of eukaryotes. *Paleobiology*, 26: 386–404.

Cavalier-Smith, T. (1999) Principles of protein and lipid targeting in secondary symbiogenesis: euglenoid, dinoflagellate, and sporozoan plastid origins and the eukaryote family tree. *Journal of Eukaryotic Microbiology*, 46: 347–366.

Chadefaud, M. (1950) Les cellules nageuses des algues dans l'embranchement des Chlorophycees. *Comptes rendus hebdomadaires des seances de l' academie des sciences*, 231: 988–990.

Chesnick, J.M., Morden, C.W., and Schmieg, A.M. (1996) Identity of the endosymbiont of *Peridinium foliaceum* (Pyrrophyta): analysis of the *rbc*LS operon. *Journal of Phycology*, 32: 850–857.

Chesnick, J.M., Kooistra, W., Wellbrock, U., and Medlin, L.K. (1997) Ribosomal RNA analysis indicates a benthic pennate diatom ancestry for the endosymbionts of the dinoflagellates *Peridinium foliaceum* and *Peridinium balticum* (Pyrrhophyta). *Journal of Eukaryotic Microbiology*, 44: 314–320.

Cho, Y.R. and Palmer, J.D. (1999) Multiple acquisitions via horizontal transfer of a group I intron in the mitochondrial cox1 gene during evolution of the Araceae family. *Molecular Biology and Evolution*, 16: 1155–1165.

Christensen, T. (1971) The gross classification of algae. In *Selected Papers in Phycology*. Vol. I. (eds. Rosowski, J.R. and Parker, B.C.). Department of Botany, University of Nebraska, Lincoln, NE, pp. 571–577.

Christensen, T. (1989) The Chromophyta, past and present. In *The Chromophyte Algae: Problems and Perspectives*. Vol. 38. (eds. Green, J.C., Leadbeater, B.S.C., and Diver, W.L.). Clarendon Press, Oxford, pp. 1–12.

Delwiche, C.F. (1999) Tracing the thread of plastid diversity through the tapestry of life. *American Naturalist*, 154: S164–S177.

Delwiche, C.F., Kuhsel, M., and Palmer, J.D. (1995) Phylogenetic analysis of *tufA* sequences indicates a cyanobacterial origin of all plastids. *Molecular Phylogenetics and Evolution*, 4: 110–128.

Delwiche, C.F., Andersen, R.A., Bhattacharya, D., Mishler, B., and McCourt Richard, M. (2004) Algal evolution and the early radiation of green plants. In *The Tree of Life* (eds. Cracraft, J. and Donoghue, M.J.). Oxford University Press, London, pp. 121–137.

Doherty, E.C. (1955) Comparative evolution and the origin of sexuality. *Systematic Zoology*, 4: 145–169, 190.

Douglas, S., Zauner, S., Fraunholz, M., Beaton, M., Penny, S., Deng, L.T., Wu, X.N., Reith, M., Cavalier-Smith, T., and Maier, U.G. (2001) The highly reduced genome of an enslaved algal nucleus. *Nature*, 410: 1091–1096.

Falkowski, P.G., Katz, M.E., Knoll, A.H., Quigg, A., Raven, J.A., Schofield, O., and Taylor, F.J.R. (2004) The evolution of modern eukaryotic phytoplankton. *Science*, 305: 354–360.

Fast, N.M., Kissinger, J.C., Roos, D.S., and Keeling, P.J. (2001) Nuclear-encoded, plastid-targeted genes suggest a single common origin for apicomplexan and dinoflagellate plastids. *Molecular Biology and Evolution*, 18: 418–426.

Fritsch, F.E. (1965) *The Structure and Reproduction of Algae*. Cambridge University Press, Cambridge, UK.

Gardner, M.J., Hall, N., Fung, E., White, O., Berriman, M., Hyman, R.W., Carlton, J.M., Pain, A., Nelson, K.E., Bowman, S., Paulsen, I.T., James, K., Eisen, J.A., Rutherford, K., Salzberg, S.L., Craig, A., Kyes, S., Chan, M.S., Nene, V., Shallom, S.J., Suh, B., Peterson, J., Angiuoli, S., Pertea, M., Allen, J., Selengut, J., Haft, D., Mather, M.W., Vaidya, A.B., Martin, D.M.A., Fairlamb, A.H., Fraunholz, M.J., Roos, D.S., Ralph, S.A., McFadden, G.I., Cummings, L.M., Subramanian, G.M., Mungall, C., Venter, J.C., Carucci, D.J., Hoffman, S.L., Newbold, C., Davis, R.W., Fraser, C.M., and Barrell, B. (2002) Genome sequence of the human malaria parasite *Plasmodium falciparum*. *Nature*, 419: 498–511.

Gibbs, S.P. (1981) The chloroplasts of some algal groups may have evolved from endosymbiotic eukaryotic algae. *Annals of the New York Academy of Sciences*, 361: 193–208.

Graham, L.E. and Wilcox, L.W. (2000) *Algae*. Prentice Hall, Upper Saddle River, NJ.

Gray, M.W. (1989) The evolutionary origins of organelles. *Trends in Genetics*, 5: 294–299.

Gray, M.W. (1999) Evolution of organellar genomes. *Current Opinion in Genetics and Development*, 9: 678–687.

Hackett, J.D., Yoon, H.S., Soares, M.B., Bonaldo, M.F., Casavant, T.L., Scheetz, T.E., Nosenko, T., and Bhattacharya, D. (2004) Migration of the plastid genome to the nucleus in a peridinin dinoflagellate. *Current Biology*, 14: 213–218.

Harper, J.T. and Keeling, P.J. (2003) Nucleus-encoded, plastid-targeted glyceraldehyde-3-phosphate dehydrogenase (GAPDH) indicates a single origin for chromalveolate plastids. *Molecular Biology and Evolution*, 20: 1730–1735.

Harper, J.T., Waanders, E., and Keeling, P.J. (2005) On the monophyly of chromalveolates using a six-protein phylogeny of eukaryotes. *International Journal of Systematic and Evolutionary Microbiology*, 55: 487–496.

Howe, C.J., Barbrook, A.C., Koumandou, V.L., Nisbet, R.E.R., Symington, H.A., and Wightman, T.F. (2003) Evolution of the chloroplast genome. *Philosophical Transactions of the Royal Society of London Series B-Biological Sciences*, 358: 99–106.

Johnson, M.D., Tengs, T., Delwiche, C.F., Oldach, D., and Stoecker, D.K. (2004) Highly divergent SSU rRNA genes found in marine ciliates *Myrionecta rubra* and *Mesodinium pulex*. *Molecular Biology and Evolution*, in press.

Ingenhousz, J. (1779) *Experiments Upon Vegetables*. P. Emsly and H. Payne, London. [Reprinted in Reed, H.S., ed. (1949) *Chronica Botanica* 11:285–396.]

Karol, K.G., McCourt, R.M., Cimino, M.T., and Delwiche, C.F. (2001) The closest living relatives of land plants. *Science*, 294: 2351–2353.

Keegstra, K. and Cline, K. (1999) Protein import and routing systems of chloroplasts. *The Plant Cell*, 11: 557–570.

Keeling, P. (2004a) A brief history of plastids and their hosts. *Protist*, 155: 3–7.

Keeling, P.J. (2004b) Diversity and evolutionary history of plastids and their hosts. *American Journal of Botany*, 91: 1481–1493.

Keeling, P.J. and Palmer, J.D. (2001) Lateral transfer at the gene and subgenic levels in the evolution of eukaryotic enolase. *Proceedings of the National Academy of Sciences of the United States of America*, 98: 10745–10750.

Kohler, S., Delwiche, C.F., Denny, P.W., Tilney, L.G., Webster, P., Wilson, R.J.M., Palmer, J.D., and Roos, D.S. (1997) A plastid of probable green algal origin in apicomplexan parasites. *Science*, 275: 1485–1489.

Laatsch, T., Zauner, S., Stoebe-Maier, B., Kowallik, K.V., and Maier, U.-G. (2004) Plastid-derived single gene minicircles of the dinoflagellate *Ceratium horridum* are localized in the nucleus. *Molecular Biology and Evolution*, 21: 1318–1322.

Margulis, L. (1968) Evolutionary criteria in thallophytes—a radical alternative. *Science*, 161: 1020–1022.

Marin, B., Nowack, E.C.M., and Melkonian, M. (2005) A plastid in the making: primary endosymbiosis. *Protist*, 156: 425–432.

Martin, W., Rujan, T., Richly, E., Hansen, A., Cornelsen, S., Lins, T., Leister, D., Stoebe, B., Hasegawa, M., and Penny, D. (2002) Evolutionary analysis of *Arabidopsis*, cyanobacterialand chloroplast genomes reveals plastid phylogeny and thousands of cyanobacterial genes in the nucleus. *Proceedings of the National Academy of Sciences of the United States of America*, 99: 12246–12251.

McFadden, G. (2001) Chloroplast origin and integration. *Plant Physiology* 125: 50–53.

McFadden, G.I., Reith, M.E., Munholland, J., and LangUnnasch, N. (1996) Plastid in human parasites. *Nature*, 381: 482.

Moreira, D. and Philippe, H. (2001) Sure facts and open questions about the origin and evolution of photo-synthetic plastids. *Research in Microbiology*, 152: 771–780.

Okamoto, N. and Inouye, I. (2005) A secondary symbiosis in progress? *Science*, 310: 287.

Palenik, B. (2002) The genomics of symbiosis: hosts keep the baby and the bath water. *Proceedings of the National Academy of Sciences of the United States of America*, 99: 11996–11997.

Palmer, J.D. (2003) The symbiotic birth and spread of plastids: how many times and whodunit? *Journal of Phycology*, 39: 4–11.

Patron, N.J., Rogers, M.B., and Keeling, P.J. (2004) Gene replacement of fructose-1,6-bisphosphate aldolase supports the hypothesis of a single photosynthetic ancestor of chromalveolates. *Eukaryotic Cell*, 3: 1169–1175.

Patterson, C. (1989) Phylogenetic relations of major groups: conclusions and prospects. In *The Hierarchy of Life*. (eds. Fernholm, B., Bremer, K., and Jornvall, H.). Elsevier Science Publishers (Biomedical Division), Amsterdam; New York, pp. 471–488.

Patterson, D.J. (1999) The diversity of eukaryotes. *American Naturalist*, 154: S96–S124.

Reumann, S., Davila-Aponte, J., and Keegstra, K. (1999) The evolutionary origin of the protein-translocating channel of chloroplastic envelope membranes: identification of a cyanobacterial homolog. *Proceedings of the National Academy of Sciences of the United States of America*, 96: 784–789.

Roger, A.J. and Silberman, J.D. (2002) Cell evolution: mitochondria in hiding. *Nature*, 418: 827–829.

Rumpho, M.E., Summer, E.J., and Manhart, J.R. (2000) Solar-powered sea slugs. Mollusc/algal chloroplast symbiosis. *Plant Physiology*, 123: 29–38.

Sánchez Puerta, M.V., Bachvaroff, T.R., and Delwiche, C.F. (2004) The complete mitochondrial genome of the haptophyte *Emiliania huxleyi* and its relation to heterokonts. *DNA Research* 11: 1–10.

Sanchez-Puerta, M.V., Lippmeier, J.C., Apt, K.E., and Delwiche, C.F. (2006) Plastid genes in a non-photosynthetic dinoflagellate. *Protist*, 158: 105–117.

Stiller, J.W., Reel, D.C., and Johnson, J.C. (2003) A single origin of plastids revisited: convergent evolution in organellar genome content. *Journal of Phycology*, 39: 95–105.

Tappan, H. (1980) *The Paleobiology of Plant Protists*. W.H. Freeman, San Francisco, CA.

Taylor, F.J.R. (1976) Flagellate phylogeny: a study in conflicts. *Journal of Protozoology*, 23: 28–40.

Tengs, T., Dahlberg, O.J., Shalchian-Tabrizi, K., Klaveness, D., Rudi, K., Delwiche, C.F., and Jakobsen, K.S. (2000) Phylogenetic analyses indicate that the 19' hexanoyloxy-fucoxanthin-containing dinoflagellates have tertiary plastids of haptophyte origin. *Molecular Biology and Evolution*, 17: 718–729.

Tyler, B.M., Tripathy, S., Zhang, X., Dehal, P., Jiang, R.H.Y., Aerts, A., Arredondo, F.D., Baxter, L., Bensasson, D., Beynon, J.L., Chapman, J., Damasceno, C.M.B., Dorrance, A.E., Dou, D., Dickerman, A.W., Dubchak, I.L., Garbelotto, M., Gijzen, M., Gordon, S.G., Govers, F., Grunwald, N.J., Huang, W., Ivors, K.L., Jones, R.W., Kamoun, S., Krampis, K., Lamour, K.H., Lee, M.-K., McDonald, W.H., Medina, M., Meijer, H.J.G., Nordberg, E.K., Maclean, D.J., Ospina-Giraldo, M.D., Morris, P.F., Phuntumart, V., Putnam, N.H., Rash, S., Rose, J.K.C., Sakihama, Y., Salamov, A.A., Savidor, A., Scheuring, C.F., Smith, B.M., Sobral, B.W.S., Terry, A., Torto-Alalibo, T.A., Win, J., Xu, Z., Zhang, H., Grigoriev, I.V., Rokhsar, D.S., and Boore, J.L. (2006) Phytophthora genome sequences uncover evolutionary origins and mechanisms of pathogenesis. *Science*, 313: 1261–1266.

van den Hoek, C., Mann, D.G., and Jahns, H.M. (1995) *Algae: An Introduction to Phycology.* Cambridge University Press, Cambridge, UK.

Yoon, H.S., Hackett, J.D., Pinto, G., and Bhattacharya, D. (2002) The single, ancient origin of chromist plastids. *Proceedings of the National Academy of Sciences of the United States of America*, 99: 15507–15512.

Yoon, H.S., Hackett, J.D., Ciniglia, C., Pinto, G., and Bhattacharya, D. (2004) A molecular timeline for the origin of photosynthetic eukaryotes. *Molecular Biology and Evolution*, 21: 809–818.

Zhang, Z., Green, B.R., and Cavalier-Smith, T. (1999) Single gene circles in dinoflagellate chloroplast genomes. *Nature*, 400: 155–159.

3 Evolution and relationships of algae: major branches of the tree of life

Thomas Cavalier-Smith

CONTENTS

ABSTRACT

Algae are oxygenic photosynthesizers other than embryophyte land plants. Despite evolutionary unity in photosynthetic machinery, their cell membranes are organized in three profoundly different ways, reflecting diverse modes of origin. Proalgae (cyanobacteria and prochlorophytes) constitute one bacterial phylum. Eualgae comprise three eukaryotic phyla that diverged after a biciliate protozoan permanently enslaved a cyanobacterium to make the first chloroplast and the ancestor of the plant kingdom: glaucophytes diverged first, then red and green algae. Other eukaryotic algae are meta-algae, formed by enslavements of eualgae by biciliate protozoan hosts that created novel genetic membranes. Most are chromophyte algae (those with chlorophyll c: dinoflagellates, cryptophytes, heterokonts, and haptophytes), which arose by one enslavement of a red alga. Numerous non-photosynthetic chromalveolates arose from chromophytes by multiple losses of photosynthesis and plastids. Chloroplast replacement occurred twice within dinoflagellates, increasing their plastid diversity. Chromalveolates may be sisters of Plantae. Other meta-algae (euglenoids, chlorarachneans) arose from cells that enslaved chlorophyte green algae and are the only algae in the protozoan infrakingdoms Excavata and Rhizaria; whether they have a photosynthetic common ancestor is unclear.

INTRODUCTION

Algae are best defined as oxygenic photosynthesizers other than embryophyte land plants. They are not a taxonomic group but an important ecologically and functionally analogous group of organisms defined both by their evolutionary origins and by the origins from them of numerous kinds of organisms that are not themselves algae. Derivatives of algae include many non-photosynthetic protists (unicellular eukaryotes) and the embryophyte land plants (Cormophyta: bryophytes and tracheophytes) that colonized the land in the Silurian period over 400 My ago.

The first algae were Cyanobacteria, which evolved from still older anoxygenic photosynthetic bacteria, about 2.8 Gy ago (Cavalier-Smith, 2006a). Their origin diversified habitats for life by adding oxygen to the atmosphere, which oxidized the surface of the earth, rusting deserts and cliffs to yellow, brown, or reddish, rather than black or green, and which generated the ozone layer that protects life against excessive ultraviolet (UV) radiation. Nearly two billion years later a cyano-bacterium was enslaved and incorporated as the first chloroplast within the cell of a biciliate protozoan to form the first eukaryote alga—the progenitor of the plant kingdom (Cavalier-Smith, 1982; Mereschkovsky, 1905).

At least four times in Earth history eukaryotic algae were similarly enslaved by and integrated into phagotrophic host cells to generate structurally even more complex algae with diverse pigments (Cavalier-Smith, 2000a). This chapter discusses these integrative processes of symbiogenesis, as Mereschkovsky (1910) first called them, and places the major algal groups within the broad evolutionary context of the "tree of life."

THE SHAPE OF THE TREE: DIVERGENCE AND SYMBIOGENETIC CELL MERGERS

Within a biparental species such as ourselves or most eukaryotic algae, "web of life" would be a more apt metaphor than a tree, as sexual cell fusion occurs frequently, making lateral links. For human, animal, or plant history, the term "family tree" is thus a marked distortion of reality by those patrilineal Eurasian societies that dominated culture during the rise of science. In sexual organisms over the short term, the stream of life actually forms a genealogical web, not a family tree. Even among species, if closely related, hybridization can occur, making microevolution reticulate, not solely divergent. On a grander scale however, the tree of life with successively diverging branches is an apt metaphor; successive bifurcations predominate, but symbiogenesis (the complete merger of symbiont and host into a single organism) causes very rare lateral fusions (which sometimes occur even in real trees). That of Figure 3.1 places algae in historical context. Although its branching order and sister and ancestor–descendant relationships are intended to be accurate, its time dimension is grossly distorted, saying almost nothing about rates of change through time. Branches on evolutionary trees are seldom evenly spaced. Often, a major innovation in body plan stimulates a rather sudden burgeoning of lineages over such a short interval that their order is very hard to resolve. Thus, if we scaled phylogenetic trees accurately in time they would often resemble a series of nested multistem bushes, not a single-trunk forest tree. I shall focus on the relationships of algae and their ancestors and descendants; only occasionally do I mention the likely timing of critical events as deduced from the fossil record and discussed in detail by Cavalier-Smith (2006a).

Figure 3.1 shows that all eukaryote algae fall in only one of the two major eukaryote branches, the bikonts, and that all four major bikont groups contain at least some algae. Even though the unikont branch is exclusively heterotrophic, some unikonts learned to cultivate algae in their own tissues to become photosynthetic superorganisms. Fungi that enslaved many kinds of algae to become lichens, and the reef-building corals that cultivate dinoflagellate algae intracellularly, are familiar examples. But although many features of the slave-owning lichen fungi and corals are adapted to their intracorporeal gardening way of life, the algae thus enslaved are not integrated into their cells and germ lines and can live independently and remain distinct organisms. They are examples of symbiosis, in a version known as helotism (slavery) rather than mutualism or parasitism, not of symbiogenesis. Photosynthetic corals and lichens are symbiotic consortia or superorganisms, not single organisms and species. These consortia are therefore not algae, even though one component can perform oxygenic photosynthesis and is not an embryophyte.

By contrast, when cyanobacteria were permanently enslaved as chloroplasts, two formerly distinct organisms were truly merged into one: genuine symbiogenesis. Symbiogenesis happened only once before, when a probably photosynthetic α–proteobacterium was enslaved to make mitochondria (during the later stages of the origin of eukaryotes) (Cavalier-Smith, 2006b). Possibly every case of symbiogenesis in the history of life involved a photosynthetic slave. The simplest criterion for distinguishing an enslaved symbiont (e.g. a dinoflagellate in a coral or a *Buchnera* bacterium in an aphid) from a former symbiont that symbiogenesis made a true cell organelle is that an organelle (e.g. a chloroplast or mitochondrion) can import numerous proteins from the rest of the cell but an enslaved symbiont (e.g. *Buchnera*) cannot (Cavalier-Smith and Lee, 1985). Thus, it was not endosymbiosis itself, but the subsequent origin of a generalized protein-import machinery that created these two novel organelles. Once a generalized protein-import mechanism evolves, the host can insert into the resulting organelle any proteins that bear the correct import signals recognized by that machinery. The imported proteins include many originally coded by the host. Conversely, copies of many symbiont genes were inevitably accidentally incorporated into the host nucleus (perhaps following accidental lysis of a symbiont); if the nuclear copies evolve the correct targeting sequences for protein import into the new organelle, and their encoded proteins can actually be imported efficiently without getting stuck in the bounding membranes, then the nuclear-coded copies can and usually do entirely replace the ancestral symbiont-located genes, which are then sooner or later inevitably lost by accidental deletions (Cavalier-Smith, 2000a). The origin of a novel organelle-specific protein-import mechanism is what integrates a symbiont as an organelle, not gene transfer itself. In addition, thousands of genes transferred to the host nucleus were retained by it without ever evolving chloroplast import sequences, so their proteins are cytosolic. Thus, both chloroplasts and the rest of the plant cell are chimaeras of host and symbiont proteins; host and symbiont are fully merged as one organism.

Symbiogenesis and symbiosis are often confused, but the distinction is very important. Intracellular symbiosis is entirely unknown in free-living bacteria, which lack evolved mechanisms for internalizing other cells. In the only case of intracellular symbiosis within a bacterium, the host (very likely with a reduced wall) is itself a symbiont within a eukaryotic cell (von Dohlen et al., 2001), making it irrelevant to the origin of mitochondria (Cavalier-Smith, 2002c). Intracellular symbiosis became very easy when the first eukaryotes evolved the capacity for phagocytosis, as the engulfed cells can frequently escape digestion and live for a period in the eukaryote host. Therefore, in all major groups of eukaryotes intracellular symbiosis has evolved many thousands of times; symbionts enter eukaryote host cells millions of times a day. Thus, symbiosis is evolutionarily easy; but symbiogenesis is not, as the requisite novel protein-import mechanisms evolved only six or seven times in the history of life. Though exceedingly rare, symbiogenesis was crucial for the evolution of all major groups of algae except cyanobacteria. The contrasting evolutionary consequences of symbiosis and symbiogenesis are well illustrated by comparing algal chloroplasts with aphid symbionts, where *Buchnera* provide amino acids to the host and are essential for its health. Even though the aphid is not healthy or fertile without a symbiont providing it with amino acids, *Buchnera* are not integrated as organelles, for protein-import machinery never evolved during the 200 My of this obligate symbiosis (van Ham et al., 2003) and the symbiont can be functionally replaced by others (Koga et al., 2003). Furthermore, despite its massive genome reduction (nearly eightfold), there is no evidence for gene transfer to the host genome, and the enslaved bacteria still retain about 580 protein-coding genes—far more than chloroplasts that never retain more than 209 and sometimes as few as 10 protein-coding genes. This degree of genome reduction by the enslaved cyanobacterium would have been impossible without a generalized protein-import mechanism.

I shall start with proalgae, the only ones not created by symbiogenesis.

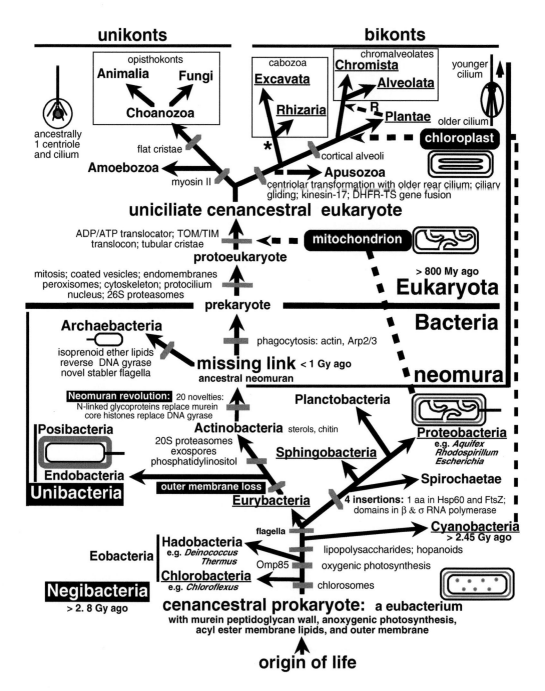

FIGURE 3.1 The tree of life showing major innovations in cell biology. Partly, largely, or entirely photosynthetic taxa are underlined. The first cells were negibacteria, with an envelope of two membranes. Negibacteria include the always oxygenically photosynthetic Cyanobacteria (proalgae), four ancestrally anoxygenically photosynthetic phyla (but with multiple losses of photosynthesis), and three purely heterotrophic phyla. Three fundamental quantum evolutionary transitions generated radically novel cell types: loss of the negibacterial outer membrane by murein hypertrophy created Posibacteria (Endobacteria and Actinobacteria) with very thick murein walls with proteins attached covalently by sortase enzymes; replacement of murein and lipoprotein by N-linked glycoproteins and of DNA gyrase by histones generated neomura from an actinobacterial ancestor; thus freed of their murein corset, one neomuran lineage evolved phagotrophy, endomembranes, and endoskeleton and enslaved an α-proteobacterium as mitochondria to become eukaryotes. Posibacteria and Negibacteria are often collectively lumped as the grade eubacteria. Early bacteria had only

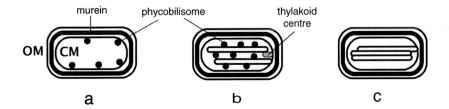

FIGURE 3.2 The three types of Cyanobacteria (proalgae) all have a murein peptidoglycan wall between the cytoplasmic membrane (CM) and the outer membrane (OM) and two membrane-embedded photosystems (PSI and PSII; the latter splits water to generate hydrogen to reduce CO_2, yielding oxygen waste). (a) The cyanobacterium *Gloeobacter* has both photosystems and the respiratory chain embedded in the CM and phycobilisomes with antenna pigments attached to it. More advanced cyanobacteria (Phycobacteria) have flattened thylakoids bearing both photosystems, with respiratory chains restricted to the cytoplasmic membrane (b, c). (b) Phycobacteria ancestrally had phycobilisomes attached to PSII and a thylakoid centre at the point where thylakoids diverge and often contact (sometimes in continuity with?) the CM. (c) Advanced phycobacteria polyphyletically lost phycobilisomes and evolved thylakoid stacking and either evolved chlorophyll b as antenna pigments (prochlorophytes; linked to PSII in *Prochloron* and PSI in *Prochlorococcus* [Bibby et al., 2003]) or replaced chlorophyll a by d in PSII (*Acaryochloris*; Chen et al., 2005).

THE PROALGAE: CYANOBACTERIA, PROCHLOROPHYTES, AND THE BACTERIAL TREE

Cyanobacteria are one of only ten bacterial phyla in my most recent classification of prokaryotes (Cavalier-Smith, 2002b, 2006c). They are the only bacteria that live by oxygenic photosynthesis and evolved three types (Figure 3.2a through Figure 3.2c). They evolved from anaerobic ancestors that carried out anoxygenic photosynthesis by a single photosystem. All photosynthetic bacteria are negibacteria, i.e. they are bounded by an envelope of two membranes: an inner cytoplasmic membrane (CM) to which ribosomes can temporarily attach by a signal recognition particle (SRP), just as occurs in the endoplasmic reticulum of eukaryotic cells, and an outer membrane (OM) separated from the CM by the periplasmic space containing the cell wall (Figure 3.3a). The OM can only grow by the export of lipids and proteins made in the CM. The cell wall of Planctobacteria (see Figure 3.1) is generally of protein, but that of all other negibacteria, including cyanobacteria, is made of the peptidoglycan murein, whose hydrophilic precursors are moved across the hydrophobic CM by covalent attachment to long isoprenol carrier molecules. Murein forms a three-dimensionally covalently bonded meshwork able to resist the immensely high internal osmotic pressures that cause eubacterial cells to grow when the cross-linking is temporarily weakened by controlled enzymatic covalent bond cleavage.

FIGURE 3.1 (CONTINUED) gliding motility, rotary flagella evolving later, independently in negibacteria and archaebacteria, but cilia evolved in the first eukaryote, an aerobic heterotroph. This split into two branches. Unikonts remained heterotrophic with the ancestral simple cytoskeleton of single microtubules diverging from the originally single centriole (later two in opisthokonts). For gliding on substrates bikonts evolved a second posterior cilium, generated by ciliary transformation over two cell cycles, and distinct anterior and posterior ciliary roots of microtubule bands (thumbnail sketch); one lineage enslaved a phagocytosed cyanobacterium to form chloroplasts and the plant kingdom. After plants diverged into three types of eualgae, another bikont enslaved a red alga (R) to make chromalveolates, and green algae were enslaved to make chlorarachnean algae (Cercozoa) within Rhizaria and Euglenia (Euglenozoa) within Excavata. Whether green algal enslavement was a unique event (asterisk) in a common ancestor of Excavata and Rhizaria or occurred twice independently is uncertain; recent multigene trees mostly show cabozoa as monophyletic (Burki and Pawlowski, 2006), but the basal branching order of bikonts is unclear—possibly cabozoa are actually sisters of chromalveolates (Burki and Pawlowski, 2006); other multigene protein and rRNA trees (Kim et al., 2006; Moreira et al., 2007) suggest that Apusozoa are the most divergent bikonts, as shown, but they might instead be related to excavates. Note that the phyla Choanozoa and Amoebozoa and infrakingdoms Excavata, Rhizaria, and Alveolata collectively comprise the basal eukaryotic kingdom Protozoa, from which the four derived kingdoms (Animalia, Fungi, Plantae, and Chromista) independently evolved.

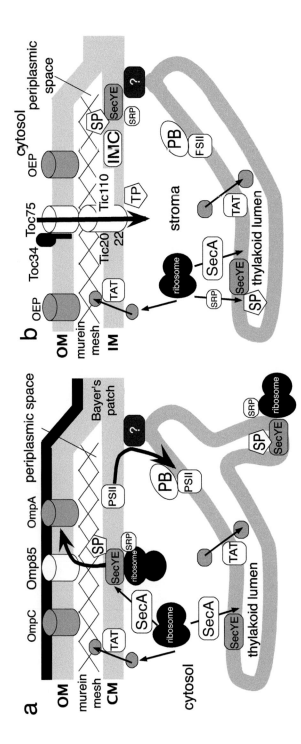

FIGURE 3.3 Envelope organization and protein targeting in typical cyanobacteria (a) and glaucophyte chloroplasts (b) showing evolutionary continuity and innovation. The outer membranes (OMs) are homologous; their proteins are β-barrel proteins absent from other membranes, but the lipopolysaccharide (LPS) of the outer leaflet (black) was replaced by phosphatidylcholine from the host when chloroplasts evolved. (a) Proteins made by membrane-bound ribosomes have N-terminal hydrophobic signals recognized by ribonucleoprotein signal recognition particles (SRPs) that dock onto membrane receptors, thus allowing the ribosomes to extrude the growing protein across the membrane through a gated SecYE channel before it folds up. Unless it is needed for attachment of the mature protein to the traversed membrane, the signal peptide is cut off by a signal peptidase (SP). In addition to this cotranslational export, free ribosomes also make proteins able to cross membranes by earlier associating with SecA chaperone and being similarly extruded through SecYE before they fold or else by folding first and then being moved across the membrane by twin arginine translocases (TAT) as native proteins (thin arrows). Cylindrical OM proteins (porins: Omps) with central pores are essential to allow nutrients to cross the OM. After synthesis and signal peptide cleavage, Omps are carried across the periplasmic space by chaperones and inserted into the OM by Omp85 (thick arrow). PSII is assembled in the CM and passes to the thylakoids by unknown mechanisms (thick arrow; ?) and phycobilisomes (PB) are added. OM lipids are made in the IM and probably exit via Bayer's patches, adhesion zones between CM and OM. (b) The key innovation that enslaved cyanobacteria to make chloroplasts was the import of novel inner membrane carriers (IMCs, e.g. triose phosphate/phosphate antiporter; Knappe et al., 2003) from the cytosol though the Toc75 channel, which evolved from Omp85, and their insertion into the inner membrane (IM; Cavalier-Smith, 2000a). Imported proteins bear N-terminal transit sequences that are recognized by a novel GTPase receptor (Toc34), which pushes them through Toc75. They are then pulled through the IM channel Tic110 by stromal chaperones using ATP energy, and a transit peptidase removes the transit sequences (thick arrow). In some proteins this exposes a subterminal signal sequence recognizable either by SRP or SecA for extrusion through SecYE into the thylakoid lumen (or out through the IM in the case of Toc75; other OM proteins (outer envelope proteins [OEPs], probably descendants of cyanobacterial porins; Vothknecht and Soll, 2005) lack transit sequences and insert directly from the cytosol. Thylakoid, IM, and OM inner leaflet lipids are galactolipids or sulpholipids, as in cyanobacteria, and are made in the chloroplast IM not the endoplasmic reticulum (ER), which makes phospholipids. Chloroplasts make all the cell's fatty acids, using stromal enzymes (all imported from the cytosol but originally of cyanobacterial origin), so many must be exported. TAT machinery is conserved in plastids but the SRP is modified by losing its RNA component and working post-translationally (Gutensohn et al., 2006).

Cell envelope evolution and the root of the tree

In cyanobacteria and most other negibacteria the main OM proteins are large cylindrical β-barrel proteins, typically with a central pore; such porins make the OM much more permeable to small hydrophilic molecules than is the CM. These β-barrel proteins pass across the periplasmic space with the help of chaperone proteins and are then inserted into the OM by a specific OM protein (Omp85) dedicated as a catalyst of their insertion (Figure 3.3).

Omp85, itself a β-barrel protein, is essential for its own insertion as well as that of all other β-barrel proteins. This mode of growth of the OM has been highly conserved in evolution and was retained by the OM of all chloroplasts, where Toc75 (a key component of their protein-import machinery) is an evolutionary homologue of negibacterial Omp85. Not only did Omp85 play a key role in the origin of chloroplasts, but understanding its early evolution is the key to correctly placing the root on the whole tree of life, because once evolved it was virtually impossible to be lost (Cavalier-Smith, 2006c). In *Escherichia coli*, deleting the Omp85 gene kills the cells because they cannot insert OM proteins. Even our own cells retain this negibacterial mechanism for inserting β-barrel proteins into the OM of mitochondria, which directly descended from the OM of the α-proteobacterium that was enslaved to make the first mitochondrion. In mitochondria the insertion catalyst is a different Omp85 homologue, Sam50, and in chloroplasts it is Toc75. Deleting Sam50 from yeast kills the cell as it can no longer insert β-barrel proteins; plants die sooner or later when Toc75 is inactivated (Hust and Gutensohn, 2006). In eukaryotes the presence of β-barrel proteins in the OM of mitochondria and chloroplasts is the strongest evidence that these organelle OMs evolved from the OM of the enslaved negibacteria, as I first argued (Cavalier-Smith, 1982, 1983), not from the food vacuole membrane of the host that first engulfed it, as previously incorrectly supposed (Schnepf, 1964; Margulis, 1970, 1981).

The fact that even the highly modified chloroplasts and mitochondria, whose genomes shrank almost to zero, retain the original negibacterial Omp85 mechanism emphasizes its indispensability. I argued that no organism in the history of life ever lost Omp85 without simultaneously losing the entire OM (Cavalier-Smith, 2006c). Loss of Omp85 alone would be lethal, as the OM would become impermeable to many nutrients if β-barrel proteins could not be inserted, thus starving the cell. This strongly indicates that the apparent absence of β-barrel proteins from the OM of Chlorobacteria must be the primitive state, not a derived one, implying that the root of the negibacteria lies between Chlorobacteria and all phyla that possess Omp85 (Cavalier-Smith, 2006a, 2006c). This probable position of the root of the tree of life (Figure 3.1) contradicts conventional, but ill-founded, assumptions that it is between neomura and eubacteria, and needs testing by studying OM proteins of Chlorobacteria. If correct, however, it implies that the simple mechanism of anoxygenic photosynthesis used by Chlorobacteria (e.g. *Chloroflexus*) is the closest to the ancestral one from which the more complex cyanobacterial machinery evolved.

Chlorobacteria and Hadobacteria (collectively designated Eobacteria) are simpler than cyanobacteria and other negibacteria in that both leaflets of their outer membrane (OM) are composed of acyl ester lipids, like the CM and eukaryote membranes. The outer leaflet of the OM of other negibacteria, including cyanobacteria, is composed instead of the very complex lipopolysaccharide (LPS) which makes it more impermeable to hydrophilic molecules. The eobacterial absence of LPS was probably the primitive state for negibacteria; I group all LPS-containing negibacteria together as glycobacteria. LPS probably evolved during a radical set of changes in the OM referred to as the glycobacterial revolution, when hopanoids (isoprenoids not requiring oxygen for their synthesis, unlike sterols) that stabilize membranes also arose, as did several macromolecular assemblies of proteins that actively import molecules (e.g. TonB) or actively export proteins (e.g. TolC) across the increasingly impermeable OM (Cavalier-Smith, 2006a, 2006c). Porins, which allow passive exchange of small molecules across the OM, probably evolved in the common ancestor of glycobacteria and Hadobacteria and may have been a prerequisite for the subsequent evolution of LPS.

Cyanobacteria are more primitive than other glycobacteria in lacking flagella. For motility they rely on gliding on surfaces by secreting slime through a large proteinaceous assemblage that penetrates both the CM and OM. I suggested that this cylindrical junctional pore complex was the evolutionary precursor of the rotary basal body of eubacterial flagella (Cavalier-Smith, 2006a, 2006c). The motor of such flagella is made of two proteins: MotA, a proton channel in the CM, and MotB that attaches MotA to the murein wall so that when MotA applies the rotary force to the basal body the flagellum rotates but MotA itself does not. MotB is an evolutionary chimaera of TonB and the porin OmpA and must have first evolved in a negibacterium as neither evolutionary precursor is ever found in posibacteria. Likewise, MotA is a homologue of a CM proton channel ExbB found in cyanobacteria and other negibacteria only. Thus all three probable precursors of flagella are present in cyanobacteria, and widely in negibacteria, but are totally absent from posibacteria. This means that eubacterial flagella almost certainly first evolved in negibacteria and that the homologous flagella of posibacteria must have evolved from them. This strong evidence for the polarization of evolution from negibacteria to posibacteria means that posibacteria evolved from negibacteria by losing the OM; such loss of the OM could have been caused mechanically by hypertrophy of murein to form the very thick cell wall of posibacteria.

This polarizing of the eubacterial tree by the evolution of Omp85/β-barrel proteins and eubacterial flagella together means that the first cells probably had an envelope of two membranes, not just one as was assumed before Blobel (1980) first suggested that negibacteria preceded posibacteria in evolution. How the double envelope originally evolved is discussed in Cavalier-Smith (2001). The rooting of the tree of life within negibacteria is also shown by trees for protein paralogues of many metabolic enzymes (Kollman and Doolittle, 2000; Peretó et al., 2004). Paralogues are functionally different but related genes produced by gene duplication. When a combined tree for both is constructed, the position of the line joining the two subtrees theoretically can locate the root of each. In practice, however, trees for different paralogues can be contradictory because of lack of resolution or systematic biases. Thus, paralogue trees for molecules such as protein synthesis elongation factors and RNA polymerases do not place the root within negibacteria but between archaebacteria and posibacteria. I have argued that this position is incorrect and is instead a tree-reconstruction long-branch artefact resulting from accelerated evolution in the ancestors of these molecules during the neomuran revolution—the major changes in protein secretion, ribosomes, and DNA handling enzymes that took place in the common ancestor of eukaryotes and archaebacteria (Figure 3.1). Many biologists (e.g. Kollman and Doolittle, 2000) make the converse assumption that the metabolic enzyme paralogue trees that typically place the root among negibacteria are the misleading ones and that the contradictory elongation factor and RNA polymerase trees are correct. In my view there is no evidence for that interpretation and many reasons for thinking that it is wrong, as explained in detail elsewhere (Cavalier-Smith, 2002b, 2006a, 2006c). These reasons include the much greater antiquity of negibacteria compared with eukaryotes as shown by palaeontology and the polarization of most of the 20 major changes during the neomuran revolution from eubacteria (negibacteria plus posibacteria) to neomura (eukaryotes plus archaebacteria) not the reverse.

A further strong argument against the root of the tree being between eubacteria and neomura comes from evolution of the proteasome, a cylindrical protein digestion chamber found not only in neomura, but also in Actinomycetales among the Actinobacteria. I argued that the 20S core of the proteasome evolved by gene duplication and differentiation into its α- and β-subunits from the simpler eubacterial protease HslV. Provided that Actinomycetales did not obtain their proteasome genes from archaebacteria by lateral transfer, which though claimed (Gille et al., 2003) has not been demonstrated, this polarization from HslV to proteasome rather than the reverse excludes the root of the tree from any clade including both neomura and Actinomycetales; thus, the root is within eubacteria, not between them and neomura as suggested by a minority of paralogue trees (Cavalier-Smith, 2006c).

Origin of oxygenic photosynthesis and thylakoids

Although it is often asserted that cyanobacteria were the first organisms to have evolved oxygenic photosynthesis, this may not be correct. Cyanobacteria are unique among the five ancestrally photosynthetic negibacterial phyla shown in Figure 3.1 in having two dissimilar photosystems. Photosystem II, which mediates water splitting and oxygen release, is related to the single photosystem possessed by purple photosynthetic bacteria (phylum Proteobacteria), whereas photosystem I is related to that of green sulphur bacteria (phylum Sphingobacteria). Yet, as Figure 3.1 indicates, Proteobacteria and Sphingobacteria are probably part of a single derived clade that diverged from the eurybacterial ancestors of Posibacteria after the origin of flagella. This means that the common ancestor of Proteobacteria and Sphingobacteria probably possessed homologues of both photosystems I and II. As there is good evidence from signature sequences and gene trees that cyanobacteria are holophyletic, not paraphyletic, this means that the photosystem duplication that generated separate complexes (I and II) took place prior to the divergence of cyanobacteria from that ancestor (i.e., below the branching point of cyanobacteria in Figure 3.1) (see Figure 3.4). If duplicated photosystems were initially selected because they allowed oxygenic photosynthesis, it follows that oxygenic photosynthesis had also evolved prior to that branch point and was secondarily lost by the ancestors of Sphingobacteria,

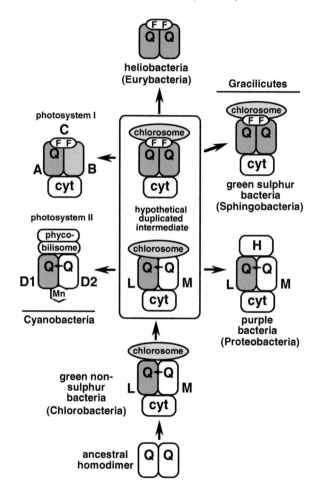

FIGURE 3.4 Hypothetical phylogeny for photosynthetic reaction centres. Prior to the last common ancestor of all extant life, the primitive reaction centre, a homodimer with two bound quinones, each donating electrons to a primitive cytochrome cc complex, evolved into the heterodimeric type found in green non-sulphur bacteria (Chlorobacteria). This was duplicated prior to divergence of cyanobacteria and gracilicutes to generate a modified homodimeric type of cytochrome bc1 complex with iron–sulphur clusters (FF). Cyanobacteria converted the two versions into photosystems I and II. Proteobacteria replaced chlorosomes in the original heterodimeric type by an H subunit with purple carotenoid, but did not retain the new duplicate with FeS clusters. By contrast, this was the only version retained by green sulphur bacteria (Sphingobacteria) and Heliobacteria, both losing the earlier heterodimeric type. (After Cavalier-Smith, T., *International Journal of Systematic and Evolutionary Microbiology*, 52, 7–76, 2002; Cavalier-Smith, T., *Biology Direct*, 1, 19, 2006.)

Proteobacteria, and Eurybacteria when they independently lost one of the two photosystems and became adapted to different anaerobic niches from those of the more primitive Chlorobacteria. This is a simpler interpretation of photosystem evolution than those that invoke lateral gene transfer of complete photosystems among bacterial phyla, which has never been adequately demonstrated (Cavalier-Smith, 2006c). If it is correct, most heterotrophic bacteria, as well as mitochondria and the host component of the eukaryote chimaera, had algal ancestors in the very distant past. Because Hadobacteria, like more derived bacteria, have more catalases for reducing poisoning by oxygen than do Chlorobacteria, and these are most likely to have evolved in the very organism that first generated oxygen, I suggested that oxygenic photosynthesis first evolved in the immediate common ancestor of Hadobacteria and Cyanobacteria (Cavalier-Smith, 2006c). If that is correct then Chlorobacteria may be the only living phylum that did not ultimately have an algal ancestry. If oxygenic photosynthesis did indeed evolve prior to the common ancestor of Cyanobacteria and Proteobacteria, then the novel feature that defines the clade Cyanobacteria is not photosystem II but the phycobilisomes attached to it. The first cyanobacteria were probably red, as some still are, not blue-green.

Gloeobacter, the most divergent cyanobacterium on sequence trees, shows that the first cyanobacteria had photosystems embedded simply in the CM, like all other photosynthetic bacteria (which like them are always negibacteria) but with phycobilisomes for absorbing a broader spectrum of light attached to the inner face of photosystem II (Figure 3.3a). All other cyanobacteria have photosystems on apparently separate membranes, the thylakoids, homologous with those of chloroplasts (Figure 3.3a). Such more advanced cyanobacteria concentrate their respiratory chain exclusively in the CM. The origin of thylakoids as specialized photosynthetic organelles was so important for cyanobacterial evolution that I placed all organisms having them in a separate subphylum, Phycobacteria, distinct from *Gloeobacter*—whose membrane topology is the same as other negibacteria (Cavalier-Smith, 2002b). I previously argued that thylakoids are a distinct genetic membrane from the CM, and that their separation involved the origin of a novel specific protein-targeting machinery so that photosynthetic enzymes are targeted to them and respiratory and cell surface proteins to the CM (Cavalier-Smith, 2000a). Their distinctness, however, is not well established, and recent evidence suggests that there may be occasional localized continuity (van de Meene et al., 2006) and that photosystem II is first targeted to the CM and then moved into thylakoids prior to the attachment of phycobilisomes (Klinkert et al., 2004). What is clear is that phycobilisomes have been secondarily lost by more than one group of cyanobacteria, and other pigments (e.g. chlorophyll b, d, and Mg-protoporphyrins) evolved instead. Those with chlorophyll b are polyphyletic and are called prochlorophytes (Lewin, 2002) on the mistaken view that they are specifically related to green algal plastids. These pigment changes are probably adaptations to oceanic low-nitrogen deserts and different light regimes.

ORIGIN AND PRIMARY DIVERGENCE OF EUKARYOTES

It is now clear that the last common ancestor of all eukaryotes was an aerobic heterotroph (with mitochondria and at least one centriole and cilium) that fed phagotrophically and lacked chloroplasts. As no bacteria feed phagotrophically, the origin of phagocytosis was central to eukaryogenesis (evolution of the eukaryote cell). I have explained elsewhere (Cavalier-Smith, 1987, 2002c, 2006a) how phagocytosis was the cause of the origin of the endomembrane system and of an internal cytoskeleton with a capacity to transport organelles by intracellular molecular motors: myosin and kinesin (of common ancestry from a bacterial GTPase) and dynein (of separate origin from a bacterial ATPase). These innovations also rapidly led to the origin of the nucleus, mitosis, and cilia, and to the enslavement of a proteobacterium to make mitochondria (Cavalier-Smith, 2006a, 2006b).

Eukaryogenesis was immediately preceded by the neomuran revolution, during which eubacterial peptidoglycan wall was replaced by a potentially more flexible wall of discrete N-linked glycoproteins (Cavalier-Smith, 2002c). Loss of the three-dimensionally cross-linked eubacterial peptidoglycan wall was an essential prerequisite for the origin of phagotrophy and therefore for the subsequent uptake of a cyanobacterium to form the first chloroplast (Cavalier-Smith, 2002a, 2002c, 2006a).

The primary split in the eukaryote tree is between unikonts and bikonts (Figure 3.1). Unikonts ancestrally had a single centriole and cilium and a simple cytoskeleton consisting of a cone of microtubules diverging from the centriole and subtending the nucleus. They were ancestrally benthic amoeboflagellates creeping on surfaces and diverged early on into two protozoan lineages: the Amoebozoa, which typically have broad pseudopods, and the Choanozoa that typically have narrow thread-like pseudopods and which gave rise independently to the animal and fungal kingdoms. Bikonts ancestrally had two centrioles and cilia and a more complex cytoskeleton consisting of separate bands of microtubules associated with their dissimilar anterior and posterior cilia. Development of bikont centrioles and cilia is spread over at least two cell cycles; the youngest centriole/cilium is the anterior one that undergoes a transformation in structure and often in beat pattern and becomes posterior in the next cell cycle and a new anterior cilium grows. This pattern of ciliary transformation does not occur in unikonts, even those that are secondarily biciliate (e.g. some myxogastrid mycetozoa where the anterior cilium is the older mature one), and is thus a shared derived character for bikonts alone. Bikonts comprise the kingdoms Plantae and Chromista, each ancestrally algal and with numerous algal classes, and three protozoan infrakingdoms (Alveolata, Rhizaria, Excavata), each with only one algal or partially algal class but numerous classes of usually phagotrophic heterotrophs.

Despite their trophic and structural heterogeneity, bikonts are probably monophyletic and holophyletic. This is supported not only by their unique pattern of ciliary transformation and distinct anterior and posterior microtubule bands, but also by the presence only in bikonts of a derived fusion between the dihydrofolate reductase and thymidine synthetase gene (separately transcribed in the ancestral bacteria and in unikonts; Stechmann and Cavalier-Smith, 2003) and the presence in bikonts only of kinesin-17, a probable ciliary kinesin (Wickstead and Gull, 2006), and ciliary gliding on surfaces. By contrast, unikont holophyly is supported by four myosin synapomorphies, including the unique presence of myosin II—the myosin used for striated muscle contraction (Richards and Cavalier-Smith, 2005). A multigene tree for 149 different proteins also shows a robust bipartition between unikonts and bikonts and the monophyly of opisthokonts (animals, fungi, Choanozoa), Plantae, chromalveolates (Chromista plus Alveolata), and Excavata (Rodríguez-Ezpeleta et al., 2005). However, large numbers of sequences are unavailable for Rhizaria and some lesser groups, so the basal branching order of bikonts is uncertain; it is unresolved on single-gene trees, which also often fail to recover the monophyly of such groups as Plantae, Chromista, chromalveolates, and excavates, which are well established from other evidence (Cavalier-Smith and Chao, 2003). The problem is that early bikont diversification appears to have been rather sudden, with numerous lineages originating almost simultaneously, and there is not enough information conserved in any single gene to show their proper branching order confidently. Similar problems are encountered within all the individual algal groups as discussed in other chapters (and for the basal branching of the eubacterial phyla; Figure 3.1, where rare cladistic characters are used to deduce topology), making it essential to use multigene trees and rare cladistic characters to establish phylogenetic history, and to not rely solely on single genes (lateral gene transfer can also make single-gene trees misleading, especially in bacteria, but is hardly ever a problem for algae).

As explained in the section below on "Red algal enslavement and the diversification of chromalveolates," the evidence that Chromista and Alveolata are sisters and that all chromophytes had a common origin by the single intracellular enslavement of a red alga, and the associated single origin of chlorophyll c2 and a shared novel protein-targeting machinery, is now very strong. I have argued that chromalveolates and Plantae are probably sisters and that cortical alveoli (smooth-membraned alveoli attached below the plasma membrane, and typically also associated with a microtubular cortical skeleton) probably evolved in their common ancestor (Figure 3.1) and thus collectively called these two major groups the corticates (Cavalier-Smith, 2003c, 2004; Cavalier-Smith and Chao, 2003), which include the vast majority of eukaryote algae. However, even if the cortical alveoli of Glaucophyta (Plantae) and Alveolata are homologous (which needs testing by sequence studies of their proteins), there is good evidence that cortical alveoli can be secondarily lost (e.g. by karyorelictid ciliates); if this happened in the ancestor of either Rhizaria or Excavata, their absence would be phylogenetically misleading, and

Plantae and chromalveolates need not necessarily be sisters as in Figure 3.1. The branching order of corticates, Rhizaria, and Excavata is even less clear and therefore shown as a trifurcation (Figure 3.1). A further complication is that there are several apparently deeply branching bikont lineages for which only one or two gene sequences are available; most of these have been grouped in the protozoan phylum Apusozoa, but both its monophyly and relationship to the Excavata and Rhizaria are uncertain. Apusozoa are probably either sisters to Excavata (Cavalier-Smith, 2003a) or sisters to all other bikont groups (Cavalier-Smith, 2002c; Cavalier-Smith and Chao, 2003). The ancestral phenotype for Rhizaria was probably a benthic biciliate, like the cercozoan *Cercomonas,* that glided on surfaces on its posterior cilium and had a marked propensity for making branching filose pseudopods. As *Amastigomonas,* within Apusozoa, has a very similar phenotype and is probably at least 500 My old (Cavalier-Smith and Chao, 2003), I suggested that a gliding benthic biciliate with a propensity for making branching pseudopods is likely to have been the ancestral bikont phenotype (Cavalier-Smith, 2002c). Both ciliary transformation and the asymmetric ciliary roots of microtubule bands found in most eukaryotic algae probably first evolved to facilitate this benthic mode of life and its associated phagotrophy (mainly feeding on small bacteria other than cyanobacteria) by a soft surfaced flagellate.

On this view, the origin of cortical alveoli was a later surface skeletal specialization that enabled corticate predators to grow large and feed as planktonic swimmers on bigger prey such as cyano-bacteria. An early corticate dwelling in the photic zone was the host that enslaved a cyanobacterium to make the first plant. Somewhat later another corticate enslaved a red alga to generate chromal-veolates. Corticates soon dominated the plankton and eventually the whole algal world after evolving external walls rather than cortical alveoli allowed some polyphyletically to become macroalgae. Thus, after two billion years of only proalgae (prokaryotic algae), successive symbio-geneses generated the modern algal diversity of eualgae (those formed by enslaving cyanobacteria) and meta-algae (those formed by enslaving eualgae), each with more genetic membranes and more complex cell biology and protein-targeting machinery than their ancestors (Cavalier-Smith, 1995).

PLASTID ORIGINS, EUALGAE, AND THE DIVERSIFICATION OF KINGDOM PLANTAE

Mereschkovsky (1910) suggested polyphyletic origins of chloroplasts from bacteria bearing different photosynthetic pigments to yield eukaryotic algae of different colours. Whilst Margulis (1970, 1981) envisaged 30 separate origins, others have argued for many fewer multiple origins, often two or three. I argued that symbiogenesis was far more difficult than these authors assumed and that all chloroplasts had a single common origin by one internalization and primary symbiogenetic enslavement of a cyanobacterium, but that there was at least one secondary transfer of preexisting chloroplasts to a different eukaryotic host cell (secondary symbiogenesis) (Cavalier-Smith, 1982). This single origin of chloroplasts to form the common ancestor of the kingdom Plantae (Glaucophyta, Rhodophyta, and Viridiplantae [Chlorophyta plus Cormophyta]) only, a novel circumscription for the kingdom estab-lished by Cavalier-Smith (1981), is now strongly supported by all available evidence.

My primary reason for arguing for a single origin was that symbiogenesis requires the evolution of complex machinery for importing over a thousand different proteins into the enslaved cell (Reumann et al., 2005). It is therefore scientifically very unparsimonious to assume multiple origins without very strong positive evidence that that was what happened. My second reason was that phylogenetic reasons then given for assuming separate origins were very superficial and did not exclude a single origin. There was, and still is among some morphologists, a tendency merely to assume that if organisms or structures are sufficiently different, they cannot have a common origin. But this assumption grossly underestimates the powers of evolution to radically change structures and thus produce very divergent descendants from one ancestor. It also underestimates the frequency with which secondary evolutionary losses remove ancestral characters and thus conceal true rela-tionships (e.g. the absence of cilia [= flagella] from red algae was widely mistakenly held to preclude a relationship with glaucophytes and other eukaryotic algae). But we now know that the red algal ancestor lost cilia (as cilia have been dozens of times in eukaryote evolution), which necessarily follows from the compelling evidence for the monophyly of both Plantae and cilia. Evolutionary loss of a process (e.g. photosynthesis) or organelle can be very easy and initiated by a single mutation provided such loss is not lethal; in such cases losses can be relatively frequent. By contrast a major innovation involving many genes is very difficult and unique; it is never exactly repeated.

Thus, enslavement of a cyanobacterium possessing both phycobilisomes and chlorophyll b by a biciliate protozoan host with cortical alveoli would have produced an ancestral chloroplast that could have generated red algae (Rhodophyta) and glaucophytes (collectively Biliphyta) by losing chlorophyll b and green algae (Chlorophyta, in the classical sense used here; Figure 3.5) by loss of phycobilisomes.

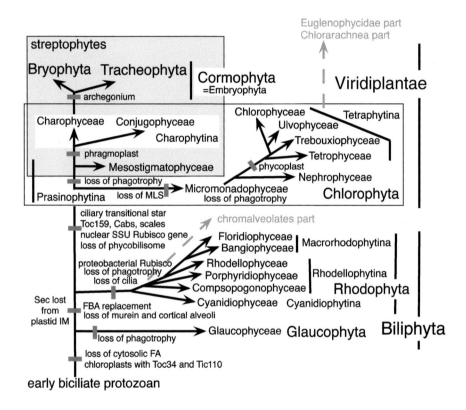

FIGURE 3.5 Phylogeny and major innovations within the kingdom Plantae. Plantae comprise a basal subkingdom Biliphyta that retains the ancestral phycobilisomes and a derived subkingdom Viridiplantae that lost them and modified the plastid protein import machinery by adding a new receptor Toc159 for importing copious amounts of chlorophyll a/b-binding proteins (Cabs). The post-Hennigian doctrine against ancestral (paraphyletic) groups is an unscientific and antiscientific idealistic philosophy that does not take account of the realities and complexities of evolutionary history (Cavalier-Smith 1998). Plantae is paraphyletic because it rightly excludes the descendants shown in grey type that form part of the cells of the three groups indicated (chromalveolates, Euglenophycidae, Chlorarachnea). I here use the older Conjugophyceae (= Zygnematophyceae) and Glaucophyta (= Glaucocystophyta); changing established untypified names to typified names (those in brackets) is undesirable instability. The classification follows Cavalier-Smith (1998) and Yoon et al. (2006); I do not think the absence of floridoside merits a separate class for Stylonematales, which I include within Porphyridiophyceae on the basis of the Golgi–ER association, even though this broader class may be paraphyletic (branching at base of Rhodellophytina is poorly resolved even on multigene trees). Tetrophyceae cl. nov. (Diagnosis: Tetraciliate cells with both scales, forming a theca, and a phycoplast; cilia quattuor, theca squamis et phycoplasta munita. Sole order Tetraselmidales) belong in Tetraphytina subphyl. nov. (diagnosis as for infraphylum Tetraphytae [Cavalier-Smith, 1998, p. 250]), not in Prasinophytina subphyl. nov. (diagnosis as for infraphylum Prasinophytae [Cavalier-Smith, 1998, p. 250]), also here raised in rank.

Thus mechanistically easy, divergent pigment adaptation from a single common ancestor simply explains the different pigments of Biliphyta and Viridiplantae. Retention of cilia, cortical alveoli, and the cyanobacterial murein wall by glaucophytes, but the loss of all three by the ancestor of red algae yielded two very different groups whose relationship was not previously suspected. As predicted by my analysis, the protein-import machinery of all three plant groups is homologous (Figure 3.3b), and proteins coded by nuclear genes of one of them can be imported *in vitro* by chloroplasts from another (Steiner and Löffelhardt, 2005; Steiner et al., 2005). Thus, the targeting signals are functionally interchangeable. However, Biliphyta have a more primitive transit sequence with an essential subterminal phenylalanine, which was lost by the ancestor of green algae when it acquired a novel import receptor (Toc159) that is specialized for import of chlorophyll-ab-binding proteins (Cabs; modified light-harvesting complexes [LHCs]) and the small subunit of Rubisco (its gene was moved to the nucleus in the ancestral green alga), two functions not needed in Biliphyta. Therefore, biliphyte transit sequences work in Viridiplantae, but not vice versa unless the missing phenylalanine is engineered into them (Steiner and Löffelhardt, 2005). This implies that the original Toc34 transit peptide receptor lost its requirement for a subterminal phenylalanine when Toc159 was added to the Toc protein translocator. The transit sequences of the phosphate translocator family are unusual in resembling those of mitochondrial presequences (Silva-Filho et al., 1997); this fits both the idea that insertion of IM carriers like triose phosphate/phosphate antiporter was the first step in chloroplast enslavement (Cavalier-Smith, 2000a) and that transit sequences may first have been recruited from preexisting mitochondrial presequences and then modified so as to reduce targeting ambiguity (Cavalier-Smith, 1982). The phosphate antiporter itself, however, evolved not from a mitochondrial protein but from an endomembrane protein lacking bacterial versions; sequence trees for it prove that red algae and Viridiplantae at least had a common origin (Weber et al. 2006).

Multigene sequence trees for chloroplast genes and nuclear genes all strongly support the monophyly of Plantae and chloroplasts (Rodríguez-Ezpeleta et al., 2005). There are also some derived gene arrangements found in plant chloroplasts that are not present in cyanobacteria that also support a common origin. Also important is the organizational and phylogenetic unity of all the LHC proteins in all chloroplasts compared with proalgae (Durnford et al., 1999; Rissler and Durnford, 2005). Although LHCs are not characterized from glaucophytes, they have a related carotenoid-binding protein (Rissler and Durnford, 2005); red and green algae and all meta-algae with plastids derived from them have LHCs with three membrane-spanning domains embedded in the thylakoid and inserted by a common mechanism involving SRP (see Figure 3.3b) (Gutensohn et al., 2006). Proalgae have related proteins with either a single domain or six (*Prochlorococcus*); thus, the three-domain proteins probably arose in the common ancestor of all Plantae from a one-domain protein (Garczarek et al., 2003).

The idea of a separate origin of red algal chloroplasts from phycobilisome-containing cyanobacteria and of green algal plastids from a prochlorophyte with chlorophyll b and no phycobilisomes is firmly refuted by the evidence for the monophyly of Plantae. The broad similarities, however, of the prochlorophyte chlorophyll-binding protein domains to those of LHCs imply that they were of symbiont, not host, origin. Furthermore, similarity in chlorophyllide a oxygenase, the enzyme that catalyses chlorophyll b synthesis, in prochlorophytes and Viridiplantae suggested that the enslaved cyanobacterium may have been one with an unusual mix of pigments, possessing both phycobilisomes and chlorophyll b, in which case Biliphyta lost chlorophyll b (Tomitani et al., 1999). Alternatively, plastids could have come from a normal cyanobacterium and chlorophyll b biosynthesis later by lateral gene transfer (Chen et al., 2005). Such transfer is easier into phagotrophic protists that often eat prey (as shown by the huge variety of genes acquired by the phagotrophic chlorarachnean algae; Archibald et al., 2003). Although Plantae are essentially non-phagotrophic, in contrast to Chromista, there is one example of phagotrophy in a micromonadophyte prasinophyte (O'Kelly, 1992), so the ancestral green alga probably retained some phagotrophic capacity and thus ability to take up prochlorophyte prey. Such accidental acquisition of genes for making chlorophyll b could have enabled them to flourish in similar niches. Firmer evidence of the precise branching position of chloroplasts and genes of cyanobacterial origin within cyanobacteria might make it possible to decide which scenario is right.

All plastids retain much of the cell division machinery of cyanobacteria, notably the FtsZ GTPase ring and proteins MinD and MinE that help position or regulate it, and others located at the cyanobacterial septum; their genes are still encoded by plastid genes in some algae but in others moved to the nucleus (Miyagishima, 2005; Miyagishima et al., 2005). In addition, however, a duplicated version of host eukaryotic dynamin was recruited as an extra, cytosolic, chloroplast division ring, probably in the ancestor of all Plantae (Miyagishima, 2005), and is thus another synapomorphy for them that helped integrate them into the host cell cycle.

Not being phagotrophs, algal Plantae seem to have acquired very few genes by lateral gene transfer independently of cellular symbiogenesis. One probable example is the red algal Rubisco that probably came from a proteobacterium, replacing the cyanobacterial version in their ancestor (Delwiche and Palmer, 1996).

RED ALGAL ENSLAVEMENT AND THE DIVERSIFICATION OF CHROMALVEOLATES

Chromalveolates comprise the classical chromophyte algae (those with chlorophyll c2) plus their numerous non-photosynthetic descendants, most of which not only lost photosynthesis but (unlike non-photosynthetic Plantae) also lost plastids altogether. Figure 3.6 indicates that of the 48 chromalveolate classes only 14 contain algae, so plastid loss has been very extensive and the closest relatives of some algal groups, notably Cryptophyceae and Peridinea, are not algae at all but phagotrophic heterotrophs traditionally treated as Protozoa. Only two phyla (Ochrophyta and Haptophyta) are predominantly algal and they are not sisters. Thus, understanding chromalveolate evolution and diversity requires a broad protistological approach. Repeated evolutionary losses of other important characters such as phagotrophy, ciliary hairs, ciliary transition region helices, photosynthetic pigments (e.g. phycobilins, fucoxanthin), nucleomorphs, or just their genomes, have also played a major role in chromalveolate evolution and in confusing taxonomists in the past about their relationships.

The idea of multiple losses of photosynthesis and that the closest relatives of many algae are heterotrophs is even older than that of symbiogenesis, going back at least to Blackman's influential proposal of autotrophic flagellates as ancestral eukaryotes (Blackman, 1900). But it fell from favour for about three decades from 1970 because of the broad popularity of the ill-founded view that chloroplast loss never occurred and that symbiogenesis is so easy that all cases where some organisms have chloroplasts and close relatives lack them are to be explained by multiple independent origins, up to 30 being casually suggested (Margulis, 1970, 1981). Recognition of the reality of evolutionary plastid loss and its much greater evolutionary ease than gain (Cavalier-Smith 1982, 1986, 1993b, 1999, 2000a, 2002a) has therefore been slow, but it is now generally accepted as a result of electron microscopic and molecular phylogenetic evidence and cell biological reasoning.

THE KINGDOM CHROMISTA

The kingdom Chromista was defined by the common membrane topology surrounding their plastids and by the presence in two of the major groups (Heterokonta, Cryptista) of bipartite or tripartite tubular rigid ciliary hairs (often called mastigonemes) (Cavalier-Smith, 1981). Both were argued to be homologous and very stable characters (Cavalier-Smith, 1986), but either could be lost, albeit rarely; recognition of membership of Chromista is only difficult when both the defining characters are lost, as in Leucocryptea (Cryptista) and a fair number of heterokonts (e.g. Opalozoa, and the bicoecean zooflagellates *Caecitellus* and *Adriamonas*). Thus, ciliary hairs of oomycetes long ago suggested they are related to heterokont algae and are not fungi (Cavalier-Smith, 2000b), so they are now placed within Heterokonta together with Hyphochytrea and the zooflagellate *Developayella* as the purely heterotrophic phylum Pseudofungi (Cavalier-Smith and Chao, 2006). It was also recognized early on that the opalozoan *Proteromonas* (saprotrophic intestinal commensal flagellate

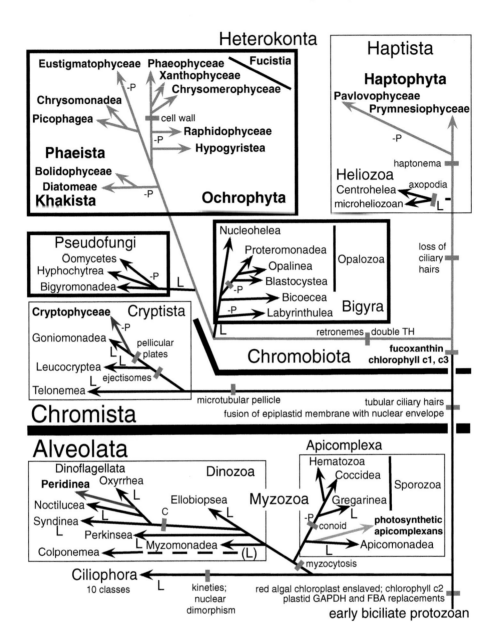

FIGURE 3.6 (Please see colour insert following page 76) Phylogeny and major innovations within chromalveolates (kingdom Chromista and protozoan infrakingdom Alveolata). Taxa that include algae are shown in grey boxes. As plastid losses (L) also occurred several times within Peridinea, chloroplasts were probably lost at least 20 times within chromalveolates. P = loss of phagotrophy (which additionally occurred within Peridinea, Prymnesiophyceae, and Chrysomonadea). The major phylogenetic uncertainty is whether Heliozoa are really sisters of Haptophyta and actually belong in Chromista rather than Protozoa; they evolved axopodia independently of Nucleohelea and pedinellids (classified as an order together with silicoflagellates, Pelagomonadales and Sarcinochrysidales, in the new subclass Alophycidae of Hypogyristea, alongside subclass Pinguiophycidae; Cavalier-Smith and Chao, 2006)—axopodia also evolved three times within Rhizaria (Figure 3.8). Phaeophyceae was recently broadened to include two subclasses, Melanophycidae, the classical brown algae, and Phaeothamniophycidae (with three orders including Schizocladiales), as Phaeista had been over-split in the past decade into a plethora of minor classes, previously incorrectly included in Chrysomonadea (= Chrysophyceae); see Cavalier-Smith and Chao (2006) for the latest more detailed classification of

of reptiles) was related to chromists because of its bipartite tubular body hairs (somatonemes) and double ciliary transition helix similar to those of Pseudofungi; but it was initially regarded as a sister group, not a member of the chromists (Cavalier-Smith, 1986). rRNA sequence trees corrected this but were misinterpreted as evidence for a heterotrophic ancestry for heterokonts (Leipe et al., 1996).

Molecular evidence is still wanting to confirm the homology of cryptist and heterokont tubular hairs. In heterokonts these hairs amplify and reverse the thrust of the ciliary beat and thus are known as retronemes and are found only on the younger anterior cilium, being lost when its centriole regrows as the smooth posterior cilium at the next cell cycle (Wetherbee et al., 1988). They play a key role in feeding by the two most important classes of phagotrophic heterokont flagellates (Chrysomonadea, Bicoecea), by wafting bacteria or other prey toward the uptake site near the ciliary base (Andersen and Wetherbee, 1992). Contrary to traditional views, Bicoecea are not specifically related to chrysomonads (Ochrophyta) but belong in the early diverging heterotrophic phylum Bigyra. Chrysomonads lost photosynthesis at least three times, but in each case (*Paraphysomonas* and the polyphyletic *Spumella*) they retained plastids and the primitive feeding mechanism in which a microtubule of the R3 root pushes upwards deforming the cell surface into a cup to catch the prey brought by the ciliary current. Andersen et al. (1999) suggested that the smooth cilium helps and that it has been lost only by chrysomonads that abandoned phagotrophy to concentrate on photosynthesis; but *Oikomonas* is probably an exception: a heterotroph with no smooth cilium (Cavalier-Smith et al., 1995/6). Possibly the feeding mechanism is subtly different in Bicoecea; it seems unlikely that in Bicoecales where the smooth cilium anchors the cell within a lorica that it can be directly involved in prey uptake, and other bicoeceans (*Siluania* and *Paramonas*) independently lost it altogether (Cavalier-Smith and Chao, 2006). Ciliary/root and trophic changes are often associated, e.g., retroneme and phagotrophy loss by the gut commensal Opalozoa, and three independent origins of axopodia for prey entrapment (superorder Actinochrysia within

FIGURE 3.6 (CONTINUED) Heterokonta, and Cavalier-Smith (2004) for overall chromalveolate evolution. Chromists comprise two subphyla, Cryptista and Chromobiota. Within Cryptista, Cryptophyceae and Goniomonadea are grouped as subphylum Cryptomonada, the only chromalveolates other than centrohelids with flat mitochondrial cristae, a secondary change from the ancestral tubular state; I here emend the other cryptist subphylum, Leucocrypta (Cavalier-Smith, 2004), by adding Telonemea, which like Leucocryptea have a two-layered microtubular cortex with some cortical alveoli (Shalchian-Tabrizi et al., 2006a), probably lost when cryptomonads evolved pellicular plates. Telonemea group with other cryptists on many but not all sequence trees (Shalchian-Tabrizi et al., 2006a) and lack ejectisomes, unlike other cryptists; unlike Leucocryptea (sisters of Cryptomonada) and goniomonads, they have tubular hairs on one cilium. Separate phyla for Leucocryptea (Okamoto and Inouye, 2005) and Telonemea (Shalchian-Tabrizi et al., 2006a) are unjustified. Chromobiotes, ancestrally with fucoxanthin, comprise infrakingdoms Heterokonta (phyla in bold boxes, ancestrally with double ciliary transition helix: TH) and Haptista, each with only one predominantly algal phylum, within which a few species only are secondarily heterotrophs (but no plastid loss has been firmly demonstrated), and at least one purely heterotrophic phylum. Retronemal thrust reversal (and coadapted ciliary roots), axopodia, and the haptonema define the three contrasting modes of prey capture that evolved divergently from the ancestral photophagotrophic chromobiote after it lost the nucleomorph genome and evolved fucoxanthin and chlorophylls c1 and c3. Among the 22 classes in Alveolata, Peridinea (not sensibly renamed Dinophyceae, as half are non-algal phagotrophs) is the only partially algal one of four in the infraphylum Dinoflagellata, collectively defined by evolution of the cingulum (C: groove for the transverse ribbon-shaped cilium with striated centrin paraxonemal rod). The subphylum Dinozoa comprises dinoflagellates plus four more diverse heterotrophic classes (comprising the paraphyletic infraphylum Protalveolata); molecular evidence is needed to position *Colponema* and see if they lost plastids independently of other heterotrophs (or even preceded the ancestral red algal enslavement!). Cavalier-Smith and Chao (2004) give a detailed classification of phylum Myzozoa, and its subphyla: Dinozoa, Apicomplexa (but excluding the then undiscovered photosynthetic apicomplexans that deserve their own class: see text).

Hypogyristea, which lost all ciliary roots and some of which lost photosynthesis; Nucleohelea and Centrohelea which independently lost cilia altogether). This emphasizes that one cannot understand chromophytes just as algae. Five classes contain algae that are simultaneously active phagotrophs: Chrysomonadea, Picophagea, Hypogyristea, Peridinea, and Prymnesiophyceae. This dual holozoic/phototrophic nutrition was the ancestral state for the whole chromalveolate clade, contradicting the artificial division between botany and zoology. It also emphasizes that morphology must not be used just as phylogenetic markers but also to interpret the adaptive significance of innovations and losses.

I originally postulated that the chromist tubular ciliary hairs reversed thrust in the ancestor of all chromists, but the discovery of somewhat less rigid tubular hairs on one of the two posterior cilia of *Telonema* (Klaveness et al., 2005), which I consider distantly related to Cryptophyceae and Leucocryptea, renders this less likely, especially as there has never been good evidence that those on anterior cilia of Cryptophyceae reverse thrust. Thus, thrust reversal is probably a synapomorphy for heterokonts alone; the original function was probably simply to increase the power of the beat. As many Myzozoa have simple ciliary hairs, it is important to see if their proteins are related to those of chromist tubular hairs; if they were, this would be another chromalveolate synapomorphy.

The branching order of the three major chromist lineages remains controversial. I have argued that haptophytes and heterokonts are sisters (collectively called chromobiotes) as they share six important synapomorphies: fucoxanthin; chlorophylls c1 and c3, and loss of phycobilins; autofluorescent posterior cilium; relocation of starch biosynthesis (ancestrally in periplastid space) to the cytosol; loss of nucleomorph genome and transfer of all essential chloroplast-protein-coding genes (around 30) to the nucleus; periplastid reticulum, which may be a relic of the nucleomorph envelope; and thylakoids stacked in threes, not two as in cryptomonads (Cavalier-Smith, 1986, 1994, 2004). Furthermore, concatenated chloroplast gene trees strongly group haptophytes and cryptophytes as sisters (Yoon et al., 2002; Rice and Palmer, 2006).

Chromobiote holophyly has been called into question by the discovery of a lateral gene transfer from a bacterium into the chloroplast that is found in haptophytes and cryptophytes only (Rice and Palmer, 2006), which might mean that they are mutually more closely related than either is to heterokonts, making it possible that the chromobiote characters mentioned above are convergent between haptophytes and heterokonts or were ancestral for all chromists but lost by cryptomonads. However, this conclusion is contradicted by the tree using 97 chloroplast genes that very strongly placed haptophytes and heterokonts as sisters (Rice and Palmer, 2006); single gene trees, unsurprisingly, were less decisive with 138 genes favouring chromobiote monophyly and only 98 (often more weakly) favouring haptophytes and cryptophytes being sisters (Rice and Palmer, 2006). The simplest way of reconciling these conflicting data would be that the bacterial gene was implanted into the ancestral chromist chloroplast and then lost by the ancestral heterokont before it lost its own version. That is a reasonably likely possibility because the bacterial gene in question encodes a ribosomal protein (Rpl36) that simply replaced a related endogenous chloroplast protein in haptophytes and cryptophytes. The simplest way of inserting the bacterial gene into the chloroplast genome would be by a single homologous recombination event between a donor circular DNA molecule bearing the gene and the chloroplast *rpl36* gene. This would necessarily yield a genome bearing both versions in tandem. If both remained in the chloroplast during the relatively short time interval needed for the mutual divergence of cryptists, haptophytes, and heterokonts, the original version could have been lost independently by cryptomonads and haptophytes and the newly inserted version by heterokonts. This involves three independent random deletions of one copy of a redundant gene, with different results among the lineages; this is no more improbable than throwing a coin three times and getting two heads and one tail (or vice versa). Persistence for a period is no problem, as these same two paralogues have persisted in some bacterial groups for far longer, probably billions of years (Rice and Palmer, 2006). Contrary to the assumption of Rice and Palmer (2006), their important data are compatible with the rather strong evidence for chromobiote monophyly, as they can be explained without having to assume two independent recombinational insertions. Only the secondary deletions, which are far more likely in the circumstances, need be independent. Thus, given the likely mechanism of insertion and the weight of other evidence for chromobiote mono-phyly, the shared *rpl36* gene replacement by lateral transfer provides much stronger evidence for a single origin for all chromistan plastids than it does against chromobiote monophyly.

PLASTID PROTEIN TARGETING AND THE UNITY OF CHROMALVEOLATES

Exactly as predicted (Cavalier-Smith, 1982, 1986), all chromistan algae use a common mechanism for import of nuclear-coded proteins into the plastid. Such proteins bear a bipartite N-terminal targeting sequence. The hydrophobic N-terminal signal sequence directs them across the ER membrane into its lumen, where the subterminal transit-like sequence directs it across both the periplastid membrane and the chloroplast envelope, as explained in Figure 3.7. After it was discovered that many sporozoan protozoa (subphylum Apicomplexa) had tiny colourless plastids surrounded by four membranes, I proposed that chromistan and dinoflagellate chloroplasts, with an unusual envelope of three membranes (Figure 3.7), were acquired in the very same red algal enslavement in a common ancestor and that chromists and alveolate Protozoa (in which both Apicomplexa and Dinoflagellata are classified as phylum Myzozoa (Cavalier-Smith and Chao, 2004) together with Ciliophora—ciliates and suctorians) are sister groups, together constituting the chromalveolates, a major clade on the eukaryote tree (Cavalier-Smith, 1999). This chromalveolate theory was rapidly confirmed by the discovery that two nuclear-encoded chloroplast proteins, glyceraldehyde 3-phosphate dehydrogenase (GAPDH) and fructose 1,6-bisphosphate aldolase (FBA), both originally encoded by the red algal nucleus, had been replaced by very different but functionally equivalent versions derived by gene duplication from the host cytosolic version of the enzymes (Harper and Keeling, 2003; Patron et al., 2004). Precisely the same replacements for both enzymes were found in all five major groups of chromalveolates furnished with plastids. This convincingly refuted the widespread alternative idea that each of these five groups had acquired plastids by independent red algal enslavements, since each would have had to have replaced both enzymes independently. Ten independent replacements, rather than only two required by the chro-malveolate theory, would require not only five independent duplications of each gene (and every time specifically of the cytosolic version rather than the also evolutionarily equally available mitochondrial version) but also five independent acquisitions of targeting sequences, and is thus so exceedingly improbable that it is now accepted that chromalveolates did indeed have a single chimaeric origin and that all their aplastidic lineages have lost plastids.

As explained in Figure 3.7, chromists also target nuclear-encoded soluble proteins into the periplastid space (PS; former red algal cytosol) across two membranes by a mechanism that is functionally analogous to that for importing nuclear-encoded chloroplast proteins across four membranes. As predicted (Cavalier-Smith, 1999), this mechanism requires bipartite leaders with N-terminal signal sequences that are cleaved off during entry into the rough ER by a signal peptide exposing the subterminal transit-like sequence that is recognised by a receptor translocon protein embedded in the PPM, moved across the PPM, and removed within the PS by a specific peptidase (Gould et al., 2006a, 2006b). Thus, it is now clear that the functional mechanism, involving a PPM translocon (Top) that recognises a transit-like peptide, is as I predicted (Cavalier-Smith, 1999), rather than by periplastid vesicles as Gibbs (1979) hypothesised.

However, the evolutionary origin of the periplastid space import machinery remains unclear. No Toc homologue has been identified, and it is possible that chromists recruited a different translocon for this purpose (e.g. the Sec61 channel through which proteins traverse ER membranes), though no periplastid compartment version of this has been detected either; if so, Top would be evolutionarily unrelated to Toc, not derived from it as previously suggested (Cavalier-Smith, 1999). Whatever its origin, Top is likely to have been so radically altered from its ancestor, to avoid harmfully recognising its original substrates, that it may be hard to identify it simply by sequence homology. It is also unclear whether the transit-like topogenic sequences really evolved from transit

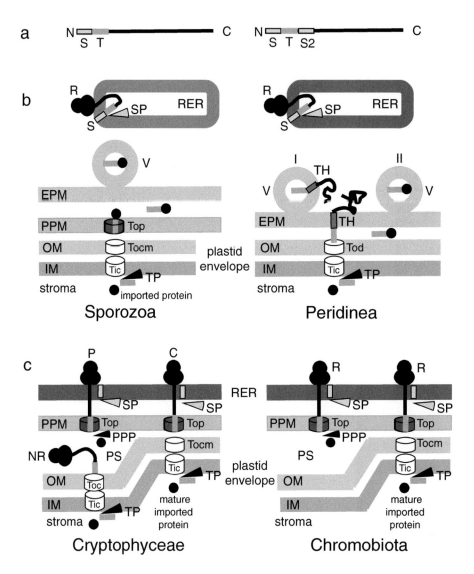

FIGURE 3.7 (Please see colour insert following page 76) Protein targeting to chromalveolate plastids. (a) All chromalveolate nuclear-coded proteins destined for the plastid stroma have bipartite N-terminal targeting sequences (left) with a signal sequence (S, pale grey box) and then a transit-peptide-like sequence (T, mid grey). Those destined for thylakoids have tripartite targeting sequences (right) with an extra signal sequence (S2) for targeting to the thylakoid membrane (after the first signal sequence is removed by signal peptidase [SP in part b] in the lumen of the rough endoplasmic reticulum [RER], and the transit-like peptide is removed by the stromal transit peptidase [TP] as shown in parts b and c). (b) In alveolates the outermost membrane around the plastids (the epiplastid membrane [EPM] derived from the food vacuole membrane that originally surrounded the phagocytosed red alga about 570 My ago) is smooth, the chloroplast proteins being made by ribosomes (R) attached by SRPs (not shown) to probably topologically separate RER cisternae, and carried to the plastids inside transport vesicles (V) suggested (but not yet demonstrated) to bud from the ER and fuse with the EPM, liberating their transported proteins into the space within. The signal peptide (S) is removed by a signal peptidase (grey segment, SP) in the RER and then digested. In dinoflagellates (right) these vesicles probably travel via the Golgi, as in euglenoids that like them have an envelope of three membranes, but in Sporozoa (left) they probably go direct to the EPM. In Sporozoa the transit-like sequence is probably recognized by receptors in the periplastid membrane (PPM, the former red algal plasma membrane) associated with a hypothetic translocase (Top) and pass through it and a channel in the plastid OM (Tocm, meaning a greatly modified Toc) to the Tic machinery of the plastid inner membrane (IM) through which it is probably

peptides, as they lack the key N-terminal phenylalanine (F) that characterizes true transit peptides in many chromalveolates (Figure 3.7); the periplastid space peptidase (PPP) that removes them (Gould et al., 2006a, 2006b) remains unidentified, so we do not know if it evolved from a transit peptidase, as postulated (Cavalier-Smith, 1999). However, it is now highly probable that nuclear-coded chloroplast proteins and PS proteins both enter the periplasmic space via Top, with discrimination between them clearly involving the presence of F in the transit peptide of plastid proteins and its absence in PS proteins (Gould et al., 2006b). In principle, discrimination could occur before

FIGURE 3.7 (CONTINUED) pulled by stromal ATPases as in plants. The transit peptide is removed by a transit peptidase (TP) and the mature protein folds up in the stroma, or is transported into thylakoids (not shown) if it has a second signal peptide or TAT signal. Tocm is proposed to have evolved from red algal Tocs transferred to the host nucleus during symbiogenesis (Cavalier-Smith, 2003b). On that theory Tocm should be drastically modified, so it is unsurprising that it has not been identified in sporozoan genomes by sequence similarity to Tocs (McFadden and van Dooren, 2004). Dinoflagellates (right) lost the PPM, so dispensed with Tops and probably modified their OM Toc/Tocm even more to become a novel translocase (Tod, d for dinoflagellate), which I suggest recognizes the transit-like sequences of both class I and class II leaders. Because class I proteins are probably trapped in the vesicle membrane by a transmembrane helix (TH) prior to fusion, they may remain thus in the EPM until pulled through Tod and Tic, which would require a greater force than for class II proteins that may be free in the intra-EPM space, but the translocation machinery could import both. (c) In the ancestral chromist the EPM fused with the ER/nuclear envelope (possibly by the transport vesicle fusing with its target as an extended tubule without ever budding) so the ancestral vesicle transport machinery was lost. In the periplastid space (PS, former red algal cytosol) Cryptophyceae (left) retain a nucleomorph (former red algal nucleus, not shown) and ribosomes that make 30 chloroplast proteins (Douglas et al., 2001). Therefore unlike other chromalveolates, they retain typical Tocs to recognize the transit sequences on these 30 proteins (which lack terminal signal sequences) and import them (left). But most cryptophyte chloroplast proteins are nuclear encoded, made by ribosomes on the RER (C), and have bipartite signals for import via putative Top/Tocm, with terminal signal sequences (pale grey box) and subterminal transit sequences (grey). Interestingly the nuclear-encoded transit sequences, though overall broadly similar to transit sequences of the nucleomorph-encoded proteins, differ systematically from them (nucleomorph ones have an N-terminal FXN, nuclear ones an FXP, motif) in keeping with the idea that Top, the hypothetical PPM translocase for plastid proteins, diverged greatly from Toc (Cavalier-Smith, 1999, 2003b), but both retain the red algal phenylalanine (F). In addition, Cryptophyceae import many proteins into the PS (e.g. those for making starch and carotenoids there; Gould et al., 2006a, 2006b) or nucleomorph (e.g. DNA polymerases), also made by ribosomes on the RER (P); as predicted (Cavalier-Smith, 2003b), these PS and nucleomorph proteins have bipartite targeting sequences (Gould et al., 2006). The fact that their transit-like sequence lacks the N-terminal F (having neither FXN nor FXP sequences) shows either that the receptor for proteins crossing the PPM differs from that for chloroplast proteins (TopI; Cavalier-Smith, 1999, 2003) and PS proteins (Top2; Cavalier-Smith, 1999, 2003) or that discrimination between plastid-destined and PS proteins takes place after entry into the PS via a common Top, and makes use of the presence or absence, respectively, of F (see text). The periplastid processing peptidase (PPP) that removes the transit-like sequence from PS proteins must differ from transit peptidases (TPs) in the chloroplast stroma. Chromobiotes (right) lost the nucleomorph genome, ribosomes, and starch from the PS, and thus also lost the OM Toc; but they still import proteins into the PS (e.g. chaperone Hsp70; Gould et al., 2006b) and thus must have kept Top and PPP. If the membranous periplastid reticulum (PR, not shown), which is located in the PS (putatively relict nucleomorph envelope membrane, perhaps retained for making PPM lipids), is topologically separate from the PPM, chromobiotes would also use Top for importing PR membrane proteins, but if the PR and PPM are continuous, or the PR buds from the PPM, PR proteins need not use that route. Modifying a nuclear-coded cryptomonad PS protein merely by changing its TP N-terminal amino acid to phenylalanine (F) makes it enter the plastid instead (Gould et al., 2006a), proving its functional similarity to chloroplast TPs. For simplicity, Tops are drawn as if they pass imported plastid proteins directly to Tocm; actually Top and Tocm need not be juxtaposed, as preproteins extruded by Top could bind PS Hsp70 chaperone and diffuse to Tocm (Gould et al., 2006b). (Modified from Cavalier-Smith, T., *Philosophical Transactions of the Royal Society of London B*, 358, 109–134, 2003b, in the light of Patron, N. J., Waller, R. F., Archibald, J. M., and Keeling, P. J., *Journal of Molecular Biology*, 348, 1015–1024, 2005, and Gould, S. B., Sommer, M. S., Kroth, P. G., Gile, G. H., Keeling, P. J., and Maier, U. G., *Molecular Biology and Evolution*, 23, 1–10, 2006.)

import into the PS (by using separate Tops, Top1 and Top2; Cavalier-Smith, 1999, 2003b) or afterwards (Gould et al., 2006b). Post-import discrimination within the PS might involve a hypothetical F-receptor directing transit-peptide-bearing proteins to Tocm (Gould et al., 2006b) or more simply if the F of transit peptides prevents PPP from removing the transit peptidase, allowing the TP to dock automatically onto Tocm without needing a hypothetical soluble F-receptor. I consider the latter view more likely as it requires neither a receptor additional to Tocm nor two separate Tops, and suggest that removal of the transit-like peptide from PS proteins, together possibly with a direct requirement by Tocm for a terminal F or similar amino acid in a transit peptide, may be sufficient for discrimination.

The demonstration that cryptophyte periplastid proteins can be correctly targeted in a diatom and that diatoms and cryptophytes have similar bipartite leaders on periplastid protein precursors (Gould et al., 2006a, 2006b) means that a common ancestral protein translocation mechanism across the PPM was in place before cryptophytes and heterokonts diverged, providing very strong corroboration for the monophyly of chromists and for their last common ancestor having been photosynthetic and all heterotrophic chromists secondarily derived, as I first argued (Cavalier-Smith, 1981, 1986). Not all cryptophyte TPs have a terminal F, and this requirement is also looser in haptophytes and dinoflagellates, suggesting different degrees of divergence from the ancestral use of F in the enslaved red alga of the first chromalveolate.

Clearly, plastid loss has been much more frequent than generally assumed (Cavalier-Smith, 2003b), generating heterotrophs from algal ancestors that often mimic either fungi (if walled) or protozoa (if unwalled) in mode of nutrition, so unsurprisingly such chromalveolates were traditionally classified thus. Given the phylogeny of Figure 3.6, plastids were lost about 20 times. The contrast with Plantae where they were never lost is marked. It probably arises because chromalveolate plastids did not become essential for host non-photosynthetic functions early in evolution, whereas those of plants did. In Plantae early loss of the host fatty acid (FA) synthetase meant that they rely entirely on that located within the plastids (of cyanobacterial origin, albeit now nuclear encoded). In chromists the host FA synthetase has been lost and the red algal plastid version retained (Ryall et al., 2003), but this dependence must have arisen after the divergence of the now photosynthetic lineages from their now non-photosynthetic relatives, which generally use the original non-cyanobacterial FA synthetase. It would be expected that the earlier the aplastidic lineages diverged, the less likely would it be that non-photosynthetic red algal genes would be retained. In keeping with this, ciliates that diverged right at the beginning of alveolate evolution from the partially plastid-bearing Myzozoa show so little evidence of this sort of their ultimately chromophyte ancestry that it was not previously suspected. By contrast the Pseudofungi, which are so closely related to Ochrophyta that they sometimes even branch within them on sequence trees based on only one or two genes, possess at least two genes of probable red algal origin (Cavalier-Smith, 2002a; Robertson and Tartar, 2006), confirming the ancient theory of their algal ancestry (De Bary, 1884). Pseudofungi apparently acquired several fungal genes by lateral transfer (Richards et al., 2006). In Coccidea (e.g. *Toxoplasma*) and Hematozoa (e.g. *Plasmodium*, the malaria parasites) the plastid is retained, despite the loss of photosynthesis hundreds of millions of years ago, because it is the only site of synthesis in the cell of fatty acids and isoprenoid lipids; the host versions have been lost, but in *Cryptosporidium*, more closely related to Gregarinea, which like it lost plastids, than to typical Coccidea, the host FA synthetase was retained, so plastid retention had no selective advantage (Cavalier-Smith, 2002a).

PHOTOSYNTHETIC DIVERSITY OF ALVEOLATES

Chromophyte evolution is also complicated by chloroplast replacement in dinoflagellates. At least twice in their history symbiogenesis replaced the normal peridinin-containing and DNA minicircle-containing plastid by one from a different group of eukaryotic algae (Saldarriaga et al., 2001). One case involved a green algal symbiont generating *Lepidodinium* and *Gymnodinium chlorophorum*; the other enslaved a haptophyte to form *Karlodinium* and *Karenia* (Shalchian-Tabrizi et al., 2006b).

Such replacements may have been especially favoured in dinoflagellates because as meta-algae they already have a complex protein-targeting machinery for protein import across more than two membranes, that might be evolutionarily co-opted for a different photosynthetic slave, and because they are one of the few groups of meta-algae that regularly eat other eukaryote algae (Cavalier-Smith, 2003b). Only some dinoflagellates have stopped being predators and are simply algal nutritionally. Though some use ordinary phagocytosis, many use other elaborate ways of getting prey, like pallial feeding or myzocytosis, where prey contents are sucked into food vacuoles minus their plasma membrane, and which probably evolved in or close to the common ancestor of all Myzozoa (named after this feeding mode that they share with primitive Apicomplexa; Cavalier-Smith and Chao, 2004), and which seems to predispose flagellates for evolving parasitism, as in the most divergent dinoflagellate class, Syndinea. Most chromistan algae lost phagocytosis, except for Actinochrysea and the more primitive chrysomonads and haptophytes, which tend to eat bacteria rather than eukaryotes. It is important not to confuse these two genuine cases of symbiogenesis with kleptoplastidy (temporarily harbouring stolen plastids, as in *Dinophysis*, which can steal cryptophyte or haptophyte plastids) or with simple obligate endosymbiosis without generalised import, as is likely to be the case for the diatoms enslaved by the common ancestor of *Kryptoperidinium foliaceum* and *Peridinium balticum* (McEwan and Keeling, 2004).

Gene replacement independently of symbiogenesis also occurred in dinoflagellates; the originally red algal two-protein proteobacterial type I Rubisco was replaced by a simpler single protein type II Rubisco from a different proteobacterium. The smaller size of this novel Rubisco probably allowed its import into the plastid more easily than for the larger type I large subunit (in *Karenia* this was replaced in turn by the haptophyte Rubisco; Yoon et al., 2005). Probably associated with the loss of the periplastid membrane (former red algal plasma membrane) in dinoflagellates alone among plastid-bearing chromalveolates, Peridinea must have evolved a novel modified mechanism of protein import. The capacity of their novel import mechanism to deal with proteins temporarily stuck in ER membranes may have made it easier for dinoflagellates to import proteins that would get stuck in envelope membranes in other chloroplasts and whose genes may therefore never have been able to be lost from chloroplasts (Cavalier-Smith, 2003b). Thus, gene transfer to the nucleus could have been much easier in dinoflagellates than in any other organisms simply because of this novel protein-import machinery, explaining why they alone have retained only 12 to 18 plastid genes. In turn, the extreme diminution of genome size could have explained their ability to survive the loss of linkage into a single genome postulated to result from mutation pressure in markedly reduced genomes; this may be why dinoflagellates alone evolved chloroplast mini-gene circles (Cavalier-Smith, 2003b; Zhang et al., 1999). If this interpretation were correct it would be another fascinating example of intracellular coevolution between different cell constituents that do not directly interact with each other. Intriguingly dinoflagellates apparently have two variants of the import mechanism (Figure 3.7). In class I after partial insertion into the ER by the signal sequence, much of the protein remains sticking out into the cytosol during passage through the Golgi, because of trapping in the membrane by a hydrophobic transmembrane helix until after the post-Golgi vesicle bearing it fuses with the epiplastid membrane and the N-terminal end engages with and is pulled in by the protein-import machinery (Nassoury et al., 2003). In proteins with class II leaders, import does not involve membrane stop-transfer sequences (Patron et al., 2005); possibly this variant mechanism is evolutionarily closer to the ancestral one for chromalveolates. It seems that throughout dinoflagellates the same set of proteins has a class I leader with transmembrane helices (e.g. Rubisco, peridinin-binding protein), and a different but conserved set of proteins has a class II leader instead (Patron et al., 2005). The fact that some genes that are transferred to the nucleus uniquely in dinoflagellates and whose proteins are therefore imported have class I and others class II leaders means that the presence of leader transmembrane helices alone cannot be the reason why dinoflagellates have transferred so many genes to the nucleus; the key feature could be a property common to the import of both types of proteins (e.g. the machinery that actually pulls them into the plastid). The unique polyuridylation of minicircle transcripts (Wang and Morse, 2006b) is another synapomorphy for Peridinea, which also gives strong support to their plastid genomes really being dramatically reduced compared with other algae. It is currently unclear whether the higher molecular weight of some dinoflagellate plastid DNAs (Wang and Morse, 2006a) is primitive or secondarily derived; while some may have been secondarily transferred to the nucleus (Laatsch et al., 2004), this is not yet convincingly demonstrated.

An exciting recent development since the above was written is the discovery of a novel group of free-living photosynthetic apicomplexans, which will deserve to be a separate algal class when they are fully characterized. Two genera of photosynthetic apicomplexans are in the hands of the isolator (R. Moore, University of Adelaide, personal communication). Their rRNA phylogeny shows them as early Apicomplexa (possibly sisters of Apicomonadea; Figure 3.6) and a formal description is in preparation detailing that one genus has a plastid containing chlorophyll a and is surrounded by four membranes (R. Moore, University of Adelaide; M. Obornik, University of South Bohemia; S. Wright, Australia Antarctic Division, personal communication). The number of membranes surrounding the plastid of the other genus has not been examined, but it too contains chlorophyll a. Neither genus has chlorophyll b or c, but the absence of chlorophyll c may reasonably be regarded as a secondary loss, so they can be treated as derived chromophytes. This more directly confirms the inference that Myzozoa were ancestrally photosynthetic, and thus further bolsters the chromalveolate theory. It also means that there are really two radically different classes of myzozoan algae, one bounded by three membranes (Peridinea) and one by four (photosynthetic apicomplexans), the ancestral state. Thus, there are really 15, not 14, classes of chromalveolate algae (and 49 chromalveolate classes in all). Detailed comparison of its plastid genome and physiology with those of Peridinea should allow one to reconstruct the nature of the algal last common ancestor of all Myzozoa, all alveolates, and all chromalveolates more confidently, though the remote possibility that they are chloroplast replacements rather than endogenous apicomplexan plastids also needs rigorous testing.

This exiting discovery and that of picobiliphytes (Not et al., 2007), which might be a new algal class related either to cryptophytes or glaucophytes, highlights how much remains to be discovered about chromalveolates. For further discussion of chromalveolate diversity and evolution see Cavalier-Smith (2004) and Cavalier-Smith and Chao (2004, 2006).

GREEN ALGAL ENSLAVEMENT: POSITION OF CHLORARACHNEAN ALGAE (CERCOZOA) WITHIN THE RHIZARIA

Chloroplasts of Chlorarachnea (Rhizaria) and Euglenia (Excavata) contain chlorophyll a and b and evolved by enslavement of a green alga. In principle it is more likely that this happened only once in a common ancestor, as this would have required only a single origin of a new protein import system. I therefore suggested that Chlorarachnea and Euglenia may have diverged from a common photosynthetic ancestor produced by such a single green algal enslavement (Cavalier-Smith, 1999). If this were true it would make Rhizaria and Excavata sisters (collectively termed "cabozoa") and mean that plastids have been lost several times within each group, as is now well established for chromalveolates. Although there is evidence that both chlorarachnean and euglenoid plastids came from the same major algal subphylum (Tetraphytina, possibly Ulvophyceae; Ishida et al., 1997) there is insufficient evidence at hand to decide between a single versus two separate origins; however, most trees in a recent analysis of up to 85 proteins show cabozoa as holophyletic (strongly for Bayesian, less decisively for maximum likelihood; Burki and Pawlowski, 2006). From Figure 3.8 it can be seen that even though Chlorarachnea is the only algal rhizarian class, and is nested among 17 heterotrophic classes, only three or four plastid losses would have occurred within Rhizaria if their ancestor was photosynthetic. Like dinoflagellates, most Chlorarachnea retain phagotrophy and many eat diverse algae. Perhaps unsurprisingly therefore they seem to have acquired a considerable variety of genes from other algae (Archibald et al., 2003), illustrating the dictum "you are what you eat" (Doolittle, 1998). They retain

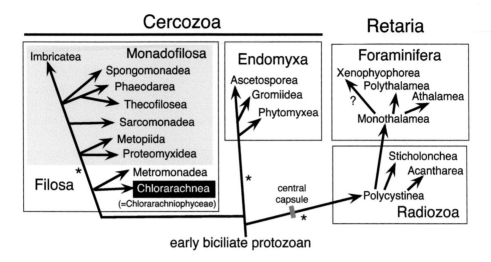

FIGURE 3.8 The protozoan infrakingdom Rhizaria comprises two phyla, Cercozoa and Retaria, with 19 classes. Chlorarachnea, the only algal class, diverged early within the cercozoan subphylum Filosa (Bass et al., 2005; Cavalier-Smith and Chao, 2003). It is unclear whether the green algal part of the chlorarachnean cellular chimaera was enslaved after it diverged from other Filosa or prior to the divergence of Rhizaria and Excavata. If Rhizaria ancestrally had plastids (Cavalier-Smith, 1999, 2000), they must have been lost three times (asterisks) if Metromonadea cl. nov. (Diagnosis as for sole order Metromonadida; Bass and Cavalier-Smith, 2004, p. 2402) belongs within Monadofilosa, as its gliding phenotype suggests. If metromonads are sisters to Chlorarachnea, as 18S rRNA trees weakly suggest, the metromonad ancestor must also have lost plastids if the cabozoan theory is correct. The euglyphid genus *Paulinella*, which has permanently enslaved a cyanobacterium intracellularly, belongs in Imbricatea. Many Retaria cultivate algae intracellularly but these are not permanently inherited. Note that Polythalamea are polyphyletic and about two distinct classes need to be segregated from them.

the nucleus of the enslaved green alga as a tiny nucleomorph (Ishida et al., 1999), whose genome is the smallest of any nuclei (Gilson et al., 2006). Unlike euglenoids they also kept the green algal plasma membrane as a periplastid membrane but the mechanism of protein translocation across it is unknown, although the preceding step probably involves vesicle fusion with the epiplastid membrane (former food vacuole membrane) as in euglenoids (Cavalier-Smith, 2003b). Chloroplast gene trees now suggest that chlorarachneans may have acquired plastids independently of euglenoids (Rogers et al., 2007); recent trees for nuclear-coded light-harvesting complex proteins fail to clarify the issue (Koziol et al., 2007).

It is sometimes implied that the enslaved cyanobacterium on which the euglyphid filose amoeba *Paulinella* (Cercozoa) depends for its food is an exception to the single primary origin of chloroplasts that formed Plantae (Marin et al., 2005). But there is currently no evidence that it evolved a generalized protein-import mechanism, the key criterion that distinguishes an obligate symbiont from an integrated organelle (Cavalier-Smith and Lee, 1985). If it lacks such a mechanism (even if it turned out to have donated a few genes to the host, for which there is no evidence), this would not be an example of symbiogenesis but simply of obligate symbiosis or cell enslavement. In that event it should be called a cyanelle, not a chloroplast. "Cyanelle" was originally invented to designate just such obligately endosymbiotic cyanobacteria that are not chloroplasts but are functionally analogous in being tapped by the host for photosynthesate (Pascher, 1929). Because glaucophyte plastids were originally incorrectly thought to be cyanelles this name was often used for them; but they ought to be called chloroplasts now that their homology is unambiguously established, and the term "cyanelle" should be reserved for such cases as the *Paulinella* cyanobacterium, which must not be called a chloroplast or organelle in the absence of evidence for integration by generalized protein import.

GREEN ALGAL ENSLAVEMENT: POSITION OF EUGLENOID ALGAE (EUGLENOZOA) WITHIN THE EXCAVATA

Because the cabozoan theory (i.e., that Rhizaria and Excavata are sisters and their common ancestor enslaved a single green algal symbiont, which was ancestral to both chlorarachnean and euglenoid plastids) has neither been refuted nor confirmed, we do not know whether excavates are ancestrally non-photosynthetic as most biologists assume (if the theory is wrong) or were ancestrally algae (if it is correct). The most direct way of testing the cabozoan theory would be to elucidate the novel post-Golgi vesicle-targeting machinery of *Euglena* (Slavikova et al., 2005) and see if it is homologous to that of chlorarachneans. What is clear is that only about half of all euglenoids are algae and also that ancestral euglenoids were phagotrophs (Figure 3.9). In contrast to chlorarachneans, which are mostly photophagotrophs, no euglenoid algae have retained phagotrophy, although

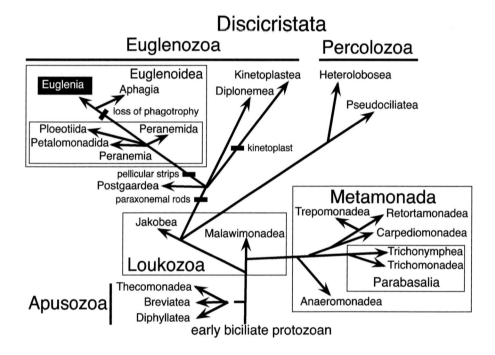

FIGURE 3.9 Relationship of euglenoid algae (subclass Euglenia) to other excavate Protozoa. Euglenia (highlighted) are the only euglenoids with chloroplasts; secondarily heterotrophic euglenians may either retain leucoplasts (*Euglena* [= *Astasia*] *longa*) or not (*Khawkinea*). If phagotrophy was lost only once in their common ancestor with the saprotrophic rhabdomonads (Aphagia), chloroplasts must have been acquired before then and lost by the rhabdomonad ancestor. Shared cyanobacteria/plastid-related genes suggest that discicristates (phyla Euglenozoa and Percolozoa) may have been ancestrally photosynthetic; analogous but less clear evidence is consistent with this being so also for the common ancestor of discicristates and Metamonada (Cavalier-Smith, 2003a). If the cabozoan theory were correct (see text and Figure 3.1) and the ancestral excavate had a plastid of green algal origin, plastids must have been lost at least eight times within excavates; the exact number is uncertain because the basal branching order within the class Euglenoidea and phylum Euglenozoa is unclear and it is uncertain whether Apusozoa are really sisters of other excavates, a possibility shown here by dashed lines, or—perhaps more likely (Moreira et al., 2007)—the deepest branching bikont phylum (Figure 3.1). At present only one plastid loss (*Khawkinea*) is proven. Excavate classification is from Cavalier-Smith (2003a); I am unconvinced that Heterolobosea and Jakobea are part of a clade that excludes Euglenozoa (Simpson et al., 2006). Loukozoa ancestrally had a ventral feeding groove and ciliary vanes, a split R1 and a singlet ciliary root and associated I, B, and C fibres (the excavate synapomorphies; Simpson, 2003) but lack the euglenozoan and metamonad synapomorphies shown.

some have relict ingestive organelles; they are best treated as an ancestrally photosynthetic subclass (Euglenia Cavalier-Smith 1993 [= Euglenophycidae: Busse and Preisfeld, 2003]; orders Euglenida and Eutreptiida) of the ancestrally phagotrophic protozoan class Euglenoidea. Arguments that early diverging euglenoids typically eat bacteria and cannot ingest eukaryotes do not, as they are sometimes mistakenly thought to (Leander, 2004), prove that the ancestral euglenoid was not photosynthetic, for on the cabozoan theory the putative single enslavement of a green alga would have been in the ancestral cabozoan well before the origin of the euglenoid pellicle. Furthermore, the basal branching order of euglenoids is not well resolved on molecular trees and it is uncertain whether the bacteria-eating Ploeotiida and Petalomonadida really preceded the eukaryote-eating Peranemida, which may be paraphyletic (von der Heyden et al., 2004). Even though bootstrap support is not strong, it is likely that Euglenia are sisters to the osmotrophic Rhabdomonadida and that there was only one loss of phagotrophy in their common ancestor. Though I once grouped them together as class Aphagea (Cavalier-Smith, 1993), I here follow Busse and Preisfeld (2003) in using this name for the osmotrophic Rhabdomonadida only, for which Aphagia (ICZN) is a preferable subclass name to their Aphagophycidae, as they are not algae. Ranking Euglenia as a class (e.g. Marin et al., 2003) unnecessarily multiplies classes and risks under-stressing fundamental euglenoid unity, as the superclass category (Cavalier-Smith, 1993a) remains little used; conversely, demoting all euglenoids to an order (Lee et al., 2002 dated 2000) devalues their distinctiveness from other Euglenozoa and internal disparity. Therefore, I now formally reduce the paraphyletic class Peranemea (Cavalier-Smith, 1993a) to subclass Peranemia, so that class Euglenoidea is now simply divided into three subclasses, respectively ancestrally phototrophic, osmotrophic, and phagotrophic. Calling the whole class euglenids, rather than the traditional euglenoids (Bütschli, 1885), risks confusion with the much narrower euglenian order Euglenida.

It has long been clear that euglenoids are related to the purely heterotrophic classes Diplonemea and Kinetoplastea with which they are grouped in the excavate phylum Euglenozoa, characterized particularly by latticed paraxonemal rods that differ between the two cilia (thinner in the younger dorsal [anterior] cilium and thicker in the older ventral cilium) and a periciliary reservoir (Cavalier-Smith, 1981; Simpson, 1997; von der Heyden et al., 2004) and by trans-splicing of most nuclear premessenger mini exams (Canaday et al., 2001). There is no justification for continued use of a separate algal phylum Euglenophyta for euglenoids alone, which conceals their true relationships and exaggerates their differences from other Euglenozoa. Nor should euglenoids be thought of primarily as algae. They are a mixture of algal, osmotrophic, and phagotrophic flagellates, seldom even slightly amoeboid, in marked contrast to ancestral bikonts. It is best to treat them entirely under the zoological code of nomenclature like other excavates, but phycologists and protozoologists alike should be encouraged to study them. There is a special need to study a greater diversity of the apparently early-diverging bacterivorous petalomonads, whose ultrastructural diversity even raises the possibility that they, and euglenoids, might be paraphyletic (Leander et al., 2001). Some molecular evidence suggests that Diplonemea may be sisters to Kinetoplastea (Simpson and Roger, 2004). If this is correct then the plicate vanes and fairly complex ingestion apparatus that diplonemids share with some euglenoids could be ancestral for the phylum. As in so many other phyla, Euglenozoa exhibit many cases of secondary simplification (e.g. reduction of the posterior cilium in many euglenoids and its loss by the parasitic trypanosomatid kinetoplastids).

For some time the typically amoeboflagellate Percolozoa (Heterolobosea plus Pseudociliatea), which have an even greater tendency toward ciliary loss, being total in many cases, have been regarded as the sister group of Euglenozoa, as both groups have discoid mitochondrial cristae, a derived character within excavates, leading to their joint designation as Discicristata. Percolozoa differ from Euglenozoa by lacking latticed paraxonemal rods and reservoir and by being ancestrally tetraciliate and by distinct Golgi dictyosomes (cisternal stacks) not being visible. Like Euglenozoa their centrioles are parallel, in marked contrast to the prototypical excavates, Jakobea and Malawimonadea, biciliates characterized by a posterior cilium in a ventral feeding groove, the scooped out appearance of which first suggested the name excavate (Simpson and Patterson, 1999).

Jakobea are closer to discicristates and Malawimonadea probably closer to the secondarily amito-chondrial phylum Metamonada, characterized by being ancestrally tetrakont, having three anterior cilia and one posterior. Thus, ancestral Percolozoa and Metamonada independently doubled their kinetids, evolving ciliary transformation over three cell cycles. (This also happened at least twice in chlorophytes, e.g. in Pyramimonadales among prasinophytes and in Tetraphytina.) Six-gene trees suggest that Jakobea may be even closer to Heterolobosea than are Euglenozoa (Simpson et al., 2006); if confirmed by more extensive data this would suggest that crista shape has undergone reversions or convergence in early excavate evolution. *Malawimonas* and Jakobea share cytoskeletal features supporting the monophyletic origin of Loukozoa, despite other cytoskeletal differences and deep molecular divergence (Cavalier-Smith, 2003a; Simpson, 2003).

The common cytoskeletal features of Loukozoa distinguish them from Apusozoa, which mostly also have a feeding groove and may therefore be distantly related to them. Apusozoa comprise the planktonic zooflagellate class Diphyllatea that feeds on algae, the mostly benthic Thecomonadea (zooflagellates with ventral groove and dense cortical plastron-like submembrane skeleton, typically bacterivorous), and the amoeboid Breviatea (Cavalier-Smith, 2003a; Walker et al., 2006). As thecomonads include both Ancyromonadida with flat cristae and Apusomonadida with tubular ones (like all other Apusozoa), which single-gene trees only group inconstantly (Cavalier-Smith and Chao, 2003), their monophyly has been uncertain, but unpublished multigene trees now unambiguously demonstrate it (Franz Lang, personal communication); our own unpublished concatenated 18S/28S rRNA trees also support thecomonad monophyly as well as *Breviata* being their sister (Berney and Cavalier-Smith, unpublished). Multigene trees may reveal whether Apusozoa are sisters of typical excavates (Figure 3.9) or whether (increasingly likely) they are the most divergent bikonts (Figure 3.1) as 18S rRNA trees, concatenated 18S/28S rRNA, and trees based on five proteins often weakly suggest (Cavalier-Smith, 2003a; Kim et al., 2006; Moreira et al., 2007). The enigmatic *Micronuclearia* resembles an *Ancyromonas* that has entirely lost cilia (Mikrjukov and Mylnikov, 2001), so may be yet another divergent apusozoan.

CONCLUDING REMARKS: ALGAL PHOTOSYNTHETIC UNITY DESPITE THEIR ORGANISMAL POLYPHYLY

The fact that eualgae arose as chimaeras of a bikont protozoan and a cyanobacterium and that at least four (five if the cabozoan theory is wrong) secondary symbiogeneses occurred to generate the diversity of meta-algae means that algae are polyphyletic as organisms, even though oxygenic photosynthesis itself clearly originated only once and is thus monophyletic. Thus, the photosynthetic components of algae have a phylogenetic unity despite the polyphyletic diversity of the cells that house them. This makes understanding algal evolutionary relationships especially complex. So also do the roughly 20 losses of plastids in chromalveolates that make chromophyte algae paraphyletic. Multiple symbiogeneses and plastid losses make it essential to consider higher algal taxonomy in conjunction with that of related heterotrophs to achieve a sound overall balance.

I started this chapter by defining algae as oxygenic photosynthesizers other than embryophyte land plants. This is more restrictive than the earlier definition that also admitted organisms that evolved from algae, but which retained plastids (Cavalier-Smith, 1995). That broader definition became undesirable when it was found that sporozoa of classes Hematozoa and Coccidea have plastids. It would make nonsense of the utility of the term "algae" to include such sporozoa in the concept of algae, still more to include also those sporozoa such as gregarines and *Cryptosporidium* that have lost plastids as well as photosynthesis. Recognition of the monophyly of chromalveolates means that there are three whole phyla (Ciliophora, Bigyra, Pseudofungi) that are heterotrophs with no trace of plastids, but that did have a distant chromophyte ancestor (Heliozoa would be a fourth if actually chromists). It makes no sense to call these phyla algae just because they had a distant and probably rather temporary algal ancestry. Thus the tradition of using *algae* to embrace any heterotrophic descendant of algae (Lemmerman, 1914; Fritsch, 1948; Silva, 1980) can no longer be justified; it is also most undesirable to use the algal suffix –phyceae for such chromistan classes

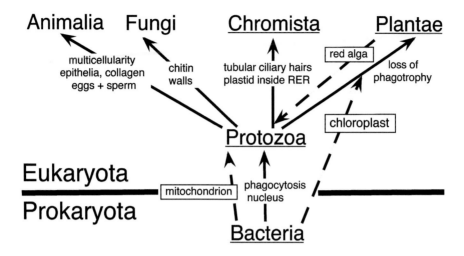

FIGURE 3.10 The six-kingdom, two-empire classification of life. The three major symbiogeneses are shown as dashed lines and the other key innovations that distinguish kingdoms are shown, except for the endoskeleton and endomembrane system of eukaryotes which give them far greater morphogenetic potential than bacteria. The four kingdoms with some algae are underlined. Macroscopic examples of animals, fungi, and plants, and the name bacteria are known to all laymen; schoolchildren can be told that brown seaweeds are examples of chromists and that slime moulds (macroscopic), amoebae, and malaria and sleeping sickness parasites are examples of protozoa, whilst some blue-green algae and actinomycetes are examples of bacteria visible to the naked eye. Everyone should learn this degree of detail and the distinction between these cellular organisms and viruses. Phylogeneticists should subdivide the ancestral kingdoms Bacteria and Protozoa to their hearts' content but not bother the general public (e.g. visitors to natural history museums) with every cladistic detail (e.g. avoid that in Figure 3.1, where kingdom Protozoa is subdivided into phyla Amoebozoa and Choanozoa (together forming subkingdom Sarcomastigota) and infrakingdoms Alveolata, Rhizaria, and Excavata (together subkingdom Biciliata; Cavalier-Smith, 2003c), and kingdom Bacteria is subdivided into its ten phyla.

(Cavalier-Smith and Chao, 1996, 2006). Likewise, heterotrophic chlorophytes (e.g. *Helicosporidium*, *Polytomella*; de Koning and Keeling, 2006; Mallet and Lee, 2006) should be regarded as parasitic or saprophytic protists that are relatively recent derivatives of algae, but not as algae themselves. By consistently excluding all secondary heterotrophs, whether they still possess leucoplasts or not, we gain a relatively simple answer to the question *what are algae*? Algae are not a taxon but an important ecologically distinctive **functional group of organisms** comprising predominantly aquatic oxygenic photosynthesizers that lack the adaptations to terrestrial existence of the archegonium and internal skeletal and typically also conductive tissues that enable cormophytes (= embryophytes) to be self-supporting in air and gather water from the soil. Figure 3.10 emphasizes their immense evolutionary and structural diversity, which is spread across four of the six kingdoms of life.

ACKNOWLEDGMENTS

I thank NERC for research grants and a Professorial Fellowship and the Canadian Institute for Advanced Research (Evolutionary Biology Programme) for fellowship support.

REFERENCES

Andersen, R. A., Van de Peer, Y., Potter, M. D., Sexton, J. P., Kawachi, M., and LaJeunesse, T. (1999) Phylogenetic analysis of the SSU rRNA from members of the Chrysophyceae. *Protist*, 150: 71–84.
Andersen, R. A. and Wetherbee, R. (1992) Microtubules of the flagellar apparatus are active in prey capture of the chrysophycean alga *Epipyxis pulchra*. *Protoplasma*, 166: 1–7.

Archibald, J. M., Rogers, M. B., Toop, M., Ishida, K., and Keeling, P. J. (2003) Lateral gene transfer and the evolution of plastid-targeted proteins in the secondary plastid-containing alga *Bigelowiella natans*. *Proceedings of the National Academy of Sciences, USA*, 100: 7678–7683.

Bass, D. and Cavalier-Smith, T. (2004) Phylum-specific environmental DNA analysis reveals remarkably high global biodiversity of Cercozoa (Protozoa). *International Journal of Systematic and Evolutionary Microbiology*, 54: 2393-2404.

Bass, D., Moreira, D., López-García, P., Polet, S., Chao, E. E., von der Heyden, S., Pawlowski, J., and Cavalier-Smith, T. (2005) Polyubiquitin insertions and the phylogeny of Cercozoa and Rhizaria. *Protist*, 156: 149–161.

Bibby, T. S., Nield, J., Chen, M., Larkum, A. W., and Barber, J. (2003) Structure of a photosystem II supercomplex isolated from *Prochloron didemni* retaining its chlorophyll a/b light-harvesting system. *Proceedings of the National Academy of Sciences, USA*, 100: 9050–9054.

Blackman, F. F. (1900) The primitive Algae and the Flagellata. *Annals of Botany (London)*, 14: 647–689.

Blobel, G. (1980) Intracellular protein topogenesis. *Proceedings of the National Academy of Sciences USA*, 77: 1496–1500.

Burki, F. and Pawlowski, J. (2006) Monophyly of Rhizaria and multigene phylogeny of unicellular bikonts. *Molecular Biology and Evolution*, 23: 1922–1930.

Busse, I. and Preisfeld, A. (2003) Systematics of primary osmotrophic euglenids: a molecular approach to the phylogeny of *Distigma* and *Astasia* (Euglenozoa). *International Journal of Systematic and Evolutionary Microbiology*, 53: 617–624.

Bütschli, O. (1885) *Dr. H. G. Bronn's Klassen und Ordnungen des Thier-Reichs. Vol. 1 Abt. II Mastigophora* p. 1016. C. F. Winter, Heidelberg.

Canaday, J., Tessier, L. H., Imbault, P., and Paulus, F. (2001) Analysis of *Euglena gracilis* alpha-, beta- and gamma-tubulin genes: introns and pre-mRNA maturation. *Molecular Genetics and Genomics*, 265: 153–160.

Cavalier-Smith, T. (1981) Eukaryote kingdoms: seven or nine? *BioSystems*, 14: 461–481.

Cavalier-Smith, T. (1982) The origins of plastids. *Biological Journal of the Linnean Society*, 17: 289–306.

Cavalier-Smith, T. (1983) Endosymbiotic origin of the mitochondrial envelope. In *Endocytobiology II*, W. Schwemmler and H. E. A. Schenk (eds), de Gruyter, Berlin, pp. 265–279.

Cavalier-Smith, T. (1986) The kingdom Chromista: origin and systematics. In *Progress in Phycological Research*, F. E. Round and D. J. Chapman (eds), Vol. 4, Biopress, Bristol, pp. 309–347.

Cavalier-Smith, T. (1987) The origin of eukaryotic and archaebacterial cells. *Annals N Y Academy of Science*, 503: 17–54.

Cavalier-Smith, T. (1993a) Kingdom Protozoa and its 18 phyla. *Microbiological Reviews*, 57: 953–994.

Cavalier-Smith, T. (1993b) The origin, losses and gains of chloroplasts. In *Origin of Plastids: Symbiogenesis, Prochlorophytes and the Origins of Chloroplasts*, R. A. Lewin (ed.), Chapman & Hall, New York, pp. 291–348.

Cavalier-Smith, T. (1994) Origin and relationships of Haptophyta. In *The Haptophyte Algae*, J. C. Green and B. S. C. Leadbeater (eds), Clarendon Press, Oxford, pp. 413–435.

Cavalier-Smith, T. (1995) Membrane heredity, symbiogenesis, and the multiple origins of algae. In *Biodiversity and Evolution*, R. Arai, M. Kato, and Y. Doi (eds), The National Science Museum Foundation, Tokyo, pp. 75–114.

Cavalier-Smith, T. (1998) A revised six-kingdom system of life. *Biological Reviews of the Cambridge Philosophical Society*, 73: 203–266.

Cavalier-Smith, T. (1999) Principles of protein and lipid targeting in secondary symbiogenesis: euglenoid, dinoflagellate, and sporozoan plastid origins and the eukaryotic family tree. *Journal of Eukaryotic Microbiology*, 46: 347–366.

Cavalier-Smith, T. (2000a) Membrane heredity and early chloroplast evolution. *Trends in Plant Science*, 5: 174–182.

Cavalier-Smith, T. (2000b) What are Fungi? In *The Mycota*, D. J. McLaughlin, E. J. McLaughlin, and P. Lemke (eds), Vol. VII Part A, Springer-Verlag, Berlin, pp. 3–37.

Cavalier-Smith, T. (2001) Obcells as proto-organisms: membrane heredity, lithophosphorylation, and the origins of the genetic code, the first cells, and photosynthesis. *Journal of Molecular Evolution*, 53: 555–595.

Cavalier-Smith, T. (2002a) Chloroplast evolution: secondary symbiogenesis and multiple losses. *Current Biology*, 12: R62–R64.

Cavalier-Smith, T. (2002b) The neomuran origin of archaebacteria, the negibacterial root of the universal tree and bacterial megaclassification. *International Journal of Systematic and Evolutionary Microbiology*, 52: 7–76.

Cavalier-Smith, T. (2002c) The phagotrophic origin of eukaryotes and phylogenetic classification of Protozoa. *International Journal of Systematic and Evolutionary Microbiology*, 52: 297–354.

Cavalier-Smith, T. (2003a) The excavate protozoan phyla Metamonada Grassé emend. (Anaeromonadea, Parabasalia, *Carpediemonas*, Eopharyngia) and Loukozoa emend. (Jakobea, *Malawimonas*): their evolutionary affinities and new higher taxa. *International Journal of Systematic and Evolutionary Microbiology*, 53: 1741–1758.

Cavalier-Smith, T. (2003b) Genomic reduction and evolution of novel genetic membranes and protein-targeting machinery in eukaryote-eukaryote chimaeras (meta-algae). *Philosophical Transactions of the Royal Society of London B*, 358: 109–134.

Cavalier-Smith, T. (2003c) Protist phylogeny and the high-level classification of Protozoa. *European Journal of Protistology*, 39: 338–348.

Cavalier-Smith, T. (2004) Chromalveolate diversity and cell megaevolution: interplay of membranes, genomes and cytoskeleton. In *Organelles, Genomes and Eukaryote Phylogeny*, R. P. Hirt and D. S. Horner (eds), CRC Press, London, pp. 75–108.

Cavalier-Smith, T. (2006a) Cell evolution and earth history: stasis and revolution. *Philosophical Transactions of the Royal Society of London B*, 361: 969–1006.

Cavalier-Smith, T. (2006b) Origin of mitochondria by intracellular enslavement of a photosynthetic purple bacterium. *Proceedings of the Royal Society of London B*, 273: 1943–1952.

Cavalier-Smith, T. (2006c) Rooting the tree of life by transition analyses. *Biology Direct*, 2006, 1: 19.

Cavalier-Smith, T. and Chao, E. E. (1996) 18S rRNA sequence of *Heterosigma carterae* (Raphidophyceae), and the phylogeny of heterokont algae (Ochrophyta). *Phycologia*, 35: 500–510.

Cavalier-Smith, T. and Chao, E. E. (2003) Phylogeny of Choanozoa, Apusozoa, and other Protozoa and early eukaryote megaevolution. *Journal of Molecular Evolution*, 56: 540–563.

Cavalier-Smith, T. and Chao, E. E (2004) Protalveolate phylogeny and systematics and the origins of Sporozoa and dinoflagellates (phylum Myzozoa nom. nov.). *European Journal of Protistology*, 40: 185–212.

Cavalier-Smith, T. and Chao, E. E. (2006) Phylogeny and megasystematics of phagotrophic heterokonts (kingdom Chromista). *Journal of Molecular Evolution*, 62: 388–420.

Cavalier-Smith, T., Chao, E. E., Thompson, C. E., and Hourihane, S. L. (1995/6) *Oikomonas*, a distinctive zooflagellate related to chrysomonads. *Archiv für Protistenkunde*, 146: 273–279.

Cavalier-Smith, T. and Lee, J. J. (1985) Protozoa as hosts for endosymbioses and the conversion of symbionts into organelles. *Journal of Protozoology*, 32: 376–379.

Chen, M., Hiller, R. G., Howe, C. J., and Larkum, A. W. (2005) Unique origin and lateral transfer of prokaryotic chlorophyll-b and chlorophyll-d light-harvesting systems. *Molecular Biology and Evolution*, 22: 21–28.

De Bary, A. (1884) *Vergleichende Morphologie und Biologie der Pilze, Mycetozoen, und Bacterien*. Engelmann, Leipzig.

de Koning, A. P. and Keeling, P. J. (2006) The complete plastid genome sequence of the parasitic green alga, *Helicosporidium* sp. is highly reduced and structured. *BMC Biology*, 4: 12.

Delwiche, C. F. and Palmer, J. D. (1996) Rampant horizontal transfer and duplication of rubisco genes in eubacteria and plastids. *Molecular Biology and Evolution*, 13: 873–882.

Doolittle, W. F. (1998) You are what you eat: a gene transfer ratchet could account for bacterial genes in eukaryotic nuclear genomes. *Trends in Genetics*, 14: 307–311.

Douglas, S., Zauner, S., Fraunholz, M., Beaton, M., Penny, S., Deng, L. T., Wu, X., Reith, M., Cavalier-Smith, T., and Maier, U. G. (2001) The highly reduced genome of an enslaved algal nucleus. *Nature*, 410: 1091–1096.

Durnford, D. G., Deane, J. A., Tan, S., McFadden, G. I., Gantt, E., and Green, B. R. (1999) A phylogenetic assessment of the eukaryotic light-harvesting antenna proteins, with implications for plastid evolution. *Journal of Molecular Evolution*, 48: 59–68.

Fritsch, F. E. (1948) *The Structure and Reproduction of the Algae*. Cambridge University Press, Cambridge.

Garczarek, L., Poupon, A., and Partensky, F. (2003) Origin and evolution of transmembrane Chl-binding proteins: hydrophobic cluster analysis suggests a common one-helix ancestor for prokaryotic (Pcb) and eukaryotic (LHC) antenna protein superfamilies. *FEMS Microbiology Letters*, 222: 59–68.

Gibbs, S. P. (1979) The route of entry of cytoplasmically synthesized proteins into chloroplasts of algae possessing chloroplast ER. *Journal of Cell Science*, 35: 253–266.

Gille, C., Goede, A., Schloetelburg, C., Presissner, R., Kloetzel, P. M., Gobel, U. B., and Frommel, C. (2003) A comprehensive view on proteasomal sequences: implications for the evolution of the proteasome. *Journal of Molecular Biology*, 326: 1437–1448.

Gilson, P. R., Su, V., Slamovits, C. H., Reith, M., Keeling, P. J., and McFadden, G. I. (2006) Complete nucleotide sequence of the chlorarachniophyte nucleomorph: nature's smallest nucleus. *Proceedings of the National Academy of Sciences, USA*, 103: 9739–9780.

Gould, S. B., Sommer, M. S., Hadfi, K., Zauner, S., Kroth, P. G., and Maier, U. G. (2006a) Protein targeting into the complex plastid of cryptophytes. *Journal of Molecular Evolution*, 62: 674–681.

Gould, S. B., Sommer, M. S., Kroth, P. G., Gile, G. H., Keeling, P. J., and Maier, U. G. (2006b) Nucleus-to-nucleus gene transfer and protein retargeting into a remnant cytoplasm of cryptophytes and diatoms. *Molecular Biology and Evolution*, 23: 1–10.

Gutensohn, M., Fan, E., Frielingsdorf, S., Hanner, P., Hou, B., Hust, B., and Klosgen, R. B. (2006) Toc, Tic, Tat et al.: structure and function of protein transport machineries in chloroplasts. *Journal of Plant Physiology*, 163: 333–347.

Harper, J. T. and Keeling, P. J. (2003) Nucleus-encoded, plastid-targeted glyceraldehyde-3-phosphate dehydrogenase (GAPDH) indicates a single origin for chromalveolate plastids. *Molecular Biology and Evolution*, 20: 1730–1735.

Hust, B. and Gutensohn, M. (2006) Deletion of core components of the plastid protein import machinery causes differential arrest of embryo development in *Arabidopsis thaliana*. *Plant Biology (Stuttgart)*, 8: 18–30.

Ishida, K., Cao, Y., Hasegawa, M., Okada, N., and Hara, Y. (1997) The origin of chlorarachniophyte plastids, as inferred from phylogenetic comparisons of amino acid sequences of EF-Tu. *Journal of Molecular Evolution*, 45: 682–687.

Ishida, K., Green, B. R., and Cavalier-Smith, T. (1999) Diversification of a chimaeric algal group, the chlorarachniophytes; phylogeny of nuclear and nucleomorph small subunit rRNA genes. *Molecular Biology and Evolution*, 16: 321–331.

Kim, E., Simpson, A. G., and Graham, L. E. (2006) Evolutionary relationships of apusomonads inferred from taxon-rich analyses of six nuclear-encoded genes. *Molecular Biology and Evolution*, 23: 2455–2466.

Klaveness, D., Shalchian-Tabrizi, K., Thomsen, H. A., Eikrem, W., and Jakobsen, K. S. (2005) *Telonema antarcticum* sp. nov., a common marine phagotrophic flagellate. *International Journal of Systematic and Evolutionary Microbiology*, 55: 2595–2604.

Klinkert, B., Ossenbuhl, F., Sikorski, M., Berry, S., Eichacker, L., and Nickelsen, J. (2004) PratA, a periplasmic tetratricopeptide repeat protein involved in biogenesis of photosystem II in *Synechocystis* sp. PCC 6803. *Journal of Biological Chemistry*, 279: 44639–44644.

Knappe, S., Flugge, U. I., and Fischer, K. (2003) Analysis of the plastidic phosphate translocator gene family in *Arabidopsis* and identification of new phosphate translocator-homologous transporters, classified by their putative substrate-binding site. *Plant Physiology*, 131: 1178–1190.

Koga, R., Tsuchida, T., and Fukatsu, T. (2003) Changing partners in an obligate symbiosis: a facultative endosymbiont can compensate for loss of the essential endosymbiont *Buchnera* in an aphid. *Proceedings in Biological Science*, 270: 2543–2550.

Kollman, J. M. and Doolittle, R. F. (2000) Determining the relative rates of change for prokaryotic and eukaryotic proteins with anciently duplicated paralogs. *Journal of Molecular Evolution*, 51: 173–181.

Koziol, A. G., Borza, T., Ishida, K., Keeling, P., Lee, R. W., and Durnford, D. G. (2007) Tracing the evolution of the light-harvesting antennae in chlorophyll a/b-containing organisms. *Plant Physiology*, 143: 1802–1816.

Laatsch, T., Zauner, S., Stoebe-Maier, B., Kowallik, K. V., and Maier, U. G. (2004) Plastid-derived single gene minicircles of the dinoflagellate *Ceratium horridum* are localized in the nucleus. *Molecular Biology and Evolution*, 21: 1318–1322.

Leander, B. S. (2004) Did trypanosomatid parasites have photosynthetic ancestors? *Trends in Microbiology*, 12: 251–258.

Leander, B. S., Witek, R. P., and Farmer, M. A. (2001) Trends in the evolution of the euglenid pellicle. *Evolution*, 55: 2215–2235.

Lee, J. J., Leedale, G., and Bradbury, P. (2002 dated 2000) *An Illustrated Guide to the Protozoa*. Society of Protozoologists, Lawrence, KS.

Leipe, D. D., Tong, S. M., Goggin, C. L., Slemenda, S. B., Pienizek, N. J., and Sogin, M. L. (1996) 16S-like rDNA sequences from *Developayella elegans*, *Labyrinthuloides haliotidis*, and *Proteromonas lacertae* confirm that the stramenopiles are a primarily heterotrophic group. *European Journal of Protistology*, 32: 449–458.

Lemmermann, E. (1914) *Flagellatae 1*. Gustav Fischer, Jena.

Lewin, R. A. (2002) Prochlorophyta—a matter of class distinctions. *Photosynthesis Research*, 73: 59–61.

Mallet, M. A. and Lee, R. W. (2006) Identification of three distinct *Polytomella* lineages based on mitochondrial DNA features. *Journal of Eukaryotic Microbiology*, 53: 79–84.

Margulis, L. (1970) *Origin of Eukaryotic Cells*. Yale University Press, New Haven, CT.

Margulis, L. (1981) *Symbiosis in Cell Evolution*. Freeman, San Francisco.

Marin, B., Nowack, E. C., and Melkonian, M. (2005) A plastid in the making: evidence for a second primary endosymbiosis. *Protist*, 156: 425–432.

Marin, B., Palm, A., Klingberg, M., and Melkonian, M. (2003) Phylogeny and taxonomic revision of plastid-containing euglenophytes based on SSU rDNA sequence comparisons and synapomorphic signatures in the SSU rRNA secondary structure. *Protist*, 154: 99–145.

McEwan, M. L. and Keeling, P. J. (2004) HSP90, tubulin and actin are retained in the tertiary endosymbiont genome of *Kryptoperidinium foliaceum*. *Journal of Eukaryotic Microbiology*, 51: 651–659.

McFadden, G. I. and van Dooren, G. G. (2004) Evolution: red algal genome affirms a common origin of all plastids. *Current Biology*, 14: 514–516.

Mereschkovsky, C. (1905) Über Natur und Ursprung der Chromatophoren im Pflanzenreiche. *Biologisches Zentralblatt*, 25: 593–604.

Mereschkovsky, C. (1910) Theorie der Zwei Plasmaarte als Grundlage der Symbiogenesis, einer neue Lehre von der Entstehung der Organismen. *Biologisches Zentralblatt*, 30: 278–303, 321–347, 353–367.

Mikrjukov, K. A. and Mylnikov, A. P. (2001) A study of the fine structure and the mitosis of a lamellicristate amoeba, *Micronuclearia podoventralis* gen. et sp. nov. (Nucleariidae, Rotosphaerida). *European Journal of Protistology*, 37: 15–24.

Miyagishima, S. Y. (2005) Origin and evolution of the chloroplast division machinery. *Journal of Plant Research*, 118: 295–306.

Miyagishima, S. Y., Wolk, C. P., and Osteryoung, K. W. (2005) Identification of cyanobacterial cell division genes by comparative and mutational analyses. *Molecular Microbiology*, 56: 126–143.

Moreira, D., von der Heyden, S., Bass, D., López-García, P., Chao, E., and Cavalier-Smith, T. (2007) Global eukaryote phylogeny: combined small- and large-subunit ribosomal DNA trees support monophyly of Rhizaria, Retaria and Excavata. *Molecular Phylogenetics and Evolution*, 44: 255–266.

Nassoury, N., Cappadocia, M., and Morse, D. (2003) Plastid ultrastructure defines the protein import pathway in dinoflagellates. *Journal of Cell Science*, 116: 2867–2874.

Not, F., Valentin, K., Romari, K., Lovejoy, C., Massana, R., Tobe, K., Vaulot, D., and Medlin, L. K. (2007) Picobiliphytes: a marine picoplanktonic algal group with unknown affinities to other eukaryotes. *Science*, 315: 253–255.

O'Kelly, C. J. (1992) Flagellar apparatus architecture and the phylogeny of "green" algae: chlorophytes, euglenoids, glaucophytes. In *Cytoskeleton of the Algae*, D. Menzel (ed.), CRC Press, Boca Raton, FL, pp. 315–345.

Okamoto, N. and Inouye, I. (2005) The katablepharids are a distant sister group of the Cryptophyta: a proposal for Katablepharidophyta divisio nova/Kathablepharida phylum novum based on SSU rDNA and beta-tubulin phylogeny. *Protist*, 156: 163–179.

Pascher, A. (1929) Über einige Endosymbiosen von Blaualgen in Einzellern. *Jahrbuch Wissenschaftliche Botanik*, 71: 386–462.

Patron, N. C., Rogers, M. B., and Keeling, P. J. (2004) Gene replacement of fructose-1,6-bisphosphate aldolase supports the hypothesis of a single photosynthetic ancestor of chromalveolates. *Eukaryotic Cell*, 3: 1169–1175.

Patron, N. J., Waller, R. F., Archibald, J. M., and Keeling, P. J. (2005) Complex protein targeting to dinoflagellate plastids. *Journal of Molecular Biology*, 348: 1015–1024.

Peret´o, J., López-García, P., and Moreira, D. (2004) Ancestral lipid biosynthesis and early membrane evolution. *Trends in Biochemical Sciences*, 29: 469–477.

Reumann, S., Inoue, K., and Keegstra, K. (2005) Evolution of the general protein import pathway of plastids (review). *Molecular Membrane Biology*, 22: 73–86.

Rice, D. W. and Palmer, J. D. (2006) An exceptional horizontal gene transfer in plastids: gene replacement by a distant bacterial paralog and evidence that haptophyte and cryptophyte plastids are sisters. *BMC Biology*, 4: 31.

Richards, T. A., Dacks, J. B., Jenkinson, J. M., Thornton, C. R., and Talbot, N. J. (2006) Evolution of filamentous plant pathogens: gene exchange across eukaryotic kingdoms. *Current Biology*, 16: 1857–1864.

Richards, T. A. and Cavalier-Smith, T. (2005) Myosin domain evolution and the primary divergence of eukaryotes. *Nature*, 436: 1113–1118.

Rissler, H. M. and Durnford, D. G. (2005) Isolation of a novel carotenoid-rich protein in *Cyanophora paradoxa* that is immunologically related to the light-harvesting complexes of photosynthetic eukaryotes. *Plant Cell Physiology*, 46: 416–424.

Robertson, D. L. and Tartar, A. (2006) Evolution of glutamine synthetase in heterokonts: evidence for endosymbiotic gene transfer and the early evolution of photosynthesis. *Molecular Biology and Evolution*, 23: 1048–1055.

Rodríguez-Ezpeleta, N., Brinkmann, H., Burey, S. C., Roure, B., Burger, G., Löffelhardt, W., Bohnert, H. J., Philippe, H., and Lang, B. F. (2005) Monophyly of primary photosynthetic eukaryotes: green plants, red algae, and glaucophytes. *Current Biology*, 15: 1325–1330.

Rogers, M. B., Gilson, P. R., Su, V., McFadden, G. I., and Keeling, P. J. (2007) The complete chloroplast genome of the chlorarachniophyte *Bigelowiella natans*: evidence for independent origins of chlorarachniophyte and euglenid secondary endosymbionts. *Molecular Biology and Evolution*, 24: 54–62.

Ryall, K., Harper, J. T., and Keeling, P. J. (2003) Plastid-derived Type II fatty acid biosynthetic enzymes in chromists. *Gene*, 313: 139–148.

Saldarriaga, J. F., Taylor, F. J., Keeling, P. J., and Cavalier-Smith, T. (2001) Dinoflagellate nuclear SSU rRNA phylogeny suggests multiple plastid losses and replacements. *Journal of Molecular Evolution*, 53: 204–213.

Schnepf, E. (1964) Zur Feinstruktur von *Geosiphon pyriforme*: ein Versuch zur Deutung cytoplasmatischer Membranen und Kompartimente. *Archiv für Microbiologie*, 49: 112–131.

Shalchian-Tabrizi, K., Eikrem, W., Klaveness, D., Vaulot, D., Minge, M. A., Le Gall, F., Romari, K., Throndsen, J., Botnen, A., Massana, R., Thomsen, H. A., and Jakobsen, K. S. (2006a) Telonemia, a new protist phylum with affinity to chromist lineages. *Proceedings of the Royal Society of London B*, 273: 1833–1842.

Shalchian-Tabrizi, K., Minge, M., Nedreklepp, J., Cavalier-Smith, T., Klaveness, D., and Jakobsen, K. S. (2006b) Combined Hsp90 and rRNA sequence phylogeny supports multiple replacements of dinoflagellate plastids. *Journal of Eukaryotic Microbiology*, 53: 217–224.

Silva, P. C. (1980) Names of classes and families of living algae with special reference to their use in the Index Nominum Genericorum (Plantarum). *Regnum Vegetabile*, 103: [iii +] 156 pp.

Silva-Filho, M. D., Wieers, M. C., Flugge, U. I., Chaumont, F., and Boutry, M. (1997) Different *in vitro* and *in vivo* targeting properties of the transit peptide of a chloroplast envelope inner membrane protein. *Journal of Biological Chemistry*, 272: 15264–15269.

Simpson, A. G. B. (1997) The identity and composition of the Euglenozoa. *Archiv für Protistenkunde*, 148: 318–328.

Simpson, A. G. (2003) Cytoskeletal organization, phylogenetic affinities and systematics in the contentious taxon Excavata (Eukaryota). *International Journal of Systematic and Evolutionary Microbiology*, 53: 1759–1777.

Simpson, A. G., Inagaki, Y., and Roger, A. J. (2006) Comprehensive multigene phylogenies of excavate protists reveal the evolutionary positions of "primitive" eukaryotes. *Molecular Biology and Evolution*, 23: 615–625.

Simpson, A. G. B. and Patterson, D. J. (1999) The ultrastructure of *Carpediemonas membranifera* (Eukaryota) with reference to the "excavate hypothesis." *European Journal of Protistology*, 35: 353–370.

Simpson, A. G. and Roger, A. J. (2004) Protein phylogenies robustly resolve the deep-level relationships within Euglenozoa. *Molecular Phylogenetics and Evolution*, 30: 201–212.

Slavikova, S., Vacula, R., Fang, Z., Ehara, T., Osafune, T., and Schwartzbach, S. D. (2005) Homologous and heterologous reconstitution of Golgi to chloroplast transport and protein import into the complex chloroplasts of *Euglena*. *Journal of Cell Science*, 118: 1651–1661.

Stechmann, A. and Cavalier-Smith, T. (2003) The root of the eukaryote tree pinpointed. *Current Biology*, 13: R665–R666.

Steiner, J. M. and Löffelhardt, W. (2005) Protein translocation into and within cyanelles (review). *Molecular Membrane Biology*, 22: 123–132.

Steiner, J. M., Yusa, F., Pompe, J. A., and Löffelhardt, W. (2005) Homologous protein import machineries in chloroplasts and cyanelles. *Plant Journal*, 44: 646–652.

Tomitani, A., Okada, K., Miyashita, H., Matthijs, H. C., Ohno, T., and Tanaka, A. (1999) Chlorophyll b and phycobilins in the common ancestor of cyanobacteria and chloroplasts. *Nature*, 400: 159–162.

van de Meene, A. M., Hohmann-Marriott, M. F., Vermaas, W. F., and Roberson, R. W. (2006) The three-dimensional structure of the cyanobacterium *Synechocystis* sp. PCC 6803. *Archives of Microbiology*, 184: 259–270.

van Ham, R. C., Kamerbeek, J., Palacios, C., Rausell, C., Abascal, F., Bastolla, U., Fernandez, J. M., Jimenez, L., Postigo, M., Silva, F. J., Tamames, J., Viguera, E., Latorre, A., Valencia, A., Moran, F., and Moya, A. (2003) Reductive genome evolution in *Buchnera aphidicola*. *Proceedings of the National Academy of Sciences, USA*, 100: 581–586.

von der Heyden, S., Chao, E. E., Vickerman, K., and Cavalier-Smith, T. (2004) Ribosomal RNA phylogeny of bodonid and diplonemid flagellates and the evolution of Euglenozoa. *Journal of Eukaryotic Microbiology*, 51: 402–416.

von Dohlen, C. D., Kohler, S., Alsop, S. T., and McManus, W. R. (2001) Mealybug β-proteobacterial endosymbionts contain γ-proteobacterial symbionts. *Nature*, 412: 433–436.

Vothknecht, U. C. and Soll, J. (2005) Chloroplast membrane transport: interplay of prokaryotic and eukaryotic traits. *Gene*, 354: 99–109.

Walker, G., Dacks, J. B., and Martin Embley, T. (2006) Ultrastructural description of *Breviata anathema*, n. gen., n. sp., the organism previously studied as "*Mastigamoeba invertens*." *Journal of Eukaryotic Microbiology*, 53: 65–78.

Wang, Y. and Morse, D. (2006a) The plastid-encoded psbA gene in the dinoflagellate *Gonyaulax* is not encoded on a minicircle. *Gene*, 371: 206–210.

Wang, Y. and Morse, D. (2006b) Rampant polyuridylylation of plastid gene transcripts in the dinoflagellate *Lingulodinium*. *Nucleic Acids Research*, 34: 613–619.

Weber, A. P. M., Linka, M., and Bhattacharya, D. (2006) Single, ancient origin of a plastid metabolite translocator family in Plantae from an endomembrane-derived ancestor. *Eukaryotic Cell*, 5: 609–612.

Wetherbee, R., Platt, S. J., Beech, P. L., and Pickett-Heaps, J. D. (1988) Flagellar transformation in the heterokont *Epipyxis pulchra* (Chrysophyceae): direct observations using image enhanced video microscopy. *Protoplasma*, 145: 47–54.

Wickstead, B. and Gull, K. (2006) A "holistic" kinesin phylogeny reveals new kinesin families and predicts protein functions. *Molecular Biology of the Cell*, 17: 1734–1743.

Yoon, H. S., Hackett, J. D., Pinto, G., and Bhattacharya, D. (2002) The single, ancient origin of chromist plastids. *Proceedings of the National Academy of Sciences, USA*, 99: 15507–15512.

Yoon, H. S., Hackett, J. D., Van Dolah, F. M., Nosenko, T., Lidie, K. L., and Bhattacharya, D. (2005) Tertiary endosymbiosis driven genome evolution in dinoflagellate algae. *Molecular Biology and Evolution*, 22: 1299–1308.

Yoon, H. S., Müller, K. M., Sheath, R. G., Ott, F. D., and Bhattacharya, D. (2006) Defining the major lineages of red algae (Rhodophyta). *Journal of Phycology*, 42: 482–492.

Zhang, Z., Green, B. R., and Cavalier-Smith, T. (1999) Single gene circles in dinoflagellate chloroplast genomes. *Nature*, 400: 155–159.

4 Classification and diatom systematics: the past, the present, and the future

David M. Williams

CONTENTS

"A dog is a mammal."
"So's a rat," Denise said.
"A rat is a vermin," Babette said.
"Mostly what a rat is," Heinrich said, "is a rodent."
"It's also a vermin."
"A cockroach is a vermin," Steffie said.
"A cockroach is an insect. You count the legs is how you know."
"It's also a vermin."

"Does a cockroach get cancer? No," Denise said. "That must mean a rat is more like a human than it is like a cockroach, even if they're both vermins, since a rat and a human can get cancer but a cockroach can't."

"In other words," Heinrich said, "She's saying that two things that are mammals have more in common than two things that are only vermins."

"Are you telling me," Babette said, "that a rat is not only a vermin and a rodent but a mammal too?"

Snow turned to sleet, sleet to rain.

Don DeLillo, 1984, pp. 124–125

Reasons for any classification are always easy to find.

Lange-Bertalot, 1993, p. 179

ABSTRACT

Biological classification is the pinnacle of comparative biology. With respect to evidence available for diatom classification, certain phases of investigation have been recognised, starting with the use of valve and frustule "shape," to the acquisition of structural data obtained with light and, subsequently, electron microscopes and, finally, with the avalanche of molecular data. In this paper I explore these phases, examining selected topics in the history of diatom classification with a view to exploring the concepts, if any, early diatom biologists utilised. Discovery of concepts may illuminate matters today, shedding some light on the "post-genomic" diatom classification. It seems, as it does for other taxa and other eras (regardless of the source of data), the main source of problems is paraphyly.

INTRODUCTION

To some, biological classification is quaint and unscientific, irrelevant to modern science (O'Hara, 1994; Felsenstein, 2004); for others it is the pinnacle of comparative biology, a representation and summary of our knowledge of the living world (Nelson & Platnick, 1981, 1984). Even so, for such a seemingly harmless subject, it has generated considerable amount of discussion, especially in the 1960s and 1970s, with reference to cladistics (Kitching et al., 1998), and again in the 1990s and 2000s. Of the many contributions, one commentator said of classification that

> During the early years of cladistics, in the 1960s and 1970s, the mismatch between cladistically reconstructed phylogenies and traditional classifications gave rise to a great deal of controversy over the relation between trees and classifications, but this controversy has today almost completely withered away. This is because more and more systematists have come to realize that in the evolutionary world the notion of classification as an object of systematics can be largely dispensed with. The point of systematics in an evolutionary world ought not to be the construction of classes, but the reconstruction of history (de Queiroz, 1988; O'Hara, 1988), and the analogy of systematics to classification is in fact a relict of the pre-evolutionary period, when living diversity was viewed ahistorically. (O'Hara, 1994, p. 14)

That viewpoint seems almost quaintly anachronistic, given that it is almost impossible to say *anything* of meaning about *any* organism without invoking some kind of classification, a name of some sorts (Stevens, 2002; Bowman, 2005). Nevertheless, there are principles to classification (Nelson, 1972, 1973), even if in the past they had not been too clearly articulated (Stevens, 1994) and in the present have become hopelessly muddled (Brummitt, 2002, 2003, for commentary see Nelson et al., 2003; Ebach & Williams, 2004). There have been efforts to establish a *Phylocode* (Cantino & de Queiroz, 2004), a new code of biological nomenclature designed to create a system of classification that purportedly reflects evolutionary relationships exactly. Although the *Phylocode* appears to be almost dead (Pickett, 2005), the relationship between classification and phylogeny remains of significance (see Williams et al., 2005), particularly in this new "genomic" age (Franz, 2005), in spite of those who speak of the irrelevance of classification (Felsenstein, 2004, p. 145).

In contrast to O'Hara, consider Platnick's words: "Back in the early days of the cladistics wars, it was fashionable in some circles to see classification as a serious problem for phylogeneticists. The problem, of course, was imaginary" (Platnick, 2001, p. 5). With respect to diatom classification and any principles that might have guided it, there appears a common belief that it has proceeded isolated from mainstream systematic thought, as if, by and large, no reasoning was ever presented for the groups so recognised: "History suggests that, in many cases, diatomists might just as well have been classifying scraps of wallpapers as diatoms. Diatom taxonomy has developed largely without a conceptual basis, using a restricted range of characters drawn from just one part of the phenotype (the valve)" (Mann, 1999, p. 482).

Darwin, rather circumstantially in the 4th edition of the *Origin*, chapter 6, "Difficulties of the Theory, Utilitarian Doctrine, How Far True: Beauty, How Acquired," posed a rather different but certainly rhetorical question: "Few objects are more beautiful than the minute siliceous cases of the diatomaceae: were these created that they might be examined and admired under the higher powers of the microscope?" (Darwin, 1866, p. 200). Darwin's comment originated from his correspondence with George Henry Kendrick Thwaites (1811–1882) (Burkhardt and Smith, 1987, pp. 131–133), an erstwhile diatomist who in 1849 moved from Bristol, UK, to Ceylon (Sri Lanka), eventually becoming director (1857–1880) of the Peradeniya Botanical Gardens (Willis, 1901). Pondering beauty is one thing (Moore, 1991), classification another.

History suggests many things but "by its nature [it is] an act of hindsight, of understanding, or understanding better, what was understood less well at the time, or of *understanding again what has been temporarily forgotten*" (Barnes, 2002, p. 9, my italics, see Ragan, 1998). Round wrote that "Awareness, even briefly, of the historical context of the systematics of the diatoms is essential to understand the problems" (1996a, p. 205). The problems of classification, that is (Kitton, 1880, 1882; Silva, 1980). Round identified certain phases of investigation, starting with the use of "shape" as a discriminatory factor through to the acquisition of structural data obtained with electron microscopes. No complete survey has ever been undertaken on the effects 50 years electron microscopy has had on the subject of diatom classification, in spite of a wealth of information (collated in Gaul et al., 1993, and Henderson & Reimer, 2003—one might consider Round et al., 1990, as a possible summary; see below). Yet pertinent comment came early in 1959 from Norman I. Hendey (1903–2004; Sims, 2005):

> There is little doubt that in the early days of electron microscopy high hopes were held by some of finding the basis for an entirely new classificatory system for the diatoms. It was assumed that the power and ability to "see" the ultimate structure of the valves must necessarily produce the information upon which a system could be devised, that not merely would sweep the classical systems away but would replace them with one that was somehow "right." (Hendey, 1959, p. 163; see also Hendey, 1958 and 1971, but see Ross, 1963, 1995, for a contrasting viewpoint.)

Round (1996a) speculated on the impact molecular data may have on diatom classification and, one might imagine, these data constituting yet another phase—and where electron microscopy has supposedly failed, who knows, molecules may succeed.

Nevertheless, Round's historical account concerns different kinds of data rather than concepts (it appears to be "without a conceptual basis," see Mann, 1999, quote above), revealing questions that appear—at least with a little thought—unanswerable: If enough data are collected then will "true" relationships be revealed? Or, if we collect enough of the "right" data, will then "true" relationships be revealed? Today, of course, the "right" data are molecular. So that raises a similar, more focused question: Will these data, like others collected before, turn out to be just as "right" or just as "wrong" (Jeffroy et al., 2006)?

I wish to pick up Round's lead (above) and spend some time looking at selected topics in the history of diatom classification but with a view to exploring the concepts, if any, those early biologists utilised. Discovery of concepts may illuminate matters today, shedding some light on the first crop of "post-genomic" diatom classifications; hence, I begin with a short discussion of classificatory groups and their use.

CLADOGRAMS, CLASSIFICATION, AND PARAPHYLY

What do invertebrates, apes, and barbarians have in common? … Invertebrates are non-vertebrate animals; apes, non-human anthropoids; barbarians, non-Greek humans, whose diverse languages, to the ears of the ancient Greeks, all sounded like "bar-bar-bar"—even perhaps like the bleating of sheep.

Nelson, 2006

Cladograms (and phylogenetic trees) relate to classification in an absolute sense (Nelson & Platnick, 1981), in spite of dissenting voices (e.g., Cavalier-Smith, 1998, figure 1 and figure 2; Brummitt, 2002). In brief, monophyletic groups correspond to nodes on a tree, paraphyletic and polyphyletic groups do not (Platnick, 1977). Many diagrams said to depict evolutionary relationships have species (hypothetical or real) at the nodes. Such diagrams may also be converted into classifications, identifying all the monophyletic groups by inclusive taxa at and on nodes (Nelson and Platnick, 1981, pp. 143–151).

Groups identified in cladograms share sets of apomorphic characters and are usually called "natural" or monophyletic. In cases where non-monophyly occurs, certain groups may be considered paraphyletic, others polyphyletic. Paraphyletic groups occur if a part of an assumed monophyletic group is missing (birds missing from reptiles). Polyphyletic groups occur when taxa either closely or distantly related, groups scattered throughout a larger, assumed monophyletic group, are named. Paraphyletic groups are non-groups, "timeless abstractions" (Brundin, 1972; Patterson, 1978, p. 220, see also Ebach & Williams, 2004).

In spite of this, paraphyletic and polyphyletic groups have been given historical explanations (Reif, 2005b). The traditional viewpoint (one that rather remarkably seems to be gaining favour again) is that paraphyletic groups are, or nearly are, equivalent to ancestors. If paraphyletic groups are non-groups, and non-groups are ancestors, then such entities can play no part in classification (or indeed phylogeny), are impossible to discover, and have no relevance beyond giving credence to particular theories of origin, whatever that theory might be. Nevertheless, it is still common to read that non-birds give rise to birds, non-humans give rise to humans (Nelson, 1989; Nelson et al., 2003), "Radial centric diatoms begot multipolar centric diatoms…" (Kooistra et al., 2003a, p. 92), betraying a belief that "Evolution is paraphyly all the way" (Brummitt, 2002, p. 40, see also Cavalier-Smith, 1998, p. 210, figure 1 and figure 2, cf. Nelson et al., 2003; Ebach & Williams, 2004; Reif, 2005a). If "evolution is paraphyly all the way," then there is a problem. Paraphyly and polyphyly are explanations for non-groupings or, more accurately, excuses for the absence of monophyly—or of recognizing it. Therefore the contradictory terms, "paraphyletic groups" and "polyphyletic groups," are more accurately termed "non-monophyletic" and should be eliminated from classifications in order to achieve progress (see essays in Cracraft & Donoghue, 2004).

INTERPRETING MORPHOLOGY

In many diatom molecular studies, morphology is simply read from the tree. A tree is acquired and the morphological data added to the branches, their "evolutionary" changes being read directly from the branches. This is irrational. Consider a contrived example. Figure 4.1 is a fabricated (untrue) diatom phylogenetic tree, with three "araphid" diatoms (*Fragilaria crotonensis* Kitton, *Fragilaria* sp., and *Centronella* M. Voigt), three raphid pennate diatoms (a species of *Lyrella* Karayeva, *Achnanthes* Bory and *Rossia* M. Voigt), and one species of "centric" diatom (*Aulacodiscus* Ehrenb.). Can morphological features be explained? Two raphid pennate diatoms—*Achnanthes* and *Rossia*—group together, but *Lyrella* is on a separate branch (Figure 4.1, node a). Thus, the origin of its raphe can be explained by two independent events, one leading to *Achnanthes* plus *Rossia* (Figure 4.1, node b), the other to *Lyrella* (Figure 4.1, node a). It might be tempting to

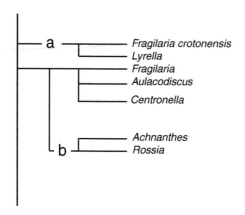

FIGURE 4.1 Hypothetical relationships among seven diatoms.

conclude that the raphe in species of *Lyrella* is different, truly different—in terms of common ancestry, not homologous—from that found in *Achnanthes* and *Rossia* because they have independent origins. Thus, we might conclude that the raphes of all diatoms are not homologous. The point here is that *any* character can be explained with reference to *any* tree, even if the tree has no basis in fact, as is the case in Figure 4.1. Morphology and molecules are independent data sources and from that perspective, require separate treatment.

CLASSIFICATION AND CHARACTERS: METHOD AND MADNESS? A SELECTION OF 19TH AND 20TH CENTURY DIATOM CLASSIFICATIONS

1. C.A. AGARDH (1785–1859)

In 1819 the entomologist William Sharp Macleay (Fletcher, 1921; Swainston, 1985; Holland, 1996) published the first part of *Horae Entomologicae*, a curious and rare but important work in the history of systematics. The second part was published in 1821 and dealt with the principles of classification (Macleay, 1819–1821; Di Gregorio, 1996). Macleay was interested in methods to discover the natural system. He believed that taxonomic groups, if represented correctly in a natural classification, would be arranged in series of interconnecting circles. He eventually understood the number to be five, connecting five particular groups together (Figure 4.2; Macleay, 1819–1821, 1825, 1830, for further discussion, especially concerning William Swainson, Macleay's most prominent interpreter and promoter, see Farber, 1985; Knight, 1985, 1986; Gardiner, 2001). As a result, Macleay's approach became known as "Quinarian," a deliberate attempt to challenge the Linnean "Dichotomous System" common at that time; Macleay was more than happy to attack his adversaries in print (Macleay, 1830).

Although Macleay and his followers' ideas were effectively eliminated from serious consideration after 1843, he did draw attention to what was understood as the central issue in systematics and classification: the discrimination of affinity, or "true relations," from analogy, or "false relations"—or as it is expressed in today's language, discriminating homology from homoplasy. Macleay's work was not ignored, with Darwin making a serious effort to understand its implications (De Beer, 1960, p. 29; Barrett, 1960, p. 256, 286; Smith, 1965, p. 100; Di Gregorio, 1981, 1982, 1996; Secord, 2000, p. 430). Indeed, Darwin acknowledged the debt in the *Origin*, a phrase that remained in all seven editions: "We can understand … the very important distinction between real affinities and analogical or adaptive resemblances. Lamarck first called attention to this distinction, and he has been ably followed by Macleay and others" (Darwin, 1859, p. 427).

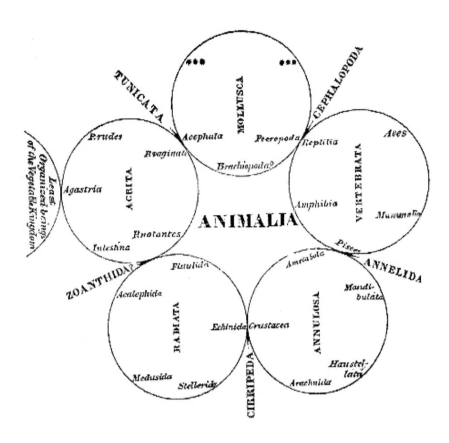

FIGURE 4.2 Quinarian diagram depicting relationships among all animals. (From Macleay, W.S., *Horae entomologicae*, S. Bagster, London, 1819–1821.)

In an 1823 paper, Macleay referred to two biologists whom he understood to have developed similar ideas independently, Carl Adolph Agardh and his student, the "father of mycology," Elias Magnus Fries (Macleay, 1823, p. 194). Macleay cited Agardh's discussion of the pitfalls and problems of separating and distinguishing analogy from affinity as significant—and Agardh was responsible for one of the first comprehensive classifications of diatoms published in the *Conspectus Criticus Diatomacearum* (Agardh, 1830–1832).

In Agardh's *Systema Algarum* (Agardh, 1823–1828; see Staflau, 1966, 1970), he classified all diatoms in one order of algae, placing the species among nine genera: *Achnanthes* Bory, *Frustulia* C.A. Agardh, *Meridion* C.A. Agardh, *Diatoma* Bory, *Fragilaria* Lyngbye, *Melosira* C.A. Agardh, *Desmidium* C.A. Agardh, *Schizonema* C.A. Agardh, and *Gomphonema* C.A. Agardh, with the genus *Diatoma* being subdivided into five groups (Agardh, 1824, in Agardh, 1823–1828). In the later *Conspectus* he rearranged the genera into three families—Cymbelleae, Styllarieae, and Fragilarieae—based on valve shape: Styllarieae included genera with cuneate (wedged-shaped) valves, Cymbelleae included genera with cymbelloid valves, and Fragilarieae included genera with rectangular valves, the latter being subdivided into two groups on the basis of their colony form. Agardh compared each of the three families with four different colony types: those with no obvious colony formation ("libera"), those attached by a stalk ("Stipitata"), those attached in chains ("In frondem composita"), and those in "cymbelloid" chains ("Fila cymbellarum frondem formantia"). In summary, Agardh (1832 in 1830–1832) presented a table contrasting valve shape with colony structure (Figure 4.3). His final classification, however, reflected frustule shape rather than colony formation. This classification

Conspectus serierum:			
Elementa	**Cymbelleæ** 1. cymbelliformis:	**Styllarieæ** 2. cuneata:	**Fragilarieæ** 3. rectangula.
1. Líbera.	1. *Cymbella.*	1. *Styllaria.*	1. *Frustulia.*
2. Stipitata.	2. *Gomphonema.*	2. *Licmophora.*	2. *Achnantes, Striatella Diatomatis, species stipitata.*
3. In frondem composita	3. *Schizonema, Berkeleya, Homœocladia, Gloiodictyon, Gloionema.*	3. *Meridion.*	3. *Diatoma, Odontella, Fragilaria, Desmidium, Melosira.*
4. Fila cymbellarum frondem formantia.	4. *Micromega, Hydrurus.*		

FIGURE 4.3 Reproduction of classification table from Agardh, 1832. (In Agardh, C.A., *Conspectus Criticus Diatomacearum,* Litteris Berlingianis, Lundae, 1830–1832.)

(derived from the table in Figure 4.3) can be represented by a branching diagram with three nodes, each node representing a family (Figure 4.4).

As noted earlier, Macleay was impressed with Agardh's approach to systematics, attempting to reconcile what he considered affinities (homologies)—in this case, shape—with what he considered analogies (non-homologies)—colony formation (see also Staflau, 1970; Ott, 1991; Woelkerling & Lamy, 1999). For Agardh, true relationships could only be understood from the characters of the frustule, colony formation providing only additional (analogous) information. Agardh, like many others of his day, was aware of the many issues in classification. The general problem of classification might be phrased: How does one discover characters of significance? Agardh had noted:

> that such a work as this cannot be free of hypotheses is self-evident. Hypotheses have always been necessary; they have never harmed the sciences, but on the contrary, have aided them even if they were found in the final analysis to be unsubstantiated. (Agardh, 1829–1830; modified from a translation in Ott, 1991, p. 304)

Agardh's approach might be considered to have some similarities with what became known as cladistics (Nelson & Platnick, 1981), if that understanding is taken to be the desire to discriminate homology from non-homology, a belief that there is but one natural system of classification to be discovered and that such a system represents the relations among organisms. The point might be contrived but is worth making:

> I take the liberty to offer you a book on Algae … [wrote Agardh in 1824 to Saint-Hilaire] … The idea that I have followed is not so much like those in present day systems that squeeze plants into a tidy frame, but rather like yours, to arrange them one nearer the other according to their greatest affinity. (in Woelkerling and Lamy, 1999, p. 78; their translation)

Agardh's classification may be regarded as an influential beginning, an early attempt to bring order to this group of algae. Yet the characters he chose—colony form and frustule shape—were to become problematic for future diatomists, as colony form is dependent on a suite of interrelated characters (spines, pore fields, etc.) and frustule shape is rather too subjective. More significantly, his approach to systematic relationships, the desire to discriminate homology from non-homology, became confused in the rush to consider all possible kinds of data.

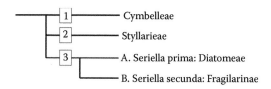

FIGURE 4.4 Branching diagram with three nodes representing table in Figure 4.3; each node represents a family.

2. F.T. KÜTZING (1807–1859)

Like Agardh, Kützing got involved with contentious topics of his day, publishing works relevant to biology in general (Kützing, 1844a). And, in his approach to diatom classification, like Agardh, he contrasted the siliceous characters of the diatom frustule with the mode of colony formation, similarly concluding that the former was of prime importance. Influential in this decision may have been the fact that Kützing had discovered their siliceous nature in 1834 (Kützing, 1844a).

Kützing (1844b) classified diatoms into three tribes, two of which were subdivided in similar ways. Tribes I and II were each split into two orders—"Astomaticae" and "Stomaticae"; he considered these tribes to be subdivided on the basis of whether they do (Stomaticae) or do not (Astomaticae) possess some kind of "opening," misinterpreted as a stomach. The third tribe was subdivided into two orders—Disciformes and Appendiculatae (Figure 4.5). Kützing's classification can be represented by a branching diagram with three nodes (Figure 4.6).

In 1846 Giuseppe Meneghini produced a lengthy critique of Kützing's work, drawing attention to the unevenness of his characters in many taxa (Meneghini, 1846, translated into English in 1853). He revised Kützing's classification, primarily at the generic level. Kützing and Meneghini's understanding of important characters differed: Meneghini (1846, 1853) suggested that *combinations* of characters were required to define groups and did not believe that any relative importance could yet be attributed to any character system: " … what value these characters have … I do not believe that we can decide in the actual state of science" (Meneghini, 1853, p. 405). Thus, Meneghini's alternative suggestion, that combinations of characters might be required to precisely identify taxa rather than search for important characters, implied that discovery of important classificatory (taxonomic) characters was irrelevant. In modern terminology, the search for homologies (defining characters) was being abandoned.

Nevertheless, Meneghini did concede that importance, albeit temporarily, may be given to "conformity of the two primary surfaces" (Meneghini, 1853, p. 384). For instance, he made various comments on the genus *Diatoma*, which he recognised as a heterogeneous mixture of unrelated organisms (Meneghini, 1853, pp. 382–384). He went on to note that the character left unifying these species (the angular connections of the filaments) was present in many other genera and hence of little significance.

In the introduction to this study, Meneghini discussed his views on natural groups and the meaning of taxonomic categories. He regarded systematic categories (at any taxonomic rank) as purely convenience groups: "The words Animal and Plant, like words in common use, as Species, Genus, Order, Class, Kingdom, do not denote any existing thing in particular. To the naturalist there exist individuals only" (Meneghini, 1853, p. 346). Of course, these thoughts allowed Meneghini to ignore any real sense of a natural classification. After all, if taxa do not exist, how can they be discovered? Meneghini's views might be considered to share certain similarities with what became known as phenetics (Sokal & Sneath, 1963), if that understanding is taken to mean abandoning the discrimination of homology from non-homology, a belief that there is not one natural system of classification but many artificial ones, the act of classification is one of imposition and that classifications represent some degree of similarity, however that concept might be conceived.

Thus far two approaches to classification were apparent: the search for important (homologies) characters or a combination of many. Each can be considered to correspond to those who wished to discover the Natural System (homologies) and those who wished to impose a classification (similarities).

3. WILLIAM SMITH (1808–1857)

William Smith offered a third view. In his *A Synopsis of the British Diatomaceae* (1853, 1856), Smith divided diatoms into two tribes, of which one was subdivided into four sub-tribes, the other five.

Tribus I.

DIATOMEAE STRIATAE. Gestreifte Diatomeen.

Lorica silicea vel laevissima vel in latere secundaria transverse striata, nec vittata nec areolata,

Ordo I. ASTOMATICAE; ostiolo medio in latere secundario nullo.

Ordo II. STOMATICAE; ostiolo medio in latere secundario.

Tribus II.

DIATOMEAE VITTATAE. Striemige Diatomeen.

Lorica silicea (in latere primario) longitudinaliter (raro transversim) vittata, laevis vel transverse striata, nec areolata.

Ordo I. ASTOMATICAE. Mundlose Diatomeen. Ostiolo medio in latere
secundario nullo.

Familia XI. LICMOPHOREAE; bacilli cuneati affixi, flabellati.

Familia XII. STRIATELLEAE; bacilli tabulati, plerumque affixi, ancipites, concatenati vel fasciati.

Ordo II. STOMATICAE. Mundführende Diatomeen. Ostiolo medio in latere secun-
dario distinctissimo magno.

Familia XIII. TABELLARIEAE; bacilli tabulati, ventricosi, plerumque affixi, concatenati vel
fasciati.

Tribus III.

DIATOMEAE AREOLATAE. Zellige Diatomeen.

Loricae siliceae latus secundarium cellulosum l. areolatum.

Anmerk. Bei manchen Formen erscheinen die zelligen Bildungen auf der Schale nur als Punkte, die oft
so klein sind, dass sie nur mit Mühe erkannt werden können. · In solchen und ähnlichen Fällen kann man die Schale
auch geglättet (laevis) nennen. Doch kommen diese Ausnahmen nur bei der Gattung Odontella vor. In andern
Fällen ist auch die Schale zellig durchbrochen (lorica perforata). — Eine mittlere Oeffnung in den Nebenseiten scheint
durchgängig zu fehlen, dagegen finden sich Oeffnungen an den Ecken, oder in der Peripherie der scheibenförmigen
Körperchen.

Ordo I. DISCIFORMES. Scheibenförmige Diatomeen.

Familia XIV. COSCINODISCEAE; latus secundarium circulare.

Familia XV. ANGULIFERAE; latus secundarium angulosum.

Ordo II. APPENDICULATAE. Mit Anhängseln versehene Diatomeen.

Familia XVI. TRIPODISCEAE; a latere secundario circulares. ·

Familia XVII. BIDDULPHIEAE; a latere secundario compressae.

Familia XVIII. ANGULATAE; a latere secundario angulosae.

Familia XIX. ACTINISCEAE; spinosae.

FIGURE 4.5 Reproduction of classification in Kützing (1844a).

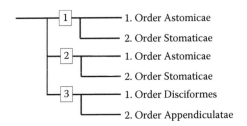

FIGURE 4.6 Branching diagram with three nodes representing classification in Figure 4.5; each node represents a family.

Each tribe and sub-tribe was characterised by some properties of the entire colony. For example, sub-tribe 3 (of Tribe 1) was characterised by "[C]onnecting membrane evanescent, or obsolete; frustules after self-division united in a compressed filament" (W. Smith, 1853, p. 7). Smith assigned to it the freshwater araphid genera *Fragilaria*, *Odontidium* Kützing, *Meridion* and *Tetracyclus* Ralfs, along with a mixture of other marine araphid and raphid genera. *Diatoma* was assigned to sub-tribe 4 (of Tribe 1), characterised by "[C]onnecting membrane subpersistent; frustules after self-division united in a zig-zag chain" (W. Smith, 1853, p. 7) and placed alongside the freshwater araphid genus *Tabellaria* Ehrenb. ex Kützing, the marine araphid genus *Grammatophora* Ehrenb., and a number of centric genera (*Amphitetras* Ehrenb., *Biddulphia* S.F. Gray, and *Isthmia* C.A. Agardh all genera with bipolar valves, that is, non-centric, centric diatoms).

Smith acknowledged that his classification may not be "natural" (in some unstated sense) but maintained that the arrangement was intended "to aid identification of species by a statement of the most obvious characters" (W. Smith, 1856, p. 39). The implication that "the most obvious characters" were invariant and, as a consequence, "important," a finding seemingly at odds with conclusions reached by earlier authors, as colony structure was not to prove useful in predicting further characters of *any* taxa and has since become minimally important for identification, as Agardh (1832) had suggested some years before.

William Smith's classification perhaps highlights an alternative approach, one that persists to the present. He stated explicitly that his classification was "to aid identification of species by a statement of the most obvious characters," in almost complete contrast to Agardh, for example, who was searching for important characters, homologies, and thus highlighting the distinction between classifications created for utility and those meant to represent nature. The former, like Smith's, are not particularly testable—in the sense of estimating or determining how further samples of characters will correspond to what is already known—such classifications are of little value in the science of systematics but are of (often exceptional) utility in identification. Clearly a classification based on a scheme to aid identification is created rather than discovered.

4. H.L. SMITH (1819–1903)

H.L. Smith's approach to classification turned out to have a lasting impact. He was convinced that the presence or absence of the raphe—a slit in the silica valve—coupled with basic valve symmetry, would prove to be decisive. He also distinguished between those diatoms that possessed a raphe and those that appeared to but on closer inspection did not. This apparent raphe-mimic, Smith called the "pseudo-raphe." A true raphe was characterised by Smith as "a true cleft, generally on the valve" (H.L. Smith, 1872, p. 2), while the pseudo-raphe was characterised as "a simple line, or blank space, without nodules" (H.L. Smith, 1872, p. 3). Smith also introduced the term "crypto-raphe" to account for species that are bilaterally symmetrical (and the few that were radially symmetrical) which bore central nodules.

The three divisions came to be known as centric, araphid, and raphid diatoms, the latter two forming the pennates: "These form the three Tribes of the Synopsis, and very seldom will any difficulty arise as to which tribe a diatom may belong to" (Figure 4.7, after H.L. Smith, 1872, pp. 4–5). Smith thus helped create today's conventional wisdom concerning the most decisive and important characters, wisdom that has only recently been challenged (Medlin et al., 1993). Smith's classification can be represented by a branching diagram with three nodes, each node representing a family (Figure 4.8).

5. ROUND, CRAWFORD, & MANN (1990)

Premolecular, or pre-genomic classification, is for the most part based on characters and criteria proposed by Friedrich Hustedt (1886–1968), first in the diatom flora of Pascher's *Die Süsswasser-flora Mitteleuropas* (Hustedt, 1930) and later in the diatom flora of Rabenhorst's *Kryptogamen-Flora von Deutschland, Österreich und der Schweiz* (Hustedt, 1927–1966). In the later more comprehensive study, Hustedt (1931 in 1927–1966) made few modifications, although he (Hustedt, 1931 in 1927–1966) did place greater emphasis on the siliceous structure of the valves and assumed a more meaningful similarity among features than is perhaps warranted. The classifications of 1930 and 1931 are essentially identical with the exception of some differing taxonomic ranks. It is evident that Hustedt's efforts were clearly directed toward artificial classifications (identification) rather than a natural classification (relationships), as he considered the identification of diatoms to be of paramount importance for progress elsewhere in diatom studies (the later schemes of Krammer and Lange-Bertalot [1991] reflecting much of Hustedt's thoughts, if not inspiration).

The most significant new classification in the late 20th century was presented in Round et al. (1990). This massive summary of diatom biology introduced many new taxa, from genus to class (Round et al., 1990, pp. 651–679). The apparent thrust was to base new taxa on as many characters as possible, statements of the positive contribution promised by a phenetic approach alarmingly naïve (Round et al., 1990, p. 119). Yet with respect to the araphid diatoms, for example, Round (1981) had earlier published a classification in which he split them into several orders based on differences in the septa of the girdle bands. The Fragilariales were characterised only by the possession of apical rimoportulae, a specialised pore of araphid and centric diatoms alike, and thus occurring in other genera excluded from the Fragilariales. No comment was made on the possibility of costae defining taxa but even earlier Round had stated:

> Certainly the Fragilarieae (containing the genera without septa on the girdle bands) of Hustedt (1959 [in 1927–1966]) is more complex than is indicated by his subdivision of it into the Diatominae and the Fragilariinae simply based on the presence in the former of massive transapical ribs and their absence in the latter. In fact, I would regard this feature as being relatively minor, and suggest that the genera *Diatoma* and *Meridion* are closer to the *Synedra* species with rimoportulae/helictoglossae than the latter is the loculate *Ardissonia*. (Round, 1979, p. 143)

Once again, emphasis was placed on what is and is not an important character—a subject Round would return to later (Round, 1996b). Round's words are similar to Meneghini's, inasmuch as Meneghini was attempting to characterise higher taxa based solely on either "striated [possessing costae] and perfectly linear frustules" or "the presence of vittae [septa]" (Meneghini, 1853). Meneghini's study was an early attempt to understand the problems of balancing the search for homologies (important characters) from combining many characters.

Round et al. provide an account of how such taxa might be related in an evolutionary sense—how these taxa are phylogenetically related (Round et al., 1990, pp. 122–123). It is difficult to understand whether the classification of Round et al. relates to the apparent phylogenetic succession, and it is not clear whether it is supposed to. What, then, is its basis? I suggest one reason: the taxa are wholly artificial such that their use, if any, is for identification not relationships and is even then virtually impossible to use in that sense. It is as if phylogeny is the subject of speculative narratives and classification the subject of learned judgement (see also Sims et al., 2006), with neither being particularly scientific.

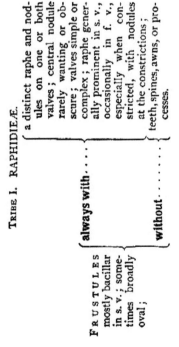

FIGURE 4.7 Reproduction of classification in H.L. Smith (1872).

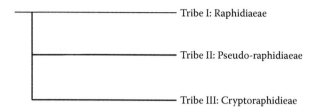

Tribe I: Raphidiaeae

Tribe II: Pseudo-raphidiaeae

Tribe III: Cryptoraphidieae

FIGURE 4.8 Branching diagram with no nodes representing the classification in Figure 4.7.

Nevertheless, the groups (taxa) proposed by H.L. Smith (1872), on the basis of the siliceous valve features, groups proposed by Pfitzer (1871) and Merezhkowsky (1901a, 1901b, 1902), on the basis of plastid distribution and auxospore behaviour, largely coincided, with the notable exception of the araphid diatoms and some nonradial centric diatoms. Convenience seemed to preserve the familiar tripartite divide, neatly encapsulated by the popularity and rigidity of the demarcations proposed in the diatom volumes of Engler & Prantl's *Die natürlichen Pflazenfamilien* (Schütt, 1896) and the later second edition (Karsten, 1928).

With very few exceptions, 20th century diatom classification almost wholly moved toward refining species-level identification rather than discovering and establishing interrelationships at all taxonomic levels. Species are seen as central to the prevailing consensus on evolutionary and ecological studies, an argument that persists, even if shrouded in modern vocabulary (cf. Lange-Bertalot, 1989; Mann, 1999).

What, then, of phylogeny and its relationship to classification?

PHYLOGENY AND CLASSIFICATION: METHOD OR MADNESS? A FURTHER SELECTION OF SOME 19TH AND 20TH CENTURY DIATOM CLASSIFICATIONS

While always an ever-present aspect of classification, visual, graphic representations of relationships in diatom studies have been few, most primarily in the form of tables (such as that of Agardh in Figure 4.3). With the introduction of genealogical diagrams (however crude), a fourth factor entered into classification: genealogy, a tradition that began with Ernst Haeckel (Williams, 2007), who indeed was the first to coin the word "phylogeny." Prior to the concept of phylogeny, the pressing issue was the discrimination of affinity from analogy, homology from non-homology, important characters and trivial characters; second, there was the alternative of ignoring the discrimination of characters into homologies and non-homologies and simply collecting a mass of data, in the belief that with enough data relationships would emerge; third, there was the issue of identification with easily observed characters. Rather than resolving these conundrums, phylogeny complicated matters.

In the following section I will discuss a few diatom examples of trees and classification before tackling the subject of post-genomic classifications.

1. Constantin S. Merezhkowsky (1855–1921)

With respect to characters and their significance, Merezhkowsky posed the following question and provided an answer, adding a fifth perspective: functionality to explain a particular character's importance:

> What is the most important character which should serve as the base for a truly natural classification? I believe that it is the presence or absence of movement, which is merely dependent on the presence or absence of a slit in the walls of the frustule [the raphe]; this character should be taken into consideration before any other. (Merezhkowsky, 1902, p. 65, an English summary of Merezhkowsky, 1901a, which includes both Russian and French text)

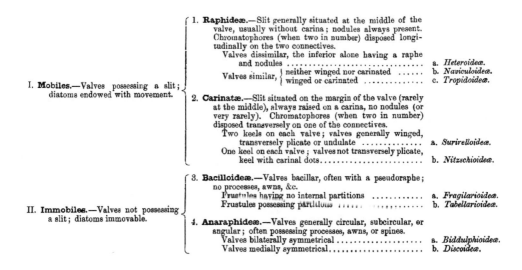

FIGURE 4.9 Reproduction of classification in Merezhkowsky (1901a, 1902).

In his classification, Merezhkowsky divided the diatoms into "two great groups," those that move, the "mobilées" (Mobiles), and those that do not, the "immobilées" (Immobiles) (Figure 4.9); H.L. Smith's raphid diatoms were (more or less) equivalent to Merezhkowsky's mobilées, and his cryptoraphids and pseudoraphids were (more or less) equivalent to his immobileés (see later and Figure 4.14). Both the mobilées (Mobiles) and immobilées themselves were subdivided into two groups, Raphideae and Carinatae for the former, Bacilloideae and Anaraphideae for the latter. Merezhkowsky's classification can be represented by a branching diagram with two nodes (Figure 4.10).

From detailed studies of diatom auxospores and plastid morphology, as well as the usual features of the valve and frustules, Merezhkowsky presented a diagram intending to depict the evolutionary relationships of the major groups (Figure 4.11, after Merezhkowsky, 1903a; a slightly different diagram was presented in Merezhkowsky, 1903b, opposite p. 204). In this tree diagram, Merezhkowsky named several (hypothetical) ancestral taxa—that is, taxa on a direct line leading to presumed descendants: *Archaideae* is the ancestor to raphid diatoms (mobilées), *Protonées* is the ancestor of the *Archaideae*, *Copuloneis* is the ancestor of *Protonées* plus Tabellarioideae, and *Urococcus* is the ancestor of all diatoms (Figure 4.11). In addition, some previously described taxa were placed in an ancestral position: Melosireae, leading to (ancestral to) the Anaraphideae, Fragilarioideae leading to (ancestral to) *Copuloneis*, *Auricula* leading to (ancestral to) the Carinatae, and *Libellus* leading to (ancestral to) Polyplacatae (Figure 4.11, enclosed in shaded boxes). Part of this diagram was included in a revised classification, with *Archaideae* added to the mobilées and the Anaraphideae re-named Centrales (Merezhkowsky, 1903b, pp. 203–204); Merezhkowsky listed *Auricula* Castracane, *Epithemia* Bréb., *Amphiprora* Ehrenb., *Amphoropsis* Grun. ex P.T. Cleve, and *Stauronella*

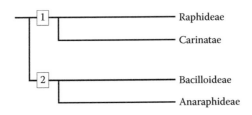

FIGURE 4.10 Branching diagram with two nodes representing the classification in Figure 4.9.

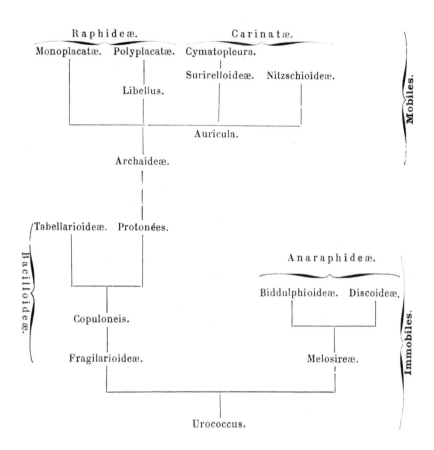

FIGURE 4.11 Merezhkowsky's genealogy, depicting the evolutionary relationships of the major groups. (Reproduced from Merezhkowsky, C., *Annales des sciences naturelles Botanique*, 17, 256, 1903. See also Merezhkowsky, C., *K morfologii diatomovkh vodorovslei = Zur Morphologie der Diatomeen = [= Morphology of diatoms]*, Kasan, Imperatorskaya Universiteta, 1903, opposite p. 204.)

Merezhkowsky as members, noting *Eunotia* Ehrenb. as *incertae sedis* (Merezhkowsky, 1902–1903, pp. 62, 157). The revised classification can be represented by a branching diagram with two nodes (Figure 4.12); Merezhkowsky's genealogy can be represented by a different branching diagram with two different nodes (Figure 4.13, taxa enclosed in boxes are the additional ancestral taxa). If the genealogy (Figure 4.11) is compared to the classification (Figure 4.12) both the Bacilloideae and the Immobiles are paraphyletic. Thus, of the four nodes in Figure 4.12 and Figure 4.13, only one—the raphid diatoms, the "mobilées—are monophyletic, the others are artificial, non-groups.

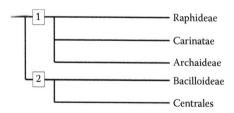

FIGURE 4.12 Branching diagram with two nodes representing Merezhkowsky's revised classification.

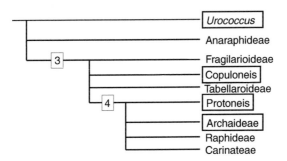

FIGURE 4 13 Branching diagram with two nodes representing the genealogy in Figure 4.11.

2. HIPPOLYTE (1851–?) & MAURICE PERAGALLO (1853–?)

In 1897 Hippolyte, the older of the two Peragallo brothers, published a preliminary discussion of diatom classification as a preamble to the forthcoming and monumental *Diatomées Marines de France*, co-authored with his brother Maurice (H. & M. Peragallo, 1897–1908). While being critical of all previous classifications, Hippolyte settled for the tripartite division of anaraphids, pseduoraphids, and raphids—not necessarily for the best of scientific reasons (Figure 4.14, H. Peragallo, 1897, pp. 16–17). In the same paper, Hippolyte included a genealogy of sorts by enclosing various diatom generic names within boxes, interconnected with each other forming a semi-reticulated, tree-like structure: centric diatoms at the base, raphid pennate diatoms at the tips, passing through various araphid genera (Figure 4.15).

It is difficult to know what exactly Hippolyte wanted to convey in this diagram, which can be interpreted in several ways. Nevertheless, following Merezhkowsky, as "genealogical" diagrams started to represent which taxa gave rise to others, more use was made of the paraphyletic basal taxa, and the more divorced they became from their graphic representation. Some years later, Hippolyte presented further views on the evolution of diatoms, proposing a revised classification primarily

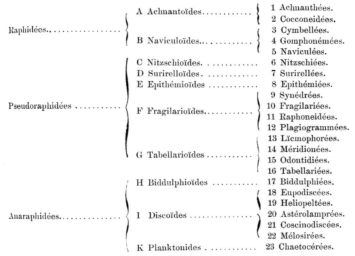

FIGURE 4.14 Reproduction of classification in Peragallo (1897).

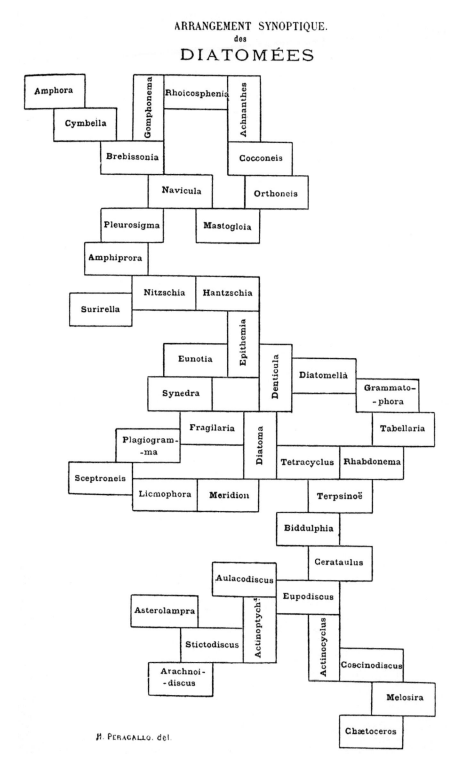

FIGURE 4.15 Reproduction of the genealogy in Peragallo (1897).

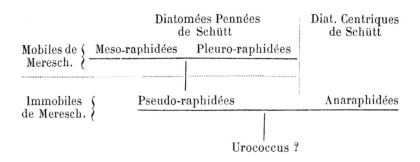

FIGURE 4.16 Reproduction of classification in Peragallo (1906).

reflecting Merezhkowsky's but recognizing the paraphyletic groups that, in his earlier diagram and those of Merezhkowsky, gave rise to other such groups (Figure 4.16, H. Peragallo, 1906).

3. ACHILLE FORTI (1878–1937)

Like the Peragallo's, Achilli Forti extended and elaborated Merezhkowsky's classification, representing certain genealogical aspects in a series of complex diagrams, linking genera via a number of reticulating lines in a network (Figure 4.17, one of a series of 11, Forti, 1911). Forti's diagrams are more like maps than trees, similar in style to those favoured by the critics of Quinarian classification (e.g., Strickland, 1844, see Stevens, 1994). That he considered a "map" of reticulating relationships rather than a tree may simply have been a sign of the times, but it seems more likely (Gola, 1938) that he regarded phylogeny as complex rather than simple, at least in its representation.

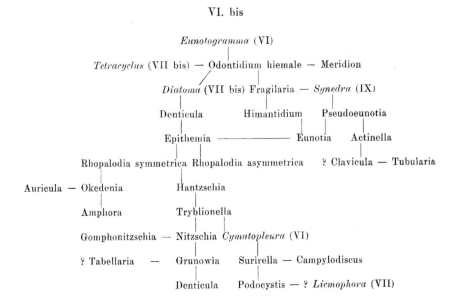

FIGURE 4.17 Reproduction of genealogy for diatoms, one of 11 in Forti (1911).

4. Reimer Simonsen (b. 1931)

In 1979 Reimer Simonsen published a critical review of diatom morphology, including a new classification and a phylogenetic diagram (Simonsen, 1979). He particularly focused on the evolution of centric diatoms, following up an earlier speculative summary (Simonsen, 1972). Alongside the new classification, he presented a full "pedigree" of diatoms—a phylogenetic tree (Figure 4.18). Simonsen (1979) added two boxes to some branches of the tree, placing them at the nodes. These were unnamed, "uncertain—not common—ancestors" (Simonsen, 1979) of all the groups distal to the box. The exact meaning of Simonsen's words is unclear, but it is evident that his intention was to add a note of doubt as to the possibility of the distal groups sharing direct common ancestry. Whatever was intended, the relationships expressed in his tree can be summarised as in Figure 4.19 (upper diagram). Here it is evident that there are (at least) four groups of centric diatoms (Figure 4.19, upper diagram, nodes 1a through d), that the pennate diatoms form a group (Figure 4.19, upper diagram, node 2), and that the araphid diatoms are more closely related to the Eunotiaceae (Figure 4.19, upper diagram, node 3). Comparison with his classification (Figure 4.19, lower diagram) suggests that the centric diatoms are monophyletic (Figure 4.19, lower diagram, node 1), and the raphid diatoms are more closely related to the Eunotiaceae rather than to the araphid diatoms (Figure 4.19, lower diagram, node 4). Thus, there is serious conflict between the evidence for the pedigree (the phylogenetic tree) and the evidence for the classification (the characters): the tree and classification, once again, are not the same.

One conclusion might be that Simonsen's classification is not supposed to represent phylogenetic relationships at all.

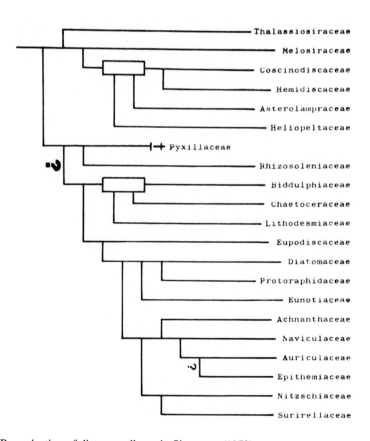

FIGURE 4.18 Reproduction of diatom pedigree in Simonsen (1979).

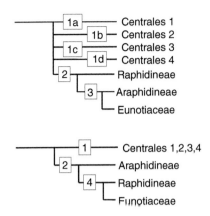

FIGURE 4.19 Upper: Branching diagram representing the diatom pedigree in Simonsen (1979) (Figure 4.18); Lower: Branching diagram representing the classification in Simonsen (1979).

Overall, it seems that the centric and araphid diatoms were always and consistently recognised as non-groups, even if appropriate terminology was not around to state it clearly. The non-groups were retained simply because little attention was paid to how to classify, as if this operation was self-evident or simply a matter of taste.

Phylogeny and classification: post-genomic clarity?

The view might be offered that from the perspective of post-genomic biology all such classifications discussed above are artificial inasmuch as they are not based on phylogenetic data—if they happen to be phylogenetic then they are so by happy accident: "Specialists have often been able to develop classifications that are highly 'natural' (phylogenetic)" (Kooistra et al., 2003a, p. 62). Coupled with this viewpoint is the implication that to be phylogenetic is to embrace the world of DNA.

It might, then, be expected that classification and phylogeny are the same, in the sense that they mean the same thing. If phylogenetic relationships are "natural," then they will be the basis for a "natural" classification. It might thus be expected that post-genomic classifications are of greater significance than those that preceded it, to be more precise, to be more "natural."

Progress in the molecular phylogeny of diatoms continues apace (Kooistra et al., 2003a; Sinninghe Damsté et al., 2004; Kooistra et al., 2004; Medlin & Kaczmarska, 2004; Sorhannus, 2004; Alverson & Theriot, 2005, Kooistra et al., 2006), yielding two landmark classifications: Medlin & Kaczmarska (2004, but see commentary in Sorhannus, 2004) and Mann (in Adl et al., 2005). Both may be amalgamated into a single branching diagram with four nodes (Figure 4.20). The first two nodes represent what is considered to be a major split, recognised as two new subdivisions—Coscinodis-cophytina and Bacillariophytina (Figure 4.20, nodes 1 and 2, respectively, Medlin & Kaczmarska, 2004, pp. 266–267)—node 3 represents the new class Mediophyceae (Figure 4.20, node 3, Medlin & Kaczmarska, 2004, based on the order Mediales Jousé & Proshkina-Lavrenko in Kryshtofovich, 1949, p. 210), and node 4 represents an emended Bacillariophyceae (Figure 4.20, node 4).

The class Mediophyceae (as emended) is said to consist of "'bi (multi) polar' and some radial centric[s] [diatoms]" including Chaetocerotales, Biddulphiales, Cymatosirales, Thalassiosirales, Triceratiales, Hemiauliales, Lithodesmiales, Toxariales, "and a suspected bipolar centric (+ Ardissoneales)" (Medlin & Kaczmarska, 2004, p. 267, additional commentary speculating on the Ardissoneales is found in Kooistra et al., 2003b, p. 196), while the Bacillariophyceae are a combination of some (most) families once described as araphid diatoms (Round et al., 1990) and all the raphid diatom families (as in Round et al., 1990; Medlin & Kaczmarska, 2004, p. 267). That is, the Fragilariophyceae (as in Round et al., 1990) is now included in the Bacillariophyceae with the exception of the few families that now are placed in the Mediophycaeae.

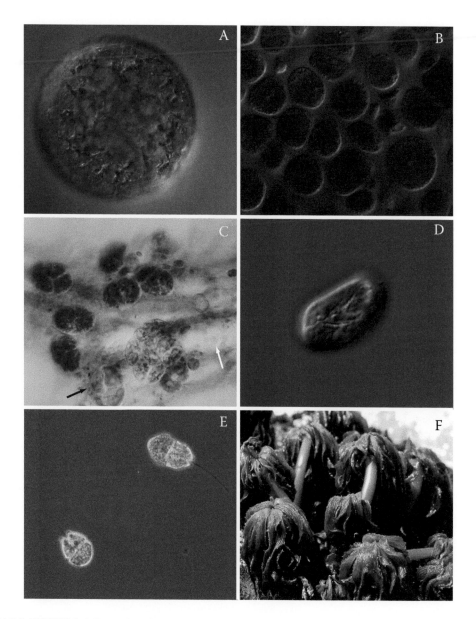

COLOR FIGURE 2.1 Examples of algal diversity. (A) *Eremosphaera*, a unicellular green alga with cells close to 1 mm in diameter (from Jyme Bog, Oneida Co., WI). (B) *Porphyridium cruentum*, a unicellular red alga; note the striking difference in colour when compared to Eremosphaera (from cultured material, UTEX 161). (C) *Glaucocystis* sp., a glaucophyte, living on the surface of the angiosperm *Ceratophyllum*, also showing the green algae *Coleochaete* and *Cosmarium* (black arrows) and plastids of *Ceratophyllum* (white arrow; from an aquatic garden at a commercial nursery, Rockville, MD). (D) *Cryptomonas* sp., a cryptomonad, and one of the organisms with secondary plastids that retain degenerate eukaryotic nuclei termed "nucleomorphs" in association with the plastid (from a seasonal freshwater wetland in Caroline County, MD). (E) *Akashiwo sanguinea*, a dinoflagellate. The trailing flagellum is visible, and the nucleus is visible as a clear region near the center of the cell, while the single plastid occupies much of the remainder of the cell (from the Cheseapeake Bay, MD). (f) *Postelsia* sp., a macroscopic brown alga that lives on exposed rocky shore lines (at Bodega Head, CA).

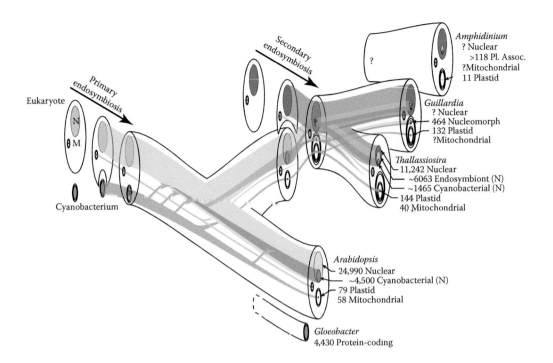

Eukaryote

Primary endosymbiosis

N

M

Cyanobacterium

Secondary endosymbiosis

?

Amphidinium
? Nuclear
>118 Pl. Assoc.
?Mitochondrial
11 Plastid

Guillardia
? Nuclear
464 Nucleomorph
132 Plastid
?Mitochondrial

Thallassiosira
11,242 Nuclear
~6063 Endosymbiont (N)
~1465 Cyanobacterial (N)
144 Plastid
40 Mitochondrial

Arabidopsis
24,990 Nuclear
~4,500 Cyanobacterial (N)
79 Plastid
58 Mitochondrial

Gloeobacter
4,430 Protein-coding

COLOR FIGURE 2.2 Gene transfer in the evolution of chloroplast genomes. One of the consequences of endosymbiosis has been the modification of both the host and endosymbiont genomes. Primary endosymbiosis (left) refers to the acquisition of a cyanobacterium by a non-photosynthetic eukaryote (already equipped with a mitochondrion). Over time, many genes have been transferred to the nuclear genome (green arrows), while others have been lost outright. Some of the transferred genes are expressed in the plastid and assist in its function. Others are now expressed in the cytosol and are not necessarily involved in plastid function. Estimates of the number of genes in each genetic compartment and their ultimate evolutionary origins are shown to the right. (Arabidopsis Genome Initiative, *Nature*, 408: 796–815, 2000. With permission.) Secondary endosymbiosis (center) refers to the acquisition of plastids from one eukaryote by a second eukaryote. As with primary endosymbiosis, substantial transfer of genes from the endosymbiont to the host has occurred, and the evolutionary history of the genomes of these organisms can be quite complex (right). (From Armbrust, E.V. et al., *Science*, 306: 79–86, 2004; Bachvaroff, T.R. et al., *Protist*, 155: 65–78, 2004; Douglas, S. et al., *Nature*, 410: 1091–1096, 2001. With permission.)

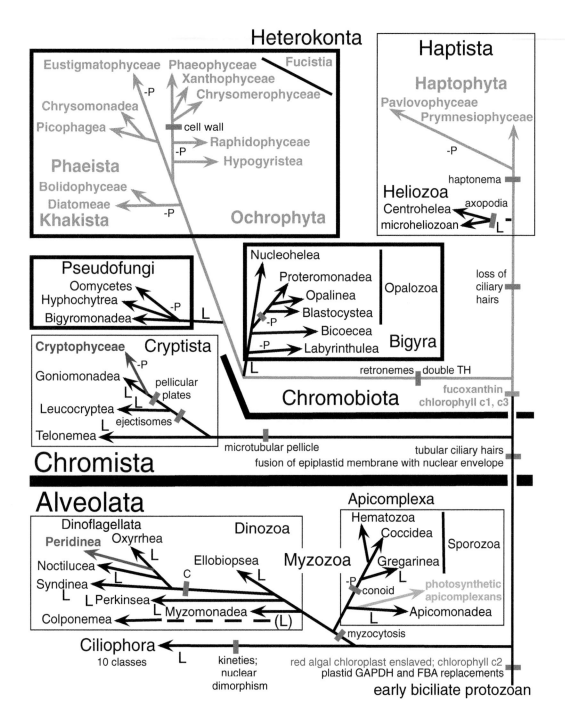

COLOR FIGURE 3.6 Phylogeny and major innovations within chromalveolates (kingdom Chromista and protozoan infrakingdom Alveolata). Phyla and classes that include algae are shown in bold and colour (orange for those ancestrally having fucoxanthin). As plastid losses (L) also occurred several times within Peridinea, chloroplasts were probably lost at least 20 times within chromalveolates. -P = loss of phagotrophy (which additionally occurred within Peridinea, Prymnesiophyceae, and Chrysomonadea). The major phylogenetic uncertainty is whether Heliozoa are really sisters of Haptophyta and actually belong in Chromista rather than Protozoa; they evolved axopodia independently of Nucleohelea and pedinellids (classified as an order

COLOR FIGURE 3.6 (CONTINUED) together with silicoflagellates, Pelagomonadales and Sarcino-chrysidales in the new subclass Alophycidae of Hypogyristea, alongside subclass Pinguiophycidae: Cavalier-Smith and Chao 2006)—axopodia also evolved three times within Rhizaria (Fig. 8). Phaeophyceae was recently broadened to include two subclasses, Melanophycidae, the classical brown algae, and Phaeothamniophycidae (with 3 orders including Schizocladiales), as Phaeista had been over-split in the past decade into a plethora of minor classes, previously incorrectly included in Chrysomonadea (=Chrysophyceae); see Cavalier-Smith and Chao (2006) for the latest more detailed classification of Heterokonta and Cavalier-Smith (2004) for overall chromalveolate evolution. Chromists comprise two subphyla, Cryptista and Chromobiota. Within Cryptista, Cryptophyceae and Goniomonadea are grouped as subphylum Cryptomonada, the only chromalve-olates other than centrohelids with flat mitochondrial cristae, a secondary change from the ancestral tubular state; I here emend the other cryptist subphylum, Leucocrypta (Cavalier-Smith 2004), by adding Telonemea, which like Leucocryptea have a two layered microtubular cortex with some cortical alveoli (Shalchian Tabrizi et al., 2006a), probably lost when cryptomonads evolved pellicular plates. Telonemea group with other cryptists on many but not all sequence trees (Shalchian-Tabrizi et al., 2006a) and lack ejectisomes, unlike other cryptists; unlike Leucocryptea (sisters of Cryptomonada) and goniomonads, they have tubular hairs on one cilium. Separate phyla for Leucocryptea (Okamoto and Inouye, 2005) and Telonemea (Shalchian-Tabrizi et al., 2006a) are unjustified. Chromobiotes, ancestrally with fucoxanthin, comprise infrakingdoms Heterokonta (phyla in bold boxes, ancestrally with double ciliary transition helix: TH) and Haptista, each with only one predominantly algal phylum, within which a few species only are secondarily heterotrophs (but no plastid loss has been firmly demonstrated), and at least one purely heterotrophic phylum. Retronemal thrust reversal (and coadapted ciliary roots), axopodia, and the haptonema define the three contrasting modes of prey capture that evolved divergently from the ancestral photophagotrophic chromobiote after it lost the nucleomorph genome and evolved fucoxanthin and chlorophylls c1 and c3. Among the 22 classes in Alveolata, Peridinea (not sensibly renamed Dinophyceae, as half are non-algal phagotrophs) is the only partially algal one of four in the infraphylum Dinoflagellata, collectively defined by evolution of the cingulum (C: groove for the transverse ribbon-shaped cilium with striated centrin paraxonemal rod). The subphylum Dinozoa comprises dinoflagel-lates plus four more diverse heterotrophic classes (comprising the paraphyletic infraphylum Protalveolata); molecular evidence is needed to position *Colponema* and see if they lost plastids independently of other heterotrophs (or even preceded the ancestral red algal enslavement!). Cavalier-Smith and Chao (2004) give a detailed classification of phylum Myzozoa, and its subphyla: Dinozoa, Apicomplexa (but excluding the then undiscovered photosynthetic apicomplexans that deserve their own class: see text).

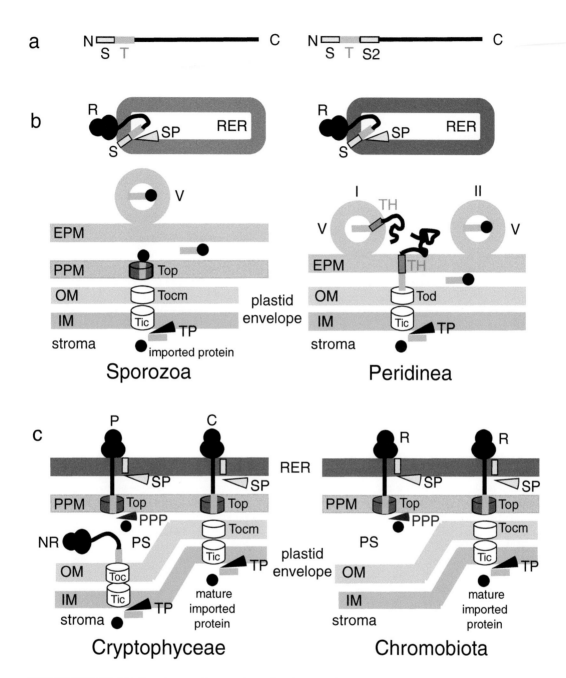

COLOR FIGURE 3.7 Protein targeting to chromalveolate plastids. (a). All chromalveolate nuclear-coded proteins destined for the plastid stroma have bipartite N-terminal targeting sequences (left) with a signal sequence (S, yellow) and then a transit-peptide-like sequence (T, green). Those destined for thylakoids have tripartite targeting sequences (right) with an extra signal sequence (S2) for targeting to the thylakoid membrane (after the first signal sequence is removed by signal peptidase (SP in b) in the lumen of the rough endoplasmic reticulum, RER, and the transit-like peptide is removed by the stromal transit peptidase (TP) as shown in (b,c)). (b). Membranes of host origin are in shades of grey; those of enslaved red algal origin are pink (former plasma membrane) or shades of green (chloroplast envelope; thylakoids not shown). In alveolates the outermost membrane around the plastids (the epiplastid membrane, EPM, derived from the food vacuole membrane that originally surrounded the phagocytosed red alga about 570 My ago) is smooth, the chloroplast proteins being made by ribosomes (R) attached by SRPs (not shown) to probably topologically separate RER cisternae and

COLOR FIGURE 3.7 (CONTINUED) carried to the plastids inside transport vesicles (V) suggested (but not yet demonstrated) to bud from the ER and fuse with the EPM, liberating their transported proteins into the space within. The signal peptide (S) is removed by a signal peptidase (yellow segment, SP) in the RER and then digested. In dinoflagellates (right) these vesicles probably travel via the Golgi, as in euglenoids that like them have an envelope of three membranes, but in Sporozoa (left) they probably go direct to the EPM. In Sporozoa the transit-like sequence is probably recognized by receptors in the periplastid membrane (PPM, the former red algal plasma membrane) associated with a hypothetic translocase (Top, red cylinder) and pass through it and a channel in the plastid OM (Tocm, meaning a greatly modified Toc) to the Tic machinery of the plastid inner membrane (IM) through which it is probably pulled by stromal ATPases as in plants. The transit peptide is removed by a transit peptidase (TP) and the mature protein folds up in the stroma, and or is transported into thylakoids (not shown) if it has a second signal peptide or TAT signal. Tocm is proposed to have evolved from red algal Tocs transferred to the host nucleus during symbiogenesis (Cavalier-Smith, 2003b). On that theory Tocm should be drastically modified, so it is unsurprising that it has not been identified in sporozoan genomes by sequence similarity to Tocs (McFadden and van Dooren, 2004). Dinoflagellates (right) lost the PPM, so dispensed with Tops and probably modified their OM Toc/Tocm even more to become a novel translocase (Tod, d for dinoflagellate), which I suggest recognizes the transit-like sequences of both class I and class II leaders. Because class I proteins are probably trapped in the vesicle membrane by a trans-membrane helix (TH, orange) prior to fusion, they may remain thus in the EPM until pulled through Tod and Tic, which would require a greater force than for class II proteins that may be free in the intra-EPM space, but the translocation machinery could import both. (c). In the ancestral chromist the EPM fused with the ER/nuclear envelope (possibly by the transport vesicle fusing with its target as an extended tubule without ever budding) so the ancestral vesicle transport machinery was lost. In the periplastid space (PS, former red algal cytosol) Cryptophyceae (left) retain a nucleomorph (former red algal nucleus, not shown) and ribosomes (NR; their rRNA is nucleomorph-encoded) that make 30 chloroplast proteins (Douglas et al., 2001). Therefore unlike other chromalveolates they retain typical Tocs to recognize the transit sequences on these 30 proteins (which lack terminal signal sequences) and import them (left). But most cryptophyte chloroplast proteins are nuclear-encoded, made by ribosomes on the RER (C), and have bipartite signals for import via putative Top/Tocm, with terminal signal sequences (yellow) and subterminal transit sequences (green). Interestingly the nuclear-encoded transit sequences, though overall broadly similar to transit sequences of the nucleomorph-encoded proteins, differ systematically from them (nucleomorph ones have an N-terminal FXN, nuclear ones an FXP, motif) in keeping with the idea that Top (red), the hypothetical PPM translocase for plastid proteins, diverged greatly from Toc (Cavalier-Smith, 1999, 2003b), but both retain the red algal phenylalanine (F). In addition Cryptophyceae import many proteins into the PS (e.g. those for making starch and carotenoids there: Gould et al., 2006a,b) or nucleomorph (e.g. DNA polymerases), also made by ribosomes on the RER (P); as predicted (Cavalier-Smith, 2003b), these PS and nucleomorph proteins have bipartite targeting sequences (Gould et al., 2006). The fact that their transit-like sequence lacks the N-terminal F (having neither FXN nor FXP sequences) shows either that the receptor for proteins crossing the PPM differs from that for chloroplast proteins (Cavalier-Smith, 1999, 2003: Top I) and PS proteins (Top2: Cavalier-Smith, 1999, 2003) or that discrimination between plastid-destined and PS proteins takes place after entry into the PS via a common Top, and makes use of the presence or absence respectively of F (see text). The periplastid processing peptidase (PPP, brown segment) that removes the transit-like sequence from PS proteins must differ from transit peptidases (TP) in the chloroplast stroma. Chromobiotes (right) lost the nucleomorph genome, ribosomes and starch from the PS, and thus also lost the OM Toc; but they still import proteins into the PS (e.g. chaperone Hsp70: Gould et al., 2006b) and thus must have kept Top and PPP. If the membranous periplastid reticulum (PR, not shown), which is located in the PS (putatively relict nucleomorph envelope membrane, perhaps retained for making PPM lipids), is topologically separate from the PPM, chromobiotes would also use Top for importing PR membrane proteins, but if the PR and PPM are continuous, or the PR buds from the PPM, PR proteins need not use that route. Modifying a nuclear-coded cryptomonad PS protein merely by changing its TP N-terminal amino acid to phenylalanine (F) makes it enter the plastid instead (Gould et al., 2006a), proving its functional similarity to chloroplast TPs. For simplicity Tops are drawn as if they pass imported plastid proteins directly to Tocm; actually Top and Tocm need not be juxtaposed, as preproteins extruded by Top could bind PS Hsp70 chaperone and diffuse to Tocm (Gould et al., 2006b). Modified from Cavalier-Smith (2003b) in the light of Patron et al. (2005) and Gould et al. (2006b).

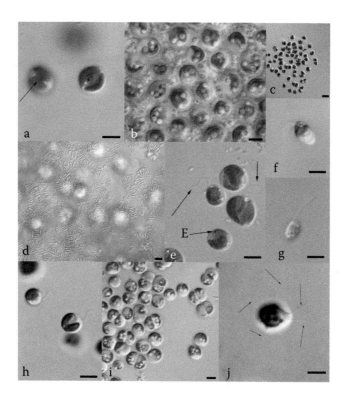

COLOR FIGURE 15.1 (a and b) *Chrysocapsa paludosa*—CCMP380: (a) Two cells in opposing view, showing the connection (arrow) between the two deeply divided chloroplast lobes. Scale bar = 5 μm. (b) Densely packed cells from an old colony showing individual gels around each cell, a chloroplast that fills only about half of each cell, and numerous globules inside and outside the cells. Scale bar = 5 μm. (c) *Naegeliella flagellifera*—CCMP280. Typical colony in culture with gelatinous hairs. Scale bar = 20 μm. (d and e) *Kremastochrysis* sp.—CCMP260: (d) Surface film of bacteria with hyponeustonic cells out of focus below the surface. (e) Flagellate cells showing the long immature flagellum (arrows) and the eyespot (E). Scale bar = 5 μm. (f) *Chrysosaccus* sp.—CCMP2717. Zoospore with an obovate shape, a small chloroplast, a long immature flagellum, a short mature flagellum, and an eyespot. Scale bar = 5 μm. (g) Unidentified chrysophyte—CCMP1161. Zoospore with oval shape, a small chloroplast, and a long immature flagellum. Scale bar = 5 μm. (h and i) *Chrysosaccus* sp.: (h) CCMP295. Vegetative cells with spherical shape, a large chloroplast that fills the cell, and an immature flagellum that beats very slowly. Scale bar = 5 μm. (i) CCMP296. Vegetative cells with spherical shape organized into a capsoid colony, cells not flagellated, several cells in division stages. (j) Unidentified chrysophyte—CCMP1161. Amoeboid cell with several fine pseudopods (arrows). Scale bar = 5 μm.

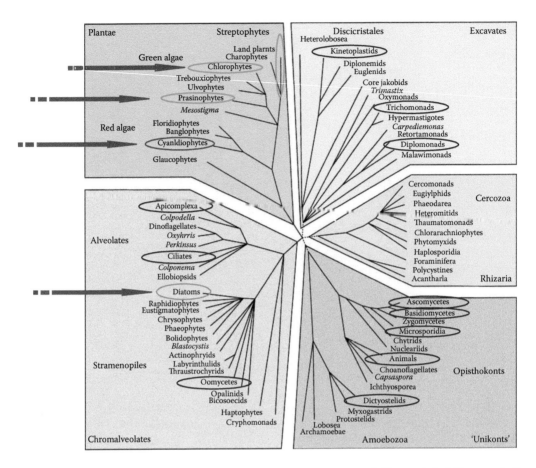

COLOR FIGURE 17.1 Taxa for which complete genome sequences are available are circled. Photosynthetic eukaryotes whose genome sequences have been completed are found within the Plantae and Chromoalevolate groups and are indicated with an arrow. (The eukaryotic phylogentic tree is reproduced with permission from Keeling, P.J., Burger, G., Durnford, D.G., Lang, B.F., Lee, R.W., Pearlman, R.E., Roger, A.J., and Gray, M.W., *Trends in Ecology and Evolution,* 20, 670–676, 2005.)

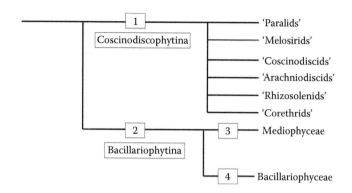

FIGURE 4.20 Branching diagram representing two recent molecular diatom classifications (Medlin & Kaczmarska, 2004, and Mann in Adl et al., 2005). Nodes 1 and 2 represent two subdivisions—Coscinodiscophytina and Bacillariophytina; node 3 represents the new class Mediophyceae; node 4 represents an emended Bacillariophyceae.

The phylogenetic tree on which this classification is apparently based can be represented by a simplified branching diagram, with 9 nodes (Figure 4.21, after Medlin & Kaczmarska, 2004, figure 3). Given the definitions of taxa above, the Coscinodiscophytina would include all taxa (and nodes) between *Paralia sol* and *Proboscia alata* (Figure 4.21, the branch leading to *Paralia sol*, nodes 2, 3, and 8, and the branch leading to *Proboscia alata*), a paraphyletic assemblage; the Mediophyceae would include all taxa (and nodes) between the Thalassiosirales and Cymatosirales (Figure 4.21, nodes 4 through 7 [excluding the pennates] and node 9), a second paraphyletic assemblage: both taxa are not natural groups, have no reality in the natural world and are imposed by convention rather than discovered (see also the comments in Sorhannus, 2004, a study that has been somewhat neglected, but see below). Curiously, Mann (in Adl et al., 2005) indicates both the Coscinodiscophytina and Mediophyceae to be paraphyletic yet retains both in the classification.

Medlin & Kaczmarska (2004, figure 1) and Sims et al. (2006, figure 1) include diagrams (trees) that show the Coscinodiscophytina and Mediophyceae to be monophyletic, yet no species names are indicated as included in each higher taxon (this is now the preferred tree; Medlin, personal communication). Yet, Sims et al. consider the possibility of a non-monophyletic Mediophyceae:

Medlin and colleagues (literature summarized by Medlin & Kaczmarska, 2004) have divided the diatoms into two groups on the basis of molecular sequence data: what initially was called "Clade 1" contains those centric diatoms with essentially radial symmetry of valve shape and structure. "Clade 2" consists

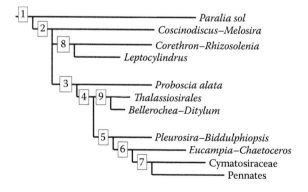

FIGURE 4.21 Phylogenetic tree for the classification in Figure 4.20; a simplified tree with nine nodes. (After Medlin, L.K. & Kaczmarska, I., *Phycologia*, 43, 245–270 [figure 3], 2004.)

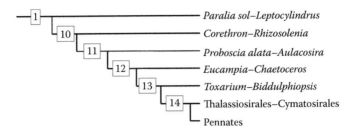

FIGURE 4.22 Phylogenetic tree, simplified with six nodes. (After Sorhannus, U., *Cladistics*, 20, 487–497, 2004.)

of two groups, the first of which contains the bi- or multipolar centrics and the radial Thalassiosirales ("Clade 2a"), and the second, the pennates ("Clade 2b") (Figs. 1, 2). Morphological and cytological support for these clades was reviewed in Medlin *et al.* (2000) and Medlin & Kaczmarska (2004). *However, whether Clade 2a is truly monophyletic is unclear (e.g. Kooistra et al. 2003a).* (Sims et al., 2006, p. 367, italics added; see also Kooistra et al., 2006, pp. 134–138, who suggest an array of different relationships with doubts toward many, as does Kooistra et al., 2003a)

Regardless of which tree is or is not preferred, none of the nine nodes in Figure 4.21 represented in the classification as recognised taxonomic groups.

Two alternatives molecular phylogenies exist, that of Sorhannus (2004) and Sinninghe Damsté et al. (2004). Sorhannus' phylogenetic tree is based on, more-or-less, the same data as used by Medlin & Kaczmarska (2004) but the resulting tree has six nodes but shares none with Medlin & Kaczmarska's tree (if a reduced version capturing all the most significant branches is considered, Figure 4.22). Of significance for classification is that neither the Coscinodiscophytina nor the Mediophyceae are found to be monophyletic, Sorhannus explicitly recognising their paraphyly (Sorhannus, 2004).

Sinninghe Damsté et al. (2004) present a third phylogenetic tree with nine nodes (Figure 4.23) and, although their data are enhanced by some additional new sequences (see supplemental material

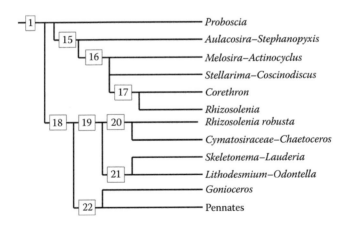

FIGURE 4.23 Phylogenetic tree simplified with nine nodes. (After Sinninghe Damsté, J.S.S., Muyzer, G., Abbas, B., Rampen, S.W., Massé, G., Allard, W.G., Belt, S.T., Robert, J.-M., Rowland, S.J., Moldowan, J.M., Barbanti, S.M., Fago, F.J., Denisevich, P., Dahl, J., Trindade, L.A.F., and Schouten, S., *Science*, 304 (5670), 584–587, 2004.)

provided with their paper), it too shares no nodes in common with either of the other two trees, nor are the Coscinodiscophytina or Mediophyceae monophyletic; Sinninghe Damsté et al. (2004) do not present a classification or comment on existing ones.

In short, the revised classification (scheme) of Medlin & Kaczmarska (2004) is (almost) identical to Merezhkowsky's (Figure 4.9 through Figure 4.13) and Peragallo's (Figure 4.14 through Figure 4.16), with the exception of shuffling a few families between different paraphyletic assemblages. All three classifications (Medlin & Kaczmarska, Merezhkowsky, Peragallo) have a paraphyletic basal group of centric diatoms, leading to an intermediate, paraphyletic assemblage of some centric and some araphid genera, terminating in a monophyletic pennate group. Progress in classification appears negligible.

Nevertheless, such is the effect of molecular data that a recent review summarised the situation thus: "Radial centric diatoms begot multipolar centrics, multipolar centrics begot pennates and the araphid pennates begot the raphid pennates" (Kooistra et al., 2003a, p. 92). Biblical allusions to one side, the meaning is clear: paraphyly is explained by ancestry in spite of the fact that it has been long acknowledged that paraphyly has no empirical reality (for recent comment see Ebach & Williams, 2004); and the explanation of Kooistra et al. (2003a) is like that of Merezhkowsky and Peragallo: the conversion of no knowledge into knowledge.

PHYLOGENY AND CLASSIFICATION: THE PERILS OF PARAPHYLY, THE POSSIBILITIES OF FOSSILS

It is beyond the task of this paper to present a revised classification, one that captures the monophyletic groups and rejects inappropriate paraphyletic groups (but see Williams & Kociolek, accepted). For one reason, there is no consensus among the three molecular phylogenetic trees on relationships within diatoms nor is there any morphological tree derived from the analysis of available data, rather than simply "mapping" characters onto the various molecular trees (Kooistra et al., 2003b, 2004; Medlin & Kaczmarska, 2004). What I will conclude with is an outline of one possible line of exploration: the relationships of pennate diatoms to all other diatoms.

The immediate sister taxon to the pennate diatoms may be one of three candidates (see Figure 4.24): (a) the Cymatosirales (Medlin & Kaczmarska, 2004), (b) the Cymatosirales plus Thalassiosirales (Sorhannus, 2004), or (c) *Gonioceros* H. Perag. in H. & M. Peragallo (Sinninghe Damsté et al., 2004).

1. THE CYMATOSIRALES AND THEIR RELATIONSHIPS

Nikolaev & Harwood (2000, 2002; Nikolaev et al., 2001) presented a classification of fossil diatoms, supporting their conclusions, for the most part, on their splendid work on the description of many new Cretaceous fossil diatoms (Gersonde & Harwood, 1990; Nikolaev, 1988, 1990; Nikolaev & Harwood,

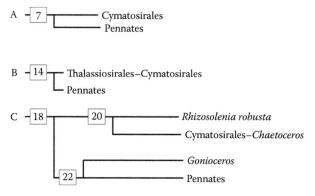

FIGURE 4.24 Three possible sister groups to the pennate diatoms: (a) the Cymatosirales (Medlin & Kaczmarska, 2004), (b) the Cymatosirales plus Thalassiosirales (Sorhannus, 2004), or (c) *Gonioceros* H. Perag. in H. & M. Peragallo (Sinninghe Damsté et al., 2004).

TABLE 4.1

Classification of Centric Diatoms According to Nikolaev and Harwood (2000) and Mann (in Adl et al., 2005)*

Nikolaev and Harwood (2000, 2002); Nikolaev et al. (2001)	Mann (in Adl et al., 2005)
Class: Centrophyceae (P)	**Subdivisions: Coscinodiscophytina (P)**
Subclasses:	"(R)—ribogroup, usually based on molecular phylogenetic analysis of rRNA genes" (Adl et al., 2005, p. 402)
Archaegladiopsophycidae	
Paraliophycidae	Paralids
Heliopeltophycidae	
	Arachniodiscids
Coscinodiscophycidae	Coscinodiscids, Melosirids
Biddulphiophycidae	
Rhizosoleniophycidae	Rhizosolenids, Corethrids

Note: P = paraphyletic.

*See text for further details

1997, 1999). While dealing primarily with centric diatoms (i.e., non-pennates, "the diatom equivalent of 'invertebrate,'" Alverson & Theriot, 2005), they place within the single class Centrophyceae six subclasses (Table 4.1, from Nikolaev et al., 2001, pp. 38–41). Their scheme can be compared with that of Mann (in Adl et al., 2005), who includes six informal taxa within the paraphyletic Coscinodiscophytina which are said to be "(R)—ribogroup[s], usually based on molecular phylogenetic analysis of rRNA genes" (Adl et al., 2005, p. 402; see Table 4.1 and Figure 4.25). While some correspondence exists between the two classifications (Figure 4.25), a more precise comparison can be made between the paraphyletic Mediophyceae and Nikolaev & Harwood's subclass Biddulphiophycidae (of unknown status, but probably paraphyletic) (Table 4.2). The two classifications have five orders in common, with a further four unique to Nikolaev & Harwood (2000) and a different four unique to Mann (in Adl et al., 2005) (Figure 4.25). Within both the Mediophyceae and Biddulphiophycidae is the Cymatosirales.

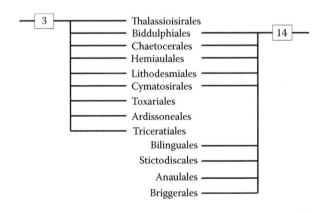

FIGURE 4.25 Comparison of the class Mediophyceae (paraphyletic) and subclass Biddulphiophycidae (probably paraphyletic; see Table 4.1).

TABLE 4.2

Comparison between the Paraphyletic Mediophyceae and Nikolaev and Harwood's Subclass Biddulphiophycidae

Nikolaev and Harwood (2000, 2002); Nikolaev et al. (2001)	Mann (in Adl et al., 2005)
Subclass: Biddulphiophycidae (?P)	Class: Mediophyceae (P)
Orders	
Biddulphiales	Biddulphiales
Hemiaulales	Hemiaulales
Chaetocerales [sic]	Chaetocerotales
Lithodesmiales	Lithodesmiales
Cymatosirales	Cymatosirales
Bilinguales*	
Briggerales*	
Stictodiscales	
Anaulales	
	Thalassioisirales
	Toxariales
	Ardissoneales
	Triceratiales

Note: P = paraphyletic.
* Indicates extinct taxa.

No classification presented so far captures the possibility of the Cymatosirales as sister to the pennate diatoms; in each classification the relationship is buried within a mass of paraphyletic taxa.

Nikolaev & Harwood's classification of the Cymatosirales has two families—Cymatosiraeae and Rutilariaceae (following Round et al., 1990). In the Cymatosiraeae there are nine genera with c. 42 living species and three extinct genera (*Bogorovia* Jousé, *Koizumia* Yanagisawa, and *Rossiella* T.V. Desik. & Maheshwari), with c. 17 species. A relationship between these extinct and living genera was suggested some years ago, by Jousé, for example: "In all probability there is some relationship between *Bogorovia*, *Campylosira* and *Cymatosira*, which three genera ought to be joined together in a separate family" (Jousé, 1973, p. 351, see also Yanagisawa, 1996, p. 275). In total, there are some 59 species of which 31 known only from fossils—around half (Hasle et al., 1983. For further details of the morphology of living taxa for this group described since Hasle et al., see Takano, 1985; Nakata, 1987; Cheng & Gao, 1993; Cheng et al., 1993; Gardner & Crawford, 1994; and footnotes to Table 4.3; numerous references are available detailing the morphology of fossil members; see Gaul et al., 1993, and Henderson & Reimer, 2003, and footnotes to Table 4.3).

The Rutilariaceae has five genera of which only *Rutilaria* Grev. has species still living (two; Ross, 1995). With the possible addition of a further four extinct genera (Table 4.3), there are a total of 59 species. Thus, almost 100% of the species in Rutilariaceae are extinct.

Finally, in 1999 Seiichi Komura described a new extinct family Parodontellaceae, with four new genera, all having characteristics of the Cymatosirales (and possibly with the Chaetocerotales), but at present their relationships are unknown (Table 4.4; Komura, 1999, see also Komura, 1997).

2. *GONIOCEROS* AND THEIR RELATIONSHIPS

According to Round et al., *Gonioceros* is placed in the order Chaetocerotales Round et Crawford, part of the Chaetocerotophycidae (Round et al., 1990, p. 127). The Chaetocerotophycidae is divided

TABLE 4.3
Genera in the Order Cymatosirales with Fossils Indicated and Numbers of Species

Nikolaev and Harwood (2000, 2002); Nikolaev et al. (2001)

Order: Cymatosirales	Species Numbers—Approximate (Fossil Numbers in Brackets)
Family: Cymatosiraceae	
Genera:	
Arçocellus	2
Brockmanniella	1
Campylosira	8 (1)
Cymatosira[1]	22 (13)
Extubocellulus	2
Leyanella	1
Minutocellulus	3
Papilliocellulus	2
Plagiogrammopsis	3
Pseudoleyanella	1
Bogorovia[*][2]	(7)
Koizumia[*][3]	(3)
Praecymatosira[4]	(1)
Rossiella[*][5]	(7)
Total:	59 (31)
Family: Rutiliariaceae	
Genera:	
Rutilaria	36 (34)
Kisseleviella[*][6]	(4)
Pseudorutilaria[*][7]	(5)
*Spinivinculum**	(3)
*Syndetocystis**	(3)
Total:	59 (57)

[1] See Schrader & Fenner (1976, p. 975, Figure 41).
[2] Gardette (1978), Desikachary et al. (1984), Yanagisawa (1995a, 1996).
[3] Yanagisawa (1994).
[4] Strelnikova in Dzinoridze et al. (1979).
[5] Yanagisawa (1995b).
[6] See Akiba (1980), Akiba & Yanagisawa (1985), and Scherer et al. (2000) for descriptions of many unnamed specimens of this genus and Harwood & Boharty (2001, p. 327) for further commentary.
[7] Harwood & Boharty (2000).

* Indicates extinct taxa.

into two orders, the Chaetocerotales and the Leptocylindrales—the Chaetocerotales having three families, two of which are monoptypic: Acanthocertaceae (with one genus *Acanthoceros* and species, *A. magdeburgense* Honigmann) and Attheyaceae (with one genus *Attheya* T. West). The third family, the Chaetocerotaceae, has, according to Round et al. (1990), three genera: *Chaetoceros* Ehrenb., *Gonioceros*, and *Bacteriastrum* Shadbolt (Figure 4.26, upper diagram). Nevertheless, *Gonioceros* may indeed be related more closely to *Attheya* than any other member of the

TABLE 4.4
Genera in the New Family *Paradontellaceae*
(Komura, 1999)

?? Order: Cymatosirales	Fossil *incertae sedis* (Species Numbers Approximate)
Family: Paradontellaceae	
Acigonium*	2
Paradontella*	4
Stylorium*	2
Thamnodiscus*	1
Total:	9

Chaetocerotaceae and a more succinct classification leads to the union of all species recognised as belonging to either *Attheya* or *Gonioceros* placed "within one genus in the family Attheyaceae … [and hence leaving] the Chaetocerotaceae containing only *Chaetoceros* and *Bacteriastrum*" (Crawford et al., 1994, see Figure 4.26, lower diagram). The classification of Round et al. (rather than the modification proposed by Crawford et al., 1994) is followed by Nikolaev & Harwood (2000, 2002; Nikolaev et al., 2001). Nevertheless, relationships within Chatocerotales are uncertain, especially as two of the families are monotypic.

3. What relationships?

It is virtually impossible to understand what systematic hypotheses are being expressed when classifications are produced largely by convention, considering group membership by default in an ever-changing sea of opinion. The three options proposed above for the direct relationships of the pennate diatoms are not evident from either Medlin & Kaczmarska (2004) or Mann (in Adl et al., 2005). The species above (regardless of higher taxon) are related in some meaningful way (see Figure 4.21 through Figure 4.23), and most probably within lurks the sister taxon of the pennate diatoms. It is also evident from above that should either the Cymatosirales or Chaetocerotales

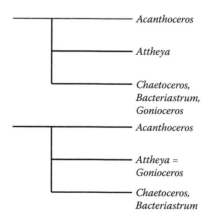

FIGURE 4.26 Two schemes of relationships among the Chaetocerotales Round et Crawford. (Upper diagram: after Round, F. E., Crawford, R.M., and Mann, D.G., *The Diatoms—Biology and Morphology of the Genera*, Cambridge University Press, Cambridge, 1990; lower diagram after Crawford, R.M., Gardner, C., and Medlin, L.K., *Diatom Research*, 9, 27–51, 1994.)

contain the sister taxon to the pennates diatoms, relevant evidence will most probably come from morphology (fossils) rather than molecules, if that evidence is not already available.

I have deliberately avoided discussing particular morphological characters as it is the relationships expressed in the various classifications and diagrams that are the focus of interest. Characters, nevertheless, are of great significance, as resolution of the problem relating to the sister group of the pennate diatoms will not, and cannot, rest solely among DNA sequence data—fossils are of crucial significance, which means without some meaningful understanding of morphology (see, for example, Hasle et al., 1983), diatom systematics cannot and will not progress.

DISCUSSION

[H]ypotheses are formulated in such a manner that they can be falsified (i.e. disproved) by new findings and—if so—can be replaced by new hypotheses.

Lange-Bertalot, 1990, p. 20

Appealing for more data (Round, 1996a) is something of a truism—since when has the accumulation of data confounded? Evidently it appears to do so in diatom systematics, if its more recent history is anything to go by. But a more general view of history suggests that it is not the data themselves that reveal relationships—species or otherwise—but the discovery of homology and taxa (relationships) and the elimination of similarity and paraphyly as a guide. While 150 years has passed, it almost seems as if Agardh's preliminary suggestions have dominated our collective efforts, and perhaps more significantly for biology in general, Agardh's general concerns with taxa and characters have never really been answered.

Nevertheless, progress does really seem to come from the elimination of paraphyly, the directing principle of cladistics (Kitching et al., 1998), a discipline understood by some as synonymous with discovering natural classification (Williams, 2004; Williams & Humphries, 2004). Cladistics aims to discover monophyletic groups by means of evidence (homologues, molecular or morphological) and eliminate paraphyletic groups not so supported (Kitching et al., 1998). Current progress in systematic biology, regardless of the organism under investigation, removal of these non-groups, the paraphyletic residue, is the primary focus of attention and their elimination understood as significant (see the essays in Cracraft & Donoghue, 2004, especially Donoghue, 2004, p. 551). This principle may be coupled with the distinction between natural and artificial classifications, where a natural classification "has an empirical connection that allows for falsification" (Nelson, 1983, p. 490). That is, data speak for or against any particular classification. Currently diatom classification has been proposed by convention rather than evidence, otherwise the phylogenetic trees found today and the resulting classifications would at least resemble each other.

But phylogeny is now supposedly the guiding principle. What, then, can be said of phylogeny? Beyond the merely simplistic ("Radial centric diatoms begot multipolar centrics, multipolar centrics begot pennates and the araphid pennates begot the raphid pennates" Kooistra et al., 2003a, p. 92), there is the more elaborate:

It is of interest that the molecular phylogeny of diatoms also favours the centric forms as ancestral ... with some molecular evidence agreeing [data] that the earliest diatoms could have been neritic ... Calibrating the molecular clock of diatoms and other Heterokontophyta from the fossil record, and applying the same nucleotide substitution rate to parts of the molecular phylogeny before the occurrence of fossil diatoms, indicates that diatoms originated close to the Permian–Triassic boundary 250 Mya, some 130 million yr earlier than the first silicified fossil diatoms ... This suggests that silicification evolved late in the evolutionary history of diatoms. It is possible that any earlier silicified diatoms have failed to be preserved for some environmental or taphonomic reasons; however, silicified organisms (sponges) have been found in phosphoritic marine sediments

> 580 million yr old … and both radiolarian and sponges were common from the Cambrian onwards. (Raven & Waite, 2004, p. 44)

Regardless of details, this is the usual stuff of phylogenies: a series of untestable stories with scant regard for the relationships expressed by the organisms themselves and the evidence upon which they are based. Thus, statements that "centric forms [are] ancestral" and "araphid pennates begot the raphid pennates" make no sense in the knowledge that centric and araphid diatoms are not groups of any kind at all. Of course, papers like those of Raven & Waite (2004) belong to the past in what might be called the narrative phase of evolutionary discourse, buried in the mid-1950s, assumed dead in the 1960s, but occasionally resurfacing (Sims et al., 2006). If systematists (molecular or otherwise) revive this tradition (accepting paraphyletic groups, viewing evolution as a narrative), then it is doomed to forever repeat the mistakes of its past masters (see above), with classification remaining that "comparatively humble and unexacting kind of science" (Medawar, 1967), which it evidently is not (Darwin, 1859, chapter 13, "Mutual Affinities of Organic Beings").

ACKNOWLEDGMENTS

I am grateful to Juliet Brodie and Jane Lewis for inviting me to present some ideas on diatom classification. Juliet Brodie and a referee provided many valuable comments and suggestions. Geraldine Reid provided support and comment. Pat Sims kindly allowed me to use some of her micrographs in the presentation of this material.

REFERENCES

Adl, S.M., Simpson, A.G.B., Farmer, M.A., Andersen, R.A., Anderson, O.R., Barta, J.R., Bowser, S.S., Brugerolle, G., Fensome, R.A., Fredericq, S., James, T.Y., Karpov, S., Kugrens, P., Krug, J., Lane, C.E., Lewis, L.A., Lodge, J., Lynn, D.H., Mann, D.G., Mccourt, R.M., Mendoza, L., Moestrup, Ø., Mozley-standridge, S.E., Nerad, T.A., Shearer, C.A., Smirnov, A.V., Spiegel, F.W., and Taylor, M.F.J.R. (2005) The new higher level classification of Eukaryotes with emphasis on the taxonomy of Protists. *Journal of Eukaryotic Microbiology*, 52: 399–451.

Agardh, C.A. (1823–1828) *Systema Algarum*, Litteris Berlingianis, Lundae.

Agardh, C.A. (1829–1830) *Lärobok i Botanik*, N.M. Thomsons Boktryckeri, Malm.

Agardh, C.A. (1830–1832) *Conspectus Criticus Diatomacearum*, Litteris Berlingianis, Lundae.

Akiba, F. (1980) A Lower Miocene diatom flora from the Boso Peninsula, Japan, and the resting spore formation of an extinct diatom, *Kisseleviella carina* Sheshukova-Poretzkaya. *Bulletin of the Technical Laboratory, JAPEX*, 23: 81–94.

Akiba, F. and Yanagisawa, Y. (1985) Taxonomy, morphology and phylogeny of the Neogene diatom zonal marker species in the middle-to-high latitudes of the North Pacific. *Initial Reports of the Deep Sea Drilling Project*, 88: 483–554.

Alverson, A.J. and Theriot, E.C. (2005) Comments on recent progress toward reconstructing the diatom phylogeny. *Journal of Nanoscience and Nanotechnology*, 5: 57–62.

Barnes, J. (2002) *Something to Declare. Essays on France*, Alfred A. Knopf, New York.

Barrett, P.H. (1960) A transcription of Darwin's first notebook on "transmutation of species." *Bulletin of the Museum of Comparative Zoology*, 122: 247–296.

Bowman, D.D. (2005) What's in a Name? *Trends in Parasitology*, 21: 267–269.

Brummitt, R.K. (2002) How to chop up a tree. *Taxon*, 51: 31–41.

Brummitt, R.K. (2003) Further dogged defense of paraphyletic taxa. *Taxon*, 52: 803–804.

Brundin, L. (1972) Evolution, casual biology, and classification. *Zoologica Scripta*, 1: 107–120.

Burkhardt, F. and Smith, S. (eds.) (1987) *The Correspondence of Charles Darwin*, Vol. 3, 1844–1846, Cambridge University Press, Cambridge.

Cantino, P.D. and de Queiroz, K. (2004) *PhyloCode: A Phylogenetic Code of Biological Nomenclature*, www.ohiou.edu/phylocode.

Cavalier-Smith, T. (1998) A revised six-kingdom system of life. *Biological Reviews*, 73: 203–266.

Cheng, Z. and Gao, Y. (1993) Nanodiatoms from Xiamen harbour (1). *Acta Phytotaxonomica Sinica*, 31: 197–200.

Cheng, Z., Gao, Y., and Lui, S. (1993) *Nanodiatoms from Fujian Coast*, China Ocean Press, Beijing.

Cracraft, J. and Donoghue, M.J. (2004) *Assembling the Tree of Life*, Oxford University Press, New York.

Crawford, R.M., Gardner, C., and Medlin, L.K. (1994) The genus *Attheya*. I. A description of four new taxa, and the transfer of *Gonioceros septentrionalis* and *G. armatas*. *Diatom Research*, 9: 27–51.

Darwin, C. (1859) *On the Origin of Species by Means of Natural Selection, or, the Preservation of Favoured Races in the Struggle for Life*, John Murray, London.

Darwin, C. (1866) *On the Origin of Species by Means of Natural Selection, or, the Preservation of Favoured Races in the Struggle for Life*, 4th ed., with additions and corrections, John Murray, London.

De Beer, G. (ed.) (1960) Darwin's notebooks on transmutation of species. *Bulletin of the British Museum (Natural History)*, 2(2): 23–186.

DeLillo, D. (1984) *White Noise*, Picador Books, London.

De Toni, G.B. (1889) [Obituary of G.G.A. Meneghini (1811–1889)]. *Notarisia*, 4; 725–732.

De Queiroz, K. (1988) Systematics and the Darwinian revolution. *Philosophy of Science*, 55: 238–259.

Desikachary, T.V., Latha, Y., and Ranjitha, Devi K.A. (1984) *Rossiella* and *Bogorovia*: two fossil diatom genea. *Palaeobotanist*, 32: 337–340.

Di Gregorio, M.A. (1981) Order or process of nature: Huxley's and Owen's different approaches to natural sciences. *History and Philosophy of the Life Sciences*, 3: 217–242.

Di Gregorio, M.A. (1982) In search of the natural system, problems of zoological classification in Victorian Britain. *History and Philosophy of the Life Sciences*, 4: 225–254.

Di Gregorio, M.A. (1996) The uniqueness of Charles Darwin: his reading of W.S. Macleay's *Horae Entomologicae*. *Historical Records of Australian Science*, 11: 103–117.

Donoghue, M.J. (2004) Immeasurable progress on the tree of life. In *Assembling the Tree of Life* (J. Cracraft and M.J. Donoghue, eds.), Oxford University Press, New York, pp. 548–552.

Dzinoridze, R.N., Jousé, A.P., and Strelnikova, N.I. (1979) Description of the diatoms. In *Istoziya mikroplanktona norvezhskogo morya [po materialom glubokovodnogo bureniya], [The History of the Microplankton of the Norwegian Sea (on the Deep Sea Drilling Materials)]*, Issledovanya fauný morei [Explorations of the Fauna of the Seas] (A.A. Strelkov and M.G. Petrushevskaya, eds.), 23 (31): 32–70.

Ebach, M.C. and Williams, D.M. (2004) Classification. *Taxon*, 53: 791–794.

Farber, P.L. (1985) Aspiring naturalists and their frustrations: the case of Williams Swainson (1789–1855). In *From Linneaus to Darwin: Commentaries on the History of Biology and Geology* (A. Wheeler and J.H. Price, eds.), Published for the Society for the History of Natural History (Special Publications Number 3), London, pp. 51–59.

Felsenstein, J. (2004) *Inferring Phylogenies*, Sinauer Associates, Sunderland, MA.

Fletcher, J.J. (1921) The society's heritage from the Macleays. *Proceedings of the Linnean Society of New South Wales*, 45: 592–629.

Forti, A. (1911) Contribuzioni diatomologische. XII. Metodo di classificazione delle Bacillariee Immobili fondato sull'affinità morfologica dei frustuli ed in relazione con l'evoluzione dell'auxospora. *Atti del Real Istituto Veneto di Scienze, Lettere ed Arti*, 71: 677–731.

Forti, A. (1926) *Achille Forti, Bibliografia (1898–1925)*, La Tipografica Veronese, Verona.

Franz, N.M. (2005) On the lack of good scientific reasons for the growing phylogeny/classification gap. *Cladistics*, 21: 495–500.

Gardette, D. (1978) Une nouvelle espece de Diatomee marine du Miocene superieur de Chypre: *Bogorovia cypriata* [A new species of diatom, *Bogorovia cypriata*, from the upper Miocene of Cyprus]. *Geobios*, 11: 761–767.

Gardiner, B. (2001) Editorial. *The Linnean*, 17(1): 1–3.

Gardner, C. and Crawford, R.M. (1994) A description of *Plagiogrammopsis mediaequatus* Gardner and Crawford, *sp. nov.* (Cymatosiraceae, Bacillariophyta) using light and electron microscopy. *Diatom Research*, 9: 53–63.

Gaul, U., Geissler, U., Henderson, M., Mahoney, R., and Reimer, C.W. (1993) Bibliography on the fine-structure of diatom frustules (Bacillariophyceae). *Proceedings of the Academy of Natural Sciences, Philadelphia*, 144: 69–238.

Gersonde, R. and Harwood, D.M. (1990) Lower Cretaceous diatoms from ODP Leg 113 Site 693 (Weddell Sea). Part I. Vegetative cells. In *Proceedings of the Ocean Drilling Program, Scientific Results*, 113 (P.F. Barker, J.P. Kennett, et al., eds.), Ocean Drilling Program, College Station, TX, pp. 365–402.

Gola, G. (1937a) Achille Forti (28.11.1878–11.2.1937). *Berichte der Deutschen Botanischen Gesellschaft,* 55: 195–208.

Gola, G. (1937b) Achille Forti (28.11.1878–11.2.1937). *Nuovo Giornale Botanico Italiano e Bolletino della Societa Botanica Italiano,* ser. 2, 44: 595–606.

Gola, G. (1937c) *Achille Forti, Bibliografia, supplemento (1925–1933).* La Tipografica Veronese, Verona.

Gola, G. (1938) Commenorzione del m.e. Dott. Achille Forti. *Atti del Real Istituto Veneto di Scienze, Lettere ed Arti,* 97: 11–35.

Harwood, D.M. and Boharty, S.M. (2000) Marine diatom assemblages from Eocene and younger Erratics, McMurdo Sound, Antarctica. *Antarctic Research Series,* 76: 73–98.

Harwood, D.M. and Boharty, S.M. (2001) Early Oligocene siliceous microfossil biostratigraphy of Cape Roberts Project Core CRP-3, Victoria Land Basin, Antarctica. *Terra Antarctica,* 8: 315–338.

Hasle, G.R., Stosch, H.A., von & Syvertsen, E.E. (1983) Cymatosiraceae, a new diatom family. *Bacillaria* 6: 9–156.

Henderson, M.V. and Reimer, C.W. (2003) Bibliography on the fine structure of diatom frustules (Bacillariophyceae). II. (+deletions, addenda and corrigenda for bibliogrphy [sic] I.). *Diatom Monographs* (W. Witkowski, ed.), 3, 372 pp.

Hendey, N.I. (1958) Electron microscope studies on some marine diatoms. *Annual Report of the Challenger Society,* 3: 21.

Hendey, N.I. (1959) The structure of the diatom cell wall as revealed by the electron microscope. *J. Quekett Microscopical Club,* 5: 147–175.

Hendey, N.I. (1971) Electron microscope studies and the classification of diatoms. In *The Micropalaeontology of Oceans: Proceedings of the Symposium Held in Cambridge from 10 to 17 September 1967* (B.M. Funnell and W.R. Reidel, eds.), Cambridge University Press, Cambridge, pp. 625–631.

Holland, J. (1996) Diminishing circles: W.S. Macleay in Sydney, 1839–1865. *Historical Records of Australian Science,* 11: 119–147.

Hustedt, F. (1927–1966) Die Kieselalgen Deutschlands, Österreichs und der Schweiz. In *Kryptogamen-Flora von Deutschlands, Österreichs und der Schweiz* (L. Rabenhorst, ed.), 7. Akademische Verlagsgesellschaft, Leipzig.

Hustedt, F. (1930) Bacillariophyta (Diatomeae). In *Die Süsswasser-flora Mitteleuropa* (A. Pascher, ed.), 10. G. Fisher, Jena.

Jeffroy, O., Brinkmann, H., Delsue, F., and Philppe, H. 2006. Phylogenomics: the beginning of incongruence? *Trends in Genetics* (in press).

Jousé, A.P. (1973) Diatoms in the Oligocene-Miocene biostratigraphic zones of the tropical areas of the Pacific Ocean. *Nova Hedwigia, Bieh.,* 45: 333–357.

Karsten, G. 1928. Bacillariophyta (Diatomaceae). In *Die Natürlichen Pflanzenfamilie* (A. Engler and K. Prantl, eds.), 2nd ed., 2, W. Englemann, Leipzig, pp. 105–303.

Kitching, I., Forey, P.L., Humphries, C.J., and Williams, D.M. (1998) *Cladistics: The Theory and Practice of Parsimony Analysis,* 2nd ed., Oxford University Press, Oxford.

Kitton, F. (1880) The early history of the Diatomaceae. *Hardwicke's Science Gossip,* 184: 77–79, 186: 133–136.

Kitton, F. (1882) Early history of the Diatomaceae. *Hardwicke's Science Gossip,* 205: 6–9.

Knight, D. (1985) William Swainson: types, circles and affinities. In *The Light of Nature* (J.D. North and J.J. Roche, eds.), Springer, New York, pp. 83–94.

Knight, D. (1986) William Swainson: Naturalist, author and illustrator. *Archives of Natural History,* 13: 275–290.

Komura, S. (1997) Additional descriptions of the new diatoms from the Miocene Nabuto formation, Central Japan. *Diatom,* 13: 63–81.

Komura, S. (1999) Barrel-shaped diatoms from the Miocene Nabuto Formation, central Japan. *Diatom,* 15: 51–78.

Kooistra, W.C.H.F, De Stefano, M., Mann, D.G., and Medlin, L.K. (2003a) The phylogeny of the diatoms. *Progress in Molecular and Subcellular Biology,* 33: 59–97.

Kooistra, W.H.C.F., De Stefano, M., Mann, D.G., Salma, N., and Medlin, L.K. (2003b) The phylogenetic position of *Toxarium,* a pennate-like lineage within centric diatoms (Bacillariophyceae). *Journal of Phycology,* 39: 185–197.

Kooistra, W.C.H.F, Forlani, G., Sterrenburg, F.A.S., and De Stefano, M. (2004) Molecular phylogeny and morphology of the marine diatom *Talaroneis posidoniae* gen. et sp. nov. (Bacillariophyta) advocate the return of the Plagiogrammaceae to the pennate diatoms. *Phycologia,* 43: 58–67.

Kooistra, W.C.H.F., Chepurnov, V., Medlin, L.K., De Stefano, M., Sabbe, K., and Mann, D.G. (2006) Evolution of the diatoms. In *Plant Genome: Biodiversity and Evolution. Vol. 2B: Lower Groups* (A.K. Sharma and A. Sharma, eds.), Oxford and IBH Publishing, New Delhi, India, and Science Publishers, Enfield, NH, pp. 117–178.

Krammer, K. and Lange-Bertalot, H. (1991) Susswasserflora von Mitteleuropa. Bacillariophyceae 3. Teil: Centrales, Fragilariaceae, Eunotiaceae. In *Süßwasserflora von Mitteleuropa* 2 (3) (H. Ettl, J. Gerloff, H. Heynig, and D. Mollenhauer, eds.), G. Fischer, Stuttgart & Jena, New York, 576 pp.

Kryshtofovich, A.N. (1949) *Diatomovi analiz [Diatom analysis]*. Botanicheskii Institut im. V.L. Komarova, Leningrad.

Kützing, F.T. (1844a) *Die Sophisten und Dialektiker die gefährlicihsten Feinde der wissenschaftlichen Botanik.* Nordhausen W. Köhn.

Kützing, F.T. (1844b) *Die kieselschaligen Bacillarien oder Diatomeen.* Nordhausen W. Köhn.

Lange-Bertalot, H. (1989) Können *Staurosirella, Punctastriata* und weitere Taxa sensu Williams & Round als Gattungen der Fragilariaceae kritischer Prüfung standhalten? *Nova Hedwigia*, 49: 79–106.

Lange-Bertalot, H. (1990) Current biosystematic research on diatoms and its implications for the species concept. *Limnetica*, 6: 13–22.

Lange-Bertalot, H. 1993. Once more: *Staurosirella, Punctastriata* etc.: a reply to Williams & Round (1992). *Nova Hedwigia*, 56: 179–182.

Macleay, W.S. (1819–1821) *Horae entomologicae*, S. Bagster, London.

Macleay, W.S. (1823) Remarks on the identity of certain general laws which have been lately observed to regulate the natural distribution of Insects and Fungi. *Transactions of the Linnean Society of London*, 14: 46–68.

Macleay, W.S. (1825) *Annulosa Javanica, or an attempt to illustrate the natural affinities and analogies of the insects.* London: Kingsbury, Parbury & Allen.

Macleay, W.S. (1830) On the dying struggle of the dichotomous system … in a letter to N.A. Vigors. *Philosophical Magazine*, 7: 431–445; 8: 53–57, 134–140, 200–207.

Mann, D.G. (1999) The species concept in diatoms. *Phycologia*, 38: 437–495.

Medawar, P.B. (1967) *The Art of the Soluble*. Methuen & Co. Ltd., London.

Medlin, L.K. and Kaczmarska, I. (2004). Evolution of the diatoms: V. Morphological and cytological support for the major clades and a taxonomic revision. *Phycologia*, 43: 245–270.

Medlin, L.K., Williams, D.M., and Sims, P.A. (1993) The evolution of the diatoms (Bacillariophyta). I. Origin of the group and assessment of the monophyly of its major divisions. *European Journal of Phycology*, 28: 261–275.

Meneghini, G. (1846) Sulla animaliti delle Diatomee e revisione organografica dei generi di Diatomee stabilite dal Kützing. *Atti del Real Istituto Veneto di Scienze, Lettere ed Arti*, 1845: 1–191.

Meneghini, G. (1853) On the animal nature of diatomeae with an organographical revision of the genera estblished by Kützing. *Botanical and Physiological Memoirs (Ray Society)*, II: 345–513.

Merezhkowsky, C. (1901a) K voprosu o Klassifikatsii diatomovkh vodorosle [On the classification of diatomic algae]. *Botanischeskaya Zapiski*, 18: 87–98.

Merezhkowsky, C. (1901b) Étude sur l'endochrome des diatomées. *Mémoires de l'Académie Impériale des Sciences de St. Petersbourg*, ser. 8, 11(6): 1–40.

Merezhkowsky, C. (1902) On the classification of diatoms. *Annals and Magazine of Natural History*, ser. 7, 9: 65–68.

Merezhkowsky, C. (1902–1903) Le types de l'endochrome chez les diatomees. *Scripta Botanica Horti Universitatis Imperialis Petropolitanae*, 21: 1–193.

Merezhkowsky, C. (1903a) Les types des auxospores chez les diatomées et leur évolution. *Annales des sciences naturelles Botanique*, 17: 225–262.

Merezhkowsky, C. (1903b) *K morfologii diatomovkh vodorovslei = Zur Morphologie der Diatomeen = [= Morphology of diatoms*], Kasan, Imperatorskaya Universiteta.

Moore, J. (1991) Deconstructing Darwinism: the politics of evolution in the 1860s. *Journal of the History of Biology*, 24: 353–408.

Müller, R.H.W. and Zaunick, R. (eds.) (1960) Friedrich Traugott Kützing 1807–1893: Aufzeichnungen und Erinnerungen. *Lebensdarstellungen deutscher Naturforscher*, 8: 1–300.

Nakata, K. (1987) The fine structure of two marine diatom species of the family Cymatosiraceae. *Bull. Tokai Reg. Fish. Res. Lab.*, 119: 41–45.

Nelson, G.J. (1972) Phylogenetic relationship and classification. *Systematic Zoology*, 21: 227–231.

Nelson, G.J. (1973) Classification as an expression of phylogenetic relationships. *Systematic Zoology*, 22: 344–359.

Nelson, G.J. (1983) Vicariance and cladistics: historical perspectives with implications for the future. In *Evolution, Time and Space: The Emergence of the Biosphere* (R.W. Sims, J.H. Price, and P.E.S. Whalley, eds.), Academic Press, London, pp. 469–492.

Nelson, G.J. (1989) Species and taxa: systematics and evolution. In *Speciation and Its Consequences* (D. Otte and J. Endler, eds.), Sinauer, Sunderland, pp. 60–81.

Nelson, G.J. (2006) Cladistics: the search for the sister group. In *Darwin's Universe: Evolution from A to Z* (Milner, R., ed.), University of California Press, Berkeley.

Nelson, G.J., Murphy, D.J., and Ladiges, P.Y. (2003) Brummitt on paraphyly: a response. *Taxon*, 52: 295–298.

Nelson, G.J. and Platnick, N.I. (1981) *Systematics and Biogeography: Cladistics and Vicariance.* Columbia University Press, New York.

Nelson, G.J. and Platnick, N.I. (1984) Systematics and evolution. In *Beyond Neodarwinism: An Introduction to the New Evolutionary Paradigm* (M.-W. Ho and P. T. Saunders, eds.), Academic Press, London, pp. 143–158.

Nikolaev, V.L. (1988) Systema klassa Centrophyceae (Bacillariophyta). *Botanischeskii Zhurnal*, 73: 486–496.

Nikolaev, V.L. (1990) The system of centric diatoms. In *Proceedings of the 10th International Diatom Symposium* (H. Simola, ed.), O. Koeltz Scientific Books, Koenigstein, Germany, pp. 17–22.

Nikolaev, V.L. and Harwood, D.M. (1997) New process, genus and family of Lower Cretaceous diatoms from Australia. *Diatom Research*, 12: 43–53.

Nikolaev, V.L. and Harwood, D.M. (1999) Taxonomy of Lower Cretaceous diatoms. In *Proceedings of the 14th International Diatom Symposium* (S. Mayama, M. Idei, and I. Koizumi, eds.), O. Koeltz Scientific Books, Koenigstein, Germany, pp. 101–111.

Nikolaev, V.L. and Harwood, D.M. (2000) Diversity and classification of centric diatoms. In *The Origin and Early Evolution of Diatoms* (A. Witkowski and J. Siemiska, eds.), W. Safer Institute of Botany, Polish Academy of Sciences, Cracow, pp. 37–53.

Nikolaev, V.L. and Harwood, D.M. (2002) Diversity and system of classification [in] centric diatoms. In *Proceedings of the 16th International Diatom Symposium* (A. Economu-Amilli, ed.), University of Athens, Greece, pp. 127–152.

Nikolaev, V.L., Harwood, D.M., and Samsonov, N.I. (2001) [*Early Cretaceous Diatoms*]. Russian Academy of Sciences, Komorov Botanical Institute.

O'Hara, R.J. (1988) Homage to Clio, or, toward an historical philosophy for evolutionary biology. *Systematic Zoology*, **37**: 142–155.

O'Hara, R.J. (1994) Evolutionary history and the species problem. *American Zoologist, 34:* 12–22.

Ott, F.D. (1991) Carl Adolph Agardh, Professor, Bishop. A translation of J.E. Areschoug's 1870 Memorial. *Archiv für Protistenkunde* 139: 297–312.

Patterson, C. (1978) Verifiability in systematics. *Systematic Zoology* 27: 218–222.

Peragallo, H. (1897) [D]iatomées marines de France. *Micrographie Préparateur*, 5: 9–17.

Peragallo, H. (1906 [1907]) Sur l'évolution des Diatomées. *Bulletin de la Station Biologique d'Arcachon*, 9: 110–124.

Peragallo, H. and Peragallo, M. (1897–1908) *Diatomées Marines de France et des districts Maritimes Voisin.* Micrographie-éditeur, Grez-sur-Loing.

Pfitzer, E. (1871) Untersuchungen uber Bau und Entwickelung der Bacillariaceen (Diatomaceen). *Botanische Abhandlungen aus dem Gebiet der Morphologie und Physiologie*. J. Hanstein, Bonn.

Pickett, K.M. (2005) The new and improved Phylocode, now with types, ranks, and even polyphyly: a conference report from the First International Phylogenetic Nomenclature Meeting. *Cladistics*, 21: 79–82.

Platnick, N.I. (1977) Paraphyletic and polyphyletic groups. *Systematic Zoology*, 26: 195–200.

Platnick, N.I. (2001) From Cladograms to classifications: the road to DePhylocode.www.systass.org/events_ archive/agm-address-dec2001.html.

Ragan, M.A. (1998) On the delineation and higher-level classification of the algae. *European Journal of Phycology*, 33: 1–15.

Raven, J.A. and Waite, A.M. (2004) The evolution of silicification in diatoms: inescapable sinking and sinking as escape? *New Phytologist* 162: 45–61.

Reif, W.-E. (2005a) Problematic issues of cladistics: 16. Taxonomic groups and the definition of paraphyly. *Neues Jahrbuch für Geologie und Palaontologie, Abhandlungen* 238: 191–229.

Reif, W.-E. (2005b) Problematic issues of cladistics: 17. Monophyletic taxa can be paraphyletic clades. *Neues Jahrbuch für Geologie und Palaontologie, Abhandlungen* 238: 313–354.

Ross, R. (1963) Ultrastructure research as an aid in the classification of diatoms. *Annals of the New York Academy of Sciences*, 108: 396–411.

Ross, R. (1995) A revision of *Rutilaria* Greville (Bacillariophyta). *Bulletin of the Natural History Museum (Botany)*, 25: 1–93.

Round, F.E. (1979) The classification of the genus *Synedra*. *Nova Hedwigia, Beiheft*, 64: 135–146.

Round, F.E. (1981) Some aspects of the origins of diatoms and their subsequent evolution. *BioSystems*, 14: 483–486.

Round, F.E. (1996a) What characters define diatom genera, species and infraspecific taxa? *Diatom Research*, 11: 203–218.

Round, F.E. (1996b) Fine siliceous components of diatom cells. *Nova Hedwigia, Bieh.*, 112: 201–213.

Round, F.E., Crawford, R.M., and Mann, D.G. (1990) *The Diatoms—Biology and Morphology of the Genera*, Cambridge University Press, Cambridge.

Sapp, J., Carrapiço, F. and Zolotonosov, M. (2002) Symbiogenesis: the hidden face of Constantin Merezhkowsky. *History and Philosophy of the Life Sciences*, 24: 413–440.

Scherer, R.P., Boharty, S.M., and Harwood, D.M. 2000. Oligocene and Lower Miocene siliceous microfossil biostratigraphy of Cape Roberts Project Core CRP 2/2A, Victoria Land Basin, Antarctica. *Terra Antarctica*, 7: 417–442.

Schlegel, H.G. (1999) Geschichte der Mikrobiologie. *Acta Hisotrica Leopoldiana*, 28, 280 pp.

Schrader, H.-J. and Fenner, J. (1976) Norwegian Sea Cenozoic biostratigraphy and taxonomy. Part 1, Norwegian Sea Cenozoic diatom biostratigraphy. *Initial Reports of the Deep Sea Drilling Project*, 38: 921–1099.

Schütt, F. 1896. Bacillariales. In *Die Natₑrlichen Pflanzenfamilie* (A. Engler and K. Prantl, eds.), 1(1b), W. Englemann, Leipzig, pp. 31–153.

Secord, J.A. (2000) *Victorian Sensation: The Extraordinary Publication, Reception, and Secret Authorship of Vestiges of the Natural History of Creation*. University of Chicago Press, London, Chicago.

Silva, P.A. (1980) Development of the concept of class and family in living algae. *Regnum Vegetabile*, 103: 4–11.

Simonsen, R. (1972) Ideas for a more natural system of the centric diatoms. *Nova Hedwigia, Beih.*, 39: 37–54.

Simonsen, R. (1979) The diatom system: ideas on phylogeny. *Bacillaria*, 2: 9–71.

Sims, P.A. (2005) Obituary. Norman Ingram Hendey (31 January 1903–30 August 2004). *Diatom Research* 20: 417–424.

Sims, P.A., Mann, D.G. & Medlin, L.K. (2006) Evolution of the diatoms: insights from fossil, biological and molecular data. *Phycologia*, 45:361-402.

Sinninghe Damsté, J.S.S., Muyzer, G., Abbas, B., Rampen, S.W., Massé, G., Allard, W.G., Belt, S.T., Robert, J.-M., Rowland, S.J., Moldowan, J.M., Barbanti, S.M., Fago, F.J., Denisevich, P., Dahl, J., Trindade, L.A.F., and Schouten, S. (2004) The rise of the Rhizosolenid diatoms. *Science*, 304 (5670): 584–587.

Smith, H.L. (1872) Conspectus of the families and genera of the Diatomaceae. *The Lens*, 1: 1–19, 72–93, 154–157.

Smith, S. (1965) The Darwin collection at Cambridge, with one example of its use: Charles Darwin and Cirripedes. *Actes du XIᵉ cong. Int. d'hist. Sci.*, 5: 96–100.

Smith, W. (1853) *A Synopsis of the British Diatomaceae*, Vol. 1, Smith and Beck, London.

Smith, W. (1856) *A Synopsis of the British Diatomaceae*, Vol. 2, Smith and Beck, London.

Sokal, R.R. and Sneath, P.H.A. (1963) *Principles of Numerical Taxonomy*, W.H. Freeman, San Francisco.

Sorhannus, U. (2004) Diatom phylogenetics inferred based on direct optimization of nuclear-encoded SSU rRNA sequences. *Cladistics*, 20: 487–497.

Staflau, F.A. (1966) Agardh's Systema Algarum. *Taxon*, 15: 267–277.

Staflau, F.A. (1970) Species Algarum. *Taxon*, 19: 630–632.

Stevens, P.F. (1994) *The Development of Biological Systematics: Antoine-Laurent de Jussieu, Nature and the Natural System*, Columbia University Press, New York.

Stevens, P.F. (2002) Why do we name organisms? Some reminders from the past. *Taxon*, 51: 11–26.

Strelnikova, N.I. (2006) Mereshkovsky K.S. and his research of diatoms (To 150th anniversary of birth). In *Voprosy Obšej Botaniki: Tradicii I Perspecktivy*, Čast 1, Kazan, Kazan State University, pp. 183–190.

Strickland, H.E. (1844) Description of a chart of the natural affinities of the insessorial order of birds. *Report of the Thirteenth Meeting of the British Association for the Advancement of Science Held at Cork in August 1843, Notices and Abstracts of Communications*, p. 69.

Swainston, A.Y. (1985) William Sharp Macleay, 1792–1865. *The Linnean*, 1(5): 11–18.

Takano, H. (1985) A new diatom from sand-flats in Mikawa Bay, Japan. *Bull. Tokai Reg. Fish. Res. Lab.*, 115: 29–37.

Williams, D.M. (2004) Homology and homologues, cladistics and phenetics: 150 years of progress. In *Milestones in Systematics* (D.M. Williams and P.L. Forey, eds.), Taylor & Francis, London, pp. 191–224.

Williams, D.M. (2007) Ernst Haeckel and Louis Agassiz: trees that bite and their geographical dimension. In *Biogeography in a Changing World* (M.C. Ebach and R. Tangey, eds.), CRC Press, Boca Raton, FL, pp. 1–59.

Williams, D.M., Ebach, M.C., and Wheeler, Q. (2005) 150 reasons for paraphyly. *Taxon*, 54: 858.

Williams, D.M. and Humphries, C.J. (2004) Homology and character evolution. In *Deep Morphology: Toward a Renaissance of Morphology in Plant Systematics* (T. Stuessy, E. Hörandl, and V. Mayer, eds.), Koeltz, Königstein, pp. 119–130.

Williams, D.M. and Kociolek, J.P. (accepted) Pursuit of a natural classification of diatoms: history, monophyly and the rejection of paraphyletic taxa. *European Journal of Phycology*.

Willis, J.C. (1901) The Royal Botanic Gardens of Ceylon, and their history. *Annals of the Royal Botanic Gardens, Peradeniya*, 1: 1–15.

Woelkerling, W.J. and Lamy, D. (1999) *Non-geniculate Coralline Red Algae and the Paris Muséum: Systematics and Scientific History.* Paris Publications Scientifiques du Muséum, A.D.A.C., Paris.

Yanagisawa, Y. (1994) *Koizumia* Yanagisawa gen. nov., a new marine fossil araphid diatom genus. *Transactions and Proceedings of the Palaeontological Society of Japan*, n.s. 176: 591–617.

Yanagisawa, Y. (1995a) Cenozoic diatom genus *Bogorovia* Jousé: an emended description. *Transactions and Proceedings of the Palaeontological Society of Japan*, n.s. 177: 21–42.

Yanagisawa, Y. (1995b) Cenozoic diatom genus *Rossiella* Desikachary et Maheshwari: an emended description. *Transactions and Proceedings of the Palaeontological Society of Japan*, n.s. 177: 1–20.

Yanagisawa, Y. (1996) Taxonomy of the genera *Rossiella*, *Bogorovia* and *Koizuma* (Cymatosiraceae, Bacillariophyceae). *Nova Hedwigia, Beiheft*, 112: 273–281.

5 The taxonomy of cyanobacteria: molecular insights into a difficult problem

Paul K. Hayes, Nermin Adel El Semary
and Patricia Sánchez-Baracaldo

CONTENTS

ABSTRACT

Cyanobacteria have been classified using features of their morphology and development, but many phenotypic characters vary as environmental conditions change, and therefore, molecular methods have also been used to describe diversity within this group. The methods used include phylogenetic reconstruction based on one or a few gene sequences, DNA fingerprinting techniques, and analysis of population genetic structures. The application of these methods has revealed previously hidden diversity in morphologically depauperate taxa, such as *Synechococcus* and *Prochlorococcus*, and has shown that diversity has been overestimated in character-rich genera such as *Nodularia* and *Microcystis*. The need for a taxonomic approach that makes use of both stable phenotypic characters and molecular markers (i.e., a polyphasic approach) is discussed, as is the demonstration of how such an approach has helped to define species boundaries in *Nodularia* and *Anabaena*.

INTRODUCTION

The cyanobacteria are impressive ecosystem engineers with an evolutionary history stretching over at least 1.5 and possibly 2.7×10^9 years (Blank, 2004). The evolution of oxygenic photosynthesis within this prokaryotic group created the conditions required for organisms that depend on aerobic respiration for their existence and led to the subsequent development of the photosynthetic eukaryotes.

Extant cyanobacteria are found in most aquatic and terrestrial habitats from the tropics to the poles. Through their metabolic activities they make quantitatively important contributions to the carbon, nitrogen, sulphur, and other biogeochemical cycles. Unlike most groups of prokaryotes, there is substantial morphological and life history variation within the cyanobacteria. They vary in complexity from simple unicellular forms, through undifferentiated filaments, to complex filamentous forms with true branching and specialised cells for nitrogen fixation (heterocysts) and perennation (akinetes). Both the unicellular and filamentous forms can live in isolation or they can aggregate into colonies, where cell and/or filament juxtaposition have been used as informative taxonomic characters. Features of cell ultrastructure, such as the arrangement of the thylakoid membranes, the structure of the cell wall, and the possession of gas vesicles, have also been used in taxonomic treatments of the group. As in all other areas of microbial taxonomy and phylogenetics, a range of molecular methods have been used in attempts to unravel the relationships within the cyanobacteria. More recently a combination of phenotypic and genotypic characters (Table 5.1) have been used, and this polyphasic approach forms the basis for the most recent broad-based description of cyanobacterial taxonomy (Castenholz, 2001). This treatment of the cyanobacteria is based almost exclusively on the features of isolates growing in pure culture, as is required under the Bacteriological Code, but the contributing authors make extensive reference to the groupings erected previously under the Botanical Code, where cultures are not required to establish types.

Attempts to describe valid cyanobacterial taxa using either phenotypic or genotypic characters, or a combination of both, are hampered by lack of an agreed species concept. On the one hand some species are proscribed by features of the phenotype and can be subdivided into ecotypes on the basis of the niche that they occupy. On the other hand are species defined on the basis of molecular characters, such as the percentage similarity across the entire genome (>70% identity in DNA–DNA reassociation studies) or at a single gene locus (>97.5% identity in ssu-rDNA nucleotide sequence) (Stackebrandt and Goebel, 1994). The problem with these approaches is that they depend on the identification of discontinuities of character distributions as a means of defining species boundaries, and they make little, if any, reference to the underlying reproductive biology of the organisms (i.e., they are somewhat removed from the original biological species concept). To illustrate the nature of these difficulties, the treatment of the Cyanobacteria in *Bergey's Manual* (Castenholz, 2001) is taken as a starting point for discussion. It recognises five Subsections within the cyanobacteria, but because of the current state of flux, descriptions within them are restricted to form-genera (form-genus being defined as an artificial taxonomic category established on the basis of morphological resemblance for organisms of obscure true relationships) (Table 5.2). Some of the groupings, both the Subsections themselves and the form-genera, are known to be polyphyletic on the basis of the analysis of single (Ishida et al., 1997; Honda et al., 1999; Turner et al., 1999; Turner et al., 2001; Litvaitis, 2002; Gugger and Hoffmann, 2004) and multiple genes

---

TABLE 5.1
Characters Used to Describe Cyanobacterial Taxa

Cell morphology	Shape, polarity, dimensions, division planes, colour, sheath, motility
Cell specialisations	Heterocysts, akinetes, baeocytes, necridia
Ultrastructure	Thylakoid arrangement, cell wall structure, internal inclusions (gas vesicles and storage granules), sheath structure
Colony/filament morphology	Shape and symmetry, trichome type (straight, helical, tapered), shape of terminal cell, terminal hairs, false/true branching, presence and distribution of specialised cells
Physiology/biochemistry	Pigmentation, range of growth temperature, pH and salinity, chemoheterotrophy, photoheterotrophy, anoxygenic photosynthesis, diazotrophy, lipid composition
Habitat/ecology	Marine/brackish/freshwater, flowing/static water, endolithic, symbiotic, incident irradiance, pH, microenvironment, associated community
Genetic characters	G + C content, DNA–DNA hybridisation, 16S rDNA sequence, other gene sequences, whole genome sequence

Source: Modified from Castenholz, 2001.

TABLE 5.2
Classification of Cyanobacteria as Described in *Bergey's Manual of Systematic Bacteriology*

Form-Genera

Subsection I (Chroococcales)	*Prochlorococcus, Prochloron, Gloeobacter, Chamaesiphon, Cyanobacterium, Cyanobium, Cyanothece, Synechococcus, Dactylococcopsis, Gloeothece, Microcystis, Synechocystis, Chroococcus, Gloeocapsa*
Subsection II (Pleurocapsales)	*Cyanocystis, Dermocarpella, Stanieria, Xenococcus, Chroococcidiopsis, Myxosarcina, "Pleurocapsa"*
Subsection III (Oscillatoriales)	*Crinalium, Starria, Spirulina, Arthrospira, Lyngbya, Microcoleus, Symploca, Leptolyngbya, Limnothrix, Prochlorothrix, Pseudanabaena, Geitlerinema, Borzia, Oscillatoria, Planktothrix, Trichodesmium, Tychonema*
Subsection IV (Nostocales)	*Anabaenopsis, Cyanospira, Aphanizomenon, Cylindrospermum, Cylindrospermopsis, Anabaena, Nostoc (1-2), Nostoc (3-5), Scytonema, Calothrix, Rivularia, Gloeotrichia, Tolypothrix, Microchaete*
Subsection V (Stigonemetales)	*Loriella, Geitleria, Iyengariella, Mastigocladopsis, Nostociopsis, Westiella, Chlorogloeopsis, Stigonema, Doliocatella, Fischerella*

Source: Castenholz, 2001.

(Sánchez-Baracaldo et al., 2005). As with other taxonomic groups, progress toward achieving a more robust taxonomy is definitely being hampered by the misidentification of strains in the literature, in culture collections, and in GenBank. The types of problems still apparent within the current scheme are illustrated below by reference to a few specific examples.

PICOCYANOBACTERIA

Within Subsection I of the Cyanobacteria (Castenholz, 2001) is a group collectively referred to as picocyanobacteria. These are unicellular organisms, <2 µm diameter, that are important and some-times dominant primary producers in both freshwater and marine ecosystems (Scanlan and West, 2002; Stockner et al., 2000). Many studies of marine picocyanobacterial systems tend to use a rather simple taxonomy, partitioning the community into just two genera: *Prochlorococcus*, for cells that lack phycobiliproteins but possess divinyl cholorophyll a and b, and *Synechococcus* for everything else, although some recent studies (e.g., Campbell et al., 2005) have included an additional group, *Crocosphaera* spp, to describe potentially diazotrophic picocyanobacteria (diaz-othrophs use nitrogenase to reduce N_2 to ammonia as source of nitrogen for growth). A similarly simple taxonomy is often used for freshwater picocyanobacteria, but here the subdivision of the community involves a partitioning into red and green forms of *Synechococcus*, with some studies erroneously assigning many green forms to *Cyanobium*. This minimalist taxonomic system is misleading. Molecular studies, based largely on the use of sequence variation within regions of the rRNA operons, suggest that *Prochlorococcus* encompasses many "ecotypes" (Rocap et al., 2002; Johnson et al., 2006), and whole genome analysis demonstrates that individual isolates differ markedly in their genome size and gene content (Hess, 2004). Marine *Synechococcus* communities can be similarly subdivided into many clades with distinct spatial distributions (Fuller et al., 2003), and freshwater picocyanobacterial communities have been shown to be even more diverse (Ernst et al., 2003; Crosbie et al., 2003; Sánchez-Baracaldo et al., personal communication).

Even in the absence of distinguishing phenotypic features, it is possible to develop methods to enumerate members of different groups within the natural environment on the basis of their differing genotypes and/or phylogenetic clades. Although not all have been used for picocy-anobacteria, the methods available include the use of group-specific fluorescent probes to label permeabilised cell suspensions that can then be counted either manually or by cytometry

TABLE 5.3
Date of Peak Abundance of Four Distinct Lineages[1] of Picocyanobacteria in Eight English Lakes

Lake	Date of Maximum Abundance (2003)			
	Clade 1	Clade 2	Clade 3	Clade 4
Priest Pot	—	—	—	—
Windermere North	16th September	5th August	—	—
Windermere South	5th August	19th August	—	—
Esthwaite Water	19th August	22nd July	—	—
Blelham Tarn	19th August	—	—	—
CWP[2] 9	30th September	8th July	—	—
CWP 123	5th August	—	—	—
CWP 124	22nd July	—	8th July	19th August

[1] Defined on the basis of well-supported clades identified using partial 16S-rDNA sequences (Sánchez-Baracaldo et al., unpublished).
[2] Cotswold Water Park.
Note: — indicates that the presence of this clade could not be detected by qPCR.

(e.g., Tobe et al., 2006), quantification of hybridisation intensity of labelled probes to DNA extracted from environmental samples (e.g., Fuller et al., 2003), or the hybridisation of labelled DNA from environmental samples to arrays of immobilised group-specific probes (e.g., Castiglioni et al., 2004). Real time polymerase chain reaction (RT PCR or qPCR) can also be used to quantify changes in community structure, for example, it has been employed to follow changes in the abundance of four clades of freshwater picocyanobacteria in several temperate lakes (Sánchez-Baracaldo et al., unpublished), where it has been found that there are significant differences in both the composition of the picocyanobacterial community in different lakes and in the timing of maximum population development (Table 5.3).

What the above examples serve to illustrate is that the picocyanobacteria represent a group where the current taxonomy leads to a substantial underestimate of community diversity. Molecular studies reveal that it is not always acceptable or necessary to simply lump all picocyanobacteria into just a few taxonomic dustbins. The different organisms clearly have distinctive ecologies and appear to occupy specific niches. It is therefore necessary to define and quantify picocyanobacterial community interactions and dynamics in order to develop a predictive understanding of both global- and local-scale primary productivity under changing environmental conditions.

TOXIC CYANOBACTERIAL BLOOMS

The development of large cyanobacterial communities in both lakes and reservoirs that are used for both recreational purposes and the supply of drinking water seems to be an increasingly common phenomenon and one that is of growing concern, particularly in the context of diminishing water supplies. The cyanobacterial genera most commonly encountered within these communities are *Microcystis*, *Anabaena*, and *Planktothrix*, all three of which include forms that accumulate toxins, such as microcystins, and forms that apparently do not (e.g., Christiansen et al., 2006). Not surprisingly, there has been a drive to differentiate between forms within these genera, largely to inform management strategies that aim to minimise the development and impact of populations of toxic cyanobacteria.

MICROCYSTIS

Microcystis is a genus of planktonic unicellular, generally gas vacuolate cyanobacteria that reproduce by binary fission in two (or possibly three) planes at right angles to one another. Cells are spherical, 3 to 6 μm in diameter, and may be united into colonial aggregates by mucilage, but are never enclosed by multilaminated sheath material (Herdman et al., 2001). In nature *Microcystis* is recognised by the fact that it forms colonies, and the appearance of the colony is one of the major characters used to delineate species: *M. aeruginosa*, *M. viridis*, *M. wesenbergii*, *M. ichthyoblabe*, *M. novacekii*, *M. flos-aquae*, *M. botrys*, and *M. panniformis*. Even though the various colony morphologies appear to be maintained within the natural environment, once brought into culture the various species of *Microcystis* tend to adopt a unicellular growth form. For this particular genus, the application of molecular methodologies has thus far not allowed the identification of species boundaries, as, for example, has been shown in the studies of Janse et al. (2004), where no consistent relationships between morphospecies and rDNA ITS type were found: Janse et al. (2004) found that colonies identified as *M. aeruginosa*, the most commonly encountered *Microcystis* morphospecies, were scattered among nine clades, *M. ichthyoblabe* was found in five clades, but *M. novacekii* was restricted to a single clade. Whole genome DNA–DNA hybridisation studies (Otsuka et al., 2001) have revealed >70% identity for pair-wise comparisons of two isolates of *M. aeruginosa* and single isolates of *M. viridis*, *M. wesenbergii*, *M. ichthyoblabe*, and *M. novacekii*, suggesting that all five morphospecies should be included in a single bacterial species. What these analyses demonstrate is that the concept of *Microcystis* as a genus is well supported, but within the genus, although there is considerable genetic variation, there are currently no well-supported species groupings. What is required to resolve the situation is the adoption of an approach that employs broad taxonomic sampling combined with information from multiple gene loci. If such an analysis reveals robust and consistent groupings, it might be possible to identify diagnostic phenotypic and genotypic characters. From the perspective of water management, however, it may not be necessary to have a robust taxonomy for *Microcystis*. What is of key interest in this context is to know whether or not a toxic population will develop in the water column, and for this it is possible to directly assay genes encoding the toxin biosynthetic enzymes, such as the *mcy* genes that encode components of the microcystin biosynthetic apparatus (e.g., Ouellette et al., 2006). This functional gene approach, however, is only possible if the toxin biosynthetic genes have been characterised and are therefore available for the design of appropriate PCR primers.

Although colony morphology does not seem to be a reliable taxonomic character in *Microcystis*, it is worthy of note that this is not the case for some other Subsection I cyanobacteria. The genus *Snowella* is not recognised in the latest edition of *Bergey's Manual* (Castenholz, 2001), but has been recognised under the Botanical Code from the distinctive nature of its colonies: the cells lie at the periphery of the colony at the ends of clearly visible radiating mucilaginous stalks (Komárek and Anagnostidis, 1999). As is the case with *Microcystis*, cultures of *Snowella* lose their characteristic colony morphology and grow either as single cells or pairs of cells that fit the description of the form-genus *Synechocystis* (Herdman et al., 2001). Analysis of ssu rDNA sequences from a small number of cultured isolates of *Snowella* (Rajaniemi-Wacklin et al., 2006) demonstrates a strong correlation between morphological and molecular data, i.e., in this case colony morphology is a good taxonomic marker for the genus, although a more extensive sampling is needed to confirm this and to address speciation within the genus.

ANABAENA

The taxonomy of *Anabaena*, like that of *Microcystis*, is also in a state of some confusion. *Anabaena* is a member of Subsection IV of the cyanobacteria: filamentous cyanobacteria that produce heterocysts to facilitate nitrogen fixation in the light and akinetes as a resting stage to overcome harsh environmental conditions. *Anabaena* and the other taxa that form blooms in fresh and brackish waters are characterised by having trichomes (chains of cells) without any basal polarity and by the fact that they have a developmental cycle that lacks structurally distinct hormogonia (motile, dispersive filaments) (Rippka et al., 2001).

All phylogenetic analyses of cyanobacteria cluster the heterocystous forms within a single well-supported clade, implying that the ability to form heterocysts arose only once within the long evolutionary history of the group (e.g., Sánchez-Baracaldo et al., 2005). What is also apparent in such analyses is that isolates of some of the form-genera encompassed within Subsection IV (e.g. *Nodularia* or *Cylindrospermopsis*) cluster together, suggesting that they represent robust taxonomic groups (Rajaniemi et al., 2005), whereas others, such as *Anabaena* and *Aphanizomenon*, are more problematic. Organisms identified as *Anabaena* on the basis of their morphology are distributed between five distinct clades, but with the majority of isolates confined within a single clade where they are intermingled with isolates of *Aphanizomenon* (Rajaniemi et al., 2005): morphologically *Anabaena* is differentiated from *Aphanizomenon* largely on the basis of the differing length/width ratio of their vegetative cells, the slight attenuation in width toward the end, and rounded to subconical cells at the ends of *Aphanizomenon* trichomes. Within the major *Anabaena/Aphanizomenon* clade there are some coherent groupings, recovered by analysis of rDNA, *rpoB*, and *rbcLX* sequences, that correspond to recognised morphospecies (e.g. *Aphanizomenon flos-aquae*), but the general picture is one where many clades comprise mixtures of *Anabaena* and *Aphanizomenon* morphospecies, with benthic species of *Anabaena* occurring at several points within the radiation.

FINDING A WAY FORWARD

The above examples confirm that there are still problems associated with the unambiguous identification of ecologically important cyanobacteria. What is needed is an approach that will allow us to assign organisms to taxonomic groupings that will stand the test of time. It seems certain that taxonomies based on just morphological and life history characters can never be guaranteed to be robust. Similarly, taxonomic affinities determined using just rDNA sequences, or other single gene loci, are unlikely to remain unchallenged. Robust taxonomies are likely to be derived using both phenotypic and genotypic characters in a polyphasic approach, but even here there may still be problems. For example, Rajaniemi et al. (2005) used both morphological and molecular data from a collection of distinct isolates in an attempt to resolve taxonomic problems within *Anabaena* and *Aphanizomenon*, but although they were clearly able to identify the problems, the solutions were not found.

Whereas most studies have, understandably, focussed on analysing one or a few distinct isolates across a broad range of diversity, some authors advocate the need to include multiple examples of apparently similar organisms. In doing this it should be possible to differentiate between the variation that occurs within a species and the variation that separates species. This is essentially an approach that uses population genetics to identify the boundaries of shared gene pools as a means of describing species. The largest-scale application of this population-based approach to taxonomy involving cyanobacteria has been a study of *Nodularia*, a close relative of *Anabaena*, from the Baltic Sea. On the basis of phenotypic characters, individual Baltic Sea *Nodularia* filaments can be assigned to one of three morphospecies: *N. spumigena*, *N. litorea*, or *N. baltica* (Komárek et al., 1993). Using a combination of allele-specific PCR, amplified fragment length polymorphisms, and single nucleotide polymorphism assays (Barker et al., 2000; Batley and Hayes, 2003), 4599 individual filaments were genotyped at a total of four gene loci: *cpcBA*-IGS, rDNA ITS, *gvpC*, and *gvpA*-IGS. From the analysis it was apparent that all of the *Nodularia* filaments had access to a shared gene pool (i.e. there was exchange of genetic information within the population) (Hayes et al., 2002), but the recognition of this was dependent on the analysis of many individual filaments. From the observation that the population had an underlying panmictic structure, it was concluded that only a single species was present. The numerical dominance of one or a few genotypes in the natural populations, however, means that undersampling would have supported the division of the *Nodularia* community into several taxa, as was the case in morphologically based studies.

A similar approach has been used to characterise a collection of clonal *Anabaena* isolates established from a series of recently created lakes at the Cotswolds Water Park in Southern England (formed following gravel extraction). Based on their filament morphology and ssu rDNA sequences isolates were assigned to either *A. solitaria* or *A. lemmermannii*. Eighteen isolates were subjected to extensive phenotypic and genotypic characterisation (El Semary, 2005): for phenotype, filament shape was recorded and both vegetative and heterocyst cell sizes were quantified, as was gas vesicle strength; for genotype, the sequences of the *rpoC1*, *gvpA*, *gvpC*, *gvpA*-IGS, *gvpAC*-IGS, *cpcBA*-IGS, and rDNA ITS were determined. The combined phenotypic/genotypic analysis suggested that the concept of *A. solitaria* is clearly defined: the phenotypic characters are confined within a fairly narrow range, isolates were closer in terms of genetic distance to each other than to other *Anabaena* isolates, and in phylogenetic analyses the isolates grouped together with high bootstrap support using most gene loci. At two loci, the rDNA ITS and the *cpcBA*-IGS, although the isolates of *A. solitaria* were still clustered, some alleles were not exclusive, i.e. they were shared with other *Anabaena* isolates. This sharing of alleles has been reported previously for other cyanobacterial taxa (Rudi et al., 1998) and might indicate that the gene pool for *A. solitaria* is not closed with some genes entering from closely related taxa. Alternatively, especially in the case of the rDNA ITS, the shared alleles may not be homologous, arising *de novo* in separate taxa due to complex genomic rearrangements that seem common at this locus (Gürtler, 1999). Analysis of *A. lemmermannii* revealed a less cohesive grouping. Individual cultured isolates differed in phenotype, with one group having very distinctive gas vesicles, and in phylogenetic analyses, where using different gene loci generated noncongruent groupings: multilocus genotyping of uncultured filaments collected from lakes confirmed the high levels of genetic diversity (El Semary, 2005). These analyses of *A. lemmermannii* need to be extended to include more isolates so as to assess whether there is a situation similar to that seen in the Baltic Sea *Nodularia*, i.e. a phenotypically plastic taxon with an open gene pool that allows frequent exchange of genetic information within the population.

CONCLUDING REMARKS

That the taxonomy of cyanobacteria is still in a state of flux and confusion should be apparent even from the limited number of examples included in this chapter. The application of molecular methodologies is starting to provide some clues about what does or does not constitute a valid taxonomic grouping, but it is clear that single gene sequences alone are insufficient. Whole genome sequencing and studies of population genetics should allow us to define the boundaries of shared gene pools, which in turn could be used to delimit species, but at present it is naive to suppose that such methodologies could be applied to more than a handful of key taxa. So what is the solution? This very much depends on the questions that need answering. Where the primary interest is in modelling global productivity, for example, then it is probably acceptable to use a very coarse taxonomic system that simply partitions the cyanobacterial community into functional groups, such as *Prochlorococcus*, *Crocosphaera*, and *Synechococcus* within the picoplankton (but see below). In a similar vein, where the primary interest is to monitor the development of particular physiological types within phytoplankton communities, such as toxin producers or nitrogen fixers, then a taxonomy based on the presence of the functional genes associated with that physiology will serve the purpose. It is only when there is a need to understand the detailed dynamics and evolution of cyanobacterial communities that a sophisticated taxonomic system is needed, and preferably one that identifies organisms to a level that defines species boundaries in terms of shared gene pools, since it is the potential genetic repertoire that will ultimately determine an organism's ability to respond to environmental challenges. Whatever system is used in generating descriptions of community structure, it is essential that authors avoid the use of generic epithets when partitioning communities into functional types. For example, *Synechococcus* and *Prochlorococcus* should not be used as shorthand for picocyanobacteria of uncertain taxonomic status. Until such a discipline is applied, the literature will become evermore confused, and erroneously labelled entries in sequence databases will continue to mislead the uninitiated.

ACKNOWLEDGMENTS

We would like to thank the Natural Environment Research Council and the Government of the Arab Republic of Egypt for financial support.

REFERENCES

Barker, G.L.A., Handley, B.A., Vacharapiyasophon, P., Stephens, J.R., and Hayes, P.K. (2000) Allele specific PCR shows that genetic exchange occurs among genetically diverse *Nodularia* (cyanobacteria) filaments in the Baltic Sea. *Microbiology UK*, 146: 2865–2875.

Batley, J. and Hayes, P.K. (2003) Development of high-throughput SNP genotyping for the analysis of *Nodularia* population structures. *Journal of Phycology*, 39: 248–252.

Blank, C.E. (2004) Evolutionary timing of the origins of mesophilic sulphate reduction and oxygenic photosynthesis: a phylogenomic dating approach. *Geobiology*, 2: 1–20.

Campbell, L., Carpenter, E.J., Montoya, J.P., Kustka, A.B., and Capone, D.G. (2005) Picoplankton community structure within and outside a *Trichodesmium* bloom in the southwestern Pacific Ocean. *Vie et Mileu*, 55: 185–195.

Castenholz, R.W. (2001) Phylum BX. Cyanobacteria. Oxygenic Photosynthetic Bacteria. In *Bergey's Manual of Systematic Bacteriology. Volume 1: The Archaea and the Deeply Branching and Phototropic Bacteria*, 2nd ed. (eds. G. Garrity, D.R. Boone, and R.W. Castenholz), Springer-Verlag, New York, pp. 474–487.

Castiglioni, B., Rizzi, E., Frosini, A., Sivonen, K., Rajaniemi, P., Rantala, A., Mugnai, M.A., Ventura, S., Wilmotte, A., Boutte, C., Grubisic, S., Balthasart, P., Consolandi, C., Bordoni, R., Mezzelani, A., Battaglia, C., and De Bellis, G. (2004) Development of a universal microarray based on the ligation detection reaction and 16S rRNA gene polymorphism to target diversity of cyanobacteria. *Applied and Environmental Microbiology*, 70: 7161–7172.

Christiansen, G., Kurmayer, R., Liu, Q., and Börner, T. (2006) Transposons inactivate biosynthesis of the nonribosomal peptide microcystin in naturally occurring *Planktothrix* spp. *Applied and Environmental Microbiology*, 72: 117–123.

Crosbie, N.D., Pöckl, M. and Weisse, T. (2003) Dispersal and phylogenetic diversity of non-marine picocyanobacteria inferred from 16S rRNA gene and *cpcBA*-intergenic spacer sequence analyses. *Applied and Environmental Microbiology*, 69: 5716–5721.

El Semary, N.A. (2005) *Anabaena* and associated bacteria: molecular approaches to studying microbial community structure and taxonomy. PhD Thesis, University of Bristol, UK.

Ernst, A., Becker, S., Wollenzien, U.I.A., and Postius, C. (2003) Ecosystem-dependent adaptive radiations of picocyanobacteria inferred from 16S rRNA and ITS-1 sequence analysis. *Microbiology*, 149: 217–228.

Fuller, N.J., Marie, D., Partensky, F., Vaulot, D., Post, A.F., and Scanlan, D.J. (2003) Clade-specific 16S ribosomal DNA oligonucleotides reveal the predominance of a single marine *Synechococcus* clade throughout a stratified water column in the Red Sea. *Applied and Environmental Microbiology*, 69: 2430–2443.

Gugger, M.F. and Hoffmann, L. (2004) Polyphyly of true branching cyanobacteria (Stigonematales). *International Journal of Systematics and Evolutionary Microbiology*, 54: 349–357.

Gürtler, V. (1999) The role of recombination and mutation in 16S-23S rDNA spacer rearrangements. *Gene*, 238: 241–252.

Hayes, P.K., Barker, G.L.A., Batley, J., Beard, S.J., Handley, B.A., Vacharapiyasophon, P., and Walsby, A.E. (2002) Genetic diversity within populations of cyanobacteria assessed by analysis of single filaments. *Antonie Van Leeuwenhok*, 81: 197–202.

Herdman, M., Castenholz, R.W., Iteman, I., Waterbury, J.B., and Rippka, R. (2001) Subsection I. (Formerly Chroococcales Wettstein 1924, emend. Rippka, Deruelles, Waterbury, Herdman, and Stanier 1979). In *Bergey's Manual of Systematic Bacteriology. Volume 1: The Archaea and the Deeply Branching and Phototropic Bacteria*, 2nd ed. (eds G. Garrity, D.R. Boone, and R.W. Castenholz), Springer-Verlag, New York, pp. 493–514.

Hess, W.R. (2004) Genome analysis of marine photosynthetic microbes and their global role. *Current Opinion in Biotechnology*, 15: 191–198.

Honda, D., Yokota, A., and Sugiyama, J. (1999) Detection of seven major evolutionary lineages in cyanobacteria based on the 16S rRNA gene sequence analysis with new sequences of five marine *Synechococcus* strain. *Journal of Molecular Evolution*, 48: 723–739.

Ishida, T., Yokota, A., and Sugiyama, J. (1997) Phylogenetic relationships of filamentous cyanobacterial taxa inferred from 16S rRNA sequence divergence. *Journal of General and Applied Microbiology*, 43: 237–241.

Janse, I., Kardinaal, W.E.A., Meima, M., Fastner, J., Visser, P.M., and Zwart, G. (2004) Toxic and nontoxic *Microcystis* colonies in natural populations can be differentiated on the basis of rRNA gene internal transcribed spacer diversity. *Applied and Environmental Microbiology*, 70: 3979–3987.

Johnson, Z.I., Zinser, E.R., Coe, A., McNulty, N.P., Malcolm, E., Woodward, S., and Chisholm, S.W. (2006) Niche partitioning among *Prochlorococcus* ecotypes along ocean-scale environmental gradients. *Science*, 311: 1737–1740.

Komárek, J. and Anagnostidis, K. (1999) *Cyanoprokaryota. 1. Teil: Chroococcales*. Gustav Fischer Verlag, Jena, 548 pp.

Komárek, J., Hübel, M., Hübel, H., and Šmarda, J. (1993) The *Nodularia* studies 2, taxonomy. *Archives for Hydrobiologie, Algological Studies*, 68: 1–25.

Litvaitis, M.K. (2002) A molecular test of cyanobacterial phylogeny: inferences from constraint analyses. *Hydrobiologia*, 468: 135–145.

Otsuka, S., Suda, S., Shibata, S., Oyaizu, H., Matsumoto, S., and Watanabe, M.M. (2001) A proposal for the unification of five species of the cyanobacterial genus *Microcystis* Kutzing ex Lemmermann 1907 under the rules of the bacteriological code. *International Journal of Systematic and Evolutionary Microbiology*, 51: 873–879.

Ouellette, A.J.A., Handy, S.M., and Wilhelm, S.W. (2006) Toxic *Microcystis* is widespread in Lake Erie: PCR detection of toxin genes and molecular characterisation of associated cyanobacterial communities. *Microbial Ecology*, 51: 154–165.

Rajaniemi, P., Hrouzek, P., Kaštovska, K., Willame, R., Rantala, A., Hoffmann, L., Komárek, J., and Sivonen, K. 2005. Phylogenetic and morphological evaluation of the genera *Anabaena*, *Aphanizomenon*, *Trichormus* and *Nostoc* (Nostocales, Cyanobacteria). *International Journal of Systematics and Evolutionary Microbiology*, 55: 11–26.

Rajaniemi-Wacklin, P., Rantala, A., Mugnai, M.A., Turicchia, S., Ventura, S., Komarkova, J., Lepisto, L., and Sivonen, K. (2006) Correspondence between phylogeny and morphology of *Snowella* spp. and *Woronichinia naegeliana*, cyanobacteria commonly occurring in lakes. *Journal of Phycology*, 42: 226–232.

Rippka, R., Castenholz, R.W., and Herdman, M. (2001) Subsection IV. (Formerly Nostocales Castenholz 1989b sensu Rippka, Deruelles, Waterbury, Herdman, and Stanier 1979). In *Bergey's Manual of Systematic Bacteriology. Volume 1: The Archaea and the Deeply Branching and Phototropic Bacteria*, 2nd ed. (eds G. Garrity, D.R. Boone, and R.W. Castenholz), Springer-Verlag, New York, pp. 562–589.

Rocap, G., Distel, L., Waterbury, J.B., and Chisholm, S.W. (2002) Resolution of *Prochlorococcus* and *Synechococcus* ecotypes by using 16S-23S ribosomal DNA internal transcribed spacer sequences. *Applied and Environmental Microbiology*, 68: 1180–1191.

Rudi, K., Skulberg, O.M., and Jakobsen, K.S. (1998) Evolution of cyanobacteria by exchange of genetic material among phyletically related strains. *Journal of Bacteriology*, 180: 3453–3461.

Sánchez-Baracaldo, P., Hayes, P.K., and Blank, C.E. (2005) Morphological and habitat evolution in the Cyanobacteria using a compartmentalization approach. *Geobiology*, 3: 145–165.

Scanlan, D.J. and West, N.J. (2002) Molecular ecology of the marine cyanobacterial genera *Prochlorococcus* and *Synechococcus*. *FEMS Microbial Ecology*, 40: 1–12.

Stackebrandt, E. and Goebel, B.M. (1994) A place for DNA-DNA reassociation and 16S rRNA sequence analysis in the present species definition in Bacteriology. *International Journal of Systematic Bacteriology*, 44: 846–849.

Stockner, J.C., Callieri, C., and Cronberg, G. (2000) Picoplankton and other non-bloom forming cyanobacteria in lakes. In *The Ecology of Cyanobacteria, Their Diversity in Time and Space*. (eds B.A. Whitton and M. Potts), Kluwer Academic Publishers, Dordrecht, pp 195–131.

Tobe, K., Eller, G., and Medlin, L.K. (2006) Automated detection and enumeration for toxic algae by solid-phase cytometry and the introduction of a new probe for *Prymnesium parvum* (Haptophyta: Prymnesiophyceae). *Journal of Plankton Research*, 28: 643–657.

Turner, S., Huang, T.C., and Chaw, S.M. (2001) Molecular phylogeny of nitrogen-fixing unicellular cyanobacteria. *Botanical Bulletin of Academia Sinica*, 42: 181–186.

Turner, S., Pryer, K.M., Miao, V.P.W., and Palmer, J.D. (1999) Investigating deep phylogenetic relationships among cyanobacteria and plastids by small submit rRNA sequence analysis. *Journal of Eukaryotic Microbiology*, 46: 327–338.

6 Molecular systematics of red algae: building future structures on firm foundations

Christine A. Maggs, Heroen Verbruggen and Olivier de Clerck

CONTENTS

ABSTRACT

Red algal systematics has a solid morphological foundation, based on analyses of female reproductive structures and post-fertilization development by Kylin and other workers. Recognition of the value of pit-plug ultrastructure was a catalyst leading to refinement of the Kylinian ordinal classification. Molecular approaches to systematics have further advanced our understanding of the red algae at every level and led to the proposal of several new orders. Species diversity in particular has traditionally been underestimated due to the presence of cryptic and pseudo-cryptic species.

A literature review covering the last two decades shows that relatively few molecular markers have been employed for studies of red algal systematics. The general trend was toward the use of multiple markers until five years ago when an increased emphasis on lower-level taxonomy (molecular identification) led to greater reliance on single markers. In the future, we predict that there will be relatively few new orders proposed, and the emphasis will change to hypothesis-driven, less descriptive molecular studies in concert with morphological, ultrastructural, and biochemical analyses.

INTRODUCTION TO THE RED ALGAE (RHODOPHYTA)

AN ANCIENT AND DIVERSE GROUP

The Rhodophyta (red algae) are eukaryotes, and the great majority of the species are marine, photosynthetic, and macroscopic. Red algae are an ancient lineage (Xiao et al., 1998; Yoon et al., 2004), including what is generally believed to be one of the oldest taxonomically resolved eukaryotic fossils, the 1.2 billion year old *Bangiomorpha pubescens* Butterfield (Butterfield, 2000). The Rhodophyta have evolved a diverse range of modifications in cellular organization and general morphology (Pueschel, 1990).

UNIQUE COMBINATION OF FEATURES

The red algae are distinguishable amongst eukaryotic lineages by a combination of biochemical and ultrastructural features. The most striking of these is that they lack flagella as well as centrioles or other 9 + 2 structures at any stage of their life histories (Pueschel, 1990; Ragan and Gutell, 1995); sex and spore dispersal therefore have to be accomplished without the benefit of flagellar propulsion. Instead, red algal sperm sport gothic-style mucilaginous appendages that affect their hydrodynamic properties and contain cell recognition proteins that attach specifically to the sessile female gametes—carpogonia (Broadwater et al., 1991; Kim et al., 1996).

Whereas green algae and plants store starch in the chloroplasts, the red algal polysaccharide reserves of floridean starch are in the cytoplasm. An important ultrastructural feature is that red algal plastids (term used for non-green chloroplasts) have unstacked thylakoid membranes and lack an encircling endoplasmic reticulum membrane. Chlorophyll *a* is the only chlorophyll, and accessory red/blue phycobilin pigments, predominantly the red-coloured phycoerythrin, occur in stalked phycobilisomes on thylakoids (van den Hoek et al., 1995). Although none of the characteristics listed here is unique to the Rhodophyta, the red algae represent the only group of organisms in which all are found, so that "in practice there is little difficulty in distinguishing what is or is not a red alga" (Ragan and Gutell, 1995).

From the early twentieth century until very recently, red algae were divided into two groups (Bangiophyceae and Florideophyceae) based on morphological, anatomical, and life-history differences (Dixon, 1973), but the taxonomic rank to which these groups were assigned fluctuated (Gabrielson et al., 1985; Garbary and Gabrielson, 1990; Murray and Dixon, 1992). In order to introduce these groups, we refer to the morphologically defined classes as Bangiophyceae or Florideophyceae "in the traditional sense"; a more recent classification is provided below in the section on molecular markers and higher-level systematics.

BANGIOPHYCEAE: RELATIVELY SIMPLE MORPHOLOGIES

The smaller class Bangiophyceae has mainly been defined rather unsatisfactorily by the absence of characters confined to the Florideophyceae (e.g. tetrasporangia, filamentous gonimoblasts) or by the presence of characters (e.g. single star-shaped plastids) that are found only in some members of the class. Compared to the florideophytes, the bangiophytes are generally morphologically simple and have traditionally been considered to contain the most primitive red algal forms (Müller et al., 2001). Their life histories are mostly poorly known but appear to be diverse (Brodie and Irvine, 2003).

The best-known genus, *Porphyra*, which is commercially cultivated for nori, displays a heteromorphic life history, the two phases of which are so morphologically dissimilar that they were linked only by culture studies (Drew, 1954). The blade-like haploid gametophytic phase gives rise to a microscopic shell-boring diploid conchocelis phase. In some species at least, meiosis takes place during germination of spores formed by the conchocelis phase (Mitman and van der Meer, 1994; Brodie and Irvine, 2003). Apart from the Bangiales, the Bangiophyceae are mostly asexual (Brodie and Irvine, 2003).

DIVERSE, COMPLEX FLORIDEOPHYCEAE

The more complex Florideophyceae (e.g. the valuable carrageen-producing "Irish Moss" *Chondrus crispus* Stackhouse) exhibit an enormous diversity of morphological structures and complicated haplo-diploid life histories. Haploid and diploid phases are morphologically closely similar if not identical (isomorphic) in some species, whereas other life histories such as that of the Japanese invasive species *Bonnemaisonia hamifera* Hariot are heteromorphic like *Porphyra*. The cryptic phase, which may be either the gametophyte (unisexual or bisexual) or the sporophyte, is crustose, filamentous, or boring, whereas the more conspicuous phase is erect, often resembling leaves or twigs. These intricate life histories involve specialized meiotic sporangia (tetrasporangia), characteristic of the Florideophyceae, that release four haploid tetraspores. Uniquely, in most Florideophyceae, the immediate product of fertilization is not the diploid sporophyte, but a hemi-parasitic diploid tissue (the "gonimoblast") surrounded by female nutritive tissue, collectively called the "cystocarp" (or carposporophyte). This stage, representing a clonal growth strategy compensating for the lack of motile sperm in the red algae (Searles, 1980), releases numerous genetically identical diploid spores that give rise to sporophytes.

ORDINAL CLASSIFICATION HAS FIRM FOUNDATIONS

FLORIDEOPHYCEAE: THE ENDURING LEGACY OF SCHMITZ AND KYLIN

Until the impact of molecular data, the ordinal classification of the Florideophyceae was still based on the monumental posthumous treatise of Kylin (1956), which incorporated his revisions of the earlier schemes of Schmitz (1889) and Oltmanns (1898). This classification depended mainly on characteristics of female reproductive anatomy before and after fertilization. Six orders of Florideophyceae (Nemalionales, Gelidiales, Cryptonemiales, Gigartinales, Rhodymeniales, and Ceramiales) were recognized (Figure 6.4; Papenfuss, 1966). In the last four of these Kylinian orders, the gonimoblast originates from the auxiliary cell, a cytoplasm-rich cell into which is injected the diploid zygotic nucleus, or a diploid nucleus produced after one or several divisions of the zygotic nucleus (Dixon, 1973). The Ceramiales was generally considered the most advanced of these orders because the auxiliary cell is produced only after fertilization. The Nemaliales was characterized by the direct development of the gonimoblast from the zygote in the absence of an auxiliary cell, and the Gelidiales had been separated from it because although auxiliary cells were present they did not initiate gonimoblasts (see Garbary and Gabrielson, 1990).

BANGIOPHYCEAE: FEWER, LESS ROBUST ORDERS

Kylin's (1956) ordinal classification of bangiophytes was less well developed than that for the florideophytes because of the scarcity of morphological characters (Dixon, 1973). In particular, the group lacks cystocarps, whereas the Schmitz–Kylin classification was erected essentially on the basis of the complex female reproductive characters of the florideophytes. The revision by Garbary et al. (1980) incorporated observations on life histories and ultrastructure in addition to new morphological characters and distinguished four orders by a combination of these traits. Erythropeltidales (features: multicellular; vegetative reproduction by monosporangia) was erected as a new order because of similarities in monosporangial formation in three families (Erythropeltidaceae, Boldiaceae, and Compsopogonaceae). The Porphyridiales circumscribed unicellular species that are free-living or held together by a mucilaginous matrix. The Rhodochaetales contained filamentous algae with pit connections and sexual reproduction by formation of a single diploid carpospore following fertilization. The most distinctive and species-rich order, the Bangiales, included all species that exhibited sexual reproduction and in which the conchocelis sporophyte generation formed pit connections.

PIT-PLUG ULTRASTRUCTURE: A CRUCIAL CATALYST

The most significant contribution to red algal systematics after Kylin's morphological synthesis and prior to the application of molecular markers came in a series of papers by Curt Pueschel, starting with Pueschel and Cole (1982), which reported on his ultrastructural studies of pit-plugs. In florideophytes and some bangiophytes, cross-wall formation after cell division is incomplete, leaving a membrane-lined pore in the central region (Pueschel, 1990). Tubular membranes appear in this region, then a homogeneously granular protein mass (the plug core) is deposited around the tubules, followed by the disappearance of the tubules. The pit-plug either consists only of the core or acquires additional features, such as carbohydrate domes and cap membranes, which are continuous with the cell membrane. Although some previous workers had doubted the possible systematic value of pit-plugs because of observed intraspecific variation (Duckett and Peel, 1978), Pueschel and Cole (1982) showed that a combination of three characters, the presence or absence of the inner and outer cap layers, and the shape of the outer cap (domed or plate-like) was useful in distinguishing among higher taxa of florideophytes. Features of pit-plugs introduced a new age of criteria for ordinal relationships in the florideophytes; two new orders (Batrachospermales and Hildenbrandiales) were segregated, and four previously described orders (e.g. the Palmariales, originally based on tetrasporangial ontogeny) were confirmed in the analyses of Pueschel and Cole (1982). Later, Trick and Pueschel (1991) demonstrated that plate-like and domed outer cap layers were chemically similar and probably homologous, and proposed that the domed state (found in the Corallinales and some Acrochaetiales) is ancestral, plate-like outer caps being derived.

MOLECULAR TOOLS AND THEIR CONTRIBUTION
TO SYSTEMATICS

PIONEER PAPERS

The first molecular phylogenetic studies that included red algae appeared in the mid-1980s. Analyses of 5S ribosomal sequences of representative prokaryotic and eukaryotic organisms found the red algae to be the most ancient eukaryote lineage (Hori et al., 1985; Lim et al., 1986; Hori and Osawa, 1987). Despite the small size and evolutionary constraints on this gene which make it inappropriate for this kind of phylogenetic analyses, trees were broadly congruent with later 18S trees (Ragan and Gutell, 1995). In Hori and Osawa's tree, the Palmariales was found to branch more basally within the Florideophyceae than the Gigartinales, Gelidiales, and Gracilariales.

Other early molecular studies used a variety of markers. Goff and Coleman (1988) employed restriction fragment length polymorphism (RFLP) analysis of plastid DNA for a geographical study of *Gracilariopsis andersonii* (Grunow) Dawson (as *Gracilaria sjoestedtii*), showing populations over a 2000 km range to be remarkably genetically homogeneous. The same technique was used to link the dissociated (apomictic) phases of the heteromorphic life history in *Gymnogongrus* (now *Ahnfeltiopsis*) (Parsons et al., 1990). The phylogenetic position of *Gracilariopsis* in the eukaryotes was determined from 18S nrDNA sequences (Bhattacharya et al., 1990). In red algae the large (*rbc*L) and small (*rbc*S) subunits of the ribulose biphosphate carboxylase/oxygenase (rubisco) gene are organised as a co-transcribed operon including a short highly variable intergenic spacer (the rubisco spacer) (Kostrzewa et al., 1990). This spacer was exploited for several early species-level studies that distinguished between populations of *Ahnfeltiopsis* (as *Gymnogongrus*) *devoniensis* (Greville) P.C. Silva et DeCew and closely related species (Maggs et al., 1992) and amongst

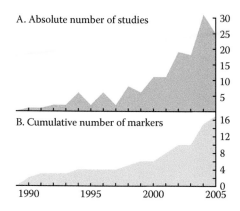

FIGURE 6.1 (A) Absolute number of red algal molecular phylogenetic studies per year through time. (B) Cumulative number of red algal molecular phylogenetic markers through time.

members of the *Gracilaria verrucosa* species complex (Destombe and Douglas, 1991). Bird et al. (1992) carried out the first molecular study at the family to species level in red algae, using using the SSU nrDNA to determine phylogenetic relationships between members of the Gracilariaceae.

MOLECULAR PHYLOGENETIC MARKERS

Since the early 1990s, the numbers both of phylogenetic studies and markers used have increased steadily (Figure 6.1A and Figure 6.1B). Phylogenetic studies have been carried out at all taxonomic levels across most of the spectrum of red algal biodiversity. Our examination of 156 red algal molecular systematic papers published between 1990 and early 2006 revealed that a wide array of nuclear ribosomal, plastid, and mitochondrial markers is currently available for red algal phylogenetics (Table 6.1). The number of molecular phylogenetic markers used has gradually increased

TABLE 6.1
Red Algal Phylogenetic Markers and Their Frequency of Use in 156 Screened Studies

Genome	Marker	Type	Reference	Frequency
Nuclear	5S	Ribosomal DNA	Hori et al. (1985)	3
	18S	Ribosomal DNA	Bhattacharya et al. (1990)	62
	28S	Ribosomal DNA	Freshwater and Bailey (1998)	21
	ITS region	Two ribosomal spacers	Steane et al. (1991)	17
	Actin	Gene	Hoef-Emden et al. (2005)	1
Mitochondrial	*cox*1	Gene	Saunders (2005)	1
	*cox*2-3	Intergenic spacer	Zuccarello et al. (1999)	11
Plastid	16S	Ribosomal DNA	Olson et al. (2005)	3
	*rbc*L	Gene	Freshwater et al. (1994)	77
	*rbc*S	Gene	Lee et al. (2001)	2
	Rubisco spacer	Intergenic spacer	Destombe and Douglas (1991)	22
	*psa*A	Gene	Yang and Boo (2004)	3
	*psa*B	Gene	Yoon et al. (2004)	1
	*psb*A	Gene	Seo et al. (2003)	4
	*psb*C	Gene	Yoon et al. (2002, 2006)	1
	*psb*D	Gene	Yoon et al. (2002, 2006)	1
	*tuf*A	Gene	Yoon et al. (2004)	1
	URP markers	Genes and spacers	Provan et al. (2004)	1

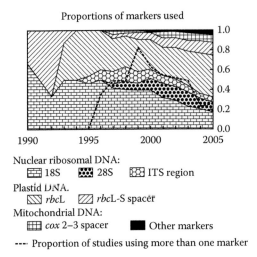

Proportions of markers used

Nuclear ribosomal DNA:
    ⊞ 18S   ▒ 28S   ⊞ ITS region
Plastid DNA.
    ◹ *rbc*L   ◺ *rbc*L-S spacer
Mitochondrial DNA:
    ⊞ *cox* 2–3 spacer   ■ Other markers
---- Proportion of studies using more than one marker

FIGURE 6.2 Use of phylogenetic markers through time, plotted as proportion of total studies. The dashed line indicates the proportion of studies using more than one marker.

with time (Figure 6.1B). Although many of the markers have been employed for studies at various taxonomic levels, plastid and mitochondrial genes are mostly exploited to resolve the relationships between species belonging to one genus or one family, at higher taxonomic levels plastid or nuclear ribosomal DNA sequences are mostly used, and spacer sequences are almost exclusively used to infer haplotype trees or networks for closely related species and populations. To evaluate the taxonomic level at which markers performed best, Verbruggen et al. (unpublished) reanalysed 104 published phylogenetic datasets using Bayesian methods. The resolution of the markers, measured as the proportion of nodes receiving posterior probability ≥0.90, was evaluated at five taxonomic levels, ranging from intraspecific nodes to nodes above the ordinal level. Whereas the resolution of plastid genes was highest at low taxonomic levels and gradually decreased toward higher ranks, nuclear rDNA markers showed the opposite trend.

There are some trends in the use of phylogenetic markers through time (Figure 6.2), such as a gradual decrease in the use of the relatively slowly evolving SSU ribosomal DNA marker in favour of faster-evolving organellar markers (*rbc*L, *rbc*L-S spacer, *cox*2-3 spacer). This trend coincides with a steady increase in the relative number of studies focussing at low taxonomic levels. The most recently developed markers reinforce the latter trend; nearly all of them are highly variable markers for use between the genus and species level (e.g. the mitochondrial *cox*1 gene and *cox*2-3 spacer, and the plastid genes *psb*A, *psa*A, and URP markers). There is also a clear trend in the number of studies using more than one marker. The use of multiple markers shows a sharp increase in the second half of the 1990s to decrease again in the 2000s (Figure 6.2). We interpret this observation as being a result of the rapid widening of the usage of molecular markers in red algal systematics. The first two decades of molecular systematic research, particularly in the pre-PCR age, involved a great deal of pioneering technical work, and researchers experimented with a wide range of different markers. Most of the recent papers are primarily taxonomically centred, and for most systematic questions, adequate data can be acquired from single markers.

HIGHER TAXONOMIC LEVELS: RELATIONSHIP BETWEEN BANGIOPHYCEAE AND FLORIDEOPHYCEAE

Two large, high-level molecular phylogenies of the red algae appeared together in 1994, utilizing the nuclear 18S ribosomal gene (Ragan et al., 1994) and the plastid-encoded gene for the large subunit of Rubisco, *rbc*L (Freshwater et al., 1994). Both of these ambitious studies suffered from various flaws, including comparatively primitive methods of data analysis, underrepresentation of bangiophyte taxa, and long-branch attraction, but they provided broad overviews of relationships among florideophyte orders, generally comparable with more recent trees. Many of the problems were overcome in later studies.

Cavalier-Smith (1998) proposed a division of the red algae into two subphyla. His Rhodellophytina, characterized by relatively simple thalli composed of small uninucleate vegetative cells, encompasses the Porphyridiales, Cyanidiales, and Compsopogonales. Cavalier-Smith's Macrorhodophytina, including the florideophytes and Bangiales, exhibit a much wider range of morphology (Goff and Coleman, 1990; Ragan and Gutell, 1995). Cavalier-Smith's classification has been adapted by Saunders and Hommersand (2004; Figure 6.3), and newer proposals are described below.

Two recent papers specifically address the subdivision of the red algae, and particularly the heterogeneous Bangiophyceae, into monophyletic subphyla and classes (Saunders and Hommersand, 2004; Yoon et al., 2006; summarized in Figure 6.3). Saunders and Hommersand (2004) present a classification based on a tree summarizing the relationships compiled from several ribosomal DNA studies. They split up the traditional bangiophycean diversity by moving the Cyanidiales into their own phylum and placing the remaining orders in three subphyla (Figure 6.3). The subphylum Metarhodophytina, with a single class Compsopogonophyceae, was erected to group the Compsopogonales, Erythropeltidales, and Rhodochaetales. A second, possibly paraphyletic, subphylum Rhodellophytina, with a single class Rhodellophyceae, contains the three porphyridialean orders. Finally, Saunders and Hommersand's third, and convincingly monophyletic, subphylum Eurhodophytina contains the Bangiophyceae s.s. (Bangiales) and the Florideophyceae.

Yoon et al. (2006) define major red algal lineages and infer their relationships on the basis of their analyses of a concatenated alignment of seven plastid protein-coding genes, 18S nuclear rDNA and 16S plastid rDNA. They argue that the Cyanidiales do not deserve recognition as a separate phylum and recognize them as a subphylum within the Rhodophyta. They support Saunders and

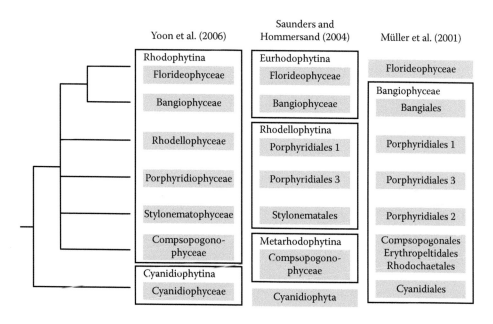

FIGURE 6.3 Phylogeny of the red algae, modified from Yoon et al. (2006), in which branches in front of nodes receiving insufficient confidence have been collapsed. Three current alternative higher-level classification schemes of the red algae into subphyla (boxes) and classes/orders (shaded) are shown, from Yoon et al. (2006, left), Saunders and Hommersand (2004, centre), and Müller et al. (2001, right). Note that in the classification of Saunders and Hommersand (2004), the Cyanidiophyceae were placed in a separate phylum.

Hommersand's Compsopogonophyceae and Bangiophyceae s.s. but raise the three porphyridialean orders to classes (Figure 6.3). In addition to the Cyanidiophytina, they recognize a second subphylum, Rhodophytina, comprising all other classes.

The relationships between the Florideophyceae and the various bangiophycean orders have been obscured until relatively recently. In earlier accounts, the sequences of the bangiophyte unicells *Dixoniella* and *Rhodella* grouped quite strongly with the solidly monophyletic Florideophyceae (Ragan and Gutell, 1995). Addition of a large number of bangiophyte sequences (Oliveira and Bhattacharya, 2000; Müller et al., 2001; Yoon et al., 2006) has repositioned *Dixoniella* and *Rhodella* away from a clade uniting the bangialean and florideophyte algae. The rather perplexing previous position of both genera in Ragan and Gutell's (1995) analyses is now attributed to long-branch attraction (Müller et al., 2001). The branching order among the major red algal lineages is still poorly resolved (Figure 6.3), even based on the nine-marker alignment of Yoon et al. (2006). As long suspected, the Bangiophyceae in the traditional sense is paraphyletic, with the Cyanidiales branching off first, and several orders falling in a polytomy. The link demonstrated between the Bangiales and Florideophyceae supports two of Magne's (1989) primary subdivisions of the Rhodophyta, the Eurhodophycidae (florideophytes + Bangiales) and the Metarhodophycidae (Erythropeltidales, Rhodochaetales, Compsopogonales), although the Porphyridiales, Magne's Archeorhodophycidae, is polyphyletic (Müller et al., 2001). Contrary to earlier beliefs, *Rhodochaete*, the single representative of the order Rhodochaetales, was not resolved as the speculated link between Bangio- and Florideophyceae but grouped closely with Erythropeltidales and Compsopogonales (Zuccarello et al., 2000).

ORDINAL SYSTEMATICS OF THE FLORIDEOPHYCEAE

Our analysis shows that there has been a slight increase in phylogenetic studies aiming to infer relationships among orders, and among families within orders, from an average of 2–3 per year in the second half of the 1990s to 5 per year in 2004 and 2005 (although the relative proportion of these studies decreases through time). The two large deep-level molecular phylogenies of the red algae that appeared together in 1994 (Freshwater et al., 1994; Ragan et al., 1994) provided an overview of relationships among florideophyte orders, generally comparable with more recent trees.

If there is one generality among ordinal studies, it is that Kylinian orders were polyphyletic, particularly with respect to taxa lacking recognizable cystocarps, and therefore are being split (Figure 6.4). The number of orders has steadily increased from six in the Kylinian system, via thirteen following the ultrastructural studies in the 1980s, to twenty-five in the present molecular phylogenetic age, and a few more families can be expected to be raised to the ordinal level in the near future.

We here give some examples of orders that have been recognized recently as a result of molecular studies, in order to analyse the contribution of molecular data to their recognition as separate orders. Separation of the family Thoreaceae from the rest of the Batrachospermales in 18S and *rbc*L sequence analyses led to the proposal of the Thoreales by Müller et al. (2002). They also carried out the most extensive survey to date of infraspecific variation in pit-plug ultrastructure, which showed that there is some variation in the degree of inflation of the outer cap. The most convincing piece of evidence supporting recognition of the new order is the discovery of unique secondary structure signatures in the 18S gene. However, analyses by Müller et al. (2002) and others show clearly that all the members of the original Nemaliales (apart from Gelidiales and Bonnemaisoniales), i.e. Batrachospermales, Thoreales, Balliales, Acrochaetiales, and Colaconematales (Harper and Saunders, 2002) form a close grouping with the Palmariales and the Balbianiales (Sheath and Müller, 1999). An argument could be made for subsuming some of these orders again into the classical Nemaliales minus Gelidiales and Bonnemaisoniales.

An elegant example of an order recognized by a combination of its molecular phylogenetic position and its morphology is the Pihiellales (Huisman et al. 2003). This order was proposed (with

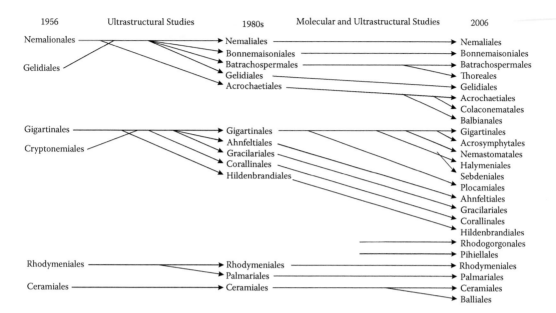

FIGURE 6.4 Changes to the ordinal classification of the florideophytes resulting from ultrastructural and molecular studies. The first column (1956) represents the Kylinian orders, the central column represents the orders recognized toward the end of the 1980s, on the basis of ultrastructural work (mostly pit-plugs), and the right column presents the currently recognized orders resulting from molecular analyses, in many cases supported by ultrastructural data.

its formal diagnosis including a GenBank code), for a previously undescribed minute endo/epiphyte growing on members of the Liagoraceae. Although these organisms were first noticed and illustrated in the mid-nineteenth century, their lack of morphological affinities with any other red alga led to a 150-year hiatus in the process of naming them. Analyses of 18S rDNA showed that the single species *Pihiella liagoraciphila* Abbott was most closely related to *Ahnfeltia*, the sole genus in the Ahnfeltiales (Maggs and Pueschel, 1989), but the extreme morphological differences and the large genetic distance between the two taxa indicate a long evolutionary divergence (Huisman et al., 2003).

Molecular phylogenetics also facilitates reconstruction of the relationships among orders. Whereas single-marker analyses left several of the deeper nodes lacking confidence, recent multi-gene approaches and more sophisticated analysis techniques have improved the statistical confidence in the relationships among orders despite some conflict among analyses and markers (Harper and Saunders, 2001; Withall and Saunders, 2006). Four main lineages of the Florideophyceae defined by Saunders and Kraft (1997) seem well supported, and a contemporary supraordinal classification for them was proposed by Saunders and Hommersand (2004). These four subclasses are also reasonably well characterized by ultrastructural characters (Table 6.2).

Species-level systematics and DNA barcoding

A large proportion of the phylogenetic studies we have screened (43%) were situated at the genus to species level. More than two-thirds of these studies have been published since 2003, indicating that species-level phylogenetics is an active field. The species-level systematic papers can roughly be subdivided into two classes. Studies belonging to the first class test species boundaries with relatively large sequence alignments in which species are represented by multiple specimens. Such studies usually also employ different types of data (morphology, interfertility) to investigate whether

TABLE 6.2
Monophyletic Supraordinal Lineages Receiving Strong Molecular Support and Defined as Subclasses by Saunders and Hommersand (2004)

Subclass	Orders	Ultrastructural Features
Hildenbrandiophycidae	Hildenbrandiales	Pit-plugs with single cap layer and membrane
Nemaliophycidae	Acrochaetiales, Balbianiales, Balliales, Batrachospermales, Colaconematales, Corallinales, Nemaliales, Palmariales, Rhodogorgonales, Thoreales	Pit-plugs with two cap layers
Ahnfeltiophycidae	Ahnfeltiales, Pihiellales	Pit-plugs naked
Rhodymeniophycidae	Acrosymphytales, Bonnemaisoniales, Ceramiales, Gelidiales, Gigartinales, Gracilariales, Halymeniales, Nemastomatales, Plocamiales, Rhodymeniales, Sebdeniales	Pit-plugs with membranes only (exception: Gelidiales)

different species concepts agree or can be reconciled. The second class of studies at the species level is of a more morphologically descriptive sort. Such studies present morphological descriptions of new species and employ molecular tools to situate the species in the genus.

The single most important conclusion to be drawn from species-level molecular phylogenetic studies is that red algal biodiversity is massively underestimated. Many studies have revealed cryptic and pseudo-cryptic species within morphologically defined species. Cryptic species, sometimes referred to as sibling species, are defined as species that are impossible to distinguish based on morphological characters (Sáez and Lozano, 2005). Pseudo-cryptic species are species that are readily distinguished morphologically once the appropriate characters are considered.

Cryptic diversity in the mangrove- and salt-marsh-dwelling species of *Bostrychia* (Rhodomelaceae) has been addressed by Joe Zuccarello and John West. Their research has shown that despite being morphologically identical, separate reproductively isolated lineages can be identified by plastid and mitochondrial haplotypes (Zuccarello and West, 2003). Moreover, some of these genuine cryptic species occur sympatrically. This situation may be found more commonly amongst the red algae when appropriate studies have been carried out.

Discoveries of pseudo-cryptic species have also been plentiful, and we will give just a few examples. In the genus *Grateloupia*, which has received much attention because of its invasive members (Verlaque et al., 2005), molecular tools unveiled several discrete entities within morphologically circumscribed species (Kawaguchi et al., 2001; De Clerck et al., 2005; Wilkes et al., 2005). These discrete entities can be morphologically recognized, and most of them have been described as separate species. A second genus exemplifying recent discovery of pseudo-cryptic entities within morphologically perceived species is *Plocamium*. Saunders and Lehmkuhl (2005) demonstrated the existence of at least eight divergent cryptic species currently included in *P. cartilagineum*, and defined morphological boundaries of four European species for which their sampling was sufficient. Likewise, Yano and coworkers (Yano et al., 2004, 2006) showed that traditional morphological species boundaries in Japanese *Plocamium* did not correspond with their molecular clusters. They could not find clear-cut morphological characters specific to their molecular entities but demonstrated that colour, bromine concentration, and cell content acidity could be used as identification clues.

A very recent trend is the use of DNA sequences as an identification tool. DNA barcoding, as this technique is commonly known, consists of a first stage in which a large database of sequences is generated from well-documented and accurately identified specimens (Newmaster et al., 2006). At a second stage, the database can be queried using sequences generated from unidentified specimens. The results of such queries allow identification of the unknown specimens. Red algal DNA barcoding is currently going through the stage of dataset creation (Saunders, 2005; Robba et al., 2006). Molecular datasets of several recent studies of species boundaries in which species are represented by multiple specimens can be used for later querying. Several more such datasets can be expected to appear in the years to come. Examples of specimen identification against a sequence database are still scarce. Rueness (2005) showed newly collected *Gracilaria* specimens from Brittany and Sweden to belong to the invasive species *G. vermiculophylla* using their ITS, *rbc*L, and *cox*2-3 spacer sequences.

When used wisely, DNA barcoding could form an incredible asset to red algal systematics. As outlined above, red algal diversity cannot always be captured using morphological characters, especially in structurally simple genera. In such cases, it is difficult or impossible to figure out which cryptic species corresponds to the type of the morpho-species. One of the great advantages of DNA barcoding is that it can be used to accurately identify cryptic species' haplotype clusters. This is exemplified by the studies of Hughey et al. (2001, 2002), in which the type specimens of several species belonging to gigartinalean genera were sequenced.

Species-level studies most commonly utilize variable plastid (*rbc*L or *rbc*L-*rbc*S spacer) and mitochondrial (*cox*2-3 spacer) markers. Recently, the mitochondrial gene *cox*1, which is used to this purpose in metazoans, has been investigated as a DNA barcoding marker and has yielded encouraging results in several red algal genera (Saunders, 2005; Robba et al., 2006). As a general rule, a single, suitable marker allows recognition of sequence clusters representing species. Fully resolving species-level phylogenies sometimes requires additional markers.

THE FUTURE OF RED ALGAL SYSTEMATICS

In the near future, molecular tools will assist red algal systematics even more than is the case today. From the technical perspective, one can expect the development of additional phylogenetic markers, the possibility of carrying out molecular analyses at very democratic prices, continuous advances in phylogenetic analysis techniques, and easier access to computer systems capable of complex analyses. As a consequence, current gaps in the red algal tree of life can be expected to be filled in relatively rapidly. It can be hoped that these efforts will be accompanied by traditional, morphological taxonomic work, as such descriptive knowledge will aid accurate identification and pave the way toward more advanced research aiming to explain the observed patterns of evolutionary diversification from genetic, physiological, and ecological perspectives. In what follows, we will sketch some perspectives for future research. However, our overview is by no means comprehensive.

MOLECULAR MARKERS

The growing amount of genomic information will facilitate further development of markers for phylogenetics and several other evolutionary questions. Currently, four plastid, three mitochondrial, and two nuclear genomes are known (Table 6.3). Furthermore, the red algal nucleus-derived sequence of the cryptomonad *Guillardia theta* D.R.A. Hill et Wetherbee nucleomorph and three large EST libraries have been determined (Table 6.3); a genome sequencing project of the first florideophyte, *Chondrus crispus* Stackhouse, has started. In our opinion, there is no urgent need for new phylogenetic markers. The existing markers allow inference of relationships at many different taxonomic levels.

The development of high-resolution molecular markers and new analytical methods allows more complex questions to be posed about the influence of dispersal on micro- and macroevolution, as red algal evolutionary studies become more hypothesis driven and ask specific questions. Such questions that will require the development of custom-tailored markers include the following: What is the precise branching order between orders X, Y, and Z? How did this or that gene family evolve in the red algae? What is the contribution of hybrid speciation to red algal diversity? Have speciation events mainly been sympatric or allopatric? How extensive is gene flow among distant populations?

TABLE 6.3
Currently Known Genomic Information*

Class	Species	Type of Information	Size	Reference
Florideophyceae	*Gracilaria tenuistipitata*	EST	3,000 seq.	P. Nyvall in GenBank
		Plastid genome	184 kb	Hagopian et al. (2004)
	Chondrus crispus	EST	4,056 seq.	Collén et al. (2006)
		Mitochondrial genome	26 kb	Leblanc et al. (1995)
Bangiophyceae	*Porphyra yezoensis*	EST	20,779 seq.	Nikaido et al. (2000)
				Asamizu et al. (2003)
	Porphyra purpurea	Plastid genome	191 kb	Reith and Munholland
				(1995)
		Mitochondrial genome	37 kb	Burger et al. (1999)
Cyanidiophyceae	*Cyanidioschyzon merolae*	Nuclear genome	16.5 Mb	Matsuzaki et al. (2004)
		Plastid genome	150 kb	Matsuzaki et al. (2004)
		Mitochondrial genome	32 kb	Matsuzaki et al. (2004)
	Galdieria sulphuraria	Nuclear genome	70% of ±11 Mb	Barbier et al. (2005)
(Cryptomonad)	*Guillardia theta*	Nucleomorph sequence	550 kb	Douglas et al. (2001)
		Plastid genome	121 kb	Douglas and Penny (1999)

*Classes as defined by Yoon et al. (2006).

What are the ecologically selective causes of speciation? Development of the hypervariable markers (e.g. single nucleotide polymorphisms (SNPs), microsatellites, nuclear introns, and spacers) needed for this kind of research will be easier with genome sequences at hand.

DEEP-LEVEL PHYLOGENETICS

Over the past 25 years, gigantic progress has been made in the delineation of red algal orders and classes. In the near future, we can expect a moderate further increase in the number of orders. The further subdivision of the Gigartinales, which remains heterogeneous, is advocated by some workers. Candidate families for recognition at the ordinal rank are Caulacanthaceae, Calosiphoniaceae, Dumontiaceae, Peyssonneliaceae, Sarcodiaceae, and Sphaerococcaceae. However, the rank at which clades are recognized is a matter of opinion, and it could be argued that a return to more inclusive orders is more practical.

Even though considerable progress has been made in establishing relationships among orders, in part thanks to molecular phylogenetics, many questions remain. In particular, the lack of confidence in nodes connecting classes and orders in molecular phylogenetic trees is troublesome (Withall and Saunders, 2006; Yoon et al., 2006). There is no silver bullet for this problem, if it is solvable at all. Confidence in phylogenetic trees depends on marker and alignment quality, appropriateness of the model used for phylogenetic reconstruction, and other factors.

The first prerequisite is that the chosen markers are suitable for resolving old divergences. Preferably, one would use DNA markers that evolve at relatively slow rates, because fast-evolving markers may show substitutional saturation at the desired taxonomic level, introducing noise into the dataset and reducing topological confidence. Of the currently used markers, 18S and 28S nuclear rDNA seem to be best suited for inference at high taxonomic levels, whereas protein-coding genes tend to deliver less resolution at deep nodes. Plastid 16S rDNA also seems to be a good candidate (Olson et al., 2005). Yoon et al. (2006) used an alignment consisting of seven plastid protein-coding genes, 16S plastid rDNA, and 18S nuclear rDNA, more than 10,000 bases in length, and were still unable to resolve fully the relationships among red algal classes, illustrating how hard and costly obtaining satisfying resolution can be. Obviously, longer alignments increase the chances of being able to resolve the branching order among a given set of taxa, but the properties of the markers in the alignment are at least as important. As more genomes become available, phylogenomic approaches toward reconstructing the red algal tree of life will gain importance (Reyes-Prieto et al., 2006).

A second issue impacting confidence in trees and their branches is alignment quality. A disappointingly small fraction of molecular phylogenetic papers specify alignment procedures, treatment of gaps and alignment ambiguities, and quality of the resulting alignment. Plastid protein-coding markers can usually be readily aligned, but this is much less the case with ribosomal DNA, especially when one tries to align sequences of highly divergent lineages, as when the focus is on ordinal relationships. Ribosomal DNA sequence alignment should ideally be based on common secondary structure of the corresponding RNA molecules (Wuyts et al., 2004). Irrespective of the marker(s) used, alignment quality and combinability of markers in multigene studies should be thoroughly examined.

The third prerequisite for obtaining highly robust phylogenetic trees is the use of appropriate methods for tree inference. Next to maximum parsimony analysis, most phylogenetic analyses are carried out in a likelihood framework, using either true likelihood approaches or Bayesian estimation. Such methods rely on models of base substitution (reviewed in Sullivan and Joyce, 2005). Hence, specification of a model appropriate for the type of data one is analysing is crucial to obtaining correct results.

In addition to those incorporated in Modeltest, specific models are available for protein-coding DNA sequences (Goldman and Yang, 1994; Shapiro et al., 2006), RNA sequences with secondary structure (Telford et al., 2005), and alignments in which rate variation across lineages is obvious (Galtier, 2001). The possibility of uncoupling models across partitions in composite alignments may also result in better model fit.

Clearly, molecular phylogenetic markers are not the only source of information that can be tapped to infer the branching order of the main red algal lineages. The emerging field of evolutionary genomics offers perspectives in this direction. In the near future, we can expect to gain information about gene order in organellar genomes and nuclear gene duplications and losses that can help to infer deep splits (green algal example: Pombert et al., 2005).

SPECIES-LEVEL SYSTEMATICS

Although considerable effort has already been made to increase our knowledge of species-level systematics using molecular data (64 studies or 43% of all published studies in the last 15 years), massive amounts of work still need to be done. As more genera are screened using molecular techniques, new species and additional cryptic and pseudo-cryptic species will be discovered. One of the crucial challenges for a stable classification lies in the reconciliation of taxa recognized in sequence alignments with traditional, morphologically defined species. Therefore, it will be necessary to integrate historic or type material into molecular systematics, as well as to continue morphological studies of living and recently collected material. Bearing this in mind, there is an urgent need for further development and perfection of ancient DNA techniques and their application to algal specimens.

The literature shows a trend toward post hoc morphological characterisation of species following their recognition using molecular data (e.g. Gurgel et al., 2003; De Clerck et al., 2005). Despite statistical analysis of morphological datasets acquired from sequenced specimens being a very powerful tool for post hoc species recognition, as exemplified by the green algal studies of Verbruggen et al. (2005a, 2005b), morphometric analysis is seldom used in concert with molecular tools in red algal systematics, with the notable exceptions of *Plocamium* (Yano et al., 2004) and *Caloglossa* (Kamiya et al., 2003). Such datasets ideally include qualitative and quantitative morphological data, and can also include ecological and physiological features that may aid species recognition (Yano et al., 2006).

Progress in techniques for databasing, querying, and evaluating DNA barcodes will facilitate data management for much of the research outlined above. DNA barcode databases will include and link to many kinds of information, including details on the morphology, geographical origin, and ecology of sequenced specimens, and provide all sorts of online tools to analyse these data (e.g. BOLD: Barcode of Life Data System; www.barcodinglife.org). Obviously, such databases and their analysis tools will be invaluable to future systematists. Hence, we are of the opinion that journals publishing integrative systematic research should investigate procedures for submission of sequence and morphological data to these digital museums of the future. Many present-day molecular markers (e.g. plastid genes, *cox*1, Rubisco, and *cox*2-3 spacers) are suitable for species-level phylogenetic inference and DNA barcoding of red algae (Verbruggen et al., unpublished). Considering that *rbc*L is the marker for which by far most sequences are available on GenBank and that it usually provides higher phylogenetic resolution at the genus to species level than spacer sequences and *cox*1, we advocate its use as the barcoding marker of choice.

SPECIATION RESEARCH

A natural next step from descriptive systematic research is trying to find out how taxa came into being. Among other things, studying speciation involves identifying and measuring reproductive isolation and looking for causes of pre- and postzygotic isolation. It encompasses analysis of geographical, ecological, and phylogenetic data, gene flow, drift, and selection in populations. For example, hybrid speciation (when two species form a hybrid that is reproductively isolated from both its parent species: Coyne and Orr, 2004) is a major mechanism of land plant diversification but has hardly been studied in red algae. Although allopatric speciation—speciation of geographically subdivided populations—has traditionally been thought to be the predominant geographic speciation mode, sympatric speciation has been documented for various twigs of the eukaryotic tree of life (e.g. Barluenga et al., 2006; Savolainen et al., 2006).

The influence of dispersal on local- and regional-scale population genetic structure is a very topical subject because of its importance for understanding speciation mechanisms. Reproductive isolation can evolve as a consequence of divergent natural selection on traits between different environments, either in sympatry or allopatry (Schluter, 2001). Confidently assessing the speciation mechanisms that have led to red algal biodiversity will demand integration of experimental systematic, genomic, and molecular cell biological research. Progress has recently been made by the discovery and characterization of "rhodobindin" gamete recognition proteins in *Aglaothamnion* (Kim and Jo, 2005). Although it is too soon to say whether these proteins actually drive speciation or diverge as a by-product of speciation, with possible further divergence through pre-mating reinforcement, discoveries such as this will hopefully spark a wide range of red algal speciation studies.

Making predictions about which red algal taxa will be used as models for speciation studies is difficult. Model taxa should ideally meet the following criteria: undemanding in laboratory culture; known life history that can be readily completed; history of molecular and interfertility testing of species boundaries; accurate information about distribution ranges; a history of genetic, cell-biological, and ecophysiological studies; scope for genetic transformation; and the existence of genomic information. We cannot think of any marine red macrophyte that currently meets all these criteria. Nonetheless, a number of taxa come to mind because they meet at least part of some of the criteria. *Aglaothamnion*, *Bostrychia*, and *Ceramium* score highly as lab rats: their rapid life histories in culture mean that they can be easily crossed and manipulated. Furthermore, rhodobindin genes have been characterized for *Aglaothamnion*. Nuclear genomic information is available for a totally different range of taxa (*Gracilaria*, *Chondrus*, and *Porphyra*), which are much more difficult to manipulate in culture.

ACKNOWLEDGMENTS

H. Verbruggen and O. De Clerck are supported by post-doctoral fellowships of the BOF (Ghent University) and the FWO-Flanders, respectively.

REFERENCES

Asamizu, E., Nakajima, M., Kitade, Y., Saga, N., Nakamura, Y., and Tabata, S. (2003) Comparison of RNA expression profiles between the two generations of *Porphyra yezoensis* (Rhodophyta), based on expressed sequence tag frequency analysis. *Journal of Phycology,* 39: 923–930.

Barbier, G., Oesterhelt, C., Larson, M.D., Halgren, R.G., Wilkerson, C., Garavito, R.M., Benning, C., and Weber, A.P.M. (2005) Comparative genomics of two closely related unicellular thermo-acidophilic red algae, *Galdieria sulphuraria* and *Cyanidioschyzon merolae*, reveals the molecular basis of the metabolic flexibility of *Galdieria sulphuraria* and significant differences in carbohydrate metabolism of both algae. *Plant Physiology,* 137: 460–474.

Barluenga, M., Stölting, K.N., Salzburger, W., Muschick, M., and Meyer, A. (2006) Sympatric speciation in Nicaraguan crater lake cichlid fish. *Nature,* 439: 719–723.

Bhattacharya, D., Elwood, H.J., Goff, L.J., and Sogin, M.L. (1990) Phylogeny of *Gracilaria lemaneiformis* (Rhodophyta) based on sequence analysis of its small subunit ribosomal RNA coding region. *Journal of Phycology,* 26: 181–186.

Bird, C.J., Rice, E.L., Murphy, C.A., and Ragan, M.A. (1992) Phylogenetic relationships in the Gracilariales (Rhodophyta) as determined by 18S rDNA sequences. *Phycologia,* 31: 510–522.

Broadwater, S.T., Scott, J.L., and West, J.A. (1991) Spermatial appendages of *Spyridia filamentosa* (Ceramiaceae, Rhodophyta). *Phycologia,* 30: 189–195.

Brodie, J.A. and Irvine, L.M. (2003) *Seaweeds of the British Isles Volume 1 Rhodophyta part 3B Bangiophycidae.* Intercept, Hampshire, UK.

Burger, G., Saint-Louis, D., Gray, M.W., and Lang, B.F. (1999) Complete sequence of the mitochondrial DNA of the red alga *Porphyra purpurea*: cyanobacterial introns and shared ancestry of red and green algae. *Plant Cell,* 11: 1675–1694.

Butterfield, N.J. (2000) *Bangiomorpha pubescens* n. gen., n. sp.: implications for the evolution of sex, multicellularity, and the Mesoproterozoic-Neoproterozoic radiation of eukaryotes. *Paleobiology,* 26: 386–404.

Cavalier-Smith, T. (1998) A revised six-kingdom system of life. *Biological Reviews,* 73: 203–266.

Collén, J., Roeder, V., Rousvoal, S., Collin, O., Kloareg, B., and Boyen, C. (2006) An expressed sequence tag analysis of thallus and regenerating protoplasts of *Chondrus crispus* (Gigartinales, Rhodophyceae). *Journal of Phycology,* 42: 104–112.

Coyne, J.A. and Orr, H.A. (2004) *Speciation.* Sinauer, Sunderland, MA.

De Clerck, O., Gavio, B., Fredericq, S., Barbara, I., and Coppejans, E. (2005) Systematics of *Grateloupia filicina* (Halymeniaceae, Rhodophyta), based on *rbc*L sequence analyses and morphological evidence, including the reinstatement of *G. minima* and the description of *G. capensis* sp nov. *Journal of Phycology,* 41: 391–410.

Destombe, C. and Douglas, S.E. (1991) Rubisco spacer sequence divergence in the rhodophyte alga *Gracilaria verrucosa* and closely related species. *Current Genetics,* 19: 395–398.

Dixon, P.S. (1973) *Biology of the Rhodophyta.* Oliver and Boyd, Edinburgh.

Douglas, S.E. and Penny, S.L. (1999) The plastid genome of the cryptophyte alga, *Guillardia theta*: Complete sequence and conserved synteny groups confirm its common ancestry with red algae. *Journal of Molecular Evolution,* 48: 236–244.

Douglas, S., Zauner, S., Fraunholz, M., Beaton, M., Penny, S., Deng, L.T., Wu, X.N., Reith, M., Cavalier-Smith, T., and Maier, U.G. (2001) The highly reduced genome of an enslaved algal nucleus. *Nature,* 410: 1091–1096.

Drew, K.M. (1954) Studies in the Bangioideae III. The life history of *Porphyra umbilicalis* (L.) Kütz. var. *laciniata* (Lightf.) J. Ag. A. The Conchocelis-phase in culture. *Annals of Botany,* 70: 184–211.

Duckett, J.G. and Peel, M.C. (1978) The role of transmission electron microscopy in elucidating the taxonomy and phylogeny of the Rhodophyta. In *Modern Approaches to the Taxonomy of Red and Brown Algae* (eds D.E.G Irvine and J.H Price), Academic Press, London, pp. 157–204.

Freshwater, D.W. and Bailey, J.C. (1998) A multigene phylogeny of the Gelidiales including nuclear large-subunit rRNA sequence data. *Journal of Applied Phycology,* 10: 229–236.

Freshwater, D.W., Fredericq, S., Butler, B.S., Hommersand, M.H., and Chase, M.W. (1994) A gene phylogeny of the red algae (Rhodophyta) based on plastid *rbc*L. *Proceedings of the National Academy of Sciences of the USA,* 91: 7281–7285.

Gabrielson, P.W., Garbary, D.J., and Scagel, R.F. (1985) The nature of the ancestral red alga: inferences from a cladistic analysis. *BioSystems,* 18: 335–346.

Galtier, N. (2001) Maximum-likelihood phylogenetic analysis under a covarion-like model. *Molecular Biology and Evolution,* 18: 866–873.

Garbary, D.J. and Gabrielson, P.W. (1990) Taxonomy and evolution. In *Biology of the Red Algae* (eds K.M. Cole and R.G. Sheath), Cambridge University Press, New York, pp. 477–498.

Garbary, D.J., Hansen, G.I., and Scagel, R.F. (1980) A revised classification of the Bangiophyceae (Rhodophyta). *Nova Hedwigia,* 33: 145–166.

Goff, L.J. and Coleman, A.W. (1988) The use of plastid DNA restriction endonuclease patterns in delineating red algal species and populations. *Journal of Phycology,* 24: 357–368.

Goff, L.J. and Coleman, A.W. (1990) DNA: microspectrofluorometric studies. In *Biology of the Red Algae* (eds. K.M. Cole and R.G. Sheath), Cambridge University Press, New York, pp. 43–72.

Goldman, N. and Yang, Z.H. (1994) Codon-based model of nucleotide substitution for protein-coding DNA sequences. *Molecular Biology and Evolution,* 11: 725–736.

Gurgel, C.F.D., Liao, L.M., Fredericq, S., and Hommersand, M.H. (2003) Systematics of *Gracilariopsis* (Gracilariales, Rhodophyta) based on *rbc*L sequence analyses and morphological evidence. *Journal of Phycology,* 39: 154–171.

Hagopian, J.C., Reis, M., Kitajima, J.P., Bhattacharya, D., and de Oliveira, M.C. (2004) Comparative analysis of the complete plastid genome sequence of the red alga *Gracilaria tenuistipitata* var. *liui* provides insights into the evolution of rhodoplasts and their relationship to other plastids. *Journal of Molecular Evolution,* 59: 464–477.

Harper, J.T. and Saunders, G.W. (2001) Molecular systematics of the Florideophyceae (Rhodophyta) using nuclear large and small subunit rDNA sequence data. *Journal of Phycology,* 37: 1073–1082.

Harper, J.T. and Saunders, G.W. (2002) A re-classification of the Acrochaetiales based on molecular and morphological data, and establishment of the Colaconematales ord. nov. (Florideophycidae, Rhodophyta). *European Journal of Phycology,* 37: 1–13.

Hoef-Emden, K., Shrestha, R.P., Lapidot, M., Weinstein, Y., Melkonian, M., and Arad, S. (2005) Actin phylogeny and intron distribution in bangiophyte red algae (Rhodoplantae). *Journal of Molecular Evolution,* 61: 360–371.

Hori, H., Lim, B.-L., and Osawa, S. (1985) Evolution of green plants as deduced from 5S rRNA sequences. *Proceedings of the National Academy of Sciences of the USA,* 82: 820–823.

Hori, H. and Osawa, S. (1987) Origin and evolution of organisms as deduced from 5S ribosomal RNA sequences. *Molecular Biology and Evolution,* 4: 445–472.

Hughey, J.R., Silva, P.C., and Hommersand, M.H. (2001) Solving taxonomic and nomenclatural problems in Pacific Gigartinaceae (Rhodophyta) using DNA from type material. *Journal of Phycology,* 37: 1091–1109.

Hughey, J.R., Silva, P.C., and Hommersand, M.H. (2002) ITS1 sequences of type specimens of *Gigartina* and *Sarcothalia* and their significance for the classification of South African Gigartinaceae (Gigartinales, Rhodophyta). *European Journal of Phycology,* 37: 209–216.

Huisman, J.M., Sherwood, A., and Abbott, I.A. (2003) Morphology, reproduction and the 18S rRNA gene sequence of *Pihiella liagoraciphila* gen. et sp. nov. (Rhodophyta), the so-called "monosporangial discs" associated with members of the Liagoraceae (Rhodophyta), and proposal of the Pihiellales ord. nov. *Journal of Phycology,* 39: 978–987.

Kamiya, M., Zuccarello, G.C., and West, J.A. (2003) Evolutionary relationships of the genus *Caloglossa* (Delesseriaceae, Rhodophyta) inferred from large-subunit ribosomal RNA gene sequences, morphological evidence and reproductive compatibility, with description of a new species from Guatemala. *Phycologia,* 42: 478–497.

Kawaguchi, S., Wang, H.W., Horiguchi, T., Sartoni, G., and Masuda, M. (2001) A comparative study of the red alga *Grateloupia filicina* (Halymeniaceae) from the northwestern Pacific and Mediterranean with the description of *Grateloupia asiatica*, sp. nov. *Journal of Phycology,* 37: 433–442.

Kim, G.H. and Jo, B.H. (2005) Cloning and characterization of a cDNA encoding a sex-specific lectin, rhodobindin, from *Aglaothamnion oosumiense* (Rhodophyta). Abstracts of papers at the Eighth International Phycological Congress, Durban, South Africa.

Kim, G.H., Lee, I.K., and Fritz, L. (1996) Cell-cell recognition during fertilization in a red alga, *Antithamnion sparsum* (Ceramiaceae, Rhodophyta). *Plant and Cell Physiology,* 37: 621–628.

Kostrzewa, M., Valentin, K., Maid, U., Radetzky, R., and Zetsche, K. (1990) Structure of the rubisco operon from the multicellular red alga *Antithamnion* spec. *Current Genetics,* 18: 465–469.

Kylin, H. (1956) *Die Gattungen der Rhodophyceen.* C.W.K. Gleerups Förlag, Lund.

Leblanc, C., Boyen, C., Richard, O., Bonnard, G., Grienenberger, J.M., and Kloareg, B. (1995) Complete sequence of the mitochondrial DNA of the rhodophyte *Chondrus crispus* (Gigartinales): gene content and genome organization. *Journal of Molecular Biology,* 250: 484–495.

Lee, S.R., Oak, J.H., Suh, Y., and Lee, I.K. (2001) Phylogenetic utility of rbcS sequences: an example from *Antithamnion* and related genera (Ceramiaceae, Rhodophyta). *Journal of Phycology,* 37: 1083–1090.

Le Gall, L. and Saunders, G.W. (2007) A nuclear phylogeny of the Florideophyceae (Rhodophyta) inferred from combined EF2, small subunit and large subunit ribosomal DNA: Establishing the new red algal subclass Corallinophycideae. *Molecular Phylogenetics and Evolution,* 43: 1118–1130.

Lim, B.-L., Kawai, H., Hori, H., and Osawa, S. (1986) Molecular evolution of 5S ribosomal RNA from red and brown algae. *Japanese Journal of Genetics,* 61: 169–176.

Maggs, C.A., Douglas, S.E., Fenety, J., and Bird, C.J. (1992) A molecular and morphological analysis of the *Gymnogongrus devoniensis* (Rhodophyta) complex in the North Atlantic. *Journal of Phycology,* 28: 214–232.

Maggs, C.A. and Pueschel, C.M. (1989) Morphology and development of *Ahnfeltia plicata* (Rhodophyta): proposal of Ahnfeltiales ord. nov. *Journal of Phycology,* 25: 333–351.

Magne, F. (1989) Classification et phylogénie des Rhodophycées. *Cryptogamie Algologie,* 10: 101–115.

Matsuzaki, M., Misumi, O., Shin-I, T., Maruyama, S., Takahara, M., Miyagishima, S.Y., Mori, T., Nishida, K., Yagisawa, F., Nishida, K., Yoshida, Y., Nishimura, Y., Nakao, S., Kobayashi, T., Momoyama, Y., Higashiyama, T., Minoda, A., Sano, M., Nomoto, H., Oishi, K., Hayashi, H., Ohta, F., Nishizaka, S., Haga, S., Miura, S., Morishita, T., Kabeya, Y., Terasawa, K., Suzuki, Y., Ishii, Y., Asakawa, S., Takano, H., Ohta, N., Kuroiwa, H., Tanaka, K., Shimizu, N., Sugano, S., Sato, N., Nozaki, H., Ogasawara, N., Kohara, Y., and Kuroiwa, T. (2004) Genome sequence of the ultrasmall unicellular red alga *Cyanidioschyzon merolae* 10D. *Nature,* 428: 653–657.

Mitman, G.G. and van der Meer, J.P. (1994) Meiosis, blade development and sex determination in *Porphyra purpurea* (Rhodophyta). *Journal of Phycology,* 30: 147–159.

Müller, K.M., Oliveira, M.C., Sheath, R.G., and Bhattacharya, D. (2001) Ribosomal DNA phylogeny of the Bangiophycidae (Rhodophyta) and the origin of secondary plastids. *American Journal of Botany,* 88: 1390–1400.

Müller, K.M., Sheath, R.G., Sherwood, A.R., Pueschel, C.M., and Gutell, R.R. (2002) A proposal for a new red algal order, the Thoreales. *Journal of Phycology,* 38: 807–820.

Murray, S.N. and Dixon, P.S. (1992) Rhodophyta: some aspects of their biology. *Oceanography and Marine Biology: An Annual Review,* 30: 1–148.

Newmaster, S.G., Fazekas, A.J., and Ragupathy, S. (2006) DNA barcoding in land plants: evaluation of *rbc*L in a multigene tiered approach. *Canadian Journal of Botany,* 84: 335–341.

Nikaido, I., Asamizu, E., Nakajima, M., Nakamura, Y., Saga, N., and Tabata, S. (2000) Generation of 10,154 expressed sequence tags from a leafy gametophyte of a marine red alga, *Porphyra yezoensis. DNA Research,* 7: 223–227.

Oliveira, M.C. and Bhattacharya, D. (2000) Phylogeny of the Bangiophycidae (Rhodophyta) and the secondary endosymbiotic origin of algal plastids. *American Journal of Botany,* 87: 482–492.

Olson, K.N., Melton, R.S., Yaudes, K.M., Norwood, K.G., and Freshwater, D.W. (2005) Characteristics and utility of plastid-encoded 16S rRNA gene sequence data in phylogenetic studies of red algae. *Journal of the North Carolina Academy of Science,* 120: 143–151.

Oltmanns, F. 1898. Zur entwicklungsgeschichte der Florideen. *Botanische Zeitung,* 56: 99–140.

Papenfuss, G.F. (1966) A review of the present system of classification of the Florideophycidae. *Phycologia,* 5: 247–255.

Parsons, T.J., Maggs, C.A., and Douglas, S.E. (1990) Plastid DNA restriction analysis links the heteromorphic phases of an apomictic red algal life history. *Journal of Phycology,* 26: 495–500.

Pombert, J.-F., Otis, C., Lemieux, C., and Turmel, M. (2005) The chloroplast genome sequence of the green alga *Pseudendoclonium akinetum* (Ulvophyceae) reveals unusual structural features and new insights into the branching order of chlorophyte lineages. *Molecular Biology and Evolution,* 22: 1903–1918.

Provan, J., Murphy, S., and Maggs, C.A. (2004) Universal plastid primers for Chlorophyta and Rhodophyta. *European Journal of Phycology,* 39: 43–50.

Pueschel, C.M. (1990) Cell structure. In *Biology of the Red Algae* (eds K.M. Cole and R.G Sheath), Cambridge University Press, New York, pp. 7–42.

Pueschel, C.M. and Cole, K.M. (1982) Rhodophycean pit plugs: an ultrastructural survey with taxonomic implications. *American Journal of Botany,* 69: 703–720.

Ragan, M.A., Bird, C.J., Rice, E.L., Gutell, R.R., Murphy, C.A., and Singh, R.K. (1994) A molecular phylogeny of the marine red algae (Rhodophyta) based on nuclear small-subunit rRNA gene. *Proceedings of the National Academy of Sciences of the USA*, 91: 7276–7280.

Ragan, M.A. and Gutell, R.R. (1995) Are red algae plants? *Botanical Journal of the Linnean Society*, 118: 81–105.

Reith, M. and Munholland, J. (1995) Complete nucleotide sequence of the *Porphyra purpurea* chloroplast genome. *Plant Molecular Biology Reporter*, 13: 333–335.

Reyes-Prieto, A., Yoon, H.S., and Bhattacharya, D. (2006) Phylogenomics and its growing impact on algal phylogeny and evolution. *Algae*, 21: 1–10.

Robba, L., Russell, S.J., Barker, G.L., and Brodie, J. (2006) Assessing the use of the mitochondrial *cox*1 marker for use in DNA barcoding of red algae (Rhodophyta). *American Journal of Botany*, 93 (in press).

Rueness, J. (2005) Life history and molecular sequences of *Gracilaria vermiculophylla* (Gracilariales, Rhodophyta), a new introduction to European waters. *Phycologia*, 44: 120–128.

Sáez, A.G. and Lozano, E. (2005) Body doubles. *Nature*, 433: 111.

Saunders, G.W. (2005) Applying DNA barcoding to red macroalgae: a preliminary appraisal holds promise for future applications. *Philosophical Transactions of the Royal Society of London. Biological Sciences*, 360: 1879–1888.

Saunders, G.W. and Hommersand, M.H. (2004) Assessing red algal supraordinal diversity and taxonomy in the context of contemporary systematic data. *American Journal of Botany*, 91: 1494–1507.

Saunders, G.W. and Kraft, G.T. (1997) A molecular perspective on red algal evolution: focus on the Florideophycidae. *Plant Systematics and Evolution*, 11 (Supplement): 115–138.

Saunders, G.W. and Lehmkuhl, K.V. (2005) Molecular divergence and morphological diversity among four cryptic species of *Plocamium* (Plocamiales, Florideophyceae) in northern Europe. *European Journal of Phycology*, 40: 293–312.

Savolainen, V., Anstett, M.C., Lexer, C., Hutton, I., Clarkson, J.J., Norup, M.V., Powell, M.P., Springate, D., Salamin, N., and Baker, W.J. (2006) Sympatric speciation in palms on an oceanic island. *Nature*, 441: 210–213.

Schluter, D. (2001) Ecology and the origin of species. *Trends in Ecology and Evolution*, 16: 372–380.

Schmitz, F. (1889) Systematische übersicht der bisher bekannten gattungen der Florideen. *Flora*, 72: 435–456.

Searles, R.B. (1980) The strategy of the red algal life history. *American Naturalist*, 115: 113–120.

Seo, K.S., Cho, T.O., Park, J.S., Yang, E.C., Yoon, H.S., and Sung, M.B. (2003) Morphology, basiphyte range, and plastid DNA phylogeny of *Campylaephora borealis* stat. nov (Ceramiaceae, Rhodophyta). *Taxon*, 52: 9–19.

Shapiro, B., Rambaut, A., and Drummond, A.J. (2006) Choosing appropriate substitution models for the phylogenetic analysis of protein-coding sequences. *Molecular Biology and Evolution*, 23: 7–9.

Sheath, R.G. and Müller, K.M. (1999) Systematic status and phylogenetic relationships of the freshwater genus *Balbiania* (Rhodophyta). *Journal of Phycology*, 35: 855–864.

Sherwood, A.R. and Presting, G.G. (2007) Universal primers amplify a 23s rDNA plastid marker in eukaryotic algae and cyanobacteria. *Journal of Phycology*, 43: 605–608.

Steane, D.A., McClure, B.A., Clarke, A.E., and Kraft, G.T. (1991) Amplification of the polymorphic 5.8S ribosomal RNA gene from selected Australian gigartinalean species (Rhodophyta) by polymerase chain-reaction. *Journal of Phycology*, 27: 758–762.

Sullivan, J. and Joyce, P. (2005) Model selection in phylogenetics. *Annual Review of Ecology Evolution and Systematics*, 36: 445–466.

Telford, M.J., Wise, M.J., and Gowri-Shankar, V. (2005) Consideration of RNA secondary structure significantly improves likelihood-based estimates of phylogeny: examples from the bilateria. *Molecular Biology and Evolution*, 22: 1129–1136.

Trick, H.N. and Pueschel, C.M. (1991) Cytochemical evidence for homology of the outer cap layer of red algal pit plugs. *Phycologia*, 30: 196–204.

van den Hoek, C., Mann, D.G., and Jahns, H.M. (1995) *Algae: An Introduction to Phycology*. Cambridge University Press, Cambridge.

Verbruggen, H., De Clerck, O., Cocquyt, E., Kooistra, W.H.C.F., and Coppejans, E. (2005a) Morphometric taxonomy of siphonous green algae: a methodological study within the genus *Halimeda* (Bryopsidales). *Journal of Phycology*, 41: 126–139.

Verbruggen, H., De Clerck, O., Kooistra, W.H.C.F., and Coppejans, E. (2005b) Molecular and morphometric data pinpoint species boundaries in *Halimeda* section *Rhipsalis* (Bryopsidales, Chlorophyta). *Journal of Phycology*, 41: 606–621.

Verlaque, M., Brannock, P.M., Komatsu, T., Villalard-Bohnsack, M., and Marston, M. (2005) The genus *Grateloupia* C. Agardh (Halymeniaceae, Rhodophyta) in the Thau Lagoon (France, Mediterranean): a case study of marine plurispecific introductions. *Phycologia*, 44: 477–496.

Wilkes, R.J., McIvor, L.M., and Guiry, M.D. (2005) Using *rbc*L sequence data to reassess the taxonomic position of some *Grateloupia* and *Dermocorynus* species (Halymeniaceae, Rhodophyta) from the north-eastern Atlantic. *European Journal of Phycology*, 40: 53–60.

Withall, R.D. and Saunders, G.W. (2006) Combining small and large subunit ribosomal DNA genes to resolve relationships among orders of the Rhodymeniophycidae (Rhodophyta): recognition of the Acrosymphytales ord. nov. and Sebdeniales ord. nov. *European Journal of Phycology*, 41: 379–394.

Wuyts, J., Perriere, G., and van de Peer, Y. (2004) The European ribosomal RNA database. *Nucleic Acids Research*, 32 (Special Issue): D101–D103.

Xiao, S., Zhang, Y., and Knoll, A. (1998) Three-dimensional preservation of algae and animal embryos in a Neoproterozoic phosphorite. *Nature*, 391: 553–558.

Yang, E.C. and Boo, S.M. (2004) Evidence for two independent lineages of *Griffithsia* (Ceramiaceae, Rhodophyta) based on plastid protein-coding *psa*A, *psb*A, and *rbc*L gene sequences. *Molecular Phylogenetics and Evolution*, 31: 680–688.

Yano, T., Kamiya, M., Arai, S., and Kawai, H. (2004) Morphological homoplasy in Japanese *Plocamium* species (Plocamiales, Rhodophyta) inferred from the Rubisco spacer sequence and intracellular acidity. *Phycologia*, 43: 383–393.

Yano, T., Kamiya, M., Murakami, A., Sasaki, H., and Kawai, H. (2006) Biochemical phenotypes corresponding to molecular phylogeny of the red algae *Plocamium* (Plocamiales, Rhodophyta): implications of incongruence with the conventional taxonomy. *Journal of Phycology*, 42: 155–169.

Yoon, H.S., Hackett, J.D., and Bhattacharya, D. (2002) A single origin of the peridinin- and fucoxanthin-containing plastids in dinoflagellates through tertiary endosymbiosis. *Proceedings of the National Academy of Sciences of the USA*, 99: 11724–11729.

Yoon, H.S., Hackett, J.D., Ciniglia, C., Pinto, G., and Bhattacharya, D. (2004) A molecular timeline for the origin of photosynthetic eukaryotes. *Molecular Biology and Evolution*, 21: 809–818.

Yoon, H.S., Müller, K.M., Sheath, R.G., Ott, F.D., and Bhattacharya, D. (2006) Defining the major lineages of red algae (Rhodophyta). *Journal of Phycology*, 42: 482–492.

Zuccarello, G.C., Burger, G., West, J.A., and King, R.J. (1999) A mitochondrial marker for red algal intraspecific relationships. *Molecular Ecology*, 8:1443–1447.

Zuccarello, G.C. and West, J.A. (2003) Multiple cryptic species: molecular diversity and reproductive isolation in the *Bostrychia radicans/B. moritziana* complex (Rhodomelaceae, Rhodophyta) with focus on North American isolates. *Journal of Phycology*, 39: 948–959.

Zuccarello, G., West, J., Bitans, A., and Kraft, G.T. (2000) Molecular phylogeny of *Rhodochaete parvula* (Bangiophycidae, Rhodophyta). *Phycologia*, 39: 75–81.

7 Systematics of the green algae: conflict of classic and modern approaches

Thomas Pröschold and Frederik Leliaert

CONTENTS

ABSTRACT

Traditionally the green algae were classified in orders or classes according to the morphological species concept. For example, monadoid species (flagellates) were summarized in the order Volvocales, coccoids in the Chlorococcales, filaments in the Ulotrichales or Chaetophorales, and siphonocladous algae in the Cladophorales or Siphonocladales. Later, a new classification was proposed based on ultrastructural investigations of the basal bodies in the flagellar apparatus and cell division. The species with basal bodies in clockwise (CW) or directly opposite (DO) orientation were classified in the class Chlorophyceae, the counterclockwise (CCW)-orientated species in the Ulvophyceae and Trebouxiophyceae (= Pleurastrophyceae). Phylogenetic analyses of nuclear-encoded SSU and ITS rDNA sequences have basically confirmed the classification based on ultra-structural characters. However, most genera and orders are polyphyletic and the relationships between many of the phylogenetic lineages remain unclear. Traditionally taxonomic approaches often depend on single or even negetative "characters" (e.g. absence of zoospore formation or pyrenoids). The authors feel that in some cases these may be given excessive "weight" and advocate the usage of polyphasic approaches (e.g. secondary structures of SSU and ITS rDNA sequences, results of crossing experiments, sporangium autolysin data, and studies of life cycles, multigene approaches, amplified fragment length polymorphism [AFLP]) for the classification of green algae. New generic and species concepts (Z- and CBC-clade concepts, biological species concept, phylogenetic concepts) can be designed for many orders and most of the classes in the Viridiplantae on the basis of this approach.

INTRODUCTION

The green algae are photosynthetic eukaryotes characterized by the presence of chloroplasts with two envelope membranes, stacked thylakoids and chlorophyll *a* and *b*. (Few genera like *Prototheca, Polytoma, Polytomella,* and *Hyalogonium* are colourless, but the cells contain leucoplasts, which secondarily lost their pigments; Pringsheim, 1963.) All green algae produce starch as the main reserve polysaccharide, which is deposited inside the plastids. The plastids arrived in a single primary endocytobiosis (endosymbiosis) event, where a cyanobacterium was taken up by a colourless eukaryote host (see reviews: Delwiche and Palmer, 1997, Delwiche, 1999, Keeling, 2004; see also chapters 2 and 3 this volume). The green algae are one of the most diverse groups of eukaryotes, showing morphological forms ranging from flagellated unicells, coccoids, branched or unbranched filaments, to multinucleated macrophytes and taxa with parenchymatic tissues (Figure 7.1). They are distributed worldwide and can be found in almost every habitat from Arctic and Antarctic regions to oceans and freshwater lakes as well as in soil from temperate and arid areas. Green algae are also found in different symbioses including lichens, protozoa, and foraminifers, or as parasites on tropical plants. There are estimated to be at least 600 genera with 10,000 species within the green algae (Norton et al., 1996). Estimations of age vary from 600 My ago (Tappan, 1980, van den Hoek et al., 1988) to 1,500 Ma (Yoon et al., 2004).

Based on the structure and pigment composition of their plastids as mentioned above, green algae and land plants are closely related. This hypothesis was put forward a long time before

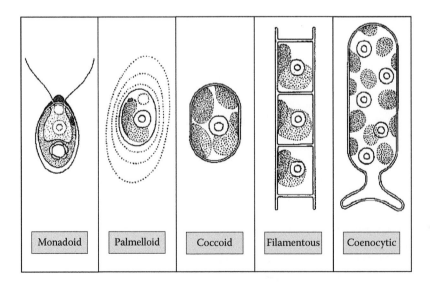

| Monadoid | Palmelloid | Coccoid | Filamentous | Coenocytic |

FIGURE 7.1 Different morphological organization in green algae (modified after Ettl, H. and Gärtner, G., *Syllabus der Boden-, Luft- und Flechtenalgen.* Gustav Fischer, Stuttgart, 1995). Parenchymatous and siphono-cladous organization are not illustrated.

molecular and ultrastructural data were available. The present contribution will focus on green algae in their narrow sense (Chlorophyta *sensu* Bremer, 1985; see also Bremer et al., 1987). The systematics of the Streptophyta (*sensu* Bremer or Charophyta *sensu* Cavalier-Smith, 1993) is discussed in detail in chapter 8 in this volume.

How are green algae classified?

The classification of green algae is a topic of many publications, especially in the context of the origin of land plants, and is still controversial. We will not repeat all these discussions here (see citations below). More than 20 classes have been described during the 250 years since the introduction of classification by Linnaeus (Silva, 1980), some of them partly contradict each other. Depending upon which characters (morphological, ultrastructural, molecular) were used for classification, the same species can belong to different classes. All classes, if correctly described, are valid according to the International Code for Botanical Nomenclature (Greuter et al., 2000). The priority rule is only valid at the generic and species level. For example, *Ulothrix zonata* Kützing and *Uronema belkae* (Mattox et Bold) Lokhorst, both belong to the Ulotrichophyceae according to their morphology (unbranched filaments), but to two different classes (*Ulothrix zonata*—Ulvophyceae; *Uronema belkae*—Chlorophyceae) according to the ultrastructural orientation of the basal body in the flagellar apparatus in their zoospores. To date a wide range of criteria and approaches have been used for classification at higher levels, which may be broadly considered under the following three different concepts discussed below.

The morphological concept

Traditionally, the green algae were classified according to the morphological species concept based on the organization level of the vegetative state as shown in Figure 7.1. The rationale of this classification was that unicellular flagellates are primitive within the green algae, and that they evolved initially into coccoid and sarcinoid chlorophytes, and later into colonial, filamentous, coenocytic, and siphonous forms. Blackman (1900), and Blackman and Tansley (1902) proposed the first subdivision into classes and orders, which was refined over almost a century by several other authors (Pascher, 1931, Fritsch, 1935, Fott, 1971, see also Round, 1963, 1984, Bold and Wynne, 1985). In the 1960s and 1970s, a modified classification of the green algae was undertaken using experimental studies of life cycles and architecture ("Bauplan") of flagellated cells (for details see Christensen, 1962; Round, 1971; Kornmann, 1973; van den Hoek and Jahns, 1978; Ettl, 1981, 1983; Ettl and Komárek, 1982). These authors proposed seven classes: Prasinophyceae, Chlamydophyceae, Chlorophyceae, Codiolophyceae, Oedogoniophyceae, Bryopsidophyceae, and Zygnematophyceae.

The ultrastructural concept

Mattox and Stewart (1984) proposed a new classification based on the ultrastructure of the basal body in flagellated cells (Figure 7.2) and cytokinesis during the mitosis (Figure 7.3). The underlying principle of their classification scheme was that the first radiation of the green algae took place at the flagellate level, resulting in a multitude of ancient lineages of flagellates, some of which then went on to give rise to non-flagellate coccoid, sarcinoid, filamentous, siphonocladous, and siphonous representatives. Five classes were proposed: The Micromonadophyceae (uni-quadriflagellates) include all the prasinophytes except the Tetraselmidales, which they transferred to the Pleurastrophyceae. This class was admittedly paraphyletic, exhibiting primitive characteristics for many of the features used to define the other classes of green algae. Likewise, the class Prasinophyceae is defined by the presence of a primitive trait (i.e., the possession of organic body scales and flagellar scales) and by the absence of more advanced traits. These primitive, ancestral traits cannot be used to characterize particular groups within the Chlorophyta, since they are common to green algae as a whole. The Pleurastrophyceae and Ulvophyceae are characterized by a counterclockwise

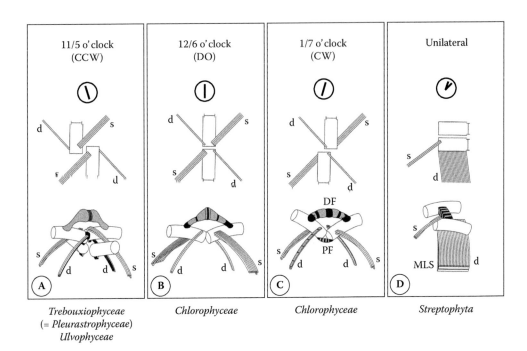

FIGURE 7.2 Different types of flagellar apparatus found among the green algae, viewed from the top (upper figure) and from the side (lower figure) (modified after Graham, L.E. and Wilcox, L.W., *Algae*. Prentice Hall, Upper Saddle River, NJ, 2000). The apparatuses generally include two or four basal bodies (shown here as rectangles or cylinders), microtubular roots (s or d), and distal (DF) or proximal (PF) connecting fibers. (A) Flagellar apparatus with cruciate roots and basal bodies displaced in counterclockwise (CCW) direction (Trebouxiophyceae and Ulvophyceae). (B) Flagellar apparatus with cruciated roots showing directly opposed (DO) displacement flagellar basal bodies (Chlorophyceae). (C) Flagellar apparatus with clockwise (CW) displaced flagellar basal bodies (Chlorophyceae). (D) Flagellar apparatus of the Streptophyta with asymmetrical (unilateral) distribution of the flagellar roots, showing the characteristic multilayered structure (MLS).

orientation of the basal body (CCW; Figure 7.2A); the Chlorophyceae are characterized by a directly opposite (DO; Figure 7.2B) or clockwise (CW; Figure 7.2C) orientation. The architecture of the flagellar apparatus in flagellated cells provides a very important taxonomic and phylogenetic character and is reviewed in detail by Melkonian (1982) and Melkonian and Surek (1995), and for the Ulvophyceae by Sluiman (1989). The Zygnematophyceae were transferred as an order to the Charophyceae based on biochemical features and development during the cell division (see Figure 7.3G through Figure 7.3H; for more details see van den Hoek et al., 1995). In addition to the phragmoplast, the Charophyceae *sensu* Mattox and Stewart are characterized by the unilateral root of their flagellar apparatus (Figure 7.2D). Most of the other classes earlier proposed by Ettl, Kornmann, and others (citations see above) were rejected and incorporated into the five classes *sensu* Mattox and Stewart.

Van den Hoek et al. (1988), who argued that the ultrastructural characters of the flagellated cell were insufficient, on their own, to characterize classes within the Chlorophyta, refined this five-class system by also taking into account the structure of the vegetative cells, mitosis and cell division, the composition of the cell walls, and the life history. These authors established seven classes (including the streptophycean algae), which combined the morphological and ultrastructural concepts: Prasinophyceae, Chlamydophyceae, Ulvophyceae, Chlorophyceae, Zygnematophyceae, Trentepohliophyceae, and Charophyceae.

The "Ulvophyceae" *sensu lato* were later subdivided into five new classes (Ulvophyceae *sensu stricto*, Cladophorophyceae, Bryopsidophyceae, Dasycladophyceae, and Trentepohliophyceae)

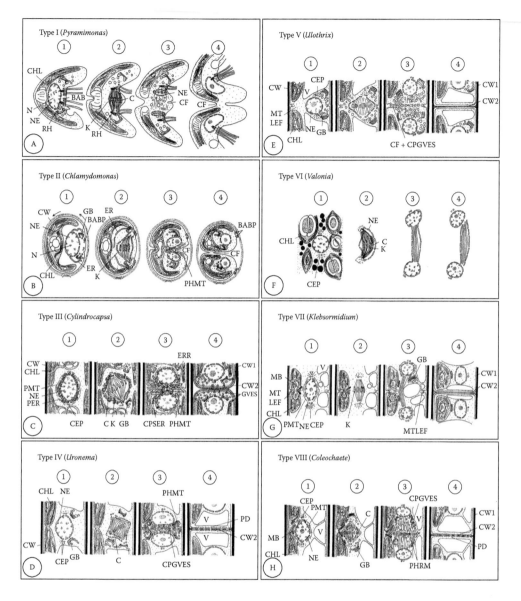

FIGURE 7.3 Different mitosis-cytokinesis types occuring among the green algae (modified after van den Hoek et al., 1988). 1. Early prophase. 2. Metaphase. 3. Late telophase. 4. Early interphase. BAB = basal body; BABP = pair of basal bodies; C = chromosome; CEP = pair of centrioles; CF = cleavage furrow; CHL = chloroplast; CPGVES = cell plate of Golgi-derived vesicles; CPSER = cell plate of smooth ER-vesicles; CW = cell wall; CW1 = old cell wall; CW2 = young cell wall; ER = endoplasmic reticulum; EER = rough endoplasmic reticulum; GB = Golgi body; GVES = Golgi-derived vesicles; K = kinetochore; MB = microbody; MTLEF = microtubules along leading edge of cleavage furrow of plasma membrane; N = nucleus; NE = nuclear envelope; PD = plasmodesma; PER = perinuclear endoplasmic reticulum; PHMT = phycoplast microtubule; PHRT = phragmoplast microtubule; PMT = perinuclear microtubules; RH = rhizoplast; V = vacuole. (A) Open mitosis with a persistent telophase spindle; cytokinesis by cleavage furrow (*Pyramimonas*). (B) Closed mitosis with a non-persistent telophase spindle; cytokinesis by a cleavage furrow in a phycoplast (*Chlamydomonas*). (C) Closed mitosis with a non-persistent spindle; cytokinesis by a cell plate of smooth ER-vesicles in a phycoplast (*Cylindrocapsa*). (D) Closed mitosis with a non-persistent telophase spindle; cytokinesis by a cell plate of Golgi-derived vesicles in a phycoplast (*Uronema*). (E) Closed mitosis with a persistent telophase spindle; cytokinesis by a cleavage furrow to which Golgi-derived vesicles are added (*Ulothrix*). (F) Closed mitosis with a prominent persistent telophase spindle, causing a typical dumbbell shape; cytokinesis does not immediately follow mitosis (*Valonia*). (G) Open mitosis with a prominent persistent telophase spindle; cytokinesis by cleavage furrow (*Klebsormidium*). (H) Open mitosis with a prominent persistent telophase spindle; cytokinesis by a cell plate of Golgi-derived vesicles in a phragmoplast (*Coleochaete*).

based on apparent differences in thallus architecture, cellular organisation, chloroplast morphology, cell wall composition, and life histories (van den Hoek et al., 1993, 1995). These classes were not validly described according to the rules of ICBN (no Latin diagnosis; not type genus designated; see Table 7.1).

TABLE 7.1
Classification of Green Algae According to the Different Concepts Using Morphological, Ultrastructural, and Molecular Approaches*

Class Author	Monophyletic clade supported in Figure 7.4 and Figure 7.5	Validly described according to ICBN	Remarks (see additional comments in Silva, 1980)
MORPHOLOGICAL CONCEPT			This concept used the morphology of the vegetative state of the organisms for classification.
Prasinophyceae Moestrup et Throndsen	–	+	Originally circumscribed and designated as a class by Christensen (1962) and was validated by Moestrup and Throndsen (1988); this class contains all flagellates with organic scales on the surface of cells, including the flagella; phylogenetically paraphyletic, containing several classes.
Chlamydophyceae Ettl	(+)	+	This class contains all monadoid and coccoid green algae with *Chlamydomonas*-like cells or zoospores, rejected by Mattox and Stewart (1984) and Deason (1984), because the Dunaliellales were not included and the subdivision in four orders is artificial, which was already mentioned by Ettl (1981); corresponds with CW-group (see Figure 7.5), needs emendations at the ordinal level.
Codiolophyceae Kornmann	+	–	This class contains all algae that form a "*Codiolum*"-stage in their life cycle; rejected by Mattox and Stewart (1984), because Ulvales are not included; invalidly described according to ICBN (no Latin diagnosis); if the Ulvales are included, monophyletic group.
Chlorophyceae Wille ex Warming	–	+	This class originally includes all green algae. Later, subsections were separated and described as other classes. Mattox and Stewart (1984) defined this class and included in it only species with CW or DO orientation (see Figure 7.2) of the basal body in their flagellated cells. *Chlorococcum* was chosen as the type genus by Christensen (1994); no solid molecular support for recognition of the monophyletic nature of the class.
Oedogoniophyceae Round	+	–	This class contains the three filamentous genera, *Oedogonium*, *Oedocladium*, and *Bulbochaete* because of their special life cycle; included as order in the Chlorophyceae by Mattox and Stewart (1984), invalidly described (no Latin diagnosis), monophyletic group.
Bryopsidophyceae Bessey	n.d.	+	In this class the two orders Bryopsidales and Halimedales, are summarized which were included in the Ulvophyceae *sensu* Mattox and Stewart (1984); no SSU rDNA sequences available.
Zygnematophyceae Round	+	–	This class summarized all conjugating algae, which were previously described as Conjugatophyceae by Engler (1892). Because *Conjugata* (= *Spirogyra*) and *Zygnema* are not synonyms, Zygnematophyceae is not a legitimate substitute for the Conjugatophyceae (see Silva, 1980); invalidly described (no Latin diagnosis), monophyletic group.

TABLE 7.1 (CONTINUED)

Classification of Green Algae According to the Different Concepts Using Morphological, Ultrastructural, and Molecular Approaches*

Class Author	Monophyletic clade supported in Figure 7.4 and Figure 7.5	Validly described according to ICBN	Remarks (see additional comments in Silva, 1980)
ULTRASTRUCTURAL CONCEPT (*sensu* **Mattox and Stewart 1984**)			This concept used only the basal body orientation of the flagellar apparatus and cytokinesis (mitosis) for classification.
Micromonadophyceae Mattox et Stewart	–	–	This class contains all prasinophytes as described above except the order Tetraselmidiales, which Mattox and Stewart (1984) transferred to the Pleurastrophyceae; invalidly described (no type genus designated), paraphyletic group.
Pleurastrophyceae Mattox et Stewart	–	–	This class contains algae with CCW basal body orientation (see Figure 7.2). Mattox and Stewart (1984) also included *Tetraselmis* in this class; invalidly described (no type genus designated), the nominal type genus *Pleurastrum* and their type species *P. insigne* belong to the *Stephanosphaera*-clade of the CW group (Friedl, 1996; see Figure 7.5), no solid molecular support for recognition of the monophyletic nature of the class.
Ulvophyceae Mattox et Stewart	–	–	This class also contains algae with CCW basal body orientation (see Figure 7.2). Besides the orders Ulvales and Ulotrichales, Mattox and Stewart (1984) also included in this class the orders Bryopsidales, Dasycladales, Trentepohliales, and Cladophorales/Siphonocladales; invalidly described (no type genus designated), the nominal type genus *Ulva* and their type species *U. lactuca* belong to the *Ulva*-clade (see Figure 7.4B), no solid molecular support for recognition of the monophyletic nature of the class.
Chlorophyceae Wille ex Warming	–	+	See above.
Charophyceae Rabenhorst	+	+	This class contains the order Charales, which is characterized by special differentiation in nodial and internodial cells, a complex life cycle, and a unilateral basal body orientation (see Figure 7.2) and phragmoplast (Figure 7.3). Originally described by Rabenhorst (1863), a defined characterization is given by Mattox and Stewart (1984) with Latin diagnosis, but without designation of the type genus; emendations necessary, monophyletic group.
ULTRASTRUCTURAL CONCEPT (*sensu* **van den Hoek et al., 1988, 1995**)			In addition to the concept presented by Mattox and Stewart (1984), the morphology, life cycle, and biochemical characters were used for this concept.
Prasinophyceae Moestrup et Throndsen	–	+	See above.
Chlamydophyceae Ettl	(+)	+	See above.
Ulvophyceae *s.str.* van den Hoek, Mann et Jahns	+	+	This class contains all species belonging to the Codiolophyceae (see above) *sensu* Kornmann (1973) including the Ulvales; invalidly described (no Latin diagnosis and no type genus designated), monophyletic group

(Continued)

TABLE 7.1 (CONTINUED)
Classification of Green Algae According to the Different Concepts Using Morphological, Ultrastructural, and Molecular Approaches*

Class Author	Monophyletic clade supported in Figure 7.4 and Figure 7.5	Validly described according to ICBN	Remarks (see additional comments in Silva, 1980)
Bryopsidophyceae Bessey	n.d.	+	See above.
Cladophorophyceae van den Hoek, Mann et Jahns	+	−	In this class are summarized the two orders Cladophorales and Siphonocladales, which were included to the Ulvophyceae *sensu* Mattox and Stewart (1984); invalidly described (no Latin diagnosis); monophyletic group.
Dasycladophyceae van den Hoek, Mann et Jahns	+	−	This class contains the order Dasycladales, which Mattox and Stewart (1984) included to the Ulvophyceae; invalidly described (no Latin diagnosis); monophyletic group.
Chlorophyceae Wille ex Warming	−	+	See above.
Zygnematophyceae Round	+	−	See above.
Trentepohliophyceae van den Hoek, Stam et Olsen	+	−	This class contains order Trentepohliales with the genera *Trentepohlia, Cephaleuros, Phycopeltis,* and *Stomatochroon,* which Mattox and Stewart (1984) included to the Ulvophyceae; invalidly described (no Latin diagnosis); monophyletic group.
Charophyceae Rabenhorst	+	+	See above.
MOLECULAR CONCEPT (*sensu* **Lewis and McCourt 2004**)			This classification is based on the molecular phylogeny of SSU rDNA or *rbc*L sequences.
Chlorophyta	+	+	This division contains all algae, their cells divide via phycoplast during cytokinesis (see Figure 7.3); solid molecular support for recognition of the monophyletic nature of the division.
Chlorophyceae Wille ex Warming	−	+	See above.
Trebouxiophyceae Friedl	−	+	This contains all taxa belonging to the class Pleurastrophyceae *sensu* Mattox and Stewart (1984) excluding the Tetraselmidales (see above) and a group of autosporic coccoid green algae; a new name was necessary, because the genus *Pleurastrum* belongs to the Chlorophyceae (Friedl, 1996); no solid molecular support for recognition of the monophyletic nature of the class.
Ulvophyceae Mattox et Stewart	−	−	See above.
Prasinophyceae Moestrup et Throndsen	−	+	See above.
Streptophyta = Charophyta	+	+	This division contains all algae, bryophytes, ferns, and higher plants, their cells divide via phragmoplast during cytokinesis (see Figure 7.3); solid molecular support for recognition of the monophyletic nature of the division.

TABLE 7.1 (CONTINUED)
Classification of Green Algae According to the Different Concepts Using Morphological, Ultrastructural, and Molecular Approaches*

Class Author	Monophyletic clade supported in Figure 7.4 and Figure 7.5	Validly described according to ICBN	Remarks (see additional comments in Silva, 1980)
Mesostigmatophyceae Marin et Melkonian	–	+	This class contains the two genera *Mesostigma* and *Chaetosphaeridium*, the latter genus belongs to the *Coleochaete*-clade (see Figure 7.4A, see also chapter 8, this volume); emendation necessary (exclusion of *Chaetosphaeridium*), then monophyletic group.
Chlorokybophyceae Bremer	+	–	This class contains few strains of a single sarcinoid species *Chlorokybus atmophyticus*; invalidly described (no Latin diagnosis); monophyletic group.
Klebsormidiophyceae van den Hoek, Mann et Jahns	–	–	Originally described by Jeffrey (1982) containing all streptophycean algae (exclusive *Coleochaete*). Van den Hoek et al. (1995) included the genera *Klebsormidium*, *Chlorokybus*, *Chaetosphaeridium*, *Coleochaete*, *Stichococcus*, and *Raphidonema* in this class, the last two genera belong to the Trebouxiophyceae, the others were transferred to other classes; invalidly described (no Latin diagnosis); monophyletic group, if confined to *Klebsormidium* and relatives (see Figure 7.4).
Zygnematophyceae Round	+	–	See above.
Coleochaetophyceae Bessey ex Woods	+	+	This class contains species of the genus *Coleochaete* and *Chaetosphaeridium*; Jeffrey (1982) described this class in the same way (invalidly without Latin diagnosis); monophyletic group.
Charophyceae Rabenhorst	+	+	See above.

Note: + = supported; – = not supported if class is revised; n.d. = no SSU rDNA sequencing data available.
*The table contains only classes that are widely accepted by the science community. For further described classes, see Silva (1980).

On the basis of the type of cell division (Figure 7.3), the green algae (inclusive of the higher plants) summarised as Viridiplantae (*sensu* Cavalier-Smith, 1981; Sluiman, 1985), or Chlorobionta (*sensu* Bremer, 1985; Bremer et al., 1987) were subdivided into two divisions—the Streptophyta (including Zygnematophyceae, Charophyceae, and higher plants), which form a phragmoplast during the cell division, and the Chlorophyta, which contain the other five classes, which mostly form a phycoplast (Pickett-Heaps, 1975).

THE MOLECULAR CONCEPT (PHYLOGENETIC CONCEPT)

In the 1990s, the application of phylogenetic analyses of molecular markers was introduced into the systematics and taxonomy of algae. Typical genetic markers used are the nuclear ribosomal operon (SSU, 5.8S, and LSU including ITS-1 and ITS-2), actin, several chloroplast genes (*rbc*L, *atp*B, and others), and mitochondrial genes (*nad*5). Phylogenetic analyses of SSU rDNA have provided support for the original suggestion, based on ultrastructural data, that there are two main lineages among the green plants (Friedl, 1997): the first one comprising charophycean algae and the land plants (the Streptophyta *sensu* Bremer, 1985), the other lineage consisting of the remaining green algae (Chlorophyta *sensu* Bremer, 1985). According to these analyses, the following lineages were revealed within the green algae (including the main publications): Prasinophyceae (e.g. Steinkötter et al., 1994; Nakayama et al., 1998; Guillou et al., 2004), Chlorophyceae (e.g. Hepperle et al., 1998; Pröschold et al., 2001; Krienitz et al., 2003; Friedl, 1997), Trebouxiophyceae (e.g. Friedl, 1995; 1997; Krienitz et al., 2004), Ulvophyceae (e.g. O'Kelly et al., 2004a, 2004b, 2004c), plus the streptophyte algae (e.g. Huss and Kranz, 1997; Marin and Melkonian, 1999; Karol et al., 2001).

At present these classes have been largely accepted by the science community; however, there is still discussion, because some lineages showed a "Long-Branch Attraction" (LBA) phenomenon *sensu* Philippe (2000) and the relationship among these lineages remains unclear. Based on the currently combined evidence (morphology, ultrastructure, and molecular phylogeny), the recognition of more classes can be revealed, and this is discussed below using the class Ulvophyceae as an example.

Mattox and Stewart (1984) and O'Kelly and Floyd (1984) defined the Ulvophyceae primarily based on ultrastructural features of the flagellar apparatus (i.e., a CCW orientation of the basal body). This class was furthermore regarded as more "advanced" in view of the fact that their vegetative thallus is non-flagellated and presumably derived from scaly green flagellates. Five orders were recognized within the Ulvophyceae (Ulotrichales, Ulvales, Siphonocladales/ Cladophorales, Dasycladales, and Caulerpales) based on modifications of the flagellar apparatus, differences in zoosporangial and gametangial structures, and life histories. This classification scheme has been largely adopted until today (Lewis and McCourt, 2004); however, molecular data (SSU rDNA and *rbc*L) have, so far, not been able to provide solid support for recognition of the monophyletic nature of the class Ulvophyceae *sensu* Mattox and Stewart (Zechman et al., 1990; Watanabe et al., 2001; López-Bautista and Chapman, 2003; see also Figure 7.4). Instead, molecular phylogenies reveal five separate, well-supported clades for which the mutual interrelationships remain ambiguous. These phylogenetic results would favour the recognition of five separate classes (Ulvophyceae *sensu stricto*, Cladophorophyceae, Bryopsidophyceae, Dasycladophyceae, and Trentepohliophyceae), as has been proposed by van den Hoek et al. (1995) based on apparent differences in thallus architecture, cellular organisation, chloroplast morphology, cell wall composition, and life histories. We refer to the following papers for molecular phylogenetic studies within each of these five lineages: Ulvophyceae (Hayden and Waaland, 2002; Friedl and O'Kelly, 2002; O'Kelly et al., 2004a, 2004b, 2004c), Cladophorophyceae (Bakker et al., 1994; Hanyuda et al., 2002; Leliaert et al., 2003), Bryopsidophyceae (Woolcott et al., 2000; Lam and Zechman, 2006), Dasycladophyceae (Olsen et al., 1994; Zechman, 2003), and Trentepohliophyceae (López-Bautista and Chapman, 2003).

Lewis and McCourt (2004) have dealt with these classification problems by only presenting a "working classification of green algae and land plants." To summarize the current state of classification, we present a phylogenetic tree of SSU rDNA sequences including representatives of all lineages available in GenBank (Figure 7.4). The clades in Figure 7.4, which form monophyletic lineages and are well-supported in bootstrap analyses, are named after a representative species or genus without ranking at higher levels. The different classification systems using morphological, ultrastructural, and molecular concepts are summarized with additional information in Table 7.1.

In the next section, we want to focus on classification at the generic and species level, because many questions need to be clarified before a new or revised classification at the higher level can be proposed. To give an example of a problematic situation, the order Ulotrichales (traditionally contains only unbranched filaments) is classified within the class Ulvophyceae based on the ultrastructural results and the molecular position of a single taxon of the genus *Ulothrix*, *U. zonata* (O'Kelly and Floyd, 1984; see also Lewis and McCourt, 2004). However, the situation is further "confused" as the type species of *Ulothrix*, *U. tenuissima* Kützing, has not been investigated employing ultrastructural and molecular methods. According to the ICBN, the name of an order must be combined and connected to the type genus and its type species.

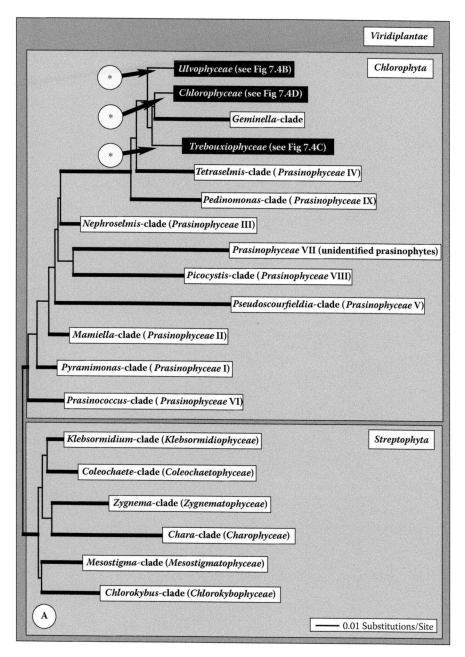

FIGURE 7.4 Molecular phylogeny of the Viridiplantae based on SSU rDNA sequence comparisons. The phylogenetic tree shown was inferred by the neighbour-joining method based on distances of 1668 aligned positions of 469 taxa using PAUP 4.0b10 (Swofford, 2002). Subsections of this dataset (parts A through D) were analyzed using the best model (TrN+I+G; Tamura and Nei, 1993) calculated by Modeltest 3.7 (Posada and Buckley, 2004; Posada and Crandall, 1998). Bootstrap percentage values (>70%) are marked by bold branches given for neighbour-joining (using TrN+I+G model; 1000 replicates). The monophyletic clades are provisionally named after a representative taxon. (A) Phylogenetic tree of the Chlorophyta and Streptophyta (67 taxa). The three classes Ulvophyceae, Chlorophyceae, and Trebouxiophyceae are represented by few taxa in this analysis (best model: TrN+I+G; I = 0.49, G = 0.58). The clades containing species of the three subsections (common branch marked by an asterisk) were not supported in bootstrap analyses; however, they were separately analyzed in parts B through D and marked in black.

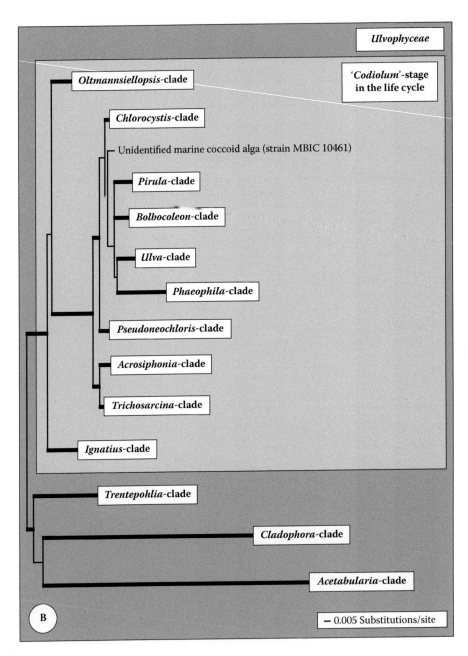

FIGURE 7.4 (CONTINUED) (B) Phylogenetic tree of the Ulvophyceae (142 taxa; best model: TrN+I+G; I = 0.35, G = 0.58).

CLASSIC VERSUS MODERN APPROACHES: PROBLEMS WITH IDENTIFICATION OF SPECIES AND GENERA

At the generic and species levels, algae (especially microalgae) are traditionally classified according to morphological and cytological characters of vegetative stages in their life cycle. Phylogenetic analyses of ribosomal genes (SSU and ITS rDNA sequences) have demonstrated that this morphological concept is artificial for most of the algal genera and needs to be revised. For example, within the Volvocales (monadoid unicells and colonies) 60 genera with 2000 species have been described on the basis of phenotypic characters. The largest genus *Chlamydomonas* contains 800 species,

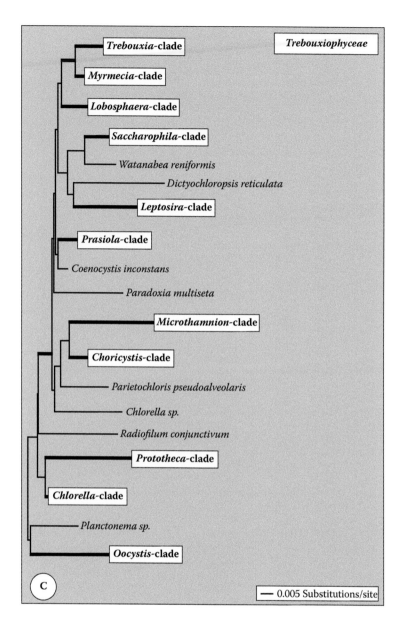

FIGURE 7.4 (CONTINUED) (C) Phylogenetic tree of the Trebouxiophyceae (77 taxa; best model: TrN+I+G; I = 0.52, G = 0.61).

which are characterized by different cell sizes and shapes, different chloroplast shapes, the number and position of pyrenoids within the chloroplast, and the position of eyespot and nucleus within the cell. However, according to phylogenetic analyses, species of the unicellular genus *Chlamydomonas* form eight monophyletic lineages within the Chlorophyceae. Strains belonging to other genera are in the same lineage as some of these "*Chlamydomonas*" (Pröschold et al., 2001; Figure 7.5).

Coleman (2000) introduced new generic and species concepts (the Z-clade and CBC-clade concepts) based on compensatory base changes (CBC) in ITS sequences ("genetic signatures" or even "DNA barcode") compared with the mating ability of species. In contrast to the biological species concept (*sensu* Mayr, 1948), strains of the same Z(ygote)-clade can at least form zygotes, though they may or may not germinate and produce fertile F1 generations. A Z-clade contains one,

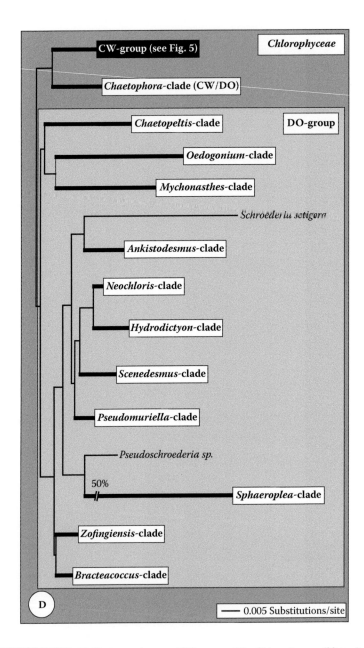

FIGURE 7.4 (CONTINUED) (D) Phylogenetic tree of DO-group of the Chlorophyceae (80 taxa). The CW-group of the Chlorophyceae is represented by few taxa in this analysis (best model: TrN+I+G; I = 0.57, G = 0.67).

or a very few, biological species, each with genetically very similar members sharing identical ITS rDNA sequences (certainly no CBCs). Therefore, ITS sequences have proven to be very good tools for evolutionary comparisons, especially at the biological species level (Coleman, 2003). For example in *Chlamydomonas allensworthii* Starr, Marner et Jaenicke, two CBC clades (which correlate with their pheromone production) and five Z-clades could be recognized (Coleman et al., 2001; Figure 7.6). This demonstrated that in the "morpho-species" *Chlamydomonas allensworthii* at least five biological species are present. These species are not correlated to their geographical origins. Furthermore, morphological differences are not observed in the strains investigated and only two different pheromones are found: lurlenol and lurlenic acid (Jaenicke and Starr, 1996),

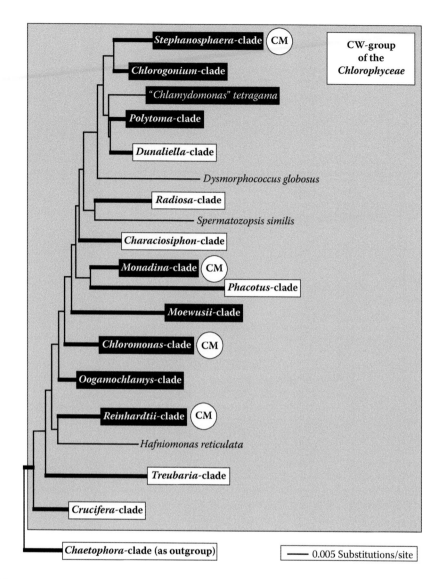

FIGURE 7.5 Molecular phylogeny of the CW-group (131 taxa) based on SSU rDNA sequence comparisons using the *Chaetophora*-clade *sensu* Pröschold et al. (2001) as outgroup. The phylogenetic tree shown was inferred by the neighbour-joining method based on distances of 1668 aligned positions calculated by the model of Tamura and Nei (Tamura and Nei, 1993; TrN+I+G; I = 0.51, G = 0.57), which was calculated as best model by Modeltest 3.7 (Posada and Buckley, 2004; Posada and Crandall, 1998). Bootstrap percentage values (>70%) are marked by bold branches given for neighbour-joining (using TrN+I+G model; 1000 replicates). The monophyletic clades are provisionally named after a representative taxon. The clades containing species of *Chlamydomonas* and *Chloromonas* (CM) are marked in black.

which also correlate with the phylogenetic analyses presented in Figure 7.6. For analysis of compensatory base changes, the secondary structures of SSU and ITS rDNA sequences have to be determined by folding, which requires accuracy of sequencing. However, this is not always available, particularly in the sequences published in GenBank prior to 1997–1998 and the advent of automatic sequencing. Therefore, for this type of analysis we recommend resequencing where more than two ambiguities or mismatches are found.

As demonstrated above (see also examples presented in Lewis and McCourt, 2004), most of the genera in the Chlorophyceae (and in the Chlorophyta in general) are polyphyletic and a revision

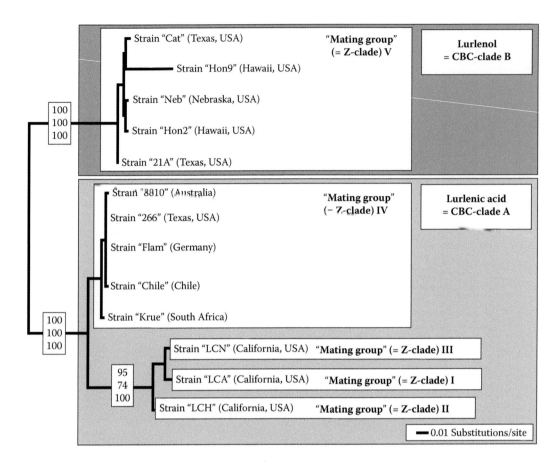

FIGURE 7.6 Molecular phylogeny of *Chlamydomonas allensworthii* based on ITS rDNA sequence comparisons. The phylogenetic tree shown was inferred by maximum likelihood method using K80+G model (Ti/Tv = 1.65; G = 0.007; Kimura, 1980) calculated as best model by Modeltest 3.7 (Posada and Buckley, 2004; Posada and Crandall, 1998). Bootstrap percentage values (>50%) are given for maximum likelihood (using K80+G; bold italic), neighbour-joining (using K80+G; bold), and unweighted maximum parsimony (not bold). Two CBC clades recognized are marked gray and correspond to the pheromone production (lurlenol and lurlenic acid). The Z-clades are highlighted.

needs to be performed that follows the regulations of the ICBN. According to the ICBN, each genus validly described has a type species, which has a type specimen deposited in a public herbarium (for macroalgae), or a type figure in a publication (for microalgae). Following these rules, a genus name can only remain for a group (clade, monophyletic lineage), which contains the type species. However, for many type species of genera (especially for microalgae), no "type material" (mostly only figures) is available as reference, thus no DNA could be extracted for phylogenetic comparisons. To solve this problem, an epitype (a new herbarium specimen for macroalgae or a cryopreserved culture for microalgae) or a neotype, when the type specimen is lost or destroyed, can be designated and deposited in a public herbarium or culture collection. For example in the genus *Chlamydomonas*, Pröschold and Silva (2007) have proposed *C. reinhardtii* P.A. Dangeard as conserved type of the genus and designated the culture strain SAG 11-32b (= CCAP 11/32A, = UTEX 90) cryopreserved as epitype, to emend an ancient crude drawing. In microalgae in particular, this procedure needs to be performed for all genera in order to link traditional and modern approaches, which are dependent on reference material that is available for the science community. This material (cultures, extracted DNA) can be provided by culture collections; however, the reference material is needed in a genetically stable condition and in a metabolic inactive state (only this is accepted by ICBN).

For microalgae, cryopreservation is the long-term solution, provided in major culture collections (Day and Brand, 2005). The long-term stability of reference material is essential for all research fields in biology; later comparisons can utilize different molecular techniques, e.g. amplified fragment length polymorphism (AFLP) as demonstrated for *Chlorella vulgaris* Beijerinck (see details in Müller et al., 2005).

TAXONOMIC REVISION OF GENERA AND SPECIES USING POLYPHASIC APPROACHES

Traditional taxonomic approaches often depend on single or even negative "characters" (e.g. absence of zoospore formation). For example, in unicellular flagellates, the genus *Chloromonas* is separated, by the absence of pyrenoids in the chloroplast, from the genus *Chlamydomonas*. Phylogenetic analyses have demonstrated that some strains of *Chloromonas* and *Chlamydomonas* can belong to the same clade and, in the case of *Chloromonas reticulata* (Goroschankin) Wille, to the same species (containing strains with or without pyrenoids; see details in Pröschold et al., 2001). Nozaki et al. (1998) have demonstrated for *Chlorogonium*, another unicellular flagellate, that the presence or absence of pyrenoids is dependent on culture conditions (autotrophic: in light in mineral medium, or heterotrophic: in dark in medium with organic compounds like yeast extract), where the algae grow. An additional example includes the species *Chlorella vulgaris* and *Micractinium pusillum* Fresenius, which are closely related according to phylogenetic analyses (Krienitz et al., 2004; Luo et al., 2005, 2006). Cultured under axenic conditions in defined culture medium, both showed smooth-walled unicells by light microscopy. However, if a grazer is added to the algal culture, in this case the rotifer *Brachionus calciflorus* Pallas, strains of *Micractinium pusillum* form colonies and cell wall spines, which are typical for this genus (Luo et al., 2005, 2006). In contrast, no phenotypic differences are observed in cultures of *Chlorella vulgaris* under the influence of a grazer.

For the green seaweeds *Ulva*, *Enteromorpha*, and *Monostroma*, a similar morphological poly-morphism is known. If *Ulva lactuca* Linnaeus is grown under axenic condition, the thalli lose the ability to produce the natural foliose morphology. Adding marine bacteria isolated from *Ulva* thalli to the axenic cultures causes the algae to grow an *Enteromorpha*-like thallus (Bonneau, 1977; Provasoli and Pintner, 1980). The influence of marine bacteria on morphogenesis has also been reported for *Ulva pertusa* Kjellman (Nakanishi et al., 1996), *Enteromorpha linza* (L.) J. Agardh, and *E. intestinalis* (L.) Nees (Fries, 1975). The phylogenetic analyses of ITS rDNA sequences have confirmed that *Ulva* and *Enteromorpha* are not distinct genera, and therefore, *Enteromorpha* was included into the genus *Ulva* (Tan et al., 1999, Hayden et al., 2003). The associated bacteria isolated from *Monostroma oxyspermum* Kützing also have a morphogenetic influence on axenic cultures of this alga as well as on *Ulva pertusa* and *U. intestinalis* (= *Enteromorpha intestinalis*). The bacteria were characterized by phylogenetic analyses as *Cytophaga-Flavobacterium-Bacteroides* (CFB) complex (Matsuo et al., 2003), and a morphogenetic inducer (Thallusin) could be isolated from these bacteria, which revealed its importance for the natural growth of the seaweeds tested (Matsuo et al., 2005). The extreme phenotypic plasticity of siphonocladalean macroalgae (e.g. *Anadyomene*, *Ernodesmis*, *Microdictyon*, *Siphonocladus*, and *Struvea*) can also be demonstrated by growing isolates under different culture conditions. Certain morphological characters, including some that are traditionally used for generic delineation, are found to be variable and dependent on environmental conditions like temperature and light regimes (e.g. the formation of blades in *Anadyomene* and *Microdictyon*). In view of this mor-phological variability in many green algae, we would like to stress the importance of field studies and comparative studies between material collected in the field and material grown in culture. In several green algae, detailed field studies, including the investigation of ecology and phenology, have been of significant assistance to the clarification of the taxonomic identity of several taxa (see for example López-Bautista et al., 2006, for the order Trentepohliales).

The major work of a taxonomic revision of genera and species regards consideration of morphological plasticity under different culture and environmental conditions. It needs also a wider context. We propose here the usage of a polyphasic approach (plasticity of phenotype and different life stages under different conditions, biochemical and physiological approaches, phylogenetic concepts, comparison of species concepts, multi-gene approach), which we demonstrate for the following examples presented below.

POLYPHASIC APPROACHES USED FOR CHARACTERIZATION OF THE GENERA *OOGAMOCHLAMYS* AND *LOBOCHLAMYS*

A group of "former" *Chlamydomonas* species were transferred to these two newly described genera by Pröschold et al. (2001), on the basis of phylogenetic analyses of SSU rDNA sequences. Both genera belong to the *Oogamochlamys*-clade (well-supported monophyletic lineage shown in Figure 7.5), and can be clearly identified by using multiple approaches, which are not causally connected. The genus *Oogamochlamys* is characterized by chloroplast morphology (parietal, massive plastids with ridges on the surface), multiple pyrenoids irregularly distributed, and homothallic protandric oogamy; the genus *Lobochlamys* has a cup-shaped chloroplast with incisions, cell wall with mucilage layer around the flagellated cells, and homo- or heterothallic isogamous sexual reproduction (Pröschold et al., 2001). The two genera, each have several unique compensatory base changes (CBC) in the secondary structure of SSU rRNA (e.g. in Helix E23_1 and E23_2 demonstrated here in Figure 7.7). Using the Z-clade concept *sensu* Coleman (2000), three species can be identified in *Oogamochlamys*: *O. gigantea* (Dill) Pröschold, Marin, Schlösser et Melkonian; *O. zimbabwiensis* (Heimke et Starr) Pröschold, Marin, Schlösser et Melkonian; and *O. ettlii* (Dill) Pröschold, Marin, Schlösser et Melkonian— furthermore, these can be characterized by morphology (Figure 7.8). The species of *Lobochlamys*, *L. segnis* (Ettl) Pröschold, Marin, Schlösser et Melkonian and *L. culleus* (Ettl) Pröschold, Marin, Schlösser et Melkonian, are characterized by different sporangia wall lytic enzymes (Figure 7.9).

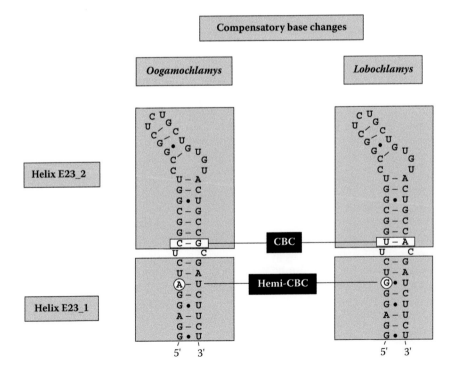

FIGURE 7.7 Comparison of Helices E23_1 and E23_2 of the SSU rRNA secondary structure between *Oogamochlamys* and *Lobochlamys*. Compensatory bases changes (Hemi-CBC and CBC) are highlighted.

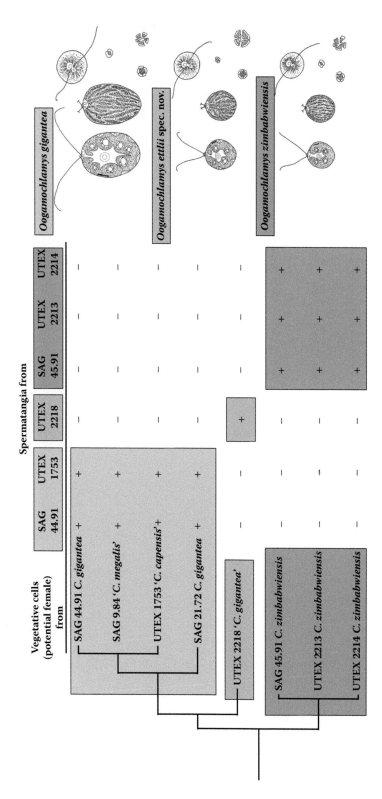

FIGURE 7.8 Molecular phylogeny of *Oogamochlamys* based on SSU rDNA sequence comparisons. The phylogenetic tree shown was inferred by the maximum-likelihood method using the TrNef+I+G (equal base frequencies, I = 0.72, G = 0.63) calculated as best model. The results of crossing experiments are marked in boxes for *O. gigantea*, *O. ettlii*, and *O. zimbabwiensis* (+ = zygote formation; − = no zygote formation). The strains named in the tree are the original designation in the culture collection. The morphological re-investigation of these strains is shown on the right.

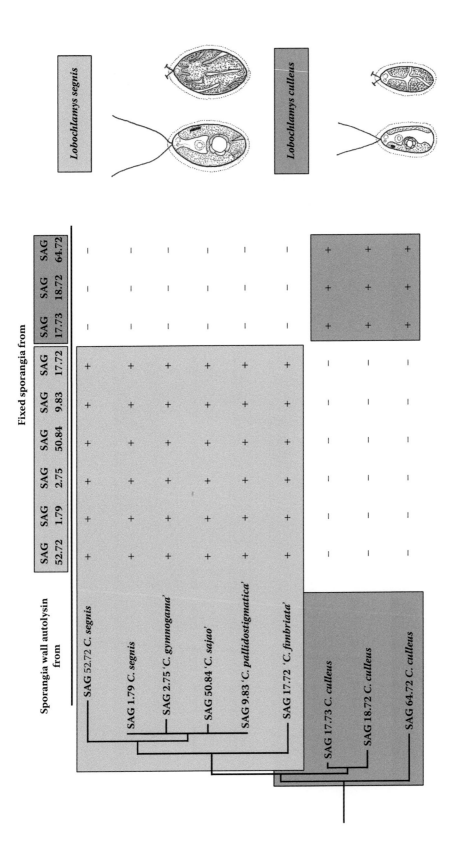

FIGURE 7.9 Molecular phylogeny of *Lobochlamys* based on SSU rDNA sequence comparisons. The phylogenetic tree shown was inferred by the maximum-likelihood method using the TrNef+I+G (equal base frequencies, I = 0.72, G = 0.63) calculated as best model. The results of biological tests using the sporangium autolysin on fixed sporangia are marked in light gray for *L. segnis* and in dark gray for *L. culleus* (+ = release of zoospores; − = no release of zoospores) according to the VLE groups 9 (*L. segnis*) and 10 (*L. culleus*; Schlösser, 1976, 1984). The strains named in the tree are the original designation in the culture collection. The morphological re-investigation of these strains shown on the right.

Schlösser (1976, 1984) determined that these enzymes (called sporangium autolysin, or vegetative lytic enzymes [VLEs]) are produced by zoospores for release from the sporangium cell wall, are stage and species specific, and can be used for classification ("autolysin concept"). Fifteen VLE groups were distinguished among 65 strains of different *Chlamydomonas* species examined (Schlösser, 1976, 1984), from which VLE group 9 correspond to *L. segnis* and 10 to *L. culleus*.

DELIMITING PHYLOGENETIC SPECIES BY A MULTI-GENE APPROACH IN *MICROMONAS* AND *HALIMEDA*

Different empirical methods for delimiting species have been described recently (Sites and Marshall, 2003). In one of these methodologies, termed the "exclusivity criterion," phylogenetic species are delimited, using genealogical concordance of multiple independent loci (Dettman et al., 2003). With this method, species are delimited based on two requirements: species are exclusive groups (those in which all members are more closely related to each other than to any organism outside of the group) and species reside at the boundary between reticulate and divergent genealogy, where unlinked genes should have concordant genealogical histories (Sites and Marshall, 2003). Within single interbreeding species (or in case of hybridisation between lineages), the mixing effects of recombination between genes would cause unlinked loci to have different genealogies, but between genetically isolated species, the extinction of ancestral alleles by genetic drift would lead to the congruence of genealogies. Hence, the transition between deep genealogical concordance and shallow genealogical discordance can be used to recognize phylogenetic species (Taylor et al., 2000). A recent paper by Slapeta et al. (2006) serves as one of the few examples in which multi-gene phylogenies are employed to delimit species in the green algae. In this study, phylogenetic analyses of four independent nuclear, plastid, and mitochondrial loci (rDNA, ß-tubulin, *rbc*L, *cox*1) has led to the recognition of numerous cryptic species in the marine prasinophyte, *Micromonas pusilla* (Butcher) Manton et Parke. Multiple semi-cryptic species were also revealed in the tropical seaweed *Halimeda* based on nuclear ribosomal and plastid DNA sequences (Verbruggen et al., 2005a, 2005b). In this study the hypothesis was formulated that reticulate evolution could have been involved in speciation within *Halimeda*, based on discordances between phylogenetic trees inferred from nuclear and plastid DNA sequences. Both studies confirm the fact that the morphospecies concept is untenable because it overlooks a large genetic and species diversity, both in green macro- and microalgae, even in the smallest eukaryotes.

These examples have clearly demonstrated that the usage of polyphasic approaches could be the answer to systematic and taxonomic questions, at least at the generic and species levels. Not all of the different methods and concepts described here for taxonomic revisions are suitable for all groups of green algae, and others need to be developed (ecological, physiological, and biochemical data). In addition, descriptions of habitat and origin especially for seaweeds should also be included. However, many of these methods examined only for a particular group can be used for other groups as well. For example, the specificity of autolysins presented here for *Chlamydomonas* and its relatives is also found in filamenteous green algae. Schlösser (1987) distinguished fragmentation autolysins in *Uronema confervicola* Lagerheim, *Radiofilum transversale* (de Brébisson) Christensen, and in an unidentified species of *Geminella*. The autolysin of *Uronema confervicola* react in bioassays on other strains of *Uronema* and strains of other genera including *Chaetophora*, *Draparnaldia*, *Fritschiella*, and *Stigeoclonium* (all branched filaments), which all belong to the *Chaetophora*-clade according to the phylogenetic analyses presented in Figure 7.4. However, no fragmentation could be observed when *Uronema* autolysin was applied to strains of the genera *Ulothrix*, *Geminella*, *Klebsormidium*, *Trentepohlia*, and others. This is particularly interesting as phylogenetic analyses positioned these taxa in other clades (see Figure 7.4).

Differentiation of genera and species based on single characters (e.g. morphology of the vegetative cell alone) is often not possible and inevitably leads to ambiguous classifications. This was recognized by several authors prior to the availability of ultrastructural and molecular phylogenetic data for characterization. For example, Kornmann and Sahling (1983) studied the life cycle of *Chlorocystis cohnii* (Wright) Reinhard (Figure 7.10) and *Halochlorococcum marinum* P. Dangeard, two marine coccoid green algae, and found a *Codiolum* stage as a zygote in sexual reproduction, which is characteristic for filamenteous and parenchymatous marine green algae (e.g. *Acrosiphonia*, *Urospora*, or *Ulva*). They concluded that both taxa are closely related to these algae and described a new order Chlorocystidales in the Codiolophyceae (= Ulvophyceae *sensu stricto*). However, Komárek and Fott (1983) integrated both species as marine representatives of the freshwater genera *Chlorochytrium* and *Spongiochloris* based on the morphology of their vegetative cells. *Chlorocystis* and *Chlorochytrium* have a cup-shaped plastid, *Halochlorococcum* and *Spongiochloris* have a reticulated chloroplast, and otherwise the cell morphology of all four is very similar. Phylogenetic analyses confirmed the placement of *Chlorocystis* and *Halochlorococcum* in the Ulvophyceae as proposed by Kornmann and Sahling (Pröschold et al., unpublished data). In contrast, *Chlorochytrium* and *Spongiochloris* on the basis of ultrastructural and molecular data belong to the Chlorophyceae (Watanabe and Floyd, 1994, Lewis, 1997; *Stephanosphaera*-clade in Figure 7.5).

CONCLUSIONS

BIODIVERSITY OF GREEN ALGAE BASED ON TAXONOMIC REVISION USING POLYPHASIC APPROACHES

As demonstrated here, polyphasic approaches can distinguish and delimit species and genera, which lead on one hand to a reduction of described species (e.g. Figure 7.9) and on the other hand to more biological species, which are morphologically identical (e.g. Figure 7.6). For example, the 800 described species of *Chlamydomonas* can be reduced to approximately 100 to 150 species using these approaches (Pröschold, unpublished data). However, Fawley et al. (2004) have shown that the biodiversity of green microalgae is much higher than expected. They isolated 273 strains with 93 distinct SSU rDNA sequences from four different sites in North Dakota and Minnesota (USA). Only four of these matched with any sequences published in GenBank. It is widely accepted that microbial diversity differs fundamentally from biodiversity of larger animals and plants. Furthermore, it has been suggested that free-living microbes have a cosmopolitan distribution (Fenchel and Finlay, 2003) and that most protistan organisms, smaller than 1 mm in size, have worldwide distribution wherever their required habitats are realised. This is a consequence of ubiquitous dispersal driven by huge population sizes, and consequently low probability of local extinction. Organisms larger than 10 mm are much less abundant, and rarely cosmopolitan (Finlay, 2002; Finlay and Fenchel, 2004). However, these hypotheses are only based on the phenotypic approach (morpho-species) and are dependent on clear identification of the microorganisms. For microalgae the hypotheses are still under discussion (Finlay and Fenchel, 2002; Coleman, 2002) and are not yet proven. Polyphasic approaches as suggested here have indicated that the biodiversity of microalgae is much higher than the described diversity according to the morpho-species concept. For example, the morpho-species *Gonium pectorale*, a member of the colonial Volvocales, contains at least five independent interbreeding biological species (Fabry et al., 1999) with little correlation to their origin. It seems that sex and the type of interbreeding (heterothally: self-sterile, need two gametes originating from different clones, or homothally: self-fertile, gametes originating from a single clone) need to be considered in the context of biogeographical distribution. Similarly in *Cladophora* (one of the largest green macro-algal genera), the number of recognized morpho-species has been reduced from over 1000 to about 100 based on comparative morphological studies performed on field, culture, and herbarium material (e.g. van den Hoek, 1963, 1982). More recently, molecular studies have revealed that some of these morphological species (e.g. *C. vagabunda* (L.) van den Hoek, which occupies one huge continuous geographic area) represent different divergent lineages that can be related to thermal ecotypes (Bakker et al., 1995; Breeman et al., 2002).

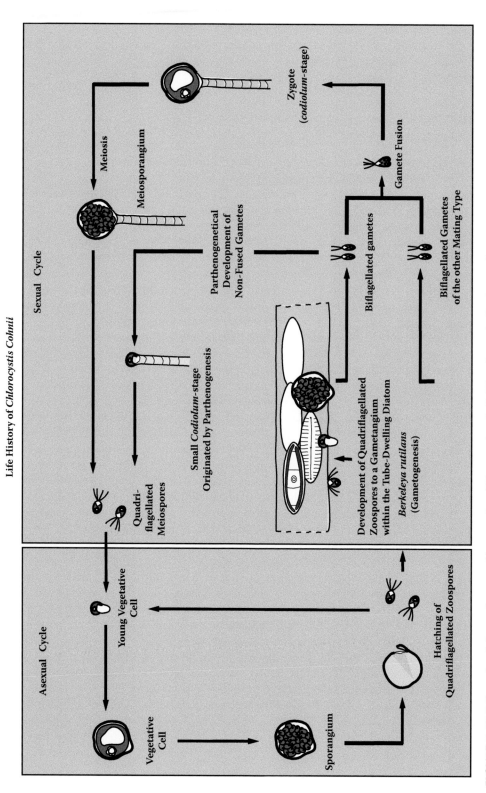

FIGURE 7.10 Life cycle of the marine coccoid green alga *Chlorocystis cohnii* (Ulvophyceae). (Redrawn after Kornmann, P. and Sahling, P.-H., *Helgoländer Meeresuntersuchungen*, 36: 1–65, 1983. With permission.)

Depending on the interpretation of these molecular and temperature tolerance data, *C. vagabunda* can be regarded as a single species, or alternatively a cryptic species complex (multiple species). Similar results were found by van der Strate et al. (2002), who demonstrated, based on ITS sequence divergence, differential microsatellite amplification and thermal ecotypes, that the Atlantic taxon *Cladophoropsis membranacea* (Hofman Bang ex C. Agardh) Børgesen consists of at least three cryptic species that have overlapping biogeographies.

RECONSTRUCTING THE GREEN ALGAL TREE BY MULTI-GENE AND WHOLE-GENOME ANALYSIS

Although the challenges associated with reconstruction of deep relationships are fundamentally different from those of shallow ones, multi-gene approaches are fast becoming the norm in both types of studies. In shallow reconstruction problems, where the complex effects of reticulation and lineage sorting are persistently present, gene genealogies of multiple independent, fast-evolving loci are analyzed to answer questions related to species concepts, speciation events, hybridisation, and gene flow in or between species (see, for example, Slapeta et al., 2006). In deep phylogenies, lineages are widely separated in time, resulting in few problems with reticulation and lineage sorting. Yet, deep phylogenetic reconstructions suffer from a number of other persistent problems. One of these is the well-known problem of long-branch attraction that can be caused by a variety of factors, including the fact that the set of lineages is relatively incomplete due to extinction events. A related problem to long-branch attraction is substitution saturation, which particularly hampers phylogenetic reconstruction of deep branches.

Phylogenetic relationships within the green algae have been based primarily on sequence analyses of the nuclear-encoded SSU rDNA. Based on this single marker, phylogenetic relationships among the major groups of green algae have been found particularly difficult to resolve, because the internal branches, grouping the different orders and classes, are generally very short relative to the subsequent evolutionary history of the group (with the exception of a few internal branches that have undergone considerable rate acceleration of the SSU rDNA, such as the ones leading to the Bryopsidales, Cladophorales, and Dasycladales). These short internal, often weakly supported branches can be attributed to a number of factors including conflict between characters due to homoplasy within a sequence, insufficient sequence length, rate variation across characters or taxa, the presence of taxa with unstable positions that may reduce support levels in the tree as a whole (Sanderson and Shaffer, 2002), or an historical signal of a rapid evolutionary radiation (Wortley et al., 2005). The factors most open to investigation in the challenge of the green algal tree reconstruction are sequence length and taxon sampling. The relative contribution of taxon number and gene number to accuracy in phylogenetic inference is still a major issue in phylogenetics and has been widely discussed (Graybeal, 1998). The general consensus today is that increasing taxon number correlates with a slight decrease in phylogenetic accuracy, while increasing sequence length or gene number has a significant positive effect on phylogenetic accuracy (Rokas and Caroll, 2005). Indeed a number of recent studies have shown that large-scale molecular sequencing projects and subsequent concatenation of multiple markers should be able to resolve the deep and problematic phylogeny of ancient eukaryotic groups, including the green algae (see for example Simpson et al., 2006).

The challenge of constructing the green algal tree by employing multi-gene or whole-genome molecular analysis techniques is still in its infancy. At present these phylogenies have been based primarily on concatenated gene sequences from organellar genomes from a relative small number of taxa (see Pombert et al., 2004, 2005, 2006, including references). Organellar genomes are particularly useful for phylogenomic reconstruction because of their relatively high gene content, condensed in comparison to nuclear genomes. Also, organellar genes are typically single-copy, in contrast to many nuclear genes that are multi-copy in nature, often having a confounding effect on phylogenetic reconstruction.

Pombert et al. (2005, 2006) analyzed the concatenated amino acid and nucleotide sequences derived from 58 protein-coding genes that are shared among the chloroplast DNAs of *Pseudendoclonium*,

Oltmannsiellopsis, Chaetosphaeridium, Chlamydomonas, Chlorella, Mesostigma, Marchantia, Nephroselmis, and *Nicotina.* As expected, their results confirmed the long-standing view in which the chlorophytes and streptophytes form two distinct lineages. In addition this whole-genome analysis supported the hypothesis that the Ulvophyceae is "sister" to the Trebouxiophyceae but could not eliminate the hypothesis that the Ulvophyceae is "sister" to the Chlorophyceae. The latter hypothesis was also supported by phylogenetic analysis of gene order data and by independent structural evidence based on shared gene losses and rearrangement break points within ancestrally conserved gene clusters. Phylogenetic analyses of seven concatenated mtDNA-encoded protein sequences also revealed a close relationship between the ulvophyte *Pseudendoclonium* and chlorophycean taxa, with the trebouxiophyte *Prototheca* occupying a basal position (Pombert et al., 2004). The difficulty in unequivocally resolving the basal divergences within the green algae, even with large numbers of concatenated genes, might be attributed to a real rapid, early radiation of the green algae, rather than to data availability. An additional potential cause is that various protein-encoded genes might have lost their phylogenetic information due to substitution saturation, plaguing the phylogenetic analysis involving deep branches in the green algal tree. The chloroplast and mitochondrial genome sequences of additional green algae will be required to provide unambiguous support for the ancient divergences within the green algae (Pombert et al., 2005). Currently the "Organelle Genome Megasequencing Program" (http://megasun.bch.umontreal.ca/ogmp/) and a research project funded through the "Assembling of the Tree of Life program" (http://ucjeps.berkeley.edu/TreeofLife/) are involved in sequencing and analysing a large number of organellar genomes, including those of various members of charophytes, trebouxiophytes, ulvophytes, and chlorophycean green algae (Monique Turmel, Université de Laval, Canada, personal communication).

OUTLOOK

HOW SHOULD WE APPROACH TAXONOMIC REVISION USING POLYPHASIC APPROACHES?

The key to answering this question is to have reference material for comparisons. In many cases this material can be provided by major culture collections, where strains should be deposited. This sounds like a relatively trivial point, but in many publications in the scientific literature no strain numbers are given and only species names have been designated. As indicated in this review, most of the genera and species are polyphyletic and without strain designations, comparisons are at least difficult and often not possible. This was shown to be the case of the "model" organism, *Chlamydomonas reinhardtii*, where the two mating types and strains of F1 generations from a single zygote isolated by G.M. Smith in 1945 were sent from one research laboratory to others over 60 years (for details, see genealogy in Pröschold et al., 2005).

Starting with reference material from a culture collection, the SSU and ITS rDNA and possibly other "marker" genes should be sequenced and phylogenetically analysed using different molecular methods (maximum likelihood, maximum parsimony, distance, and Bayesian). In the case of SSU and ITS rDNA, the secondary structures should be elucidated for detection of compensatory base changes and synapomorphies. Therefore, for this type of analysis we recommend resequencing where more than two ambiguities or mismatches are found. Using these data as a starting point, all strains of a clade should then be studied using the polyphasic approaches as described above. For these approaches, experience and expertise are necessary, which can be provided only by utilizing the knowledge of a "taxonomic college" (scientific experts of a certain groups of algae and protozoa). Finally, with the input of these experts the strains of particular groups of green algae can be revised and then be cryopreserved, if possible, for long-term stability. In addition, the extracted DNA of these strains could be preserved. This process has been initiated at the Culture Collection of Algae and Protozoa (CCAP, Dunbeg by Oban, Scotland), and it is hoped this approach will in due course involve a wide spectrum of phycological taxonomists and other culture collections.

We are confident that the reconstruction of the ancient divergences within the green algae will be greatly facilitated by the ongoing sequencing efforts of a large number of genes (including complete plastid and mitochondrial genomes, and a variety of novel nuclear markers) of a wide range of green algae. This will undoubtedly get us closer to understanding the "true" phylogeny of the green algae. In the short term we believe that the SSU rDNA phylogeny presented here could be easily complemented with complete LSU rDNA sequences (although this gene has been found difficult to amplify in some green algal groups, for example, the Bryopsidales). This gene has proved to be more variable and phylogenetically more informative than the SSU rDNA in several eukaryotic groups including green algae (see, for example, Buchheim et al., 2001; Shoup and Lewis, 2003; Leliaert et al., 2003). LSU rDNA has furthermore the advantage over SSU rDNA in that it contains three hypervariable regions (see Wuyts et al., 2001), which make it suitable to be phylogenetic informative at the species level.

ACKNOWLEDGMENTS

We thank Annette W. Coleman (Brown University, Providence, RI, USA) and John G. Day (CCAP) for correcting the English and for critical discussions.

REFERENCES

Bakker, F.T., Olsen, J.L., Stam, W.T., and van den Hoek, C. (1994) The *Cladophora* complex (Chlorophyta): new views based on 18S rRNA gene sequences. *Molecular Phylogenetics and Evolution*, 3: 365–382.

Bakker, F.T., Olsen, J.L., and Stam, W.T. (1995) Global phylogeography in the cosmopolitan species *Cladophora vagabunda* (Chlorophyta) based on nuclear rDNA internal transcribed spacer sequences. *European Journal of Phycology*, 30: 197–208.

Blackman, F. (1900) The primitive algae and the Flagellata. *Annals of Botany*, 14: 647–688.

Blackman, F.E. and Tansley, A.G. (1902) A revision of the classification of the green algae. *New Phytology*, 1: 17–24, 67–77, 89–96, 114–120, 133–144, 163–168, 189–192, 213–220, 258–264.

Bold, H.C. and Wynne, M.J. (1985) *Introduction to the Algae. Structure and Reproduction.* Prentice Hall, Englewood Cliffs, NJ.

Bonneau, E.R. (1977) Polymorphic behavior of *Ulva lactuca* (Chlorophyta) in axenic culture. 1. Occurence of *Enteromorpha*-like plants in haploid clones. *Journal of Phycology*, 13: 133–140.

Breeman, A.M., Oh, Y.S., Hwang, M.S., and van den Hoek, C. (2002) Evolution of temperature responses in the *Cladophora vagabunda* complex and the *C. albida/sericea* complex (Chlorophyta). *European Journal of Phycology*, 37: 45–58.

Bremer, K. (1985) Summary of green plant phylogeny and classification. *Cladistics*, 1: 369–385.

Bremer, K., Humphries, C.J., Mishler, B.D., and Churchill, S.P (1987) On cladistic relationships in green plants. *Taxon*, 36: 339–349.

Buchheim, M.A., Michalopulos, E.A., and Buchheim, J.A. (2001) Phylogeny of the Chlorophyceae with special reference to the Sphaeropleales: a study of 18S and 26S rDNA data. *Journal of Phycology*, 37: 819–835.

Cavalier-Smith, T. (1981) Eukaryote kingdoms: seven or nine? *Biosystems*, 14: 461–481.

Cavalier-Smith, T. (1993) Kingdom Protozoa and its 18 phyla. *Microbiological Review*, 57: 953–994.

Christensen, T. (1962) Systematik Botanik, Alger. In *Botanik* (eds. T.W. Böcher, M. Lange, and T. Sorensen), Munksgaard, København, pp. 1–178.

Christensen, T. (1994) Typification of the class name Chlorophyceae. *Taxon*, 43: 245–246.

Coleman, A.W. (2000) The significance of a coincidence between evolutionary landmarks found in mating affinity and a DNA sequence. *Protist*, 151: 1–9.

Coleman, A.W. (2002) Microbial eukaryote species. *Science*, 297: 337.

Coleman, A.W. (2003) ITS2 is a double-edged tool for eukaryote evolutionary comparisons. *Trends in Genetics*, 19: 370–375.

Coleman, A.W., Jaenicke, J., and Starr, R.C. (2001) Genetics and sexual behavior of the pheromone producer *Chlamydomonas allensworthii* (Chlorophyceae). *Journal of Phycology*, 37: 345–349.

Day, J.G. and Brand, J.J. (2005) Cryopreservation methods for maintaining microalgal cultures. In *Algal Culturing Techniques* (ed. R.A. Andersen), Elsevier Academic Press, Oxford, pp. 165–187.

Deason, T.R. (1984) A discussion of the class Chlamydophyceae and Chlorophyceae and their subordinate taxa. *Plant Systematics and Evolution*, 146: 75–86.

Delwiche, C.F. (1999) Tracing the thread of plastid diversity through the tapestry of life. *American Naturalist*, 154: S164–S177.

Delwiche, C.F. and Palmer, J.D. (1997) The origin of plastids and their spread via secondary symbiosis. *Plant Systematics and Evolution*, 11 (Supplement): 53–86.

Dettman, J.R., Jacobson, D.J., and Taylor, J.W. (2003) A multilocus geneological approach to phylogenetic species recognition in the model eukaryote *Neurospora*. *Evolution*, 57: 2703–2720.

Engler, A. (1892) *Syllabus der Vorlesungen über specielle und medicinisch-pharmceutische Botanik. Eine Übersicht über das Gesamte Pflanzenreich mit Berücksichtigung der Medicinal und Nutzpflanzen.* Gebrüder Bornträger, Berlin.

Ettl, H. (1981) Die Klasse Chlamydophyceae, eine natürliche Gruppe der Grünalgen (Chlorophyta) (Systematische Bemerkungen zu den Grünalgen I). *Plant Systematics and Evolution*, 137: 107–126.

Ettl, H. (1983) Chlorophyta I (Phytomonadina). In *Süßwasserflora von Mitteleuropa, Band 9* (eds. H. Ettl, J. Gerloff, H. Heynig, and D. Mollenhauer), Gustav Fischer, Stuttgart.

Ettl, H. and Gärtner, G. (1995) *Syllabus der Boden-, Luft- und Flechtenalgen.* Gustav Fischer, Stuttgart.

Ettl, H. and Komárek, J. (1982) Was versteht man unter dem Begriff "kokkale Grünalgen"? (Systematische Bemerkungen zu den Grünalgen II). *Archiv für Hydrobiologie, Supplement*, 60 (*Algological Studies*, 41): 345–374.

Fabry, S., Köhler, A., and Coleman, A.W. (1999) Intraspecies analysis: comparison of ITS sequence data and gene intron sequence data with breeding data for a worldwide collection of *Gonium pectorale*. *Journal of Molecular Evolution*, 48: 94–101.

Fawley, M.W., Fawley, K.P., and Buchheim, M.A. (2004) Molecular diversity among communities of freshwater microchlorophytes. *Microbial Ecology*, 48: 489–499.

Fenchel, T. and Finlay, B.J. (2003) Is microbial diversity fundamentally different from biodiversity of larger animals and plants? *European Journal of Phycology*, 39: 486–490.

Finlay, B.J. (2002) Global dispersal of free-living microbial eukaryote species. *Science*, 296: 1061–1063.

Finlay, B.J. and Fenchel, T. (2002) Microbial eukaryote species—response. *Science*, 297: 337.

Finlay, B.J. and Fenchel, T. (2004) Cosmopolitan metapopulations of free-living microbial eukaryotes. *Protist*, 155: 237–244.

Fott, B. (1971) *Algenkunde*, Fischer, Jena.

Friedl, T. (1995) Inferring taxonomic positions and testing genus level assignments in coccoid green lichen algae—a phylogenetic analysis of 18S ribosomal-RNA sequences from *Dictyochloropsis reticulata* and from members of the genus *Myrmecia* (Chlorophyta, Trebouxiophyceae cl. nov.). *Journal of Phycology*, 31: 632–639.

Friedl, T. (1996) Evolution of the polyphyletic genus *Pleurastrum* (Chlorophyta): inferences from nuclear-encoded ribosomal DNA sequences and motile cell ultrastructure. *Phycologia*, 35: 456–469.

Friedl, T. (1997) The evolution of the green algae. *Plant Systematics and Evolution*, 11 (Supplement): 87–101.

Friedl, T. and O'Kelly, C.J. (2002) Phylogenetic relationships of green algae assigned to the genus *Planophila* (Chlorophyta): evidence from 18S rDNA sequence data and ultrastructure. *European Journal of Phycology*, 37: 373–384.

Fries, L. (1975) Some observations on morphology of *Enteromorpha linza* (L.) J.Ag. and *Enteromorpha compressa* (L.) Grev. in axenic culture. *Botanica Marina*, 18: 251–253.

Fritsch, F.E. (1935) *The Structure and Reproduction of the Algae. Volume I. Introduction, Chlorophyceae. Xanthophyceae, Chrysophyceae, Bacillariophyceae, Cryptophyceae, Dinophyceae, Chloromonineae, Euglenineae, Colourless Flagellata.* Vol. I, Cambridge University Press, Cambridge.

Graham, L.E. and Wilcox, L.W. (2000) *Algae.* Prentice Hall, Upper Saddle River, NJ.

Graybeal, A. (1998) Is it better to add taxa or characters to a difficult phylogenetic problem? *Systematic Biology*, 47: 9–17.

Greuter, W., McNeill, J., Barrie, F.R., Burdet, H.-M., Demoulin, V., Filgueiras, T.S., Nicolson, D.H., Silva, P.C., Skog, J.E., Trehane, P., Turland, N.J., and Hawksworth, D.L. (2000) International Code of Botanical Nomenclature (St. Louis Code). *Regnum Vegetabile*, 138. Koeltz Scientific Books, Königstein.

Guillou, L., Eikrem, W., Chretiennot-Dinet, M.J., Le Gall, F., Massana, R., Romari, K., Pedros-Alio, C., and Vaulot, D. (2004) Diversity of picoplanktonic prasinophytes assessed by direct nuclear SSU rDNA sequencing of environmental samples and novel isolates retrieved from oceanic and coastal marine ecosystems. *Protist*, 155: 193–214.

Hanyuda, T.I., Wakana, S., Arai, K., Miyaji, Y., Watano, K., and Ueda, K. (2002) Phylogenetic relationships within Cladophorales (Ulvophyceae, Chlorophyta) inferred from 18S rRNA gene sequences with special reference to *Aegagropila linnaei*. *Journal of Phycology*, 38: 564–571.

Hayden, H.S. and Waaland, J.R. (2002) Phylogenetic systematics of the Ulvaceae (Ulvales, Ulvophyceae) using chloroplast and nuclear DNA sequences. *Journal of Phycology*, 38: 1200–1212.

Hayden, H.S., Blomster, J., Maggs, C.A., Silva, P.C., Stanhope, M.J., and Waaland, J.R. (2003) Linnaeus was right all along: *Ulva* and *Enteromorpha* are not distinct genera. *European Journal of Phycology*, 38: 277–294.

Hepperle, D., Nozaki, H., Hohenberger, S., Huss, V.A.R., Morita, E., and Krienitz, L. (1998) Phylogenetic position of the Phacotaceae within the Chlamydophyceae as revealed by analysis of 18S rDNA and *rbc*L sequences. *Journal of Molecular Evolution*, 47: 420–430.

Huss, V.A.R. and Kranz, H.D. (1997) Charophyte evolution and the origin of land plants. *Plant Systematics and Evolution*, 11 (Supplement): 103–114.

Jaenicke, L. and Starr, R.C. (1996) The lurlenes, a new class of plastoquinone-related mating pheromones from *Chlamydomonas allensworthii* (Chlorophyceae). *European Journal of Biochemistry*, 241: 581–585.

Jeffrey, C. (1982) Kingdoms, codes and classification. *Kew Bulletin*, 37: 403–416.

Karol, K.G., McCourt, R.M., Cimino, M.T., and Delwiche, C.F. (2001) The closest living relatives of land plants. *Science*, 194: 2351–2353.

Keeling, P.J. 2004 Diversity and evolutionary history of plastids and their hosts. *American Journal of Botany*, 91: 1481–1493.

Kimura, M. (1980) A simple method for estimating evolutionary rate of base substitutions through comparative studies of nucleotide sequences. *Journal of Molecular Evolution*, 16: 111–120.

Komárek, J. and Fott, B. (1983) Chlorophyceae (Grünalgen) Ordnung: Chlorococcales. In *Das Phytoplankton des Süsswassers, Band 7 (1. Hälfte)* (ed. G. Huber-Pestalozzi), Schweizerbart, Stuttgart, pp. 1–1044.

Kornmann, P. (1973) Codiolophyceae, a new class of Chlorophyta. *Helgoländer Wissenschaftliche Meeresuntersuchungen*, 25: 1–13.

Kornmann, P. and Sahling, P.-H. (1983) Meeresalgen von Helgoland: Ergänzung. *Helgoländer Meeresuntersuchungen*, 36: 1–65.

Krienitz, L., Hegewald, E., Hepperle, D., and Wolf, M. (2003) The systematics of coccoid green algae: 18S rRNA gene sequence data versus morphology. *Biologia*, 58: 437–446.

Krienitz, L., Hegewald, E., Hepperle, D., Huss, V.A.R., Rohrs, T., and Wolf, M. (2004) Phylogenetic relationship of *Chlorella* and *Parachlorella* gen. nov. (Chlorophyta, Trebouxiophyceae). *Phycologia*, 43: 529–542.

Lam, D.W. and Zechman, F.W. (2006) Phylogenetic analyses of the Bryopsidales (Ulvophyceae, Chlorophyta) based on RUBISCO large subunit gene sequences. *Journal of Phycology*, 42: 669–678.

Leliaert, F., Rousseau, F., De Reviers, B., and Coppejans, E. (2003) Phylogeny of the Cladophorophyceae (Chlorophyta) inferred from partial LSU rRNA gene sequences: is the recognition of a separate order Siphonocladales justified? *European Journal of Phycology*, 38: 233–246.

Lewis, L.A. (1997) Diversity and phylogenetic placement of *Bracteacoccus* Tereg (Chlorophyceae, Chlorophyta) based on 18S ribosomal RNA gene sequence data. *Journal of Phycology*, 33: 279–285.

Lewis, L.A. and McCourt, R.M. (2004) Green algae and the origin of land plants. *American Journal of Botany*, 91: 1535–1556.

Lopez-Bautista, J.M. and Chapman, R.L. (2003) Phylogenetic affinities of the Trentepohliales inferred from small-subunit rDNA. *International Journal of Systematic and Evolutionary Microbiology*, 53: 2099–2106.

Lopez-Bautista, J.M., Rindi, F., and Guiry, M.D. (2006) Molecular systematics of the subaerial green algal order Trentepohliales: an assessment based on morphological and molecular data. *International Journal of Systematic and Evolutionary Microbiology*, 56: 1709–1715.

Luo, W., Krienitz, L., Pflugmacher, S., and Walz, N. (2005) Genus and species concept in *Chlorella* and *Micractinium* (Chlorophyta, Chlorellaceae): genotype versus phenotypical variability under ecosystem conditions. *Verhandlungen der Internationalen Vereinigung für theoretische und angewandte Limnologie*, 29: 170–173.

Luo, W., Pflugmacher, S., Pröschold, T., Walz, N., and Krienitz, L., (2006) Genotype versus phenotype variability in *Chlorella* and *Micractinium* (Chlorophyta, Trebouxiophyceae). *Protist*, 157: 315–333.

Marin, B. and Melkonian, M. (1999) Mesostigmatophyceae, a new class of streptophyte green algae revealed by SSU rRNA sequence comparisons. *Protist*, 150: 399–417.

Matsuo, Y., Suzuki, M., Kasai, H., Shizuri, Y., and Harayama, S. (2003) Isolation and phylogenetic characterization of bacteria capable of inducing differentiation in the green alga *Monostroma oxyspermum*. *Environmental Microbiology*, 5: 25–35.

Matsuo, Y., Imagawa, H., Nishizawa, M., and Shizuri, Y. (2005) Isolation of an algal morphogenesis inducer from a marine bacterium. *Science*, 307: 1598.

Mattox, K.R. and Stewart, K.D. (1984) Classification of the green algae: a concept based on comparative cytology. In *The Systematics of Green Algae*, (eds. D.E.G. Irvine and D.M. John), The Systematics Association Special Volume 27, Academic Press, London, pp. 29–72.

Mayr, E. (1948) The bearing of the new systematics on genetical problems: the nature of species. *Advances in Genetics*, 2: 205–237.

Melkonian, M. (1982) Structural and evolutionary aspects of the flagellar apparatus in green algae and land plants. *Taxon*, 31: 255–265.

Melkonian, M. and Surek, B. (1995) Phylogeny of the Chlorophyta—congruence between ultrastructural and molecular evidence. *Bulletin de la Societe Zoologique de France—Evolution et Zoologie*, 120: 191–208.

Moestrup, Ø. and Throndsen, J. (1988) Light and electron microscopical studies on *Pseudoscourfieldia marina*, a primitive scaly green flagellate (Prasinophyceae) with posterior flagella. *Canadian Journal of Botany*, 66: 1415–1434.

Müller, J., Friedl, T., Hepperle, D., Lorenz, M., and Day, J.G. (2005) Distinction between multiple isolates of *Chlorella vulgaris* (Chlorophyta, Trebouxiophyceae) and testing for conspecificity using amplified fragment length polymorphism and its rDNA sequences. *Journal of Phycology*, 41: 1236–1247.

Nakanishi, K., Nishijima, M., Nishimura, M., Kuwano, K., and Saga, N. (1996) Bacteria that induce morphogenesis in *Ulva pertusa* (Chlorophyta) grown under axenic conditions. *Journal of Phycology*, 32: 479–482.

Nakayama, T., Marin, B., Kranz, H.D., Surek, B., Huss, V.A.R., Inouye, I., and Melkonian, M. (1998). The basal position of scaly green flagellates among the green algae (Chlorophyta) is revealed by analyses of nuclear-encoded SSU rRNA sequences. *Protist*, 149: 367–380.

Norton, T.A., Melkonian, M., and Andersen, R.A. (1996). Algal biodiversity. *Phycologia*, 35: 308–326.

Nozaki, H., Ohta, N., Morita, E., and Watanabe, M.M. (1998) Toward a natural system of species in *Chlorogonium* (Volvocales, Chlorophyta): a combined analysis of morphological and *rbc*L gene sequence data. *Journal of Phycology*, 34: 1024–1037.

O'Kelly, C.J. and Floyd, G.L. (1984) Correlations among patterns of sporangial structure and development, life histories, and ultrastructural features in the Ulvophyceae. In *The Systematics of Green Algae* (eds D.E.G. Irvine and D.M. John), The Systematics Association Special Volume 27, Academic Press, London, pp. 121–156.

O'Kelly, C.J., Bellows, W.K., and Wysor, B. (2004a) Phylogenetic position of *Bolbocoleon piliferum* (Ulvophyceae, Chlorophyta): evidence from reproduction, zoospore and gamete ultrastructure, and small subunit rRNA gene sequences. *Journal of Phycology*, 40: 209–222.

O'Kelly, C.J., Wysor, B., and Bellows, W.K. (2004b) *Collinsiella* (Ulvophyceae, Chlorophyta) and other ulotrichalean taxa with shell-boring sporophytes form a monophyletic clade. *Phycologia*, 43: 41–49.

O'Kelly, C.J., Wysor, B., and Bellows, W.K. (2004c) Gene sequence diversity and the phylogenetic position of algae assigned to the genera *Phaeophila* and *Ochlochaete* (Ulvophyceae, Chlorophyta). *Journal of Phycology*, 40: 789–799.

Olsen, J., Stam, W.T., Berger, S., and Menzel, D. (1994) 18S rDNA and evolution in the Dasycladales (Chlorophyta): modern living fossils. *Journal of Phycology*, 30: 850–856.

Pascher, A. (1931) Systematische Übersicht über die mit Flagellaten in Zusammenhang stehenden Algenreihen und Versuch einer Einreihung dieser Algenstämme in die Stämme des Pflanzenreiches. *Beihefte zum Botanischen Centralblatt*, 48 (Abteilung II, 2): 317–332.

Philippe, H. (2000) Opinion: long branch attraction and protist phylogeny. *Protist*, 151: 307–316.

Pickett-Heaps, J.D. (1975) *Green Algae. Structure, Reproduction and Evolution in Selected Genera*. Sinauer, Sunderland, MA.

Pombert, J.F., Otis, C., Lemieux, C., and Turmel, M. (2004) The complete mitochondrial DNA sequence of the green alga *Pseudoendoclonium akinetum* (Ulvophyceae) highlights distinctive evolutionary trends in the Chlorophyta and suggests a sister-group relationship between the Ulvophyceae and Chlorophyceae. *Molecular Biology and Evolution*, 21: 922–935.

Pombert, J.F., Otis, C., Lemieux, C., and Turmel, M. (2005) The chloroplast genome sequence of the green alga *Pseudendoclonium akinetum* (Ulvophyceae) reveals unusual structural features and new insights into the branching order of chlorophyte lineages. *Molecular Biology and Evolution*, 22: 1903–1918.

Pombert, J.F., Lemieux, C., and Turmel, M. (2006) The complete chloroplast DNA sequence of the green alga *Oltmannsiellopsis viridis* reveals a distinctive quadripartite architecture in the chloroplast genome of early diverging ulvophytes. *BMC Biology*, 4:3.

Posada, D. and Buckley, T.R. (2004) Model selection and model averaging in phylogenetics: advantages of the AIC and Bayesian approaches over likelihood ratio tests. *Systematic Biology*, 53: 793–808.

Posada, D. and Crandall, K.A. (1998) Modeltest: testing the model of DNA substitution. *Bioinformatics*, 14: 817–818.

Pringsheim, E.G. (1963) *Farblose Algen—Ein Beitrag zur Evolutionsforschung.* Gustav Fischer, Stuttgart.

Pröschold, T., Harris, E.H., and Coleman, A.W. (2005) Portrait of a species: *Chlamydomonas reinhardtii.* *Genetics*, 170: 1601–1610.

Pröschold, T., Marin, B., Schlösser, U.G., and Melkonian, M. (2001) Molecular phylogeny and taxonomic revision of *Chlamydomonas* (Chlorophyta). I. Emendation of *Chlamydomonas* Ehrenberg and *Chloromonas* Gobi, and description of *Oogamochlamys* gen. nov. and *Lobochlamys* gen. nov. *Protist*, 152: 265–300.

Pröschold, T. and Silva, P.C. (2007) Proposal to change the listed type of *Chlamydomonas* Ehrenb., nom. cons. (*Chlorophyta*). *Taxon* 56: 595–596.

Provasoli, L. and Pintner, I.J. (1980) Bacteria induced polymorphism in an axenic laboratory strain of *Ulva lactuca* (Chlorophyceae). *Journal of Phycology*, 16: 196–201.

Rabenhorst, L. (1863) *Kryptogamenflora von Sachsen, der Oberlausitz, Thüringen und Nordböhmen, mit Berücksichtigung der benachbarten Länder. I. Abtheilung: Algen im weiteren Sinne, Leber- und Laubmoose.* E. Kummer, Leipzig.

Rokas, A. and Carroll, S.B. (2005) More genes or more taxa? The relative contribution of gene number and taxon number to phylogenetic accuracy. *Molecular Biology and Evolution*, 22: 1337–1344.

Round, F.E. (1963) The taxonomy of the Chlorophyta. *British Phycological Bulletin*, 2: 224–245.

Round, F.E. (1971) The taxonomy of the Chlorophyta. II. *British Phycological Journal*, 6: 235–264.

Round, F.E. (1984) The systematics of the Chlorophyta: an historical review leading to some modern concepts (the taxonomy of the Chlorophyta III). In *The Systematics of Green Algae* (eds D.E.G. Irvine and D.M. John), The Systematics Association Special Volume 27, Academic Press, London, pp. 1–28.

Sanderson, M.J., and Shaffer, H.B. (2002) Troubleshooting molecular phylogenetic analysis. *Annual Review of Ecology and Systematics*, 33: 49–72.

Schlösser, U.G. (1976) Entwicklungsstadien- und sippenspezifische Zellwand-Autolysine bei der Freisetzung von Fortpflanzungszellen in der Gattung *Chlamydomonas. Berichte der Deutsche Botanischen Gesellschaft*, 89: 1–56.

Schlösser, U.G. (1984) Species-specific sporangium autolysins (cell-wall dissolving enzymes) in the genus *Chlamydomonas.* In *The Systematics of Green Algae* (eds D.E.G. Irvine and D.M. John), The Systematics Association Special Volume 27, Academic Press, London, pp. 409–418.

Schlösser, U.G. (1987) Action of cell wall autolysins in asexual reproduction of filamentous green algae: evidence and species specificity. In *Algal Development: Molecular and Cellular Aspects* (eds. W. Wiessner, D.G. Robinson, and R.C. Starr), Springer, Berlin Heidelberg, Germany, pp. 75–80.

Shoup, S. and Lewis, L.A. (2003) Polyphyletic origin of parallel basal bodies in swimming cells of Chlorophycean green algae (Chlorophyta). *Journal of Phycology*, 39: 789–796.

Silva, P.C. (1980) Names of classes and families of living algae: with special reference to their use in the Index Nominum Genericorum (Plantarum). *Regnum Vegetabile*, 103: 1–156.

Simpson, A.G.B., Inagaki, Y., and Roger, A.J. (2006) Comprehensive multigene phylogenies of excavate protists reveal the evolutionary positions of "primitive" eukaryotes. *Molecular Biology and Evolution*, 23: 615–625.

Sites, J.W. and Marshall J.C. (2003) Delimiting species: a Renaissance issue in systematic biology. *Trends in Ecology and Evolution*, 18: 462–470.

Slapeta, J., López-García, P., and Moreira, D. (2006) Global dispersal and ancient cryptic species in the smallest marine eukaryotes. *Molecular Biology and Evolution*, 23: 23–29.

Sluiman, H.J. (1985) A cladistic evaluation of the lower and higher green plants (Viridiplantae). *Plant Systematics and Evolution*, 149: 217–232.

Sluiman, H.J. (1989). The green algal class Ulvophyceae. An ultrastructural survey and classification. *Cryptogamic Botany*, 1: 83–94.

Steinkötter, J., Bhattacharya, D., Semmelroth, I., Bibeau, C., and Melkonian, M. (1994) Prasinophytes form independent lineages within the Chlorophyta—evidence from ribosomal-RNA sequence comparison. *Journal of Phycology*, 30: 340–345.

Swofford, D.L. (2002) PAUP*. Phylogenetic Analysis Using Parsimony (*and Other Methods). Version 4. Sinauer Associates, Sunderland, MA.

Tamura, K. and Nei, M. (1993) Estimation of the number of nucleotide substitutions in the control region of mitochondrial DNA in humans and chimpanzees. *Molecular Biology and Evolution*, 10: 512–526.

Tan, I.H., Blomster, J., Hansen, G., Leskinen, E., Maggs, C.A., Mann, D.G., Sluiman, H.J., and Stanhope, M.J. (1999). Molecular phylogenetic evidence for a reversible morphogenetic switch controlling the gross morphology of two common genera of green seaweeds, *Ulva* and *Enteromorpha*. *Molecular Biology and Evolution*, 16: 1011–1018.

Tappan, H. (1980) *Palaeobiology of Plant Protists*. Freeman, San Francisco.

Taylor, J.W., Jacobson, D.J., Kroken, S., Kasuga, T., Geiser, D.M., Hibbett, D.S., and Fisher, M.C. (2000) Phylogenetic species recognition and species concepts in fungi. *Fungal Genetics and Biology*, 31: 21–32.

van den Hoek, C. (1963) *Revision of the European Species of Cladophora*. Brill, Leiden.

van den Hoek, C. (1982) A taxonomic revision of the American species of *Cladophora* (Chlorophyceae) in the North Atlantic Ocean and their geographic distribution. *Verhandelingen der Koninklijke Nederlandse Akademie van Wetenschappen, Afdeling Natuurkunde, Tweede Reeks*, 78: 1–236.

van den Hoek, C. and Jahns, H.M. (1978) *Algen. Einführung in die Phykologie*. Thieme, Stuttgart.

van den Hoek, C., Mann, D.G., and Jahns, H.M. (1993) *Algen. Einführung in die Phykologie*. Thieme, Stuttgart.

van den Hoek, C., Mann, D.G., and Jahns, H.M. (1995) *Algae. An Introduction to Phycology*. Cambridge University Press, Cambridge.

van den Hoek, C., Stam, W.T., and Olsen, J.L. (1988). The emergence of a new chlorophytan system, and Dr. Kornmann's contribution thereto. *Helgoländer Meeresuntersuchungen*, 42: 339–383.

van der Strate, H.J., Boele-Bos, S.A., Olsen, J.L., van de Zande, L., and Stam, W.T. (2002) Phylogeographic studies in the tropical seaweed *Cladophoropsis membranacea* (Chlorophyta, Ulvophyceae) reveal a cryptic species complex. *Journal of Phycology*, 38: 572–582.

Verbruggen, H., De Clerck, O., Kooistra, W.H.C.F., and Coppejans, E. (2005a) Molecular and morphometric data pinpoint species boundaries in *Halimeda* section *Rhipsalis* (Bryopsidales, Chlorophyta). *Journal of Phycology*, 41: 606–621.

Verbruggen, H., De Clerck, O., Schils, T., Kooistra, W.H.C.F., and Coppejans, E. (2005b) Evolution and phylogeography of *Halimeda* section *Halimeda* (Bryopsidales, Chlorophyta). *Molecular Phylogenetics and Evolution*, 37: 789–803.

Watanabe, S. and Floyd, G.L. (1994) Ultrastructure of the flagellar apparatus of the zoospores of the irregularly shaped coccoid green-algae *Chlorochytrium lemnae* and *Kentrosphaera gibberosa* (Chlorophyta). *Nova Hedwigia*, 59: 1–11.

Watanabe, S., Kuroda, N., and Maiwa, F. (2001) Phylogenetic status of *Helicodictyon planctonicum* and *Desmochloris halophila* gen. et comb. nov. and the definition of the class Ulvophyceae (Chlorophyta). *Phycologia*, 40: 421–434.

Woolcott, G.W., Knoller, K., and King, R.J. (2000) Phylogeny of the Bryopsidaceae (Bryopsidales, Chlorophyta): cladistic analyses of morphological and molecular data. *Phycologia*, 39: 471–481.

Wortley, A.H., Rudall, P.J., Harris, D.J., and Scotland, R.W. (2005) How much data are needed to resolve a difficult phylogeny? Case study in Lamiales. *Systematic Biology*, 54: 697–709.

Wuyts, J., van de Peer, Y., and Wachter, R.D. (2001) Distribution of substitution rates and location of insertion sites in the tertiary structure of ribosomal RNA. *Nucleic Acids Research*, 29: 5017–5028.

Yoon, H., Hackett, S.J., Ciniglia, C., Pinto, G., and Bhattacharya, D. (2004) A molecular timeline for the origin of photosynthetic eukaryotes. *Molecular Biology and Evolution*, 21: 809–818.

Zechman, F.W. (2003) Phylogeny of the Dasycladales (Chlorophyta, Ulvophyceae) based on analyses of RUBISCO large subunit (*rbc*L) gene sequences. *Journal of Phycology*, 39: 819–827.

Zechman, F.W., Theriot, E.C., Zimmer, E.A., and Chapman, R.L. (1990) Phylogeny of the Ulvophyceae (Chlorophyta): cladistic analysis of nuclear-encoded rRNA sequence data. *Journal of Phycology*, 26: 700–710.

8 In the shadow of giants: systematics of the charophyte green algae

John D. Hall and Charles F. Delwiche

CONTENTS

ABSTRACT

Charophyte green algae are those organisms most closely related to land plants. The group has at least five major lineages: the Charophyceae, Coleochaetophyceae, Zygnematophyceae, Klebsormidiophyceae, Chlorokybophyceae, and probably the Mesostigmatophyceae. These organisms are briefly introduced and their relative phylogenetic positions discussed. Current systematic understanding of the groups is discussed as well as the potential role of genomic studies in the systematics of charophyte green algae. Genomic studies are beginning to elucidate the order of ancient branching events in the lineage; however, greater molecular and broader taxon sampling will be required to resolve some relationships. In addition to deep nodes, molecular phylogenetic investigations of populations and species of all the lineages are wanting. Continued investigation and greater sampling will provide more insight into the evolution of these organisms and early land plant evolution.

INTRODUCTION

Green algae are one of the most diverse groups of organisms on Earth both structurally and in terms of number of described species. They occupy almost every habitat and are the algal relatives of one of the most species-rich lineages, the land plants. Although land plants evolved from green algae, there are major diversifications in the green algae that preceded the invasion of land and radiation of embryophytes. One major distinction is between the Chlorophyta and the Charophyta (Charophyceae *sensu* Mattox and Stewart, 1984). Land plants are members of the Charophyta and their closest algal relatives are here called the charophytes. The charophytes are only a small part of total green algal diversity, but their evolutionary history gives direct insight into the evolution of plants. Nomenclature used here follows Lewis and McCourt (2004; see Table 1). In this paper, "charophyte" refers to those algae most closely related to land plants (not just the Charophyceae *sensu stricto*). When referring to Charophyceae *sensu stricto*, we use either Characeae or their common name, stoneworts.

HISTORIC PERSPECTIVE

The charophytes, in the sense discussed here, have been recognized as a group only since the 1980s. They include their namesake Characeae (Charophyceae *sensu stricto*) and a collection of seemingly disparate lineages: the Coleochaetophyceae, the conjugating green algae (Zygnematophyceae), the Klebsormidiophyceae, *Chlorokybus atmophyticus* Geitler (Chlorokybophyceae), and probably, *Mesostigma* (Mesostigmatophyceae). Some charophytes are quite large, such as members of the Characeae, and have been known for several hundreds of years: Linnaeus named some species of *Chara,* which had previously been described as aquatic forms of *Equisetum* (Wood and Imahori, 1965); others, such as *Spirogyra,* were among the first algae discovered by means of Van Leeuwenhoek's microscopes in the late 17th century (Leeuwenhoek, 1674). Even the relationship between green algae and land plants had long been supposed as implied by discussions of microscopic vegetables (Ingenhousz, 1779). Bower (1908) described the plant-like characteristics of many algae, including some today recognized as charophytes. It was not until the latter part of the 20th century that science and technology (in this case, electron microscopy) converged to provide evidence for the relationships among green algae and their affinity to land plants. The groups currently thought to belong to the charophyte lineage, with the exception of the polemic *Mesostigma viride* Lauterborn, were recognized by Mattox and Stewart (1984) in their systematic treatment of the green algae based on comparative cytology.

Charophyte systematics, however, did not begin in the 20th century. Members of these lineages were known in the 19th century and earlier. Two groups in particular, the Characeae and the conjugating green algae (Zygnematophyceae), include several hundreds or thousands of named species, respectively. Both of these have a long history of independent systematic investigation, and several monographs are dedicated to their taxonomy and distribution. Pringsheim (1860) investigated *Coleochaete,* and contemporary authors have critically studied this genus and described new species (Cimino and Delwiche, 2002; Delwiche et al., 2002; Szymańska, 1989). All the charophytes are commonly (and historically) included in local florulas, though rarely as a group unto themselves. Even today, valuable systematic data are often published as part of local or regional florulas. Systematic investigation, therefore, proceeds on many fronts: higher-level classification of the lineages and families as well as the population and species levels and molecular, genomic, cytological, and morphological methods are used.

WHY SHOULD ONE STUDY THE SYSTEMATICS OF CHAROPHYTES?

The charophyte green algae hold a unique phylogenetic position as the closest extant relatives of terrestrial plants (in fact, embryophytes could be more correctly treated as a specialized charophyte lineage). Understanding of relationships within and among these lineages continues to provide insight into the evolution of land plants and their occupation of terrestrial habitats. Charophytes are also of systematic interest because they inhabit environments that are greatly affected by humans. Charophytes are primarily freshwater organisms (although a few occur in brackish pools) and many charophytes are most common in oligotrophic waters. These habitats are particularly impacted by human activities. There is a great need for monitoring and investigation of freshwater biodiversity, as estimates of the rate of extinction among freshwater organisms is very high (Leidy and Moyle, 1998; Watanabe, 2005). Few data are available on the conservation status of most charophycean taxa. Available data on Characeae and conjugating green algae indicate that local extinctions have occurred and global extinction may be likely. Few countries (including the United States) have biotic inventories that would reveal local extinctions, so the real loss of global charophyte biodiversity is unknown. Those countries with biotic inventories indicate that many charophyte taxa are threatened, endangered, or already locally extinct (Adam, 2004; Krause, 1984; Németh, 2005; Siemińska, 1986; Stewart and Church, 1992; Watanabe, 2005). Charophyte systematics must continue with some haste if we are to record the global diversity and natural distribution of these important organisms.

WHICH GREEN ALGAE BELONG TO THE CHAROPHYTA?

The charophytes constitute one of the two primary lineages of green algae and are distinguished from their relatives, the chlorophytes, by a number of distinct if not immediately obvious features. Mattox and Stewart (1984) separated the charophyte algae from other green algae based on the presence of a multilayer structure at the base of the two flagella which insert subapically on the asymmetric flagellate cells. It is now known that multilayer structures occur in other groups of algae besides the charophytes, but the subapical insertion of two similar flagella is distinctive to the group. The Zygnematophyceae lack flagella and were included in the group because of their persistent mitotic spindle (Mattox and Stewart, 1984). Other characteristics were later discovered that indicated that the conjugating green algae are closely related to other charophytes, but these are not diagnostic of all charophytes. It is important to note that the charophytes are only a monophyletic group if land plants are considered among their ranks (Bhattacharya et al., 1998; Karol et al., 2001).

The classification provided by Lewis and McCourt (2004) assigns class status to the major lineages of charophytes, including the embryophytes (land plants). Higher-level nomenclature is still a source of disagreement among systematists, as are the relationships among the charophyte lineages. However, most authors do agree that the charophyte algae are different from other green algae (Chlorophyta), that they are closely related to land plants, and that most of the taxa proposed by Mattox and Stewart (1984) belong to this group. The inclusion of *Mesostigma viride* is less certain but seems very likely (see discussion below). The placement of embryophytes as a monophyletic lineage, deeply embedded within the charophytes, is not controversial, but the nomenclatural implications of this biological fact are highly so.

The number of charophyte species reported in the literature is not known with certainty. The best estimates are: Mesostigmatophyceae, 2; Chlorokybophyceae, 1 (Geitler, 1942b); Klebsormidiophyceae, 19 (Ettl and Gärtner, 1995; Lokhorst, 1996); Zygnematophyceae, 4000 to 13,000 (Gerrath, 1993; Hoshaw et al. 1990); Coleochaetophyceae, 22 (Szymańska, 2003; Thompson, 1969); and Charophyceae, 395 (Wood and Imahori, 1965). These estimates should be seen as minimum, provisional, good faith estimates taken from the literature, and should be interpreted with caution.

Species estimates among charophytes are difficult: inclusion of certain genera in the group is not certain and estimates are strongly biased by the treatment of varieties, particularly among the stoneworts and conjugating green algae. The total diversity of all other charophytes falls well within the uncertainty of estimates for the conjugating green algae alone. That said, estimates for the number of conjugating green algae range from about 3000 desmids (Gerrath, 1993) and 800 filamentous Zygnematales (Kadlubowska, 1984) to 10,000 to 12,000 placoderm desmids excluding the filamentous and unicellular Zygnematales (Hoshaw et al., 1990). Differences in these estimates are due in part to the inaccessibility of the relevant literature: no one is certain how many taxa have been described, much less how many are synonymous. Another major factor in the variability of estimates is the treatment of varieties. Taxonomists of the Desmidiaceae (Zygnematophyceae) continue to use subspecies, varieties, and forms, a practice abandoned in many other algal groups. This means that any one species name can have as many as several dozen subspecific taxa associated with it. The relationship between a desmid "species" and a biological species is not clear. If the varieties of conjugating green algae were treated as species, their number would likely approach the 12,000 species estimate of Hoshaw et al. (1990). Estimates of the number of species of charophytes are consequently subject to considerable interpretation.

It might be expected that with the number of papers dedicated to, in particular, the taxonomy of the conjugating green algae, there would be a clear understanding of their diversity or at least the number of species on Earth. Unfortunately, information on their numbers and distribution is limited by the geographic location of investigators, accessibility of study sites, and time available for investigation. A limitation unique to the charophytes, to a greater or lesser degree, includes a tendency for some charophyte species to be overlooked in general floristic studies: *Coleochaete* grows attached to substrata, such as rocks and aquatic plants, which are frequently not collected; many desmids are benthic and do not appear in plankton studies; and smaller species of conjugating green algae and *Klebsormidium* are often mistaken for chlorophytes or xanthophytes in floristic surveys. As with many microscopic taxa, apparent distributions may be primarily a function of the geography of the investigators. It is also important to note that estimates reflect only the number of described taxa; the actual number of living charophytes in the world could be much greater. Vast portions of the world have not been investigated and new species are frequently described from even the best-studied regions (e.g. Coesel, 2002; Szymańska, 2003).

GENERAL SYSTEMATIC STUDIES OF CHAROPHYTES

Since the time the charophytes were formally recognized as a distinct group of green algae, few studies have investigated the relationships among these disparate lineages. Many of these studies have focused on two important questions: which taxon is most closely related to land plants and how might the ancestor of the charophytes (and possibly all green algae) have appeared? If these questions could be confidently answered, one could use the characteristics of the extant charophytes to make inferences about the evolution of land plants and charophyte algae. Systematic investigation, particularly molecular systematics, is necessary because the fossil record of these early diverging lineages is very poor; only the stoneworts (Charophyceae *sensu stricto*) have a well-documented fossil record (Feist and Grambast-Fessard, 1990). Few molecular systematic investigations have focused on the relationships within the charophyte lineages.

Nearly every lineage of charophytes, or combination of lineages, has been proposed as the most closely related to land plants at one time or another (Huss and Kranz, 1995; Karol et al., 2001; Kranz et al., 1995; Turmel et al., 2002a; Turmel et al., 2002b). However, most studies have very limited molecular or taxonomic sampling or both. Lemieux et al. (2000), based on a dataset of 53 genes and eight taxa, recovered *Mesostigma viride* at the base of the Viridiplantae clade. Other comparable phylogenies based on fewer genes support the placement of *Mesostigma* as the basal-most lineage of the Charophyta (Bhattacharya et al., 1998; Karol et al., 2001). In a study using 72 mitochondrial genes and six taxa, Turmel et al. (2003) found a sister relationship between stoneworts and embryophytes. One molecular study (Karol et al., 2001) exhibits both substantial molecular sampling (four genes) and a broad taxon sampling (40 taxa). This study found the stoneworts sister to land plants and the remaining lineages to be a paraphyletic assemblage with *Mesostigma* the earliest diverging member of the group (Figure 8.1). The relationship between stoneworts and land plants was strongly

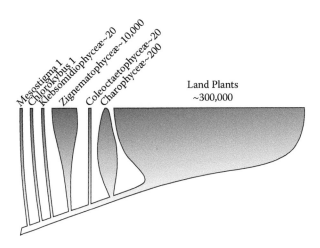

FIGURE 8.1 Diagram of the branching order of the lineages of Charophyta based on Karol et al. (2001). Line width at top of diagram is approximately proportional to the number of described species for the group. Charophyceae and the land plants are the only lineages for which there are ancient fossils. These fossils indicate that the Charophyceae were probably more diverse in the past than they are at present. For the other lineages of charophyte algae, their prehistoric diversity is almost completely unknown.

supported, as was the placement of *Mesostigma* as the basal most charophyte, although the relationships among the other lineages of the charophytes received less support (Karol et al., 2001). It is worth noting, however, that several studies (particularly those using rDNA data) have found the stoneworts to be very distantly related to the land plants and, generally, suggest that the lineage most closely related to land plants is one or an assemblage of the other charophyte classes (Huss and Kranz, 1995; Kranz et al., 1995; Peterson et al., 2003; Turmel et al., 2002a). One study based on 76 chloroplast genes found the conjugating green algae (Zygnematophyceae) to be the sister lineage to land plants with stoneworts being only distantly related (Turmel et al., 2006). Evolution of the charophyte lineages is not completely understood.

SYSTEMATICS OF THE CHAROPHYTE LINEAGES

Mesostigma

The earliest diverging lineage of the Charophyta may be *Mesostigma viride* Lauterborn (Figure 8.2A). This organism was originally classified among the prasinophyte green algae (scaly unicellular flagellates). On the basis of root configuration, Melkonian (1989) noted the affinity of *M. viride* to the charophytes. This was confirmed by analyses of actin sequences (Bhattacharya et al., 1998) that indicated *Mesostigma* is the basal-most lineage of the Charophyta. *Mesostigma viride* is unicellular and covered in minute scales. Cells are circular and compressed with a concave inner surface (Melkonian, 1989). Unlike other charophytes, *Mesostigma viride* has two multilayer structures (MLSs) and an eyespot (Rogers et al., 1981). *Mesostigma viride* is known only from freshwater habitats and has been infrequently reported from the wild. Although its phylogenetic position is not known with certainty, a number of studies suggest that *Mesostigma viride* is either very ancient or very divergent (probably both) compared to other green algae: it has an unusual complement of photosynthetic pigments (Stabenau and Winkler, 2005; Yoshii et al., 2003), and Stabenau and Winkler (2005) suggest that the microbodies found in *Mesostigma* may be ancestral to leaf peroxisomes and glyoxysomes since it has enzymes normally associated with both.

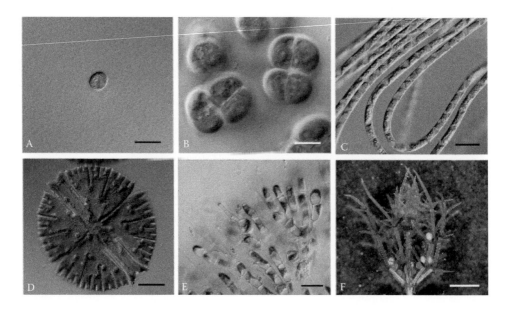

FIGURE 8.2 Micrographs of representatives of the major lineages of the Charophyta. (A) *Mesostigma viride*, scale bar = 20 μm; (B) *Chlorokybus atmophyticus*, scale bar = 10 μm; (C) *Klebsormidium flaccidum*, scale bar = 25 μm; (D) *Micrasterias rotata*, scale bar = 50 μm; (E) *Coleochaete pulvinata*, scale bar = 25 μm; (F) *Chara tomentosa*, scale bar = 5 mm.

Some molecular phylogenetic studies place *Mesostigma* at the base of the Charophyta (Bhattacharya et al., 1998; Karol et al., 2001; Marin and Melkonian, 1999), while others find it to be sister to both the charophytes and chlorophytes (Lemieux et al., 2000; Turmel et al., 2002b; Turmel et al., 2002c). Evidence in favor of including *Mesostigma* in the Charophyta is mounting: *Mesostigma*, the other charophytes, and land plants seem to share a unique duplication of the *GapA/GapB* gene (Peterson et al., 2006). Regardless of its position, *Mesostigma* is critical to our understanding of evolution of the green algae. At least one other species of *Mesostigma* has been described (*M. grande* Korshikov), but this species is not available in culture and, consequently, has not been thoroughly investigated.

Chlorokybus atmophyticus

Chlorokybus atmophyticus Geitler is the sole known representative of its lineage, the Chlorokybophyceae. The species was discovered growing among mosses in a park in Vienna, Austria (Geitler, 1942b). It is unique among the charophyte algae, with the possible exception of some conjugating green algae, in its sarcinoid growth habit (i.e., it grows as small packets of cells enveloped in a common mucilaginous matrix; see Figure 8.2B). Its cell division, as well as production of autospores and swarmers, has been documented (Geitler, 1942a; Rieth, 1972). Sexual reproduction has not been reported. Zoospores of *C. atmophyticus* have a single multilayer structure and two subapical flagella that are covered in scales (Rieth, 1972; Rogers et al., 1980), a structure that is consistent with other flagellate charophyte cells. Although its distribution is unknown, it has been found in subaerial habitats in Europe and Russia (Ettl and Gärtner, 1995; Geitler, 1942b; Rieth, 1972).

Many molecular studies place *Chlorokybus atmophyticus* near the base of the Charophyta (Bhattacharya and Medlin, 1998; Karol et al., 2001). Phylogenetic analyses of 18S rDNA and chloroplast *rbc*L sequences suggest that *Spirotaenia* (normally considered to be a member of the Zygnematophyceae) may form a monophyletic group with *Chlorokybus* (Gontcharov and Melkonian, 2004). This relationship is certainly unexpected since *Spirotaenia* is known to lack flagella and to conjugate during sexual reproduction, characteristics typical of the conjugating green algae. The analysis showed only moderate statistical support for the group and could not rule out other placements of *Spirotaenia*. However, those data did reject a sister relationship between *Spirotaenia* and the conjugating green algae. The relationships of *Chlorokybus* and *Spirotaenia* to other charophyte algae certainly merit further investigation.

KLEBSORMIDIOPHYCEAE

The Klebsormidiophyceae, particularly *Klebsormidium* spp. (Figure 8.2C), are among the most ubiquitous charophytes. The two genera most commonly included in the group (*Klebsormidium* and *Entransia*) are typically unbranched filamentous algae that reproduce by fragmentation of the filaments and the release of (presumably) biflagellate zoospores (Cook, 2004; Lockhorst, 1996). Although structures consistent with zoosporangia were observed in *Entransia fimbriata* Hughes, no flagellate cells were observed (Cook, 2004). *Klebsormidium* typically has a single parietal chloroplast that partly encircles the cell with a single, embedded pyrenoid. Filaments are sometimes attached by holdfasts or interrupted by "H" pieces (Lockhorst, 1996). *Entransia* has one or two parietal chloroplasts that are deeply, irregularly lobed (see Cook, 2004, for discussion of morphology). Inclusion of other genera, such as *Stichococcus* and *Raphidonema*, remains uncertain; some species of *Stichococcus* have been transferred to *Klebsormidium* (Ettl and Gärtner, 1995), while others clearly belong to the Chlorophyta (Lewis and Lewis, 2005.)

The genus name *Klebsormidium* was created to accommodate *Hormidium sensu* Klebs. The taxon *Hormidium* was abandoned because of synonymy and general confusion (Silva et al., 1972). Since that time, a total of about seventeen species have been described or transferred to the genus. Taxonomy of the genus is based on morphological characters, such as filament width, cell wall surface characteristics, and chloroplast shape (Lockhorst, 1996). *Klebsormidium* species are common in freshwater habitats but also occur in many subaerial habitats including desert crusts, urban walls, and freshwater seeps (Johansen et al., 2004; Lewis and Fletchner, 2002). *Entransia* is clearly distinct from *Klebsormidium,* seems to be rare, and occurs infrequently in ponds, bogs, or seeps (Cook, 2004; Hughes, 1948; Prescott, 1966).

Entransia was originally thought to belong to the conjugating green algae (Zygnematophyceae), although it is not known to conjugate. Only one of the two species has been investigated, *E. fimbriata*, which is thought to reproduce by zoospores (Cook, 2004). Sexual reproduction has not been reported. Molecular phylogenetic investigations by McCourt et al. (2000) placed *E. fimbriata* outside the conjugating green algae, and Karol et al. (2001) found it to be most closely related to *Klebsormidium*, a relationship supported by ultrastructural data (Cook, 2004). Very few *Klebsormidium* isolates have been studied using molecular systematic methods (Novis, 2006) and more work certainly remains.

ZYGNEMATOPHYCEAE

The Zygnematophyceae are the most species-rich clade of the Charophyta (excepting land plants). They are commonly referred to as the conjugating green algae because of their unusual mode of sexual reproduction by fusion of non-flagellate gametes. Zygnematophytes may be unicellular, colonial, or filamentous, depending on the species. Historically, the filamentous species with smooth walls (Zygnemataceae) and the unicellular forms were treated separately, although they were thought to be closely related to one another long before the evolutionary process was understood (Ralfs, 1848). The current family-level classification of the conjugating green algae is based primarily on differences in cell wall structure (Mix, 1972). The class is often divided into two orders, the Zygnematales and the Desmidiales, with two and four families, respectively. The species number in the thousands, with the majority belonging to the Desmidiaceae (Gerrath, 1993).

The placement of the Zygnematophyceae within the Charophyta is unclear. Most molecular studies place these algae sister to a clade with Coleochaetophyceae, stoneworts, and land plants, or as part of a lineage sister to land plants and including other charophyte classes. Consistent with the hypothesis that the conjugating green algae are more closely related to land plants than *Klebsormidium* and *Chlorokybus* is the presence of a phragmoplast-like microtubule array in some Zygnematophyceae (Fowke and Pickett-Heaps, 1969). Additionally, the Zygnematophyceae, Coleochaetophyceae, Characeae, and land plants share similar cellulose synthesizing rosettes (Tsekos, 1999). However, one study using many chloroplast genes (but relatively few taxa) places the conjugating green algae sister to land plants (Turmel et al., 2006).

Molecular phylogenetic studies suggest that the two traditional families of the Zygnematales, the Zygnemataceae and the Mesotaeniaceae, are not monophyletic with respect to one another (Gontcharov et al., 2003; McCourt et al., 2000; McCourt et al., 1995). It also appears that many of the genera in the traditional family Mesotaeniaceae may not be monophyletic (Gontcharov et al., 2003, 2004). The order Zygnematales may not be monophyletic but rather consist of two lineages in paraphyly, one containing *Spirogyra* and *Sirogonium*, and the other containing most of the remaining Zygnematales (Gontcharov et al., 2003; McCourt et al., 2000). The lineage that best corresponds to the classical Desmidiales (Figure 8.2D) almost certainly includes organisms formerly classified among the Mesotaeniaceae (Gontcharov et al., 2003; McCourt et al., 2000; Park et al., 1996). One study (Gontcharov and Melkonian, 2004) indicates that at least one genus classified among the conjugating green algae, *Spirotaenia*, may not be part of the main line of zygnematophyte evolution.

Taxonomy within the Zygnematophyceae has relied heavily on general morphology and fine structure, particularly cell wall structures (Desmidiaceae), or spore wall ornamentation (Zygnemataceae). Molecular phylogenetic methods may prove valuable for infrageneric phylogeny within the group, as indicated by the few studies that investigated species relationships (Denboh et al., 2001; Gontcharov and Melkonian, 2005; Lee, 2001; Nam and Lee, 2001). However, most of these studies did not test difficult relationships, and different molecules and methods may be necessary for investigating closely related species.

COLEOCHAETOPHYCEAE

The Coleochaetophyceae are branched, filamentous epiphytes. Four genera, *Coleochaete*, *Chaetosphaeridium*, *Chaetotheke*, and *Awadhiella*, are thought to belong to this group (Bourrelly, 1990; Delwiche et al., 2002). The organisms may be epiphytic, endophytic, or loosely attached to submerged vascular plants, Characeae, or other suitable substrate, and are occasionally found free floating. Sexual reproduction is oogamous in all species for which sexual reproduction has been described. Zoospores, meiospores, and sperm are biflagellate with an MLS and lateral, subapical insertion of the flagella. Thalli are composed of compact or loosely branched filaments. *Chaetosphaeridium* is occasionally reported as unicellular, although this is true only in early developmental stages and mature plants form filaments (Thompson, 1969), albeit often with widely spaced cells. About 16 species of *Coleochaete* have been described (Szymańska, 2003), and at least four species of *Chaetosphaeridium* (Thompson, 1969). A single species of *Awahdiella* is known but it is extremely rare and its phylogenetic placement largely speculative. *Chaetotheke* is more common but is difficult to recognize and it has received little study. A number of other genera have been classified among the Coleochaetophyceae. Although many of these taxa can probably be referred to the Chlorophyta, others may legitimately belong to this lineage (Bourrelly, 1990; Printz, 1964).

The relationship between *Coleochaete* (Figure 8.2E) and *Chaetosphaeridium* has been assumed for some time, on the basis of their unusual sheathed hairs and similar chloroplast structure. Some phylogenetic analyses of 18S rDNA sequences indicate that the two may not be closely related, and place *Chaetosphaeridium* in a monophyletic group with *Mesostigma viride* at the base of the charophyte tree (Marin and Melkonian, 1999). Our own analyses indicate that rDNA data provide weak support for such a placement, but analyses of *rbcL*, *atpB*, and *nad5* consistently show the Coleochaetophyceae as a monophyletic group, albeit often with modest bootstrap support (Cimino and Delwiche, 2002; Delwiche et al., 2002; Karol et al., 2001). This position is consistent with morphological and cytological characteristics (Delwiche et al., 2002; Karol et al., 2001). The phylogenetic placement of *Awahdiella* and *Chaetotheke* is unknown, although the filament structure, hairs, and chloroplasts are similar to those of *Coleochaete* and *Chaetosphaeridium*.

The most recent treatment of *Chaetosphaeridium* is that of Thompson (1969), which is not a full monograph. The sparse information on *Chaetosphaeridium* is probably not because the organism is rare, but because it is easily overlooked and more difficult to isolate than other charophytes. *Coleochaete* has been the subject of more comprehensive systematic studies. The first monograph listed four species (Pringsheim, 1860), and a number of studies have revised and added to the work (Delwiche et al., 2002; Printz, 1964; Szymańska, 1989, 2003). Several new species were recently described including a previously unrecognized group characterized by incomplete envelopment of the zygote following fertilization, or incomplete "cortication" (Szymańska, 1988, 1989). Studies of endophytic strains akin to *C. nitellarum* Jost suggest that much of *Coleochaete* diversity remains undescribed (Cimino and Delwiche, 2002). Certainly, further investigation will reveal still more species and, very likely, a greater structural diversity than is currently recognized.

CHAROPHYCEAE *SENSU STRICTO* (THE STONEWORTS)

The stoneworts, or Charophyceae *sensu stricto,* are the most plant-like in appearance among the charophyte algae: They are macroscopic green algae with whorls of branches at nodes. Thalli may be monoecious or dioecious, depending on the species (Corillion, 1972; Wood and Imahori, 1965). All species are oogamous with motile sperm produced in complex antheridia. Oogonia and antheridia are surrounded by sterile jacket cells. Fertilized eggs (zygotes) develop a thick covering of sporopollenin. Zygotes and thalli may become impregnated with calcium carbonate (Wood and Imahori, 1965). These sporopollenin-encrusted spores (called gyrogonites) are well preserved in the fossil record, which extends back in excess of 380 million years (Feist and Grambast-Fessard, 1991). Six genera in a single family represent the extant Characeae. The two most common genera are *Chara* (Figure 8.2F) and *Nitella*. Fossil structural diversity is greater than extant diversity, however, and many families are known only from the fossil record. Taxonomy within the group has been greatly affected by Wood and Imahori (1965), who produced the most recent global monograph of the Characeae.

In their monograph, Wood and Imahori (1965) divide the family Characeae into two tribes—the Chareae (*Chara, Lamprothamnium, Nitellopsis*, and *Lychnothamnus*) and the Nitelleae (*Nitella* and *Tolypella*). Both seem to be monophyletic (Sanders et al., 2003), or the Nitellae may be paraphyletic (McCourt et al., 1999; McCourt et al., 1996). The genera, although represented by very few species in most analyses, seem to be monophyletic with the possible exception of *Chara*, which may include *Lamprothamnium*, based on 18S rDNA sequences and a broad taxon sampling within *Chara* (Meiers et al., 1999). Systematic investigation within the two largest genera, *Chara* and *Nitella*, is wanting. Molecular investigations of *Nitella* subgenus *Tieffallenia* suggest that some sections may be artificial and that mesospore membrane fine structure may be a valuable taxonomic characteristic, at least in subgenus *Tieffallenia* (Sakayama et al., 2004a, 2004b; 2005). Although Wood and Imahori (1965) reduced many described species to varieties, they also provided a list of the "species" described at that time, which they termed "microspecies." The molecular investigations as well as other morphological studies seem to favor the "microspecies" concept (Wood and Imahori, 1965), particularly in the genus *Chara* (Corillion, 1972; Krause, 1997). Regardless of the treatment of varieties, there are probably a few hundred described, extant stoneworts.

FUTURE DIRECTIONS AND THE ROLE OF GENOMICS IN CHAROPHYTE SYSTEMATICS

INSIGHTS FROM PUBLISHED GENOMIC DATA

Some charophyte organellar genomic data have been published, including the complete chloroplast genome of *Zygnema*, *Staurastrum* (Turmel et al., 2005), *Chaetosphaeridium* (Turmel et al., 2002a), *Chara* (Turmel et al., 2006), and *Mesostigma* (Lemieux et al., 2000), as well as the mitochondrial genome of *Chara* (Turmel et al., 2003), *Chaetosphaeridium* (Turmel et al., 2002b), and *Mesostigma* (Turmel et al., 2002c). These studies provide insight into the evolution of chloroplast and mito-chondrial genomes. In particular, complete organellar genomes provide information about gene content, gene order, and trends in evolution of these genomes. The value of these sequences for systematic investigation will increase as more genome-scale data are collected from other charo-phyte algae. As of the beginning of 2006, no charophyte nuclear genome has been published (excluding embryophytes).

Besides organellar genomes, two expressed sequence tag (EST) surveys have been published. The first investigated the expressed mRNAs of members of the *Closterium peracerosum–strigosum–littorale* complex (Sekimoto et al., 2003), and related studies identified a pheromone that induces sexual cell division (Fukumoto et al., 2003; Tsuchikane et al., 2003). The second survey sequenced more than 10,000 ESTs from *Mesostigma viride* and recovered transcripts of genes important for cellular processes such as translation and transcription, signaling, and metabolism, although the majority have unknown functions (Simon et al., 2006). Such studies may provide the phylogenetic data necessary to resolve the branching order of the charophytes. However, it is important to remember the value of this information outside pure systematics. Genomic studies provide valuable information about the biology of these organisms and how their ancestors may have evolved to give rise to the complex metabolic pathways and gene families found in land plants. This information is critical to our understanding and interpretation of evolution within the charophytes as well as the origin and evolution of land plants.

FUTURE SYSTEMATIC INVESTIGATIONS

As noted above, the branching order of the deeper nodes (classes) remains uncertain. Molecular phylogenetic analyses and continued cytological observations may provide the data needed to answer these questions. Many published molecular datasets are very limited in character or taxon sampling, both of which affect inference of relationships. However, molecular investigations of currently available taxa may not be enough to resolve the branching order of the charophyte tree. Evolution within the lineages will remain uncertain until more taxa are available for investigation. Within the Zygnematophyceae, for instance, the known structural diversity has yet to be probed. Relationships among orders, families, and genera of the conjugating green algae remain poorly resolved. Fewer than half of the known genera of Zygnematophyceae are available for molecular investigation. Molecular phylogenetic analyses of the Klebsomidiophyceae remain in their early stages, and many relationships remain unclear in the Coleochaetophyceae and Characeae. Future studies may depend on a broader sampling of taxa as well as more sizable molecular datasets.

None of the charophyte lineages has been comprehensively surveyed by molecular methods. Species relationships remain poorly understood in all but the Coleochaetophyceae. Published work on the Characeae and Zygnematophyceae has only begun to address species relationships using molecular methods, and population-level studies remain few. Although a number of outstanding morphological, mating, and AFLP analyses have been published (Grant and Proctor, 1972; Griffiin and Proctor, 1964; Mannschreck et al., 2002; Meiers et al., 1999), very few studies have addressed the structure of charophyte populations using molecular sequence techniques. How populations interact as well as the distribution of charophyte algae are generally unknown. Even though there are scattered reports for Characeae and Zygnematophyceae, little is known about global population status, and local surveys suggest that many populations may be severely pressured (Krause, 1984; Németh, 2005; Siemińska, 1986; Stewart and Church, 1992; Watanabe, 2005).

Besides the ordering and documentation of the known species, systematists are also concerned with the discovery of new species. A number of extensive floristic studies exist, but as is often the case for widespread and taxonomically difficult taxa, these are strongly biased by geographic location of investigators and accessibility of study sites. The physically smaller species in particular have been poorly documented. It is difficult to imagine that *Chlorokybus* and *Klebsormidium* are represented by the relatively few described species, particularly since *Klebsormidium* thrives in a wide range of habitats. A great many species are likely to be hidden in such unlikely places as university fountains and garden walls, as was the case for *Chlorokybus* (Geitler, 1942).

Another strong bias in current charophyte systematics is the "microbial bias." Except for the Characeae, nearly all charophytes used in molecular phylogenetic studies have been cultured from the wild. This almost certainly introduces biases in the investigation analogous to those encountered when surveying Bacteria and Archea by culturing. As a complement to traditional culture-based methods, molecular sequence data provide an independent means of investigating diversity that eliminates the culture bias, though it introduces others. Molecular environmental studies of even well-characterized habitats have received a lot of attention because many new lineages have been discovered (López-García et al., 2001; Venter et al., 2004). Relating these sequences to the organisms from which they came, however, is difficult. Not just new species, but new kinds of organisms with potentially different life histories and metabolic pathways may yet be found.)

Systematic investigation of the charophyte green algae will continue to provide insight into the diversity and evolution of these exceptional organisms. Future studies will, hopefully, embrace new techniques and technologies and use them to answer fundamental systematic questions. Much remains to be investigated at all levels and in all charophyte lineages.

REFERENCES

Adam, P. (2004) *Nitella partita* (a form of algae) endangered species listing. New South Wales National Parks and Wildlife Service.

Bhattacharya, D. and Medlin, L. (1998) Algal phylogeny and the origin of land plants. *Plant Physiology*, 116: 9–15.

Bhattacharya, D., Weber, K., An, S.S., and Berning-Koch, W. (1998) Actin phylogeny identifies *Mesostigma viride* as a flagellate ancestor of the land plants. *Journal of Molecular Evolution*, 47: 544–550.

Bourrelly, P. (1990) *Les Algues Vertes*. Société Nouvelle des éditions Boubée, Paris.

Bower, F.O. (1908) *The Origin of a Land Flora*. Macmillan and Co., London.

Cimino, M.T. and Delwiche, C.F. (2002) Molecular and morphological data identify a cryptic species complex in endophytic members of the genus *Coleochaete* Brébisson (Charophyta: Coleochaetaceae). *Journal of Phycology*, 38: 1213–1221.

Coesel, P.F.M. 2002. New, intriguing desmid taxa from the Netherlands. *Algological Studies*, 104: 69–79.

Cook, M.E. (2004) Structure and asexual reproduction of the enigmatic charophycean green alga *Entransia fimbriata* (Klebsormidiales, Charophyceae). *Journal of Phycology*, 40: 424–431.

Corillion, R. (1972) *Les Charophycées de France et d'Europe Occidental*. Otto Koeltz Verlag, Angers.

Delwiche, C.F., Karol, K.G., Cimino, M.T., and Sytsma, K.J. (2002) Phylogeny of the genus *Coleochaete* (Coleochaetales, Charophyta) and related taxa inferred by analysis of the chloroplast gene *rbc*L. *Journal of Phycology*, 38: 394–403.

Denboh, T., Hendrayanti, D., and Ichimura, T. (2001) Monophyly of the genus *Closterium* and the order Desmidiales (Charophyceae, Chlorophyta) inferred from nuclear small subunit rDNA data. *Journal of Phycology*, 37: 1063–1072.

Ettl, H. and Gärtner, G. (1995) *Syllabus der Boden-, Luft- und Flechtenalgen*. Gustav Fischer Verlag, Stuttgart, Jena and New York.

Feist, M. and Grambast-Fessard, N. (1990) The genus concept in Charophyta: evidence from Palaeozoic to recent. In *Calcareous Algae and Stromatolites* (ed. R. Riding), Springer-Verlag. Berlin, pp. 189–201.

Fowke, L.C. and Pickett-Heaps, J.D. (1969) Cell division in *Spirogyra* I. Mitosis. *Journal of Phycology*, 5: 240–259.

Fukumoto, R.-H., Fujii, T., and Sekimoto, H. (2003) Cloning and charcterization of a cDNA encoding a sexual cell division-inducing pheromone from a unicellular green alga *Closterium ehrenbergii* (Chlorophyta). *Journal of Phycology*, 39: 931–936.

Geitler, L. (1942a) Morphologie, Entwicklungsgeschicte und Systematik neuer bemerkenswerter atmophytischer Algen aus Wien. *Flora*, 36: 1–29.

Geitler, L. (1942b) Neue luftlebige Algen aus Wien. *Osterreichische Botanische Zeitschrift*, 41: 49–51.

Gerrath, J.F. (1993) The biology of desmids: a decade of progress. *In Progress in Phycological Research Volume 9* (eds F.E. Round and D.J. Chapman), Biopress Ltd., Bristol, pp. 79–192.

Gontcharov, A.A., Marin, B., and Melkonian, M. (2003) Molecular phylogeny of conjugating green algae (Zygnemophyceae, Streptophyta) inferred from SSU rDNA sequence comparisons. *Journal of Molecular Evolution*, 56: 89–104.

Gontcharov, A.A., Marin, B., and Melkonian, M. (2004) Are combined analyses better than single gene phylogenies? A case study using SSU rDNA and *rbc*L sequence comparisons in the Zygnematophyceae (Streptophyta). *Molecular Biology and Evolution*, 21: 612–624.

Gontcharov, A.A. and Melkonian, M. (2004) Unusual position of *Spirotaenia* (Zygnematophyceae) among streptophytes revealed by SSU rDNA and *rbc*L sequence comparisons. *Phycologia*, 43: 105–113.

Gontcharov, A.A. and Melkonian, M. (2005) Molecular phylogeny of *Staurastrum* Meyen ex Ralfs and related genera (Zygnematophyceae, Streptophyta) based on coding and noncoding rDNA sequence comparisons. *Journal of Phycology*, 41: 887–899.

Grant, M.C. and Proctor, V.W. (1972) *Chara vulgaris* and *C. contraria*: patterns of reproductive isolation for two cosmopolitan species complexes. *Evolution*, 26: 267–281.

Griffin, D.G. and Proctor, V.W. (1964) A population study of *Chara zeylanica* in Texas, Oklahoma, and New Mexico. *American Journal of Botany*, 51: 120–124.

Hoshaw, R.W., McCourt, R.M., and Wang, J.-C. (1990) Phylum Conjugaphyta. In *Handbook of Protoctista* (eds L. Margulis, J. Corliss, M. Melkonian, and D. Chapman), Jones and Bartlett Publishers, Boston, MA, pp. 119–131.

Hughes, E.O. (1948) New fresh-water Chlorophyceae from Nova Scotia. *American Journal of Botany*, 35: 424–427.

Huss, V.A.R. and Kranz, H.D. (1995) Charophyte evolution and the origin of land plants. In *Origins of Algae and Their Plastids* (ed. D. Bhattacharya), Springer, Vienna and New York, pp. 103–114.

Ingenhousz, J. (1779) Experiments upon vegetables. *Chronica Botanica*, 11 (5/6): 285–396.

Johansen, J.R., Lowe, R., Gomez, S.R., Kociolek, J.P., and Makosky, S.A. (2004) New algal species records for the Great Smoky Mountains National Park, U.S.A., with an annotated checklist of all reported algal species for the park. *Algological Studies*, 111: 17–44.

Kadlubowska, J.Z. (1984) *Conjugatophyceae I. Zygnemales*. Gustav Fischer Verlag, Stuttgart.

Karol, K.G., McCourt, R.M., Cimino, M.T., and Delwiche, C.F. (2001) The closest living relatives of land plants. *Science*, 294: 2351–2353.

Kranz, H.D., Miks, D., Siegler, M.-L., Capesius, I., Sensen, C.W., and Huss, V.A.R. (1995) The origin of land plants: phylogenetic relationships among charophytes, bryophytes, and vascular plants inferred from complete small-subunit ribosomal RNA gene sequences. *Journal of Molecular Evolution*, 41: 74–84.

Krause, W. (1984) Rote Liste der Armleuchteralgen (Charophyta). In *Rote Liste der gefährdeten Tiere und Pflanzen in der Bundesrepublik Deutschland* (eds J. Blab, E. Nowak, and W. Trautmann), Kilda-Verlag, Berlin, pp. 184–187..

Krause, W. (1997) *Charales (Charophyceae)*. Gustav Fischer, Jena, Stuttgart, Lübeck, Ulm.

Lee, O.-M. (2001) The nucleotide sequences variability in ITS and 5.8S regions of the nuclear rDNA among *Cosmarium* species. *Algae*, 16: 129–136.

Leeuwenhoek, A.V. (1674) *Letters to the Royal Society of London*. Swets & Zeitlinger, Amsterdam.

Leidy, R.A. and Moyle, P.B. (1998) Conservation status of the world's fish fauna: an overview. In*Conservation Biology* (eds. P.L. Fiedler and P.M. Kareiva), Chapman & Hall. New York, pp. 187–227.

Lemieux, C., Otis, C., and Turmel, M. (2000) Ancestral chloroplast genome of *Mesostigma viride* reveals an early branch of green plant evolution. *Nature*, 403: 649–652.

Lewis, L.A. and Flechtner, V.R. (2002) Green algae (Chlorophyta) of desert microbiotic crusts: diversity of North American taxa. *Taxon*, 51: 443–451.

Lewis, L.A. and Lewis, P.O. (2005) Unearthing the molecular phylodiversity of desert soil green algae (Chlorophyta). *Systematic Biology*, 54: 936–947.

Lewis, L.A. and McCourt, R.M. (2004) Green algae and the origin of land plants. *American Journal of Botany*, 91: 1535–1556.

Lockhorst, G.M. (1996) *Comparative Taxonomic Studies on the Genus Klebsormidium (Charophyceae) in Europe*. Gustav Fischer Verlag, Stuttgart, Jena, New York.

López-García, P., Rodriguez-Valera, F., Pedrós-Alió, C., and Moreira, D. (2001) Unexpected diversity of small eukaryotes in deep-sea Antarctic plankton *Nature*, 409: 603–607.

Mannschreck, B., Fink, T., and Melzer, A. (2002) Biosystematics of selected *Chara* species (Charophyta) using amplified fragment length polymorphism. *Phycologia*, 41: 657–666.

Marin, B. and Melkonian, M. (1999) Mesostigmatophyceae, a new class of streptophyte green algae revealed by SSU rRNA sequence comparisons. *Protist*, 150: 399–417.

Mattox, K.R. and Stewart, K.D. (1984) Classification of the green algae: a concept based on comparative cytology. In *Systematics of the Green Algae* (eds D.E.G. Irvine and D.M. John), Academic Press, New York, pp. 29–72.

McCourt, R.M., Karol, K.G., Bell, J., Helm-Bychowski, M., Grajewska, A., Wojciechowski, M.F., and Hoshaw, R.W. (2000) Phylogeny of the conjugating green algae (Zygnematophyceae) based on *rbc*L sequences. *Journal of Phycology*, 36: 747–758.

McCourt, R.M., Karol, K.G., Casanova, M.T., and Feist, M. (1999) Monophyly of genera and species of Characeae based on *rbc*L sequences, with special reference to Australian and European *Lychnothamnus barbatus* (Characeae: Charophyceae). *Australian Journal of Botany*, 47: 361–369.

McCourt, R.M., Karol, K.G., Guerlesquin, M., and Feist, M. (1996) Phylogeny of extant genera in the family Characeae (Charales, Charophyceae) based on *rbc*L sequences and morphology. *American Journal of Botany*, 83: 125–131.

McCourt, R.M., Karol, K.G., Kaplan, S., and Hoshaw, R.W. (1995) Using *rbc*L sequences to test hypotheses of chloroplast and thallus evolution in the conjugating green algae (Zygnematales, Charophyceae). *Journal of Phycology*, 31: 989–995.

Meiers, S.T., Proctor, V.W., and Chapman, R.L. (1999) Phylogeny and biogeography of *Chara* (Charophyta) inferred from 18S rDNA sequences. *Australian Journal of Botany*, 47: 347–360.

Melkonian, M. (1989) Flagellar apparatus ultrastructure in *Mesostigma viride* (Prasinophyceae). *Plant Systematics and Evolution*, 164: 93–122.

Mix, M. (1972) Die Feinstruktur der Zellwande bei Mesotaeniaceae und Gonatozygaceae mit einer vergleichenden Betrachtung de verschiedenen Wandentypen der Conjugatophyceae und uber deren systematischen Wert. *Archiv fur Mikrobiologie*, 81: 197–220.

Nam, M. and Lee, O.-M. (2001) A comparative study of morphological characters and sequences of *rbc*L gene in *Staurastrum* (Desmidiaceae). *Algae*, 16: 363–367.

Novis, P.M. (2006) Taxonomy of *Klebsormidium* (Klebsormidiales, Charophyceae) in New Zealand streams and the significance of low pH habitats. *Phycologia*, 45: 293–301.

Németh, J. (2005) Red list of algae in Hungary. *Acta Botanica Hungarica*, 47: 379–417.

Park, N., Karol, K.G., Hoshaw, R.W., and McCourt, R.M. (1996) Phylogeny of *Gonatozygon* and *Genicularia* (Gonatozygaceae, Desmidiales) based on *rbc*L sequences. *European Journal of Phycology*, 31: 309–313.

Peterson, J., Brinkmann, H., and Cerff, R. (2003) Origin, evolution, and metabolic role of a novel glycolytic GAPDH enzyme recruited by land plant plastids. *Journal of Molecular Evolution*, 57: 16–26.

Peterson, J., Teich, R., Becker, B., Cerff, R., and Brinkmann, H. (2006) The *GapA/B* gene duplication marks the origin of Streptophyta (charophytes and land plants). *Molecular Biology and Evolution*, 23: 1109–1118.

Prescott, G.W. (1966) Algae of the Panama Canal and its Tributaries-II. Conjugales. *Phykos*, 5: 1–49.

Pringsheim, N. (1860) Beitrage zur Morphologie und Systematik der Algen. *Jahrbücher für Wissenschaftliche Botanik*, 2: 1–38.

Printz, H. (1964) *Die Chaetophoralean der Binnengewässer* Verlag Dr W. Junk, Den Haag.

Ralfs, J. (1848) *British Desmidieae*. Richard and John E. Taylor, London.

Rieth, A. (1972) Über *Chlorokybus atmophyticus* Geitler 1942. *Archiv för Protistenkund*, 114: 330–342.

Rogers, C.E., Domozych, D.S., Stewart, K.D., and Mattox, K.R. (1981) The flagellar apparatus of *Mesostigma viride* (Prasinophyceae): multilayered structures in a scaly green flagellate. *Plant Systematics and Evolution*, 138: 247–258.

Rogers, C.E., Mattox, K.R., and Stewart, K.D. (1980) The zoospore of *Chlorokybus atmophyticus*, a charophyte with sarcinoid growth habit. *American Journal of Botany*, 67: 774–783.

Sakayama, H., Hara, Y., Arai, S., Sato, H., and Nozaki, H. (2004a) Phylogenetic analyses of *Nitella* subgenus *Tieffallenia* (Charales, Charophyceae) using nuclear ribosomal DNA internal transcribed spacer sequences. *Phycologia*, 43: 672–681.

Sakayama, H., Hara, Y., and Nozaki, H. (2004b) Taxonomic re-examination of six species of *Nitella* (Charales, Charophyceae) from Asia, and phylogenetic relationships within the genus based on *rbc*L and *atp*B gene sequences. *Phycologia*, 43: 91–104.

Sakayama, H., Miyaji, K., Nagumo, T., Kato, M., Hara, Y., and Nozaki, H. (2005) Taxonomic reexamination of 17 species of *Nitella* subgenus *Tieffallenia* (Charales, Charophyceae) based on internal morphology of the oospore wall and multiple DNA marker sequences. *Journal of Phycology*, 41: 195–211.

Sanders, E.R., Karol, K.G., and McCourt, R.M. (2003) Occurence of *mat*K in a *trn*K group II intron in charophyte green algae and phylogeny of Characeae. *American Journal of Botany*, 90: 628–633.

Sekimoto, H., Tanabe, Y., Takizawa, M., Ito, N., Fukumoto, R.H., and Ito, M. (2003) Expressed sequence tags from the *Closterium peracerosum-strigosum-littorale* complex, a unicellular charophycean alga, in the sexual reproduction process. *DNA Research*, 10: 147–153.

Siemińska, J. (1986) Red list of threatened algae in Poland. In *Lista roślin wymierających i zagrożonych w Polsce.* (eds K. Zarzyckiego and W. Wojewody), Państwowe Wydawnictwo Naukowe, Warszawa, pp. 31–44.

Silva, P.C., Mattox, K.R., and Blackwell, W.H., Jr. (1972) The generic name *Hormidium* as applied to green algae. *Taxon*, 21: 639–645.

Simon, A., Glockner, G., Felder, M., Melkonian, M., and Becker, B. (2006) EST analysis of the scaly green flagellate *Mesostigma viride* (Streptophyta): implications for the evolution of green plants (Viridiplantae). *BMC Plant Biology.*

Stabenau, H. and Winkler, U. (2005) Glycolate metabolism in green algae. *Physiologia Plantarum*, 123: 235–245.

Stewart, N.F. and Church, J.M. (1992) *Red Data Books of Britain and Ireland.* The Joint Nature Conservation Committee, Peterborough.

Szymańska, H. (1988) *Coleochaete sieminskiana* Szym. sp. nov. (Chlorophyta)—a new species from Poland. *Nova Hedwigia*, 46: 143–147.

Szymańska, H. (1989) Three new *Coleochaete* species (Chlorophyta) from Poland. *Nova Hedwigia*, 49: 435–446.

Szymańska, H. (2003) *Coleochaete spalikii* Szymanska sp. nov. (Charophyceae, Chlorophyta)—a new member of the *Coleochaete sieminskiana* group. *Nova Hedwigia*, 76: 129–135.

Thompson, R.H. (1969) Sexual reproduction in *Chaetosphaeridium globosum* (Nordst.) Klebahn (Chlorophyceae) and description of a species new to science. *Journal of Phycology*, 5: 285–290.

Tsekos, I. (1999) The sites of cellulose synthesis in algae: diversity and evolution of cellulose-synthesizing enzyme complexes. *Journal of Phycology*, 35: 635–355.

Tsuchikane, Y., Fukumoto, R.H., Akatsuka, S., Fujii, T., and Sekimoto, H. (2003) Sex pheromones that induce sexual cell division in the *Closterium peracerosum-strigosum-littora* complex (Charophyta). *Journal of Phycology*, 39: 303–309.

Turmel, M., Ehara, M., Otis, C., and Lemieux, C. (2002a) Phylogenetic relationships among streptophytes as inferred from chloroplast small and large subunit rRNA gene sequences. *Journal of Phycology*, 38: 364–375.

Turmel, M., Otis, C., and Lemieux, C. (2002b) The chloroplast and mitochondrial genome sequences of the charophyte *Chaetosphaeridium globosum*: insights into the timing of the events that restructured organelle DNAs within the green algal lineage the led to land plants. *Proceedings of the National Academy of Science*, 99: 11275–11280.

Turmel, M., Otis, C., and Lemieux, C. (2002c) The complete mitochondrial DNA sequence of *Mesostigma viride* identifies this green alga as the earliest green plant divergence and predicts a highly compact mitochondrial genome in the ancestor of all green plants. *Molecular Biology and Evolution*, 19: 24–38.

Turmel, M., Otis, C., and Lemieux, C. (2003) The mitochondrial genome of *Chara vulgaris*: insights into the mitochondrial DNA architecture of the last common ancestor of green algae and land plants. *The Plant Cell*, 15: 1888–1903.

Turmel, M., Otis, C., and Lemieux, C. (2005) The complete chloroplast DNA sequences of the charophycean green algae *Staurastrum* and *Zygnema* reveal that the chloroplast genome underwent extensive changes during the evolution of the Zygnematales. *BMC Biology* 3.

Turmel, M., Otis, C., and Lemieux, C. (2006) The chloroplast genome sequence of *Chara vulgaris* sheds new light into the closest green algal relatives of land plants. *Molecular Biology and Evolution*, 23: 1324–1338.

Venter, J.C., Remington, K., Heidelberg, J.F., Halpern, A.L., Rusch, D., Eisen, J.A., Wu, D., Paulsen, I., Nelson, K.E., Nelson, W., Fouts, D.E., Levy, S., Knap, A.H., Lomas, M.W., Nealson, K., White, O., Peterson, J., Hoffman, J., Parsons, R., Baden-Tillson, H., Pfannkoch, C., Rogers, Y.H., and Smith, H.O. (2004) Environmental genome shotgun sequencing of the Sargasso Sea. *Science*, 304: 66–74.

Watanabe, M.M. (2005) Cultures as a means of protecting biological resources: *ex situ* conservation of threatened algal species. In *Algal Culturing Techniques* (ed. R. Anderson), Elsevier Academic Press, New York, pp. 419–428.

Wood, R.D. and Imahori, K. (1965) *A Revision of the Characeae*. Verlag von J. Cramer, Weinheim.

Yoshii, Y., Takaichi, S., Maoka, T., and Inouye, I. (2003) Photosynthetic pigments composition in the primitive green alga *Mesostigma viride* (Prasinophyceae): phylogenetic and evolutionary implications. *Journal of Phycology*, 39: 270–576.

9 The chlorarachniophytes: evolution and classification

Ken-ichiro Ishida, Akinori Yabuki and Shuhei Ota

CONTENTS

ABSTRACT

The chlorarachniophytes are one of the most evolutionarily interesting algal groups. Their cells have small organelles, called nucleomorphs, which provide us with direct evidence for the lateral transfer of plastids through a secondary endosymbiosis. Advances in molecular phylogenetics have proved that the chlorarachniophytes originated from a cercozoan protist that engulfed a green alga and retained it as a plastid. The focus of chlorarachniophyte research has now shifted to tackling the question "how did the endosymbiont become an organelle?" In addition to this evolutionary research, the recognition of the chlorarachniophytes as a group and their diversity has also progressed in the past decade. Currently, five genera and nine species make up the chlorarachniophytes and several new taxa are waiting to be described. Remarkable diversity in life cycle and ultrastructure and fascinating cell behaviour are beginning to be revealed.

INTRODUCTION

In a bright shallow lagoon of a coral reef, among the bushes of tropical green seaweeds, such as *Halimeda* and *Caulerpa* species, microscopic filose amoebae with green plastids are crawling around. In the middle of the Atlantic Ocean, among the drifting seaweeds, tiny green single-flagellated plankton swim. These are members of the chlorarachniophytes. Among the many groups of algae, the chlorarachniophytes are particularly interesting and enigmatic due to their unique evolutionary history, cellular structure, and genetic organization. Research has revealed that the chlorarachniophytes are one of the groups that acquired plastids via a secondary endosymbiosis in which an eukaryotic alga was engulfed and retained by a protozoan host (Hibberd and Norris, 1984; Ludwig and Gibbs, 1989; McFadden et al., 1994). The chlorarachniophytes are, like cryptophytes, unique among those "secondary algae" in having the vestigial nucleus of the endosymbiotic alga, the ancestor of their plastids, in each of their plastids (Hibberd and Norris, 1984; Ludwig and Gibbs, 1989; McFadden et al., 1994). This vestigial nucleus is called the nucleomorph and provides direct evidence for the secondary endosymbiotic origin of chlorarachniophytes (McFadden et al., 1994; Gilson and McFadden, 1996). It makes the chlorarachniophytes a very important group of microorganisms that hold clues to understanding the cellular and plastid evolution of photosynthetic eukaryotes. Much recent research on the chlorarachniophytes, therefore, has focused on the origin and the genome of the nucleomorph (McFadden et al., 1995; Gilson and McFadden, 1996; Ishida et al., 1997; Van de Peer and De Wachter, 1997; Ishida et al. 1999, Gilson and McFadden 1999; Gilson et al., 2006), and several reviews have been published on evolution of plastids (Gilson and McFadden, 1997; McFadden, 1999, McFadden, 2001, Archibald and Keeling, 2002; Archibald, 2005). However, the diversity and evolution within the chlorarachniophytes has been neglected. In this chapter, the use of molecular data as well as ultrastructure and morphology in understanding the diversity and evolution of this interesting group of algae will be emphasized while outlining their current status. The fundamental process of the acquisition of plastids in the chlorarachniophytes will also be discussed.

GENERAL FEATURES OF THE CHLORARACHNIOPHYTES

The chlorarachniophytes are a group of unicellular algae (or photosynthetic protists) that have green plastids with chlorophylls *a* and *b*, the same chlorophyll composition as that of green algae and euglenophytes (Hibberd and Norris, 1984; Sasa et al., 1992). Unlike the green algae and euglenophytes, however, the majority of chlorarachniophyte species are amoeboid with thin thread-like pseudopodia (filopodia) (Figure 9.1a). Their ultrastructure, especially with regard to plastid structure, is also very different from those of the green algae and euglenophytes. Therefore, the chlorarachniophytes are recognized as an independent phylum in most algal classification schemes (Hibberd and Norris, 1984).

The chlorarachniophytes are the only algal group that has never been found in fresh water; therefore, their distribution seems to be strictly marine. In the marine environment, however, they appear to be widely distributed throughout tropical and temperate regions of the globe (Figure 9.2). They have been found from various kinds of marine habitats, including sandy beaches, rocky shores, tide pools, oceanic surface seawater, on seaweeds, on sand grains, and at the bottom of shallow seas.

Considering the small numbers of chlorarachniophyte species that are known, their morphological and life style diversity is remarkable. A chlorarachniophyte cell can be an amoeba with filose pseudopodia (filopodia) (Figure 9.1a), a coccoid cell with a cell wall (Figure 9.1b), or a

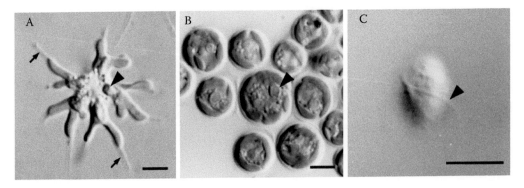

FIGURE 9.1 Three types of chlorarachniophyte cell. Bars = 5 μm. (A) Amoeboid cell of *Gymnochlora stellata*, showing several filopodia (arrows) and conspicuously projecting pyrenoids (arrowhead). (B) Coccoid cells of *Lotharella globosa*, showing a projecting pyrenoid (arrowhead). (C) Flagellated cell (zoospore) of *L. globosa* with a single flagellum (arrowhead) wrapping around the cell. (From Ishida, K. and Hara, Y., *Phycologia*, 33: 321–331. With permission.)

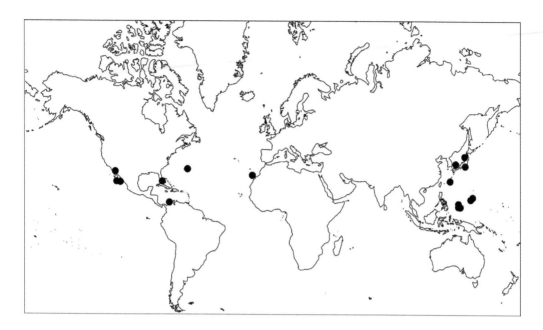

FIGURE 9.2 Map of the world showing sites (black dots) where chlorarachniophytes have been collected.

flagellate with a single flagellum (Figure 9.1c). The amoeboid cells of the species *Chlorarachnion reptans* are connected with each other by fusion of their filopodia (reticulopodia), while amoeboid cells of other species are solitary (Geitler, 1930; Hibberd and Norris, 1984; Calderon-Saenz and Schnetter, 1989; Ishida et al., 1996; Ishida et al., 2000). Cells of a flagellated species, *Bigelowiella natans*, are very small, about 5 μm in diameter, and recognized as plankters (Moestrup and Sengco, 2001), and an amoeboid species, *Gymnochlora stellata*, has multinucleated giant cells, which grow up to 100 μm in diameter, as part of the life cycle (Kaneda, unpublished data). *Bigelowiella longifila* is known to form suspendable mucilaginous colonies (Ota et al., 2007).

The structure of the chlorarachniophyte plastid is unique. Each plastid has a pyrenoid that conspicuously projects toward the centre of the cell (Hibberd and Norris, 1984; Ishida et al., 1996). This makes it easy for us to distinguish the chlorarachniophytes from the other chlorophyll-*b*-containing algae, the chlorophytes and the euglenophytes, under a light microscope. Under the transmission electron microscope (TEM) (Figure 9.3), the projecting pyrenoid is usually capped with a cytoplasmic vesicle, called a capping vesicle, which is filled with a storage product, a ß-1,3 glucan, not starch (McFadden et al., 1997b). Unlike other algal plastids, the chlorarachniophyte plastid is surrounded by four smooth membranes (Hibberd and Norris, 1984). No direct connection between the outermost membrane and other endomembrane systems has been reported. One of the most remarkable features of the plastid is that it is accompanied by a small nucleus-like structure, called the nucleomorph, which is present in the small space between the inner two and outer two membranes, called the periplastidial compartment (Figure 9.4) (Hibberd and Norris, 1984).

The ultrastructure of chlorarachniophyte mitochondria is also different from those of other chlorophyll-*b*-containing algae. The mitochondrial cristae are tubular in the chlorarachniophytes, while mitochondria of chlorophytes and euglenophytes are plate-like and fan-shaped cristae, respectively (Hibberd and Norris, 1984). The cell walls of coccoid cells are usually multilayered (Ishida and Hara, 1994; Ota et al., 2005).

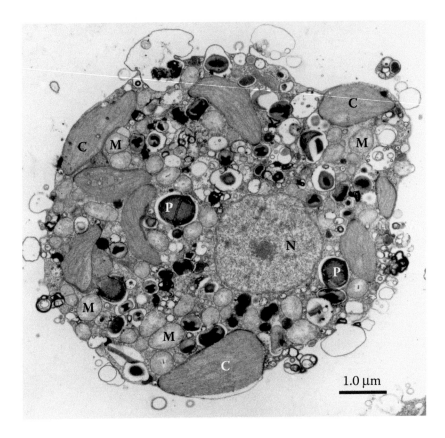

FIGURE 9.3 Transmission electron micrograph of a *Lotharella amoebiformis* cell showing a central nucleus (N), several plastids (C), pyrenoids (P), and mitochondria (M).

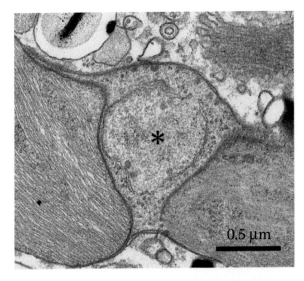

FIGURE 9.4 Transmission electron micrograph of a *Lotharella amoebiformis* nucleomorph*.

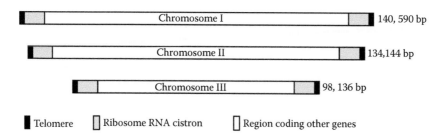

FIGURE 9.5 Structure of nucleomorph chromosomes showing that both ends of each chromosome are capped by telomeres and ribosomal RNA cistrons. The size of each chromosome in the *Bigelowiella natans* nucleomorph is also shown.

THE NUCLEOMORPH

Nucleomorphs have been found only in chlorarachniophytes and cryptophytes, another major group of algae whose origin is different from that of chlorarachniophytes (McFadden and Gilson, 1995). The chlorarachniophyte nucleomorph was first discovered by Hibberd and Norris (1984) from *Chlorarachnion reptans*, the first species in the phylum Chlorarachniophyta. Since then, it has been confirmed that all chlorarachniophyte species examined by electron microscopy possess nucleomorphs (Ishida and Hara, 1994; Ishida et al., 1996; Ishida et al., 2000; Dietz et al., 2003; Moestrup and Sengco, 2001; Ota et al., 2005).

The nucleomorph is the reduced nucleus of the ancient alga that was engulfed by a protozoan host and became the chlorarachniophyte plastid (Ludwig and Gibbs, 1987; Ludwig and Gibbs, 1989; McFadden et al., 1994). It is a tiny organelle surrounded by two membranes with pores or gaps in several places, and found in the periplastidial compartment of each chlorarachniophyte plastid (Hibberd and Norris, 1984; Ludwig and Gibbs, 1987; Ishida et al., 2000). Like a eukaryotic nucleus, it contains DNA (Ludwig and Gibbs, 1989) and divides, though it does not form a spindle (Ludwig and Gibbs, 1989). Gilson and McFadden (1996) found that the genome of the nucleomorph consists of three linear chromosomes (Figure 9.5), which are capped by telomeres at the both ends, like other eukaryotic chromosomes (see also Gilson and McFadden, 1997; McFadden et al., 1997a). Recently, the whole nucleomorph genome sequence has been reported from the tiny flagellated chlorarachniophyte *Bigelowiella natans* (Gilson et al., 2006). The total size of the genome is only 373 kb, just two to three times larger than the ordinary plastid genomes of land plants (Gilson et al., 2006). It encodes 17 putative plastid-targeted protein genes and about 350 housekeeping genes (Gilson et al., 2006). Many introns were also found in the genome and they are usually 18 to 21 bp in length, the smallest introns known (Gilson and McFadden, 1996).

The nucleomorph is present in the chlorarachniophytes, probably because it still codes the 17 plastid protein genes, which have not transferred to the host nucleus. Comparison with a cryptomonad's nucleomorph genome suggests that no single plastid gene is responsible for the retention of nucleomorphs in both algal groups (Gilson et al., 2006). The nucleomorphs may still be in the process of reduction, and may be lost in the future.

SECONDARY ENDOSYMBIOSIS GENERATED THE CHLORARACHNIOPHYTES

The presence of a nucleomorph confirms that the chlorarachniophytes originated from a secondary endosymbiosis in which a eukaryotic protozoan host engulfed a photosynthetic eukaryote and kept it as a photosynthetic organelle, the plastid. The origins of the plastid and the host component have been the focus of much chlorarachniophyte research.

Molecular phylogenetic analyses using proteins and genes encoded by plastid DNA indicated that a green alga was the origin of the plastid, as expected from its colour (McFadden et al., 1995;

Ishida et al., 1997). The nucleomorph-encoded 18S rRNA gene trees also gave the same results (Van de Peer and De Wachter, 1997; Ishida et al., 1999). It is, therefore, widely accepted that the origin of chlorarachniophyte plastids is from a green alga. However, which green alga is still unclear, because the position of chlorarachniophyte plastids in green algal phylogenetic trees is different depending on the molecule used for constructing the tree and the taxa included in the analysis (McFadden et al., 1995; Ishida et al., 1997; Van de Peer and De Wachter, 1997; Ishida et al., 1999). Besides the chlorarachniophytes, euglenophytes are also known to have acquired plastids from a green alga via a secondary endosymbiosis. Cavalier-Smith (1999) suggested that chlorarachniophytes and the euglenophytes share a common origin of plastids, but other evidence indicates the plastids were acquired from separate green algal endosymbionts (Ishida et al., 1997; Ishida et al., 1999).

Molecular phylogenetic analyses using nuclear-encoded genes (small subunit ribosomal [SSU] and large subunit [LSU] rRNA) and proteins (Actin and alpha-tubulin) clearly indicate that the origin of the chlorarachniophyte host component was a cercozoan protist (Bhattacharya et al., 1995; Van de Peer and De Wachter, 1997; Keeling et al., 1998; Ishida et al., 1999, Keeling, 2001). The Cercozoa is a recently established protistan group containing cercomonads, testate filose amoebae, and several other amoeboflagellates (Cavalier-Smith, 1993; Cavalier-Smith and Chao, 2003). Because there is no evidence for an alternate possibility, it is accepted that the protist that engulfed a green alga and became a chlorarachniophyte was a cercozoan protist. At present, the molecular phylogenetic trees are the only clear evidence for the cercozoan origin of chlorarachniophyte host component, and other kinds of evidence such as ultrastructure have not confirmed this (Moestrup and Sengco, 2001). This is probably because of the paucity of studies other than molecular phylogenetic analyses that have been conducted on the cercozoan protists.

At present, we can conclude from the evidence to date that the chlorarachniophytes were created from a secondary endosymbiosis between a cercozoan host and a green algal endosymbiont.

HOW THE GREEN ALGAL ENDOSYMBIONT WAS INTEGRATED AS A PLASTID

It is easy to say that chlorarachniophyte plastids evolved from a green algal endosymbiont engulfed by a cercozoan host, but how the endosymbiont became the plastids in the chlorarachniophyte cell is not clear. It was probably not a simple process. Archibald et al. (2003) demonstrated that, in the chlorarachniophytes, at least 20% of 78 examined nuclear-encoded plastid protein genes did not come directly from the green algal endosymbiont that evolved to the plastids, but from different sources, such as red algae, streptophytes, and bacteria, via lateral gene transfers (LGTs). Establishment of a protein-targeting system for the nuclear-encoded plastid proteins must have been essential in the process of plastid acquisition (McFadden, 1999; Cavalier-Smith, 2000). The targeting sequences of nuclear-encoded plastid proteins have been studied, and this predicted that each targeting sequence is composed of a typical ER signal sequence and a putative transit peptide that is different from those of other photosynthetic eukaryotes (Rogers et al., 2004). These reports indicate that the acquisition of plastids was a complex process that required a major modification and reorganization of the cellular system. The chlorarachniophytes are an excellent model for investigating the process of incorporation of a photosynthetic endosymbiont as a plastid in the process of a secondary endosymbiosis.

LIFE CYCLE DIVERSITY

Chlorarachniophytes show amazing diversity in life cycles, as illustrated in Figure 9.6. Basically, there are three cell forms (amoeboid, coccoid, and flagellate; Figure 9.1a through Figure 9.1c) in chlorarachniophytes and, dependent on the genera and species, the arrangement and transition of cell forms in a life cycle and what type of cell is vegetative are different. The type species, *Chlorarachnion reptans*, has all of the three cell types in its life cycle in which the amoeboid cell

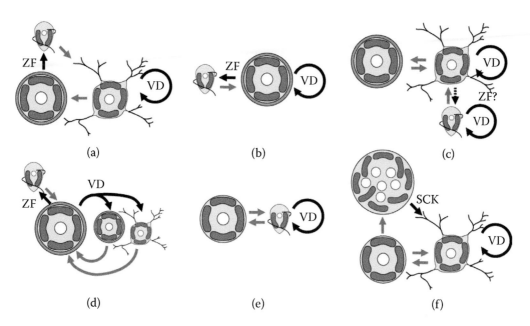

(a)   (b)   (c)

(d)   (e)   (f)

FIGURE 9.6 Schematic drawings of life cycles of (a) *Lotharella amoebiformis*, (b) *L. globosa*, (c) *Bigelowiella longifila*, (d) *L. vacuolata*, (e) *B. natans*, and (f) *Gymnochlora stellata* ZF: zoospore formation, VD: vegetative division, SCK: synchronous cytokinesis.

proliferates and constitutes the main stage (Hibberd, 1990). The life cycle of *Lotharella globosa* is composed of a vegetative coccoid cell stage and a non-vegetative flagellate cell (zoospore) stage, and lacks the amoeoid cell (Ishida and Hara, 1994). On the other hand, the life cycle of *Gymnochlora stellata* consists of a vegetative amoeboid stage and a non-vegetative wall-less coccoid stage, while it lacks the flagellate cell (Ishida et al., 1996). Recently, some cells in an old culture of this species were found to form multinuclear giant cells of up to 100 μm in diameter (Kaneda et al., unpublished data). A tiny planktonic species, *Bigelowiella natans*, has a life cycle with a main vegetative flagellate stage and minor non-vegetative wall-less cell stages (Moestrup and Sengco, 2001). This diversity of life cycle probably reflects the diversity of habitat and lifestyle, a diversity that has been created through evolution from a heterotrophic cercozoan protist to the photosynthetic chlorarachniophytes.

Sexual reproduction has only been observed in *Chlorarachnion reptans* (Grell, 1990) and *Cryptochlora perforans* (Beutlich and Schnetter, 1993). In *Chlorarachnion reptans*, an amoeboid "gamete" penetrates into a coccoid "gamete" and fuses to produce a slightly large zygote (Grell, 1990). However, how the "gametes" are produced and when meiotic division takes place has not been observed. In *Cryptochlora perforans*, two types of amoeboid gametes fuse to produce a large diploid coccoid cell, which can grow by mitotic cell division or can undergo meiosis to produce haploid zoospores (Beutlich and Schnetter, 1993).

BEHAVIOUR OF DAUGHTER CELLS

Recently, some interesting behaviour in chlorarachniophyte cells has been revealed by time-lapse video microscopy (Ota et al., 2005; Ota et al., 2007). In a few species, a vegetative cell division produces two daughter cells that show different movements (Dietz et al., 2003; Ota et al., 2005; Ota et al., 2007). In one of the coccoid species, *Lotharella vacuolata*, vegetative cell division takes place in a coccoid cell surrounded by a multilayered cell wall. After cell division, one of the daughter cells transforms into an amoeboid cell that subsequently hatches from the parental cell and becomes a new coccoid cell, while the other daughter cell stays in the parental cell

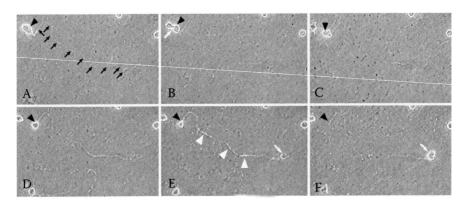

FIGURE 9.7 Time-lapse video sequence showing a post-cell-division behaviour of daughter cells in *B. longifila*. (A) Amoeboid cell (arrowhead) with a long filopodium (arrows). (B) THE cell is dividing. The cleavage of the dividing cell (white arrow) is seen. An arrowhead shows the daughter cell that inherits the long filopodium from the parent cell. (C) A scene just after the cell division. (D) The daughter cell (arrowhead) with the long filopodium that is about to undergo migration. (E) The daughter cell (arrowhead) is migrating to the other end (white arrow) of the long filopodium by transporting its contents through the filopodium (white arrowhead). (F) The daughter cell that completed the migration. The place where the cell body was (arrowhead) is now the distal end of the filopodium, and the place where the distal end of filopodium was (white arrow) now became the cell body.

(Ota et al., 2005). In another species, *Bigelowiella longifila*, which is a planktonic species with both amoeboid and flagellate vegetative stages, an amoeboid cell has basically only one long filopodium. When this amoeboid cell divides, one daughter cell inherits the filopodium and migrates by transporting all the cell contents to the distal end of the filopodium (Figure 9.7), and the other daughter cell stays and extends a new filopodium (Ota et al., 2007). It also appears that this "cell-content transportation" is the only way of migration for the amoeboid cells of this species (Ota et al., 2007). Dietz and Schnetter (1996) demonstrated, using inhibitors for microfilaments, that microtubules might be involved in bidirectional particle transport in the filopodia of a chlorarachniophyte, *Cryptochlora perforans*. How and why have these unusual and curious behaviours evolved? What is their meaning in the life of the chlorarachniophytes? In order to answer these questions, detailed observations of cellular behaviour in a wide variety of cercozoan organisms will be required.

CLASSIFICATION OF THE CHLORARACHNIOPHYTES

The first species of the phylum Chlorarachniophyta, *Chlorarachnion reptans* Geitler, was described from an enrichment culture of siphonous green algae collected from the Canary Islands, although it was first classified within the Xanthophyceae, Heterokontophyta (Geitler, 1930). In 1966, Norris rediscovered *Chlorarachnion reptans* from a coast of Mexico and established a clonal culture (Norris, 1967). The phylum Chlorarachniophyta was established in 1984, i.e. just over twenty years ago, based on the unusual combination of ultrastructural characteristics and pigment composition obtained from the cultured *Chlorarachnion reptans* (Hibberd and Norris, 1984). Since then, four genera: *Cryptochlora* Calderon-Saenz et Schnetter (Calderon-Saenz and Schnetter, 1987), *Lotharella* Ishida et Y. Hara (Ishida et al., 1996), *Gymnochlora* Ishida et Y. Hara (Ishida et al., 1996), and *Bigelowiella* Moestrup (Moestrup and Sengco, 2001), and eight species: *Cryptochlora perforans* Calderon-Saenz et Schnetter (Calderon-Saenz and Schnetter, 1987), *L. globosa* (Ishida et Y. Hara) Ishida et Y. Hara (Ishida and Hara, 1994; Ishida et al., 1996), *L. amoebiformis* Ishida et Y. Hara (Ishida et al., 2000), *L. polymorpha* Dietz et al. (Dietz et al., 2003), *L. vacuolata* S. Ota et Ishida (Ota et al., 2005),

---

TABLE 9.1
Classification of the phylum Chlorarachniophyta

Phylum Chlorarachniophyta Hibberd et Norris 1984

Class Chlorarachniophyceae Hibberd et Norris 1984

Genus 1. *Chlorarachnion* Geitler 1930
  Species *C. reptans* Geitler 1930
Genus 2. *Cryptochlora* Calderon-Saenz et Schnetter 1987
  Species *C. perforans* Calderon-Saenz et Schnetter 1987
Genus 3. *Gymnochlora* Ishida et Y. Hara 1996
  Species *G. stellata* Ishida et Y. Hara 1996
Genus 4. *Lotharella* Ishida et Y. Hara 1996
  Species 1. *L. globosa* (Ishida et Y. Hara) Ishida et Y. Hara 1996
  Species 2. *L. amoebiformis* Ishida et Y. Hara 2000
  Species 3. *L. polymorpha* Dietz et al. 2003
  Species 4. *L. vacuolata* Ota et Ishida 2005
Genus 5. *Bigelowiella* Moestrup 2001
  Species 1. *B. natans* Moestrup 2001
  Species 2. *B. longifila* S. Ota et Ishida 2007

G. stellata Ishida et Y. Hara (Ishida et al., 1996), *B. natans* Moestrup (Moestrup and Sengco, 2001), and *B. longifila* S. Ota et Ishida (Ota et al., 2007), have been described (See Table 9.1).

The genera of chlorarachniophytes are classified based mainly on the ultrastructural characteristics of the pyrenoid and nucleomorph and the types of cell that constitute the vegetative stage in the life cycle (Ishida et al., 1996; Moestrup and Sengco, 2001). *Chlorarachnion* is characterized mainly by the deep and wide longitudinal invagination of the periplastidial compartment into the pyrenoid and the nucleomorph embedded in the invagination (Ishida et al., 1996). The nucleomorph of other genera is localized in the swelling of periplastidial compartment beside the stem of the projecting pyrenoid. *Lotharella* has a pyrenoid that is longitudinally divided into two halves by the thin invagination of the periplastidial compartment (Ishida et al., 1996). The pyrenoid of *Gymnochlora* has no invagination of the periplastidial compartment. Instead, there are many tubular invaginations of the innermost plastid envelope membrane (Ishida et al., 1996). *Bigelowiella* is not defined by the pyrenoid ultrastructure but mainly by the fact that flagellate cells constitute the vegetative stage in the life cycle (Moestrup and Sengco, 2001). *Cryptochlora* was described based only on the light microscopic morphology and defined mainly by the solitary nature of the amoeboid cell (Calderon-Saenz and Schnetter, 1987). This appears to be an ancestral characteristic of the chlorarachniophytes; however, the criterion of this genus may need to be reexamined using a wider range of characteristics including ultrastructure and molecular data.

At present, three of the five genera are monospecific, and species diversity is recognized only in *Lotharella* and *Bigelowiella* (Ishida et al., 2000; Dietz et al., 2003; Ota et al., 2005; Ota et al., 2007). Species of these genera are classified based on morphology and the pattern of the life cycle. *Lotharella globosa*, the type species of *Lotharella*, is a coccoid species, in which coccoid cells constitute the main vegetative stage and zoospores are occasionally released. No amoeboid cell has been reported (Ishida and Hara, 1994). *L. polymorpha* and *L. vacuolata* appear to be similar to each other. Both have a coccoid cell stage as the main vegetative stage of the life cycle. Unlike *L. globosa*, they also have amoeboid cells in the life cycle. Those two species are mainly distinguished from each other by the presence or absence of "heliozoan-like cells" and the structure of the cell wall. Another species, *L. amoebiformis*, is characterized by dominance of the amoeboid cell stage in the life cycle. Coccoid cells and flagellate cells are rare and do not proliferate (Ishida et al., 2000). The type species of *Bigelowiella*, *B. natans*, is characterized by the fact that the flagellate cell stage is the only vegetative stage of the life cycle (Moestrup and Sengco, 2001). Although amoeboid cells with no flagellum and walled cells (cysts) are also present, they are only seen in old cultures and do not proliferate (Moestrup and Sengco, 2001). *B. longifila* is distinguished from *B. natans* by the presence of a vegetative amoeboid stage as well as the vegetative flagellate stage in the life cycle (Ota et al., 2007). The amoeboid cells of *B. longifila* usually have a single very long flagellum and form mucilaginous suspendable colonies (Ota et al., 2007).

PHYLOGENY OF CHLORARACHNIOPYTES

The phylogeny of the chlorarachniophytes has been investigated by molecular phylogentic analyses using nuclear-encoded small subunit ribosomal RNA (SSU rRNA) genes (Ishida et al., 1999; Yabuki et al., unpublished data). The latest analysis recognized six well-supported clades: namely, *Bigelowiella*, BC52, *Chlorarachnion*, *Lotharella globosa*, *Gymnochlora* and *L. amoebiformis* clades (Figure 9.8). Members of *Lotharella* were split into two distinct clades, indicating that taxonomic reexamination is probably needed on this genus. Besides the two *Lotharella* clades, those clades seem to correspond well to the genera of chlorarachniophytes (Yabuki et al., unpublished data).

Not only for taxonomic purposes but also for understanding how a heterotrophic protozoan evolved into many diverse photosynthetic organisms, elucidating the phylogenetic relationships of the chlorarachniophytes is important. The phylogenetic relationship among the six clades is still arguable, since the branch leading to the chlorarachniophyte cluster is somewhat long (Figure 9.8). This is probably because the closest cercozoan relative of the chlorarachniophytes remains to be discovered. In order to obtain a reliable topology of molecular phylogeny in the chlorarachniophytes, it will be necessary to recognize further diversity in the colourless cercozoans, of which we know only a tiny fraction.

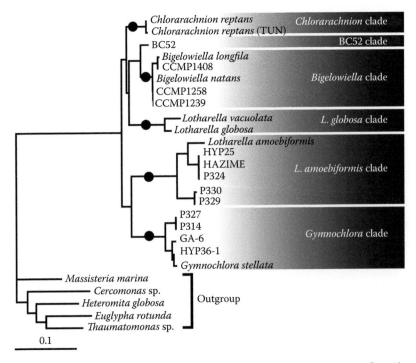

FIGURE 9.8 Maximum likelihood tree of nuclear-encoded SSU rRNA gene sequences from the chlorarachniophytes and a few colourless cercozoans (as outgroup). Black dots indicate chlorarachniophyte subclades that are supported by 100% of bootstrap value.

ACKNOWLEDGMENTS

Authors would like to thank Jane Lewis for valuable comments and help in improving the use of the English language in this manuscript.

REFERENCES

Archibald, J.M. (2005) Jumping genes and shrinking genomes — probing the evolution of eukaryotic photosynthesis with genomics. *International Union of Biochemistry and Molecular Biology: Life*, 57: 539–547.

Archibald, J.M. and Keeling, P.J. (2002) Recycled plastids: a "green movement" in eukaryotic evolution. *Trends in Genetics*, 18: 577–584.

Archibald, J.M., Rogers, M.B., Toop, M., Ishida, K., and Keeling, P.J. (2003) Lateral gene transfer and the evolution of plastid-targeted proteins in the secondary plastid-containing alga *Bigelowiella natans*. *Proceedings of the National Academy of Sciences of the United States of America*, 100: 7678–7683.

Beutlich, A. and Schnetter, R. (1993) The life cycle of *Cryptochlora perforans* (Chlorarachniophyta). *Botanica Acta*, 106: 441–447.

Bhattacharya, D., Helmchen, T., and Melkonian, M. (1995) Molecular evolutionary analyses of nuclear-encoded small subunit ribosomal RNA identify an independent rhizopod lineage containing the Euglyphina and the Chlorarachniophyta. *Jounal of Eukaryotic Microbiology*, 42: 65–69.

Calderon-Saenz, E. and Schnetter, R. (1987) *Cryptochlora perforans*, a new genus and species of algae (Chlorarachniophyta), capable of penetrating dead algal filaments. *Plant Systematics and Evolution*, 158: 69–71.

Calderon-Saenz, E. and Schnetter, R. (1989) Morphology, biology, and systematics of *Cryptochlora perforans* (*Chlorarachniophyta*), a phagotrophic marine alga. *Plant Systematics and Evolution*, 163: 165–176.

Cavalier-Smith, T. (1993) Kingdom protozoa and its 18 phyla. *Microbiological Reviews*, 57: 953–994.

Cavalier-Smith, T. (1999) Principles of protein and lipid targeting in secondary symbiogenesis: euglenoid, dinoflagellate, and sporozoan plastid origins and the eukaryote family tree. *Journal of Eukaryotic Microbiology*, 46: 347–366.

Cavalier-Smith, T. (2000) Membrane heredity and early chloroplast evolution. *Trends in Plant Science*, 5: 174–182.

Cavalier-Smith, T. and Chao, E.E. (2003) Phylogeny and classification of phylum Cercozoa (Protozoa). *Protist*, 154: 341–358.

Dietz, C., Ehlers, K., Wilhelm, C., Gil-Rodríguez, M.C., and Schnetter, R. (2003) *Lotharella polymorpha* sp. nov. (Chlorarachniophyta) from the coast of Portugal. *Phycologia*, 42: 582–593.

Dietz, C., and Schnetter, R. (1996) Arrangement of F-actin and microtubules in the pseudopodia of *Cryptochlora perforans* (Chlorarachniophyta). *Protoplasma*, 193: 82–90.

Geitler, L. (1930) Ein grünes Filarplasmodium und andere neue Protisten. *Archiv für Protistenkunde*, 69: 615–636.

Gilson, P.R. and McFadden, G.I. (1996) The miniaturized nuclear genome of a eukaryotic endosymbiont contains genes that overlap, genes that are cotranscribed, and the smallest known spliceosomal introns. *Proceedings of the National Academy of Sciences of the United States of America*, 93: 7737–7742.

Gilson, P.R. and McFadden, G.I. (1997) Good things in small packages: the tiny genomes of chlorarachniophyte endosymbionts. *Bioessays*, 19: 167–173.

Gilson, P.R. and McFadden, G.I. (1999) Molecular, morphological and phylogenetic characterization of six chlorarachniophyte strains. *Phycological Research*, 47: 7–19.

Gilson, P.R, Su, V., Slamovits, C.H., Reith, M.E., Keeling, P.J., and McFadden, G.I. (2006) Complete nucleotide sequence of the chlorarachniophyte nucleomorph: nature's smallest nucleus. *Proceedings of the National Academy of Sciences of the United States of America*, 103: 9566–9571.

Grell, K.G. (1990) Some light microscope observations on *Chlorarachnion reptans* Geitler. *Archiv für Protistenkunde*, 138: 271–290.

Hibberd, D.J. (1990) Phylum Chlorarachnida. In *Handbook of Protoctista* (eds L. Margulis, J.O. Corliss, and M. Melkonian), Jones & Bartlett Publishers, Boston, pp. 288–292.

Hibberd, D.J. and Norris, R.E. (1984) Cytology and ultrastructure of *Chlorarachnion reptans* (Chlorarachniophyta divisio nova, Chlorarachniophyceae classis nova). *Journal of Phycology*, 20: 310–330.

Ishida, K., Cao, Y., Hasegawa, M., Okada, N., and Hara,Y. (1997) The origin of chlorarachniophyte plastids, as inferred from phylogenetic comparisons of amino acid sequences of EF-Tu. *Journal of Molecular Evolution*, 45: 378–384.

Ishida, K., Green, B.R., and Cavalier-Smith, T. (1999) Diversification of a chimaeric algal group, the chlorarachniophytes: phylogeny of nuclear and nucleomorph small-subunit rRNA genes. *Molecular Biology and Evolution*, 16: 321–331.

Ishida, K. and Hara, Y. (1994) Taxonomic studies on the Chlorarachniophyta. I. *Chlorarachnion globosum* sp. nov. *Phycologia*, 33: 351–358.

Ishida, K., Ishida, N., and Hara, Y. (2000) *Lotharella amoeboformis* sp. nov.: a new species of chlorarachniophytes from Japan. *Phycological Research*, 48: 221–229.

Ishida, K., Nakayama, T., and Hara, Y. (1996) Taxonomic studies on the Chlorarachniophyta. II. Generic delimitation of the chlorarachniophytes and description of *Gymnochlora stellata* gen. et sp. nov. and *Lotharella* gen. nov. *Phycological Research*, 44: 37–45.

Keeling, P.J. (2001) Foraminifera and Cercozoa are related in actin phylogeny: two orphans find a home? *Molecular Biology and Evolution*, 18: 1551–1557.

Keeling, P.J., Deane, J.A., and McFadden, G.I. (1998) The phylogenetic position of alpha- and beta-tubulins from the *Chlorarachnion* host and *Cercomonas* (Cercozoa). *Journal of Eukaryotic Microbiology*, 45: 561–570.

Ludwig, M. and Gibbs, S.P. (1987) Are the nucleomorphs of cryptomonads and *Chlorarachnion* the vestigial nuclei of eukaryotic endosymbionts? *Annals of the New York Academy of Science* 503: 198–211.

Ludwig, M. and Gibbs, S.P. (1989) Evidence that the nucleomorphs of *Chlorarachnion reptans* (Chlorarachniophyceae) are vestigial nuclei: morphology, division and DNA-DAPI fluorescence. *Journal of Phycology*, 25: 385–394.

McFadden, G.I. (1999) Endosymbiosis and evolution of the plant cell. *Current Opinion in Plant Biology*, 2: 513–519.

McFadden, G.I. (2001) Primary and secondary endosymbiosis and the origin of plastids. *Journal of Phycology*, 37: 951–959.

McFadden, G.I., Gilson, P.R., Hofmann, C.J.B., Adcock, G.J., and Maier, U.-G. (1994) Evidence that an amoeba acquired a chloroplast by relating part of an engulfed eukaryotic alga. *Proceedings of the National Academy of Sciences of the United States of America*, 91: 3690–3694.

McFadden, G.I. and Gilson, P.R. (1995) Something borrowed, something green: lateral transfer of chloroplasts by secondary endosymbiosis. *Trends in Ecology and Evolution*, 10: 12–17.

McFadden, G.I., Gilson, P.R., and Waller, R.F. (1995) Molecular phylogeny of chlorarachniophytes based on plastid rRNA and *rbc*L sequences. *Archiv für Protistenkunde*, 145: 231–239.

McFadden, G.I., Gilson, P.R., and Hofmann, C.J.B. (1997a) Division Chlorarachniophyta. In *Origins of Algae and Their Plastids* (ed. D. Bhattacharya), Springer-Verlag, Wien, pp. 175–185.

McFadden, G.I., Gilson, P.R., and Sims, I.M. (1997b) Preliminary characterization of carbohydrate stores from chlorarachniophytes (Division: Chlorarachniophyta). *Phycological Research*, 45: 117–167.

Moestrup, Ø. and Sengco, M. (2001) Ultrastructural studies on *Bigelowiella natans*, gen. et sp. nov., a chlorarachniophyte flagellate. *Journal of Phycology*, 37: 624–646.

Norris, R.E. (1967) Micro-algae in enrichment cultures from Puerto Peñasco, Sonora, Mexico. *Bulletin of the Southern California Academy of Science*, 66: 233–250.

Ota, S., Ueda, K., and Ishida, K. (2005) *Lotharella vacuolata* sp. nov., a new species of chlorarachniophyte algae, and time-lapse video observations on its unique post-cell division behavior. *Phycological Research*, 53: 275–286.

Ota, S., Ueda, K., and Ishida, K. (2007) Taxonomic study of *Bigelowiella longifila* sp. nov. (Chlorarachniophyta) and a time-lapse video observation on the unique migration of amoeboid cells. *Journal of Phycology*, 43: 333–343.

Rogers, M.B., Archibald, J.M., Field, M.A., Li, C., Striepen, B., and Keeling, P.J. (2004) Plastid-targeting peptides from the chlorarachniophyte *Bigelowiella natans*. *Journal of Eukaryotic Microbiology*, 51: 529–535.

Sasa, T., Takaichi, S., Hatakeyama, N., and Watanabe, M.M. (1992) A novel carotenoid ester, Loroxanthin dodecenoate, from *Pyramimonas parkeae* (Prasinophyceae) and a chlorarachniophyte alga. *Plant Cell and Physiology*, 33: 921–925.

Van de Peer, Y. and De Wachter, R. (1997) Evolutionary relationships among the eukaryotic crown taxa taking into account site-to-site rate variation in 18S rRNA. *Journal of Molecular Evolution*, 45: 619–630.

10 Molecular systematics of Haptophyta

Bente Edvardsen and Linda K. Medlin

CONTENTS

ABSTRACT

The predominantly marine algal division Haptophyta comprises at present about 300 species and includes all organisms with an appendage called a haptonema. Data from environmental gene libraries indicate, however, that there are a large number of additional, small (<3 μm) unknown haptophytes in the ocean. Molecular phylogenetic analyses based on nuclear and chloroplast genes and spacers in combination with morphological data have resulted in recent taxonomic changes within the division, but have also shown that further revisions are warranted. Our aim here is to review current knowledge of molecular phylogeny, genetic variation, and biodiversity in Haptophyta, and also highlight some significant gaps.

INTRODUCTION

Haptophytes occur in all seas and may be major components of the nanoplankton (Thomsen et al., 1994; Marchant and Thomsen, 1994). They are important primary producers, and some species, such as members of the genera *Emiliania*, *Gephyrocapsa*, *Phaeocystis*, *Chrysochromulina*, and *Prymnesium*, may form extensive blooms with major biogeochemical, ecological, or economical impact. Our aim is to review our current knowledge of molecular phylogeny, genetic variation, biodiversity, and genomics of Haptophyta, and also point out some major gaps.

GENERAL CHARACTERISTICS AND TAXONOMY OF HAPTOPHYTA

The Haptophyta comprises, at present, about 80 genera and 300 species (Jordan and Green, 1994; Jordan et al., 2004). Most species are marine, but a few thrive in freshwater (ten were listed by Preisig, 2003). Most members are unicellular, planktonic biflagellates, but palmelloid, coccoid, amoeboid, filamentous, colonial, and benthic forms also occur (review by Hibberd, 1980). Nearly all haptophytes (except for members of the family Papposphaeraceae and the genus *Ericiolus*) are photosynthetic, but phagotrophy appears to be common in some genera (e.g. *Chrysochromulina*, review by Jones et al., 1994). In most species, at least one stage in the life cycle possesses two flagella that are similar in form and have no tubular hairs. Between the flagella is a unique organelle, called a haptonema, which differs structurally from the flagella (Figure 10.1; Figure 4 in Jordan et al., 1995). The length of the haptonema varies, and it has been secondarily lost in a few species. It can sometimes coil or bend, but not beat. The haptonema can attach to a substratum and may be involved in food handling (Inouye and Kawachi, 1994). The flagellar apparatus in Haptophyta consists of the two flagella, haptonema and two basal bodies with associated microtubular flagellar roots and fibrous structures. It is complex

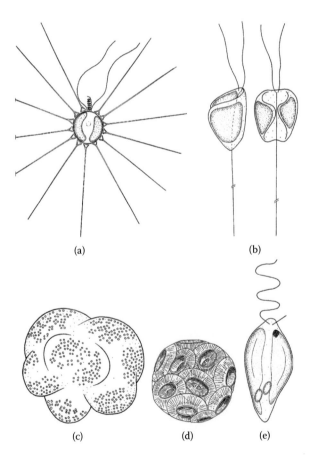

(a)   (b)

(c)   (d)   (e)

FIGURE 10.1 Haptophytes: (a) *Chrysochromulina hirta*, (b) *Chrysochromulina alifera*, (c) *Phaeocystis pouchetii* colony, (d) *Coccolithus pelagicus*, and (e) *Pavlova gyrans*. (From Throndsen, J., Hasle, G.R., and Tangen, K., *Norsk kystplanktonflora*, Almater Forlag AS, Oslo, 2003. With permission.)

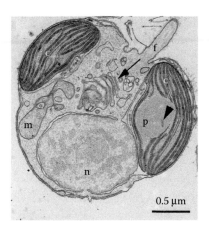

FIGURE 10.2 An electron micrograph of a thin section of a cell of *Chrysochromulina fragaria* with posterior nucleus (n), two chloroplasts (c), a pyrenoid (p) with a part of transversing thylacoid (arrowhead), mitochondria (m), flagellum (f), and a Golgi apparatus (arrow). (From Eikrem, W. and Edvardsen, B., *Phycologia*, 38, 149–155, 1999. With permission.)

and unique in Haptophyta because of the involvement of the haptonema, and has been described in previous reviews (Green and Hori, 1994; Pienaar, 1994; Jordan et al., 1995; Inouye, 1997; Billard and Inouye, 2004). The cells are typically covered by one to several layers of organic scales and coccolithophorids also have calcified scales, coccoliths. Species identification within Haptophyta is largely based on scale morphology and often requires electron microscopy. Cell coverings of Haptophyta were reviewed by Leadbeater (1994). Haptophytes generally possess 1 or 2 yellow-brown chloroplasts bounded by four chloroplast membranes and usually with a pyrenoid (Figure 10.2). The lamellae are composed of three thylakoids, and a girdle lamella is always absent. The pigment composition may also be a phylogenetic marker within Haptophyta (Van Lenning et al., 2003).

The taxonomic history of Haptophyta was reviewed by Green and Jordan (1994). For many years, the haptophytes were included in the class Chrysophyceae, but in 1962 Christensen erected a new class, Haptophyceae, within the division Chromophyta (Christensen, 1962). Hibberd (1976) suggested a separate division for the haptophytes and introduced the typified name Prymnesiophyceae for the class and Prymnesiophyta for the division. However, the descriptive name Haptophyta was already used in the literature and became the valid name of the division. Haptophyta was divided into two classes (Patelliferea and Pavlovea) by Cavalier-Smith (1993). Edvardsen and co-workers gave the class Pavlovea a formal description and renamed it Pavlovophyceae following botanical nomenclature (Edvardsen et al., 2000). They also proposed to divide Prymnesiophyceae into four orders (see below). A division of Prymnesiophyceae into six orders has also been proposed (Young and Bown, 1997; Jordan et al., 2004).

PHYLOGENY OF HAPTOPHYTA INFERRED FROM MOLECULAR AND MORPHOLOGICAL DATA

POSITION OF HAPTOPHYTA IN THE EUKARYOTIC TREE

Eukaryota may be divided into five (Keeling et al., 2005) or eight (Baldauf, 2003) major lineages. Haptophyta is a monophyletic group and not closely related to any other organisms and was not classified in any of the eight lineages by Baldauf (2003). In the eykaryotic tree by Keeling et al. (2005), it was placed in the lineage chromalveolates together with alveolates, cryptophytes, and heterokonts. These groups have members that contain a plastid derived from secondary endosymbiosis with a red alga. It has been proposed that the plastid originated from a single endosymbiontic event in their common ancestor (Cavalier-Smith, 1999, 2004), which requires that both the plastids and the host lineages share a common ancestor. However, all current molecular evidence supporting the chromalveolate hypothesis is based on plastid-related genes (plastid genes or nuclear genes coding for proteins that are targeted to the plastid and are assumed to have been transferred from the plastid, e.g., Patron et al., 2004; Yoon et al., 2004, 2005; Bachvaroff et al., 2005; Li et al., 2006). Because the plastids in the chromalveolates may have been derived from more than a single event, the evidence for the monophyly of the chromalveolates would be stronger if the host lineages also were shown to be related. A phylogenetic analysis by Harper et al. (2005) based on six nuclear-encoded genes for cytoplasmic proteins from all major lineages showed strong support for a clade embracing heterokonts and alveolates and weak support for a common ancestry for haptophytes and cryptophytes, and chromalveolates appeared paraphyletic. There is currently no molecular evidence from nuclear-related genes for the entire Chromalveolata, and the position of Haptophyta in the eukaryotic tree remains uncertain.

PHYLOGENY WITHIN HAPTOPHYTA

Molecular clock analysis based on SSU rDNA data has estimated the origin of Haptophyta to be about 1000 million years ago (Medlin et al., 1997). An analysis of six plastid genes supported this estimate (Yoon et al., 2004). The molecular phylogeny within Haptophyta has been inferred from plastid-encoded *rbc*L (Fujiwara et al., 1994, 2001; Daugbjerg and Andersen, 1997; Inouye 1997) and nuclear-encoded SSU rDNA data (Medlin et al., 1997; Simon et al., 1997; Edvardsen et al., 2000; Sáez et al., 2004). These phylogenies (except for Fujiwara et al., 1994) show the same major clades and in general support the systematic schemes based on morphological data. The inferred phylogeny by Edvardsen et al. (2000) combined SSU rDNA phylogeny with morphological and utrastructural data and resulted in several taxonomic changes. There was a clear divergence of Haptophyta into two main clades with distance of 12% that warranted separation of the two clades at the class level. The clades corresponded with the classes Prymnesiophyceae and Pavlovophyceae. The long branch before the class divergence suggests that the common ancestors have likely gone extinct. The two classes show clear morphological differences supporting the separation (Edvardsen et al., 2000, see below). The SSU rDNA molecular clock estimated the divergence between the two classes to be 400 to 500 million years ago (Medlin et al., 1997).

Characteristic features for members of Pavlovophyceae were described by Green (1980) and Edvardsen et al. (2000). They typically have two unequal (anisokont) flagella that are inserted laterally (Figure 10.1e). The long, anterior flagellum beats in an S-shaped wave and may be covered by thin non-tubular hairs and knob scales, and the posterior flagellum is shorter, smooth, sometimes with distal attenuation, and is vestigial in the genus *Rebecca*. The haptonema is modified and always short. No flat scales are produced, but knob scales may be present on the cell body, anterior flagellum, or haptonema. A stigma (eyespot) may be present, and some produce paramylon as reserve metabolite (Figure 10.1e). Another characterising feature is an elongate invagination of the plasmalemma that forms a closed canal (Green, 1980). The flagellar apparatus and the mitotic process also differ from members of Prymnesiophyceae. Members of Pavlovophyceae have a fibrous flagellar root passing deep into the cell that acts as a microtubular organizing centre during mitosis. This structure has not been observed in Prymnesiophyceae (Green, 1980; Green and Hori, 1988, 1994; Hori and Green, 1994). Members of Prymnesiophyceae, by contrast, typically have two smooth, more or less equal flagella inserted apically or subapically. The haptonema may be conspicuous. No stigma is present, and generally the cells are covered by organic plate scales and/or coccoliths (see reviews by Hibberd, 1980; Inouye, 1997).

PHYLOGENY WITHIN PAVLOVOPHYCEAE

The class Pavlovophyceae currently consists of one order, Pavlovales, four genera, and 12 species (Jordan et al., 2004). Their phylogenetic relationship was examined by Van Lenning et al. (2003) with a parallel analysis of SSU rDNA and pigment signatures. Three pigment groups were found based on the presence or absence of two chlorophyll c forms and one unknown diadinoxanthin-like xanthophyll. The SSU rDNA tree was congruent with the classification based on pigment type and also with ultrastructural data previously produced by Green (1980). Morphological characters that appeared phylogenetically important were presence or absence, placement and nature of the stigma, pyrenoid and flagella. Parsimony analysis of pigment evolution suggested a gradual simplification of the light-harvesting antenna during evolution. The genus *Pavlova* appeared paraphyletic, and a taxonomic revision is warranted (Van Lenning et al., 2003). The authors proposed to transfer *Pavlova lutheri* (Droop) Green and *P. virescens* Billard (and potentially *P. calceolata* van der Veer and *P. noctivaga* (Kalina) van der Veer et Leewis not yet examined genetically) to a new genus, but other solutions discussed were to transfer these species to the genus *Diacronema,* or transfer *Diacronema vlkianum* Prauser emend. Green et Hibberd to the genus *Pavlova.*

PHYLOGENY WITHIN PRYMNESIOPHYCEAE

In the SSU rDNA analysis by Edvardsen et al. (2000) several major clades were recognized within Prymnesiophyceae. The branching order was not well resolved, but the individual clades were strongly supported. The major, well-supported clades were clade A, consisting of *Phaeocystis* spp., clades B1 and B2 containing non-mineralized taxa in the order Prymnesiales, clade C embracing the coccolithophorids, and clades D and E that included sequences from environmental samples representing not yet identified picoplankton taxa. Our most recent haptophyte SSU rDNA phylogenetic study (Figure 10.3; Edvardsen et al., in prep.) included 133 haptophyte taxa. Phylogenies within clades A through C and unifying features are described below.

Clade A corresponding to the order Phaeocystales

Members are characterized by swimming cells with two equal flagella, a short stiff haptonema, and cell body covered by small organic plate scales. Some may produce filamentous star-like structures that are ejected from the cells. A complex life cycle is known in some species, which includes a colonial stage, where the cells are encapsulated in a gelatinous matrix (Figure 10.1c). Lange et al. (2002) analysed 16 strains representing six *Phaeocystis* species in coding and non-coding rDNA regions and in the *rbc*L-*rbc*S spacer. All species were well separated by SSU rDNA. The colony-forming species *P. pouchetii* (Hariot) Lagerheim, *P. globosa* Scherffel, and *P. antarctica* Karsten grouped together in a clade and were clearly separated from two newly described species from the Mediterranean Sea, *P. cordata* Zingone et Chrétiennot-Dinet and *P. jahnii* Zingone et Chrétiennot-Dinet and a not yet described species (Lange et al., 2002). The large variation in the ITS rDNA region among and within strains of *P. globosa* may suggest that *P. globosa* is a species complex (Lange et al., 2002). *Phaeocystis scrobiculata* Moestrup is believed to represent a seventh species (Moestrup, 1979), but this has not yet been examined by molecular methods. Other as yet undescribed species are reviewed in Medlin and Zingone (2007).

Clades B1 and B2 corresponding to the order Prymnesiales

The phylogeny within Prymnesiales based on rDNA and morphology was examined by Edvardsen et al. (in prep.). Clade B1 includes members of the genera *Prymnesium, Platychrysis, Imantonia,* and some *Chrysochromulina* species, as well as a newly described species, *Hyalolithus neolepis* Yoshida (Yoshida et al., 2006; Figure 10.3). All members of clade B1 have irregular to sphaeroid cell shape and possess a haptonema that usually is of the same length as the flagella or shorter, has seven microtubuli in the free part, and usually cannot coil. The haptonema in *C. ericina* Parke et Manton and *C. hirta* Manton is an exception, being longer than the flagella and able to coil (Figure 10.1a). A compound R1 flagellar root may be present. Most members of clade B2, in contrast, are saddle shaped (see Figure 10.1b; although cells of *C. leadbeateri* Estep, Davis, Hargraves et Sieburth are usually rounded). They have a long and coiling haptonema, with six or seven

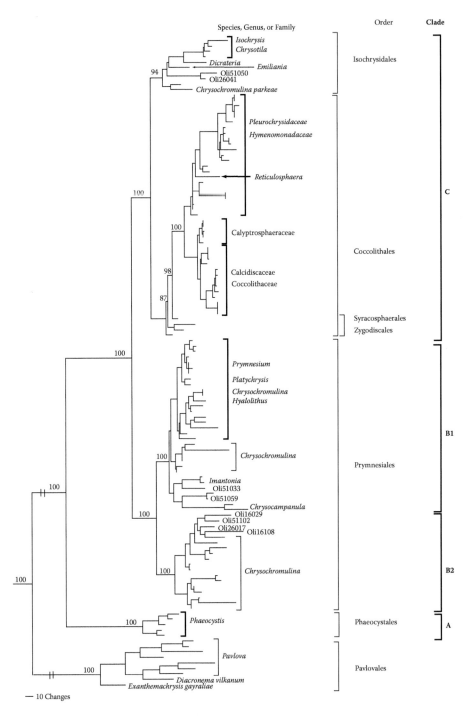

FIGURE 10.3 SSU rDNA phylogenetic tree of haptophytes including 133 taxa based upon a MrBayes analysis. The tree was rooted with *Coscinodiscus radians* and *Gonyaulax spinifera* which later were pruned from the tree. Support values for the branches (posterior probability) are presented at the internal nodes. One hundred base changes were cut out from the basal branches. Oli16029-Oli51059 refers to environmental clones that represent small picoeukaryotes from the Pacific Ocean (Moon-van der Staay et al., 2000). *Chrysochromulina* species belonging to clade B1: *C. brevifilum, C. chiton, C. ericina, C. fragaria, C.* cf. *herdlensis, C. hirta, C. kappa, C. minor, C. polylepis, C.* aff. *polylepis* PLY 200, and to the B2 clade: *C. acantha, C. campanulifera, C. cymbium, C.* cf. *ephippium, C. leadbeateri, C. parva, C. rotalis, C. scutellum, C. simplex,* and *C. throndsenii.* Information on the sequences is available on request.

microtubules in the free part, and they usually have a simple R1 flagellar root. Some have cup-formed scales. In our phylogenetic tree (Figure 10.3), all species with saddle-shaped cells form a monophyletic group (clade B2). This group includes the type species *C. parva* Lackey and therefore has priority to the name *Chrysochromulina*. Members of *Platychrysis* fall within the clade of *Prymnesium,* and should be transferred to this genus. The dominant life cycle stage of *Platychrysis* is non-motile, but the flagellated cells resemble those of *Prymnesium parvum* N. Carter. Strains of *Imantonia* group together in a clade and should retain their genus name (Figure 10.3). *Chrysoch-romulina spinifera* (Fournier) Pienaar et Norris, was originally described under the genus name *Chrysocampanula* (Fournier, 1971). It possesses a combination of morphological features (e.g. non-coiling haptonema, heterodynamic flagella, and elaborate spiny scales; Pienaar and Norris, 1979) and a SSU rDNA sequence (Figure 10.3) that deviates from the rest of the Prymnesiales, thereby warranting the reinstatement of the genus. *Chrysochromulina* species within clade B1 fall into several subclades (Figure 10.3) and should probably be transferred to several new genera or to the genus *Prymnesium.* At present, there are no characters that can be used alone to delineate various genera here, so a combination of characters have to be used. Further ultrastructural and molecular data (e.g. of LSU rDNA or *rbc*L) are probably required in order to resolve the needed taxonomic changes.

Clade C including the coccolithophorids

In the phylogenetic analysis by Edvardsen et al. (2000), only a limited number (10) of coccolitho-phorid taxa were included, and two subclades were recognized, corresponding to the orders Coc-colithales and Isochrysidales.

Bown et al. (2004) presented a coccolithophorid phylogeny based on morphological observa-tions of coccoliths and data from stratigraphic databases. This phylogeny suggested the division of the living coccolihophorid taxa into four major groups, which correspond to the orders proposed by Jordan et al. (2004): Zygodiscales, Syracosphaerales, Coccolithales, and Isochrysidales. In a phylogenetic analysis by Sáez et al. (2004) based on SSU rDNA, more coccolithophorid taxa were included, also from the two groups Zygodiscales and Syracosphaerales. In this analysis, members of Isochrysidales were separated from the remaining coccolithophorids, but the analysis could not conclude on any subdivision of the order Coccolithales, because the number of taxa from some clades was too low. In Figure 10.3 we have included 46 coccolithophorid taxa from all four orders according to Jordan et al. (2004). Also this phylogeny showed a major divergence between members of Coccolithales and Isochrysidales, and members of Syracosphaerales and Zygodiscales were placed in several clades basal within a Coccolithales clade (Figure 10.3). A final decision on whether or not to subdivide the Coccolithales depends on further molecular data and possibly also ultra-structure from more coccolithophorid species brought into culture.

The orders Coccolithales and Isochrysidales were emended and described in Edvardsen et al. (2000). Members of Isochrysidales produce unique long-chain saturated alkenones (review by Stoll and Ziveri, 2004) that may be a unifying feature. They are also characterized by having a thin membranous sheet underneath the plasmalemma, in the peripheral ER (Van der Wal et al., 1985), which has never been observed elsewhere (Inouye, 1997). The haptonema is rudimentary or absent. The coccolith bearing cells of *Emiliania huxleyi* (Lohmann) Hay et Mohler lack underlayer scales, and coccoliths lack microfibrillar base plates (review by Paasche, 2002). *Isochrysis* has secondarily lost the ability of coccolith formation and bears organic scales similar to those of flagellated cells of *Emiliania huxleyi.* All members of Coccolithales, however, have simple organic underlayer scales, and in addition the cells produce either hetero- or holococcoliths (Figure 10.1d). They have a conspicuous or reduced haptonema and a bulging pyrenoid. Coccolithophorids from all groups typically have a compound R1 flagellar root.

Slight calcification is also found in *Chrysochromulina parkeae* Green et Leadbeater (R. Andersen, personal communication in Sáez et al., 2004). In the SSU rDNA phylogeny shown here (Figure 10.3) and by Sáez et al. (2004), *C. parkeae* falls within the coccolithophorid clade supporting the hypothesis that calcification is a synapomorphy for this group and has evolved only once within Haptophyta. Molecular clock analysis based on SSU rDNA haptophyte sequences and calibrated by the coccolithophorid fossil record has estimated coccolith formation to have arisen about 220 million years ago (Medlin et al., 1997; Sáez et al., unpublished). This corresponds with the evolution of Haptophyta hypothesized by de Vargas and Probert (2004). A molecular clock based on the plastid-encoded *tuf*A gene was also produced to estimate the divergence between morphotypes of five coccolithophorid species (Sáez et al., 2003). The molecular clock estimates supported pseudocryptic speciation events to have occurred between 0.3 and 13 million years ago.

Phylogenetically relevant characters in Haptophyta

Phylogenies based on the *rbc*L gene (Fujiwara et al., 2001) supported the separation of Haptophyta into two classes and of coccolithophorids into two major orders, as well as the paraphyly of *Chrysochromulina*, but the number of available *rbc*L sequences are presently too low to provide further information compared to the SSU rDNA phylogeny. Morphological and ultrastructural features that appear to be phylogenetically relevant in Haptophyta are cell shape, nature of the haptonema, calcification, scale investment, and possibly pyrenoid type and flagellar apparatus. Descriptions of the flagellar apparatus are available for only about 40 haptophyte species, and its value as phylogenetic character is not yet fully elucidated (review by Green and Hori, 1994; Nakayama et al., 2005, and references therein). There is overlap in some of these features between the various subgroups, which makes it difficult to divide this class based on morphology only. Numerous taxa have gone extinct, features have been lost (coccolith formation) or reduced (haptonema, flagella) or arisen by convergent evolution (spiny organic scales) and can be primitive homologies that obscure the true phylogeny. DNA sequences have here provided a valuable and objective framework.

INTRASPECIFIC GENETIC VARIATION

The SSU rRNA gene has proven to be a valuable phylogenetic marker within Haptophyta down to species level. When measuring genetic variation below the species level, analysis of non-coding regions (e.g. ITS rDNA) or fingerprinting techniques (such as AFLP, microsatellites) are usually more appropriate. The phylogeny among closely related *Phaeocystis* species and genetic variation within *P. antarctica* were inferred from ITS1 rDNA sequences (Lange et al., 2002). In general the relationship of strains within a species is related to their geographical origin, and this was also the case for *P. antarctica* in the Antarctic. ITS1 rDNA was also analysed by Larsen and Medlin (1997) in order to determine the genetic variation within and between *Prymnesium parvum* N. Carter and *Prymnesium parvum* f. *patelliferum* (Green, Hibberd et Pienaar) A. Larsen. Their analyses showed that the strains were related by geographic origin rather than by morphotype, supporting the view that they are conspecific. Eight *Chrysochromulina polylepis* Manton et Parke isolates from Skagerrak were, however, all identical in the ITS1 rDNA region, even though the variation between *Chrysochromulina* species was so large that reliable alignment was not possible (Edvardsen and Medlin, 1998). Genetic variation has been extensively studied in populations of *Emiliania huxleyi* (Table 10.1). Large morphological and physiological differences between clones have suggested large genetic variation, but sequence comparisons of both coding and non-coding DNA regions did not suffice to reveal separate genetic taxa. Various fingerprinting techniques have, however, shown that there is large genetic diversity within this species, even within a bloom. Iglesias-Rodrígues et al. (2006) used microsatellite primers to assess the genetic variation among 85 isolates of *E. huxleyi*. Their results suggested high genetic diversity on a global scale and low gene flow between Norwegian fjord strains and strains from North East Atlantic. None of their other global isolates of *E. huxleyi* could be assigned to these two large populations in the northern hemisphere. Many haptophyte

TABLE 10.1
Studies on Genetic Variation in *Emiliania huxleyi*

Genetic Marker	Result	Reference
SSU rDNA	Morphotypes A and B identical	Medlin et al. (1994, 1996)
rbcL-rbcS spacer	Morphotypes A and B identical	Medlin et al. (1994, 1996)
RAPD/AFLP	Extensive genetic variation	Barker et al. (1994), Medlin et al. (1994, 1996)
Coccolith associated protein mRNA and DGGE	Genetic variation differentiated A and B morphotypes	Schroeder et al. (2005)
Microsatellites	Extensive genetic variation	Iglesias-Rodríguez et al. (2006)

species have been considered as cosmopolitan, but some morphologically defined species appear to have a more restricted geographical distribution when examined by molecular markers and may be cryptic species (Medlin et al., 2000). High intraspecific genetic variation within haptophyte taxa suggests that sexual reproduction must be widespread to maintain such diversity and may possibly be an adaptation to a variable environment.

LIFE CYCLES

Within Haptophyta, heteromorphic life histories are frequent and usually include an alternation between motile and non-motile stages (review by Billard, 1994). Yet, sexual reproduction with syngamy and meiosis has been directly observed only in a small number of species (Billard, 1994; Houdan et al., 2004). Haploid and diploid stages, each capable of vegetative cell divisions, have, however, been found in some species, which suggests a sexual haplo-diploid life cycle (Vaulot et al., 1994; Green et al., 1996). Among the coccolithophorids, a number of cases are now known where a heterococcolithophorid is joined in a haplo-diploid life cycle with a holococcolithophorid (Geisen et al., 2002, 2004; Houdan et al., 2004). Over 70 holococcolithophorid species have been described, and they are all expected to be the haploid stage in a haplo-diploid life cycle with a heterococcolithophorid (Billard, 1994; Billard and Inouye, 2004). In *Emiliania huxleyi,* the haploid stage is a flagellated cell covered by organic scales only (Klaveness, 1972; Green et al., 1996), although in Medlin et al. (1996), haploid stages with heterococcoliths were also documented by flow cytometry. A haplo-diploid life cycle is found also in non-mineralized genera, such as in *Phaeocystis, Chrysochromulina,* and *Prymnesium.* In *C. polylepis* the life cycle comprises two distinct flagellated cell types that differ in scale morphology, cell size, toxicity, growth preferences, and DNA content, but are identical in coding and non-coding rDNA regions (Paasche et al., 1990; Edvardsen and Paasche, 1992; Edvardsen and Vaulot, 1996; Edvardsen and Medlin, 1998). However, syngamy and meiosis have never been observed. A similar life cycle is found in *Prymnesium parvum* (Larsen and Medlin, 1997; Larsen and Edvardsen, 1998) and in four additional *Chrysochromulina* species (Edvardsen, 2006). A haplo-diploid life cycle is thus present in all four orders of Prymnesiophyceae (Phaeocystales, Vaulot et al., 1994; Prymnesiales, Edvardsen and Vaulot, 1996; Larsen and Edvardsen, 1998; Isochrysidales, Green et al., 1996; and Coccolithales, Houdan et al., 2004), suggesting that this is an autapomorphic state for this class. Whether or not members of Pavlovophyceae also have a haplo-diploid life cycle is unknown.

BIODIVERSITY REVEALED BY MOLECULAR TOOLS

Because of the small cell size of haptophytes, species identification usually has to be verified by electron microscopy or molecular methods, and new species are discovered and described all the time. Moon-van der Staay et al. (2000) produced a clone library from environmental samples which

TABLE 10.2
Current Genomic Data of Haptophyta

Genomic Data	Haptophyte species	Institute	Reference
Nuclear genome	*Emiliania huxleyi*	DOE-JGI	www.jgi.doe.gov
Mitochondrial genome	*Emiliania huxleyi*	University of Maryland, College Park	Sánchez Puerta et al. (2004)
Plastid genome	*Emiliania huxleyi*	University of Maryland, College Park	Sánchez Puerta et al. (2005)
EST	*Emiliania huxleyi*	California State University, San Marcos	Wahlund et al. (2004a, 2004b)
EST	*Isochrysis galbana*	University British Columbia	Patron et al. (2006)
EST	*Pavlova lutheri*	University British Columbia	Patron et al. (2006)
EST	*Chrysochromulina polylepis*	Alfred-Wegener Institute	John et al. (2004)
EST	*Prymnesium parvum*	Alfred-Wegener Institute	Beszteri et al. (2006)

Note: EST = Expressed sequence tag library.

was sequenced and analysed phylogenetically. They found two new clades not strongly related to any other known clades (not included in Figure 10.3), probably representing novel haptophyte taxa, in addition to taxa belonging to Prymnesiales and Isochrysidales (Figure 10.3). All environmental clones came from a 3-μm filtered water sample from the Pacific Ocean, and thus represent picoeukaryotes. These data suggest that there are many haptophyte species in the ocean awaiting a formal description. Recently, a new species belonging to one of the clades with exclusively environmental clones, Clade E, was described by Nakayama et al. (2005). This species, named *Chrysoculter rhomboideus* Nakayama, Yoshida, Noël, Kawachi et Inouye, produces non-mineralized scales only.

GENOMIC DATA

Genomic data have up to now almost exclusively been obtained from *Emiliania huxleyi*, but expressed sequence tag libraries (fragments of the genes that are expressed) have also been produced from a few additional haptophytes (Table 10.2). Genomic sequencing of the entire mitochondrion and plastid of *E. huxleyi* has been done, and sequencing of the nuclear genome is underway (Table 10.2). The discovery of all new genes in *E. huxleyi* will certainly promote the analysis of these genes in other haptophytes, and this is expected to shed further light on the evolution of Haptophyta.

CONCLUSIONS

In general, molecular data have supported systematic schemes based on morphological and ultrastructural information, at the same time providing new information on phylogenetic relationships, genetic variation, life cycle relationships, and biodiversity. Molecular data combined with morphological, physiological, biochemical, and ecological information will increase our understanding of haptophyte evolutionary, biological, and ecological processes.

ACKNOWLEDGMENTS

We would like to thank Drs Alberto Sáez, Wenche Eikrem, John Green, and Ian Probert for the collaboration and for providing unpublished data, and Dr Masanobu Kawachi for a SSU rDNA sequence of *Chrysochromulina parkeae*. We also thank Dr Wenche Eikrem and Prof. Eystein Paasche for constructive comments on the manuscript.

REFERENCES

Bachvaroff, T.R., Sanchez-Puerta, M.V., and Delwiche, C.F. (2005) Chlorophyll *c*-containing plastid relationships based on analyses of a multigene data set with all four chromalveolate lineages. *Molecular Biology and Evolution*, 22: 1772–1782.

Baldauf, S.L. (2003) The deep roots of eukaryotes. *Science*, 300: 1703–1706.

Barker, G.L.A., Green, J.C., Hayes, P.K., and Medlin, L.K. (1994) Preliminary results using the RAPD analysis to screen bloom populations of *Emiliania huxleyi* (Haptophyta). *Sarsia*, 79: 301–306.

Beszteri, S., Tillmann, U., Freitag, M., Glöckner, G., Cembella, A.D., and John, U. (2006). Molecular physiology of the toxigenic haptophyte *Prymnesium parvum*. 12th International Conference on Harmful Algae, Programme and Abstracts, Copenhagen, Denmark, 4–8 September, pp. 134–135.

Billard, C. (1994) Life cycles. In *The Haptophyte Algae* (eds J.C. Green and B.S.C. Leadbeater), Oxford University Press, Clarendon, New York, pp. 167–186.

Billard, C. and Inouye, I. (2004) What is new in coccolithophore biology? In *Coccolithophores. From Molecular Processes to Global Impact* (eds H.R. Thierstein and J.R. Young), Springer-Verlag, Berlin, pp. 1–29.

Bown, P.R., Lees, J.A., and Young, J.R. (2004) Calcareous nannoplankton evolution and diversity through time. In *Coccolithophores. From Molecular Processes to Global Impact* (eds H.R. Thierstein and J.R. Young), Springer-Verlag, Berlin, pp. 481–508.

Cavalier-Smith, T. (1993) Kingdom Protozoa and its 18 phyla. *Microbiological Reviews*, 57: 953–994.

Cavalier-Smith, T. (1999) Principles of protein and lipid targeting in secondary symbiogenesis: euglenoid, dinoflagellate, and sporozoan plastid origins and the eukaryote family tree. *Journal of Eukaryotic Microbiology*, 46: 347–366.

Cavalier-Smith, T. (2004) Chromalveolate diversity and cell megaevolution: interplay of membranes, genomes and cytoskeleton. In *Organelles, Genomes and Eukaryote Phylogeny* (eds R.P. Hirt and D.S. Horner), CRC Press, Boca Raton, FL, pp. 75–108.

Christensen, T. 1962. Alger. In *Systematisk Botanik* (eds T.W. Böcher, M. Lange, and T. Sørensen), Munksgaard, Copenhagen, pp. 1–178. (In Danish).

Daugbjerg, N. and Andersen, R.A. (1997) Phylogenetic analyses of the *rbc*L sequences from haptophytes and heterokont algae suggest their chloroplasts are unrelated. *Molecular Biology and Evolution* 14: 1242–1251.

de Vargas, C. and Probert, I. (2004) New keys to the past: current and future DNA studies in coccolithophores. *Micropaleontology*, 50 (Supplement no. 1): 45–54.

Edvardsen, B. (2006) Haplo-diploid life cycles in the genus *Chrysochromulina* (Haptophyta). 12th International Conference on Harmful Algae, Programme and Abstracts, Copenhagen, Denmark, 4–8 September 2006, pp. 166–167.

Edvardsen, B., Eikrem, W., Green, J.C., Andersen, R.A., Moon-van der Staay, S.Y., and Medlin, L.K. (2000) Phylogenetic reconstructions of the Haptophyta inferred from 18S ribosomal DNA sequences and available morphological data. *Phycologia*, 39: 19–35.

Edvardsen, B. and Medlin, L. (1998) Genetic analyses of authentic and alternate forms of *Chrysochromulina polylepis* (Haptophyta). *Phycologia*, 37: 275–283.

Edvardsen, B. and Paasche, E. (1992) Two motile stages of *Chrysochromulina polylepis* (Prymnesiophyceae): morphology, growth and toxicity. *Journal of Phycology*, 28: 104–114.

Edvardsen, B. and Vaulot, D. (1996) Ploidy analysis of the two motile forms of *Chrysochromulina polylepis* (Prymnesiophyceae). *Journal of Phycology*, 32: 94–102.

Edvardsen, B., Eikrem, W., Sáez, A.G., Probert, I., Green, J., and Medlin, L.K. A ribosomal DNA phylogeny and a morphological revision set the basis for a new taxonomy of Prymnesiales (Haptophyta). (In prep.)

Eikrem, W. and Edvardsen, B. (1999) *Chrysochromulina fragaria* sp. nov. (Prymnesiophyceae), a new haptophyte flagellate from Norwegian waters. *Phycologia*, 38: 149–155.

Fournier, R.O. (1971) *Chrysocampanula spinifera* gen. et sp. nov., a new marine haptophyte from the Bay of Chaleurs, Quebec. *Phycologia*, 10: 89–92.

Fujiwara, S., Sawada, M., Someya, J., Minaka, N., Kawachi, M., and Inouye, I. (1994) Molecular phylogenetic analysis of *rbc*L in the Prymnesiophyta. *Journal of Phycology*, 30: 863–871.

Fujiwara, S., Tsuzuki, M., Kawachi, M., Minaka, N., and Inouye, I. (2001) Molecular phylogeny of the Haptophyta based on the *rbc*L gene and sequence variation in the spacer region of the RUBISCO operon. *Journal of Phycology*, 37: 121–129.

Geisen, M., Billard, C., Broerse, A.T.C., Cros, L., Probert, I., and Young, J.R. (2002) Life-cycle associations involving pairs of holococcolithophorid species: intraspecific variation or cryptic speciation? *European Journal of Phycology*, 37: 531–550.

Geisen, M., Young, J.R., Probert, I., Sáez, A.G., Baumann, K.-H., Sprengel, C., Bollmann, J., Cros, L., de Vargas, C., and Medlin, L.K. (2004) Species level variation in coccolithophores. In *Coccolithophores. From Molecular Processes to Global Impact* (eds H.R. Thierstein and J.R. Young), Springer-Verlag, Berlin, pp. 327–366.

Green, J.C. (1980) The fine structure of *Pavlova pinguis* Green and a preliminary survey of the order Pavlovales (Prymnesiophyceae). *British Journal of Phycology*, 15: 151–191.

Green, J.C., Course, P.A., and Tarran, G.A. (1996) The life-cycle of *Emiliania huxleyi*: a brief review and a study of relative ploidy levels analysed by flow cytometry. *Journal of Marine Systems*, 9: 33–44.

Green, J.C. and Hori, T. (1988) The fine structure of mitosis in *Pavlova* (Prymnesiophyceae). *Canadian Journal of Botany*, 66: 1497–1509.

Green, J.C. and Hori, T. (1994) Flagella and flagellar roots. In *The Haptophyte Algae* (eds J.C. Green and B.S.C. Leadbeater), Clarendon Press, Oxford, pp. 47–71.

Green, J.C. and Jordan, R.W. (1994) Systematic history and taxonomy. In *The Haptophyte Algae* (eds J.C. Green and B.S.C. Leadbeater), Clarendon Press, Oxford, pp. 1–21.

Harper, J.T., Waanders, E., and Keeling, P.J. (2005) On the monophyly of chromalveolates using a six-protein phylogeny of eukaryotes. *International Journal of Systematic and Evolutionary Microbiology*, 55: 487–496.

Hibberd, D.J. (1976) The ultrastructure and taxonomy of the Chrysophyceae and Prymnesiophceae (Haptophyceae): a survey with some new observations on the ultrastructure of the Chrysophyceae. *Botanical Journal of the Linnean Society* 72: 55–80.

Hibberd, D.J. (1980) Prymnesiophytes (= Haptophytes). In *Developments in Marine Biology, Vol. 2. Phytoflagellates* (ed E.R. Cox), Elsevier North Holland, New York, pp. 273–317.

Hori, T. and Green, J.C. (1994) Mitosis and cell division. In *The Haptophyte Algae* (eds J.C. Green and B.S.C. Leadbeater), Clarendon Press, Oxford, pp. 91–109.

Houdan, A., Billard, C., Marie, D., Not, F., Sáez, A.G., Young, J.R., and Probert, I. (2004) Holococcolithophore-heterococcolithophore (Haptophyta) life cycles: flow cytometric analysis of relative ploidy levels. *Systematics and Biodiversity*, 1: 453–465.

Iglesias-Rodríguez, M.D., Schofield, O.M., Batley, J., Medlin, L.K., and Hayes, P.K. (2006) Intraspecific genetic diversity in the marine coccolithophore *Emiliania huxleyi* (Prymnesiophyceae): the use of microsatellite analysis in marine phytoplankton population studies. *Journal of Phycology*, 42: 526–536.

Inouye, I. (1997) Systematics of haptophyte algae in Asia-Pacific waters. *Algae (The Korean Journal of Phycology)*, 12: 247–261.

Inouye, I. and Kawachi, M. (1994) The haptonema. In *The Haptophyte Algae* (eds J.C. Green and B.S.C. Leadbeater), Clarendon Press, Oxford, pp. 73–89.

John, U., Mock, T., Valentin, K., Cembella, A.D., and Medlin, L.K. (2004) Dinoflagellates come from outer space, but haptophytes and diatoms do not. In *Harmful Algae 2002* (eds. K.A. Steidinger, J.H. Landsberg, C.R. Tomas, and G.A. Vargo), Florida Fish and Wildlife Conservation Commission and Intergovernmental Oceanographic Commission of UNESCO, St. Petersburg, FL, pp. 428–430.

Jones, H.L.J., Leadbeater, B.S.C., and Green, J.C. (1994) Mixotrophy in haptophytes. In *The Haptophyte Algae* (eds J.C. Green and B.S.C. Leadbeater), Clarendon Press, Oxford, pp. 247–263.

Jordan, R.W., Cros, L., and Young, J.R. (2004) A revised classification scheme for living haptophytes. *Micropaleontology*, 50 (Supplement no. 1): 55–79.

Jordan, R.W. and Green, J.C. (1994) A check-list of the extant Haptophyta of the world. *Journal of the Marine Biological Association of the United Kingdom* 74: 149–174.

Jordan, R.W., Kleijne, A., Heimdal, B.R., and Green, J.C. (1995) A glossary of the extant Haptophyta of the world. *Journal of the Marine Biological Association of the United Kingdom* 75: 769–814.

Keeling, P.J., Burger, G., Durnford, D.G., Lang, B.F., Lee, R.W., Pearlman, R.E., Roger, A.J., and Gray, M.W. (2005) The tree of eukaryotes. *Trends in Ecology and Evolution*, 20: 670–676.

Klaveness, D. (1972) *Coccolithus huxleyi* (Lohm.) Kamptn. II. The flagellate cell, aberrant cell types, vegetative propagation and life cycles. *British Phycological Journal*, 7: 309–318.

Lange, M., Chen, Y.-Q., and Medlin, L.K. (2002) Molecular genetic delineation of *Phaeocystis* species (Prymnesiophyceae) using coding and non-coding regions of nuclear and plastid genomes. *European Journal of Phycology*, 37: 77–92.

Larsen, A. and Edvardsen, B. (1998) Relative ploidy levels in *Prymnesium parvum* and *P. patelliferum* (Haptophyta) analyzed by flow cytometry. *Phycologia*, 37: 412–424.

Larsen, A. and Medlin, L.K. (1997) Inter- and intraspecific genetic variation in twelve *Prymnesium* (Haptophyceae) clones. *Journal of Phycology*, 33: 1007–1015.

Leadbeater, B.S.C. (1994) Cell coverings. In *The Haptophyte Algae* (eds. J.C. Green and B.S.C. Leadbeater), Clarendon Press, Oxford, pp. 23–46.

Li, S., Nosenko, T., Hackett, J.D., and Bhattacharya, D. (2006) Phylogenomic analysis identifies red algal genes of endosymbiotic origin in the chromalveolates. *Molecular Biology and Evolution*, 23: 663–674.

Marchant, H.J. and Thomsen, H.A. (1994) Haptophytes in polar waters. In *The Haptophyte Algae* (eds. J.C. Green and B.S.C. Leadbeater), Oxford University Press, Clarendon, New York, pp. 209–228.

Medlin, L.K., Barker, G.L.A., Baumann, M., Hayes, P.K., and Lange, M. (1994) Molecular biology and systematics. In *The Haptophyte Algae* (eds. J.C. Green and B.S.C. Leadbeater), Clarendon Press, Oxford, pp. 393–411.

Medlin, L.K., Barker, G.L.A., Campbell, L., Green, J.C., Hayes, P.K., Marie, D., Wrieden, S., and Vaulot, D. (1996) Genetic characterisation of *Emiliania huxleyi* (Haptophyta). *Journal of Marine Systems*, 9: 13–31.

Medlin, L.K., Kooistra, W.H.C.F., Potter, D., Saunders, G.W., and Andersen, R.A. (1997) Phylogenetic relationships of the "golden algae" (haptophytes, heterokont chromophytes) and their plastids. In *Origins of Algae and Their Plastids* (ed. D. Bhattacharya), Springer-Verlag, New York, pp. 187–219.

Medlin, L.K., Lange, M., Edvardsen, B., and Larsen, A. 2000. Cosmopolitan haptophyte flagellates and their genetic links. In *The Flagellates. Unity, Diversity and Evolution* (eds. B.S.C. Leadbeater and J.C. Green), The Systematics Association Special Volume Series 59, Taylor and Francis, London, pp. 288–308.

Medlin, L.K. and Zingone, A. 2007. A taxonomic review of the genus *Phaeocystis*. *Biogeochemistry*, 10.1007/10533–007, 9087–1.

Moestrup, Ø. (1979) Identification by electron microscopy of marine nanoplankton from New Zealand, including the description of four new species. *New Zealand Journal of Botany*, 17: 61–95.

Moon-van der Staay, S.Y., van der Staay, G.W.M., Guillou, L., Vaulot, D., Claustre, H., and Medlin, L.K. (2000) Abundance and diversity of prymnesiophytes in the picoplankton community from the equatorial Pacific Ocean inferred from 18S rDNA sequences. *Limnology and Oceanography*, 45: 98–109.

Nakayama, T., Yoshida, M., Noël, M.-H., Kawachi, M., and Inouye, I. (2005) Ultrastructure and phylogenetic position of *Chrysoculter rhomboideus* gen. et sp. nov. (Prymnesiophyceae), a new flagellate haptophyte from Japanese coastal waters. *Phycologia*, 44: 369–383.

Paasche, E. (2002) A review of the coccolithophorid *Emiliania huxleyi* (Prymnesiophyceae), with particular reference to growth, coccolith formation, and calcification-photosynthesis interactions. *Phycologia*, 40: 503–529.

Paasche, E., Edvardsen, B., and Eikrem, W. (1990) A possible alternate stage in the life cycle of *Chrysochromulina polylepis* Manton & Parke (Prymnesiophyceae). *Nova Hedwigia Beiheft*, 100: 91–99.

Patron, N.J., Rogers, M.B., and Keeling, P.J. (2004) Gene replacement of fructose-1,6-bisphosphate aldolase supports the hypothesis of a single photosynthetic ancestor of chromalveolates. *Eukaryotic Cell*, 3: 1169–1175.

Patron, N.J., Waller, R.F., and Keeling, P.J. (2006) A tertiary plastid uses genes from two endosymbionts. *Journal of Molecular Biology*, 357: 1373–1382.

Pienaar, R.N. (1994) Ultrastructure and calcification of coccolithophores. In *Coccolithophores* (eds A. Winter and W.G. Siesser), Cambridge University Press, Cambridge, pp. 13–37.

Pienaar, R.N. and Norris, R.E. (1979) The ultrastructure of the flagellate *Chrysochromulina spinifera* (Fournier) comb. nov. (Prymnesiophyceae) with special reference to scale production. *Phycologia*, 18: 99–108.

Preisig, H.R. (2003) Phylum Haptophyta (Prymnesiophyta). In *The Freshwater Algal Flora of the British Isles. An Identification Guide to Freshwater and Terrestrial Algae* (eds D.M. John, B.A. Whitton, and A.J. Brook), Cambridge University Press, Cambridge, pp. 211–213.

Sáez, A.G., Probert, I., Geisen, M., Quinn, P., Young, J.R., and Medlin, L.K. (2003) Pseudo-cryptic speciation in coccolithophores. *Proceedings of the National Academy of Sciences of the United States of America*, 100: 7163–7168.

Sáez, A.G., Probert, I., Young, J.R., Edvardsen, B., Eikrem, W., and Medlin, L.K. (2004) A review of the phylogeny of the Haptophyta. In *Coccolithophores. From Molecular Processes to Global Impact* (eds H.R. Thierstein and J.R. Young), Springer-Verlag, Berlin, pp. 251–269.

Sánchez Puerta, M.V., Bachvaroff, T.R., and Delwiche, C.F. (2004) The complete mitochondrial genome sequence of the haptophyte *Emiliania huxleyi* and its relation to heterokonts. *DNA Research*, 11: 1–10.

Sánchez Puerta, M.V., Bachvaroff, T.R., and Delwiche, C.F. (2005) The complete plastid genome sequence of the haptophyte *Emiliania huxleyi*: a comparison to other plastid genomes. *DNA Research*, 12: 151–156.

Schroeder, D.C., Biggi, G.F., Hall, M., Davy, J., Martínez, J.M., Richardson, A.J., Malin, G., and Wilson, W.H. (2005) A genetic marker to separate *Emiliania huxleyi* (Prymnesiophyceae) morphotypes. *Journal of Phycology*, 41: 874–879.

Simon, N., Brenner, J., Edvardsen, B., and Medlin, L.K. (1997) The identification of *Chrysochromulina* and *Prymnesium* species (Haptophyta, Prymnesiophyceae) using fluorescent or chemiluminescent oligonucleotide probes: a means for improving studies on toxic algae. *European Journal of Phycology*, 32: 393–401.

Stoll, H.M. and Ziveri, P. (2004) Coccolithophorid-based geochemical paleoproxies. In *Coccolithophores. From Molecular Processes to Global Impact* (eds H.R. Thierstein and J.R. Young), Springer Verlag, Berlin, pp. 529–562.

Thomsen, H.A., Buck, K.R., and Chavez, F.P. (1994) Haptophytes as components of marine phytoplankton. In *The Haptophyte Algae* (eds J.C. Green and B.S.C. Leadbeater), Clarendon Press, Oxford, pp. 187–208.

Throndsen, J., Hasle, G.R., and Tangen, K. 2003. *Norsk kystplanktonflora*. Almater Forlag AS, Oslo, 341 pp.

Van der Wal, P., Leunissen-Bijvelt, J.J.M., and Verkleij, A.J. (1985) Ultrastructure of the membranous layers enveloping the cell of the coccolithophorid *Emiliania huxleyi*. *Journal of Ultrastructure Research*, 91: 24–29.

Van Lenning, K., Latasa, M., Estrada, M., Sáez, A.G., Medlin, L., Probert, I., Véron, B., and Young, J. (2003) Pigment signatures and phylogenetic relationships of the Pavlovophyceae (Haptophyta). *Journal of Phycology*, 39: 379–389.

Vaulot, D., Birrien, J.-L., Marie, D., Casotti, R., Veldhuis, M.J.W., Kraay, G.W., and Chrétiennot-Dinet, M.-J. (1994) Morphology, ploidy, pigment composition, and genome size of cultured strains of *Phaeocystis* (Prymnesiophyceae). *Journal of Phycology*, 30: 1022–1035.

Wahlund, T.M., Hadaegh, A.R., Clark, R., Nguyen, B., Fanelli, M., and Read, B.A. (2004a) Analysis of expressed sequence tags from calcifying cells of marine coccolithophorid (*Emiliania huxleyi*). *Marine Biotechnology*, 6: 278–290.

Wahlund, T.M., Zhang, X., and Read, B.A. (2004b) Expressed sequence tag profiles from calcifying and non-calcifying cultures of *Emiliania huxleyi*. *Micropaleontology*, 50 (Supplement no. 1): 145–155.

Yoon, H.S., Hackett, J.D., Ciniglia, C., Pinto, G., and Bhattacharya, D. (2004) A molecular timeline for the origin of photosynthetic eukaryotes. *Molecular Biology and Evolution*, 21: 809–818.

Yoon, H.S., Hackett, J.D., Van Dolah, F.M., Nosenko, T., Lidie, K.L., and Bhattacharya, D. (2005) Tertiary endosymbiosis driven genome evolution in dinoflagellate algae. *Molecular Biology and Evolution*, 22: 1299–1308.

Yoshida, M., Noël, M.-H., Nakayama, T., Naganuma, T., and Inouye, I. (2006) A haptophyte bearing siliceous scales: ultrastructure and phylogenetic position of *Hyalolithus neolepis* gen. et. sp. nov. (Prymnesiophyceae, Haptophyta). *Protist*, 157: 213–234.

Young, J.R. and Bown, P.R. (1997) Cenozoic calcareous nannoplankton classification. *Journal of Nanoplankton Research*, 19: 36–47.

11 Decrypting cryptomonads: a challenge for molecular taxonomy

Federica Cerino and Adriana Zingone

CONTENTS

ABSTRACT

Cryptomonads have a set of unique morphological and biological features and a fascinating and intricate evolutionary history. Due to their complexity, this group still harbours many mysteries. Different characters were used in the past to derive a unified taxonomic scheme (e.g. colour, cell shape, periplast structure, morphology of the furrow/gullet system, and the phycobilin type). However, views on the phylogenetic value of these characters have changed drastically in the light of results of molecular analyses that have also highlighted the importance of heteromorphic life cycle in this group. Conventional characterization methods have been shown to have limited phylogenetic capacity, whereas the application of new techniques is constrained by the little molecular information so far available for this group of species. The future challenge to fill the current gap in the knowledge of cryptomonads is to combine molecular analyses with improved observational and cultivation techniques.

INTRODUCTION

Cryptomonads are a relatively small group of unicellular flagellates comprising about 200 photosynthetic species and a few colourless members, which are either aplastidic or leucoplast-bearing taxa. They are ubiquitous in freshwater, brackish, and marine environments, where they may form large populations and play an important role in the trophic webs, probably in relation with their good food value due to the high content of polyunsaturated fatty acids, wax esters, and polysaccharides (Klaveness, 1988, 1988–1989; Novarino, 2003, and references therein). Cryptomonads are also found in extreme habitats, including soil (Paulsen et al., 1992), groundwater (Novarino et al., 1994), and snow (Javornický and Hindà, 1970), as well as in ikaite (calcium carbonate hexahydrate) columns in Norwegian fjords (Kristiansen and Kristiansen, 1999). Distinct ecological patterns reflecting their requirements have been reported for single species both in lakes (e.g. Sommer, 1982; Moustaka-Gouni, 1995) and in the marine environment (Cerino and Zingone, 2006), showing that cryptomonads do not all belong to a homogeneous functional group. Therefore, a significant loss of information derives from the common procedure to pool them as "undetermined cryptomonads" or even with other flagellated microalgae in routine phytoplankton counts. However, the development of an appropriate knowledge on cryptomonad distribution, ecology, and biogeography is hampered by the current state of taxonomy and the difficulty of identification of the species of this group.

The name "cryptomonads," literally meaning "hidden unicellular organisms," was first chosen by Ehrenberg (1831) to designate some inconspicuous members of phytoplankton. No name has ever been so prophetic. Because of their relevance in the phylogeny of algae, cryptomonads have probably been studied more than other small flagellate groups, although even now, after about 180 years from their first report, many aspects of their biology and taxonomy remain hidden.

One of the reasons for the mystery of cryptomonads resides in the extreme complexity of these organisms and of their evolutionary history, based on which they have been judged as the most complex cells among eukaryotic microbes (Cavalier-Smith, 2002). Despite the variety of morphological and biochemical characters available, phylogeny and classification schemes of the group have changed frequently over time and have been further weakened by the discovery of heteromorphic life cycles. For this reason, cryptomonads more than any other algal group have benefited from the advent of molecular analyses, which can be applied to four distinct genome sources in this case—namely, the nuclear, nucleomorph, chloroplast, and mitochondrion DNA. Indeed, molecular investigations on a limited number of taxa have already drastically changed views on the phylogenetic values of phenetic characters and have laid the foundations for a more natural classification of the group.

In this chapter, we summarise current knowledge of the morphology, systematics, and evolutionary history of cryptomonads, detailing the unique features of this fascinating group of microalgae and highlighting the recent progress made in our understanding of the group and their classification stimulated in recent years by molecular investigations.

GENERAL CHARACTERS

The complexity of cryptomonads partially resides in their morphology (Figure 11.1 [1] through [11]). In no other unicellular microalga can so many different microstructures, accessory elements, and unique features be recognised, most of which leave much to the imagination when speculating on their function.

Unlike other flagellated microalgal species that often lack distinct identification characters in light microscopy, organisms in this group are easily recognised as cryptomonads by their typical asymmetric ellipsoidal, drop- or bean-like shape (Klaveness, 1985) (Figure 11.1 [2] through [5]). The cell shape and the flagellar emergence allow the definition of a ventral side, where a unique complex of structures is located (Figure 11.1 [1]). At the anterior end of the complex there is the *vestibulum*, which continues posteriorly with an open groove, the furrow, and/or with an invagination, the gullet, which is closed posteriorly. Because in different genera there may be either a furrow, or a gullet, or a combination of both, the organization of the furrow/gullet complex was attributed a diagnostic value as well as a phylogenetic importance in the past (Butcher, 1967; Clay et al., 1999). The complicated nature of the ventral surface of the cells is enhanced by the presence of a series of accessory structures varying among the genera and the species, which can be appreciated only in the electron microscope. These include the vestibular ligule (Figure 11.1 [6]), a finger-like projection emerging from the left side of the *vestibulum* and extending over the discharge site of a contractile vacuole (Kugrens et al., 1986); the vestibular plate (Figure 11.1 [7]), a semicircular extension located within the margin of the *vestibulum* (Hill and Wetherbee, 1986; Hill, 1991); a furrow plate lining the furrow along its left margin; accessory folders or lips bordering the furrow

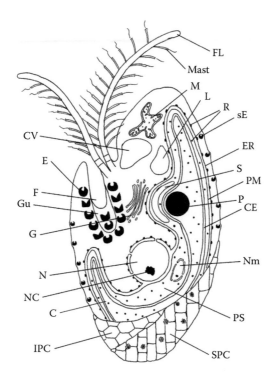

FIGURE 11.1 Cryptomonad cells. (1) Drawing of a cryptomonad cell. C: chloroplast; CE: chloroplast enve-lope; CV: contractile vacuole; E: big ejectosome; ER: rough endoplasmic reticulum; F: furrow; FL: flagellum; G: Golgi apparatus; Gu: gullet; IPC: inner periplast component; L: lysosome; M: mitochondrion; Mast: mastigoneme; N: nucleus; NC: nucleolus; Nm: nucleomorph; PM: periplastid membrane; P: pyrenoid; PS: periplastid space; R: ribosome; S: starch; sE: small ejectosome; SPC: superficial periplast component.

rim; and the stoma (Figure 11.1 [8]), a persistent oval opening at the posterior end of the furrow (Kugrens et al., 1986). In light microscopy, button-like organelles, the ejectosomes, are easily visible aligned along the furrow/gullet complex (Figure 11.1 [1], [2], and [9]). These are special explosive structures, consisting of long coiled-up ribbons (Hovasse et al., 1967; Mignot et al., 1968) that can be suddenly discharged ventrally, making the cell jump in the opposite direction. The only other algae having ejectosomes, although different in structure (Kugrens et al., 1994), are the katablepharids, a group of heterotrophic flagellates once classified as cryptophytes but recently proven to be a distant sister group (Okamoto and Inouye, 2005).

First electron microscopical observations revealed a peculiar structure for the cell envelope, consisting of a proteinaceous periplast sandwiching the cell membrane with an inner periplast component (IPC) and a superficial periplast component (SPC) (Gantt, 1971; Hibberd et al., 1971) (Figure 11.1 [10]). The periplast structure with either discrete polygonal plates or a homogeneous sheet (Figure 11.1 [4] and [5]) has been widely used for identification and classification (Gantt, 1971; Hibberd et al., 1971; Faust, 1974; Santore, 1977; Brett and Wetherbee, 1986; 1996a, b, c; Wetherbee et al., 1986, 1987; Kugrens and Lee, 1987; Novarino and Lucas, 1993; Brett et al., 1994; Perasso et al., 1997; Clay et al., 1999), yet molecular and life-cycle investigations have shown that this character has been overestimated in some cases (see below).

The two unequal flagella emerging from the *vestibulum* also have very peculiar characteristics. Distinct from Heterokonta in which mastigonemes are typically tripartite and only found on the anterior flagellum, in cryptomonads they are usually on both flagella and are bipartite (Hibberd et al., 1971), with several kinds of alternative arrangements (Kugrens et al., 1987). In addition to mastigonemes, tiny organic scales having a seven-sided (heptagonal) rosette pattern may also be

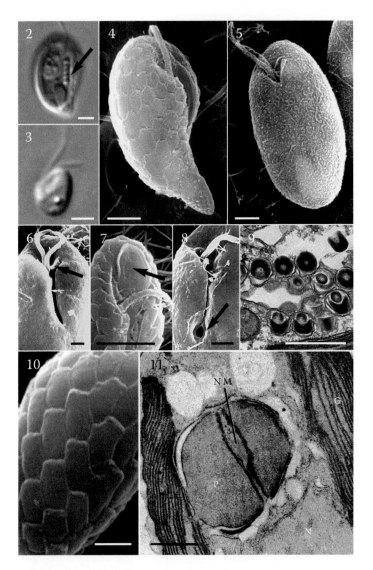

FIGURE 11.1 (CONTINUED) (2) Light micrograph of *Plagioselmis prolonga* Butcher ex Novarino, Lucas & Morrall. Note the row of ejectosomes (arrow). (3) Light micrograph of *Hemiselmis* sp. (4) Scanning electron micrograph of *Plagioselmis prolonga*. (5) Scanning electron micrograph of *Storeatula major* Hill. (6) Scanning electron micrograph of *Cryptomonas platyuris* Skuja. Detail of the vestibular ligule (arrow). (7) Scanning electron micrograph of *Proteomonas sulcata* Hill & Wetherbee (haplomorph). Detail of the vestibular plate (arrow). (8) Scanning electron micrograph of *Cryptomonas tetrapyrenoidosa* Skuja. Detail of the stoma (arrow). (9) Transmission electron micrograph of *Cryptochloris* sp. Detail of ejectosomes. (10) Scanning electron micrograph of the periplast of *Plagioselmis prolonga*. (11) Longitudinal thin section of *Pyrenomonas ovalis* Kugrens, Clay and Lee. Note the nucleomorph located in a pyrenoid invagination (C: chloroplast; N: nucleus; NM: nucleomorph; P: pyrenoid). Scale bars: Sections 2–8: 2μm; 9–11: 1μm. (Sections 2, 3, 5, and 7 from Cerino, F. and Zingone A. (2006). A survey of cryptomonad diversity and seasonality at a coastal Mediterranean site. *European Journal of Phycology*, 41: 363–378. Figures 6, 9, 16 and 40, Taylor & Francis Group. Sections 6 and 8 from Kugrens, P., Lee, R.E. and Andersen, R.A. (1986). Cell form and surface patterns in *Chroomonas* and *Cryptomonas* cells (Cryptophyta) as revealed by scanning electron microscopy, *Journal of Phycology*, 22: 512–522. Figures 11 and 6, Blackwell Publishing. Section 11 from Kugrens, P., Clay, B.L. and Lee, R.E. (1999). Ultrastructure and systematics of two new freshwater red cryptomonads, *Storeatula rhinosa*, sp. nov. and *Pyrenomonas ovalis*, sp. nov. *Journal of Phycology*, 35: 1079–1089; Figure 20, Blackwell Publishing).

associated with the flagellar surface (Pennick, 1981; Santore, 1983; Lee and Kugrens, 1986). The flagellar transitional region is another distinctive feature of cryptomonads, showing two or more plate-like partitions just below the point where the two central axonemal microtubules appear (Hibberd, 1979). In some species a longitudinal microtubular root, the rhizostyle, is present (Roberts et al., 1981). In some cases the rhizostyle is made of microtubules with wings (keeled rhizostyle), which begin near the basal body and extend posteriorly into the cell, terminating past the nucleus; in other cases it lacks wings (non-keeled rhizostyle) and terminates anterior to the nucleus. Its structure (keeled or not) and relation to other components of the flagellar apparatus vary among species (Clay et al., 1999), and also among different morphs within a single species (Hill and Wetherbee, 1986; Hoef-Emden and Melkonian, 2003).

The plastidial complex has a special structure showing vestiges of the evolutionary history of the cryptomonads. In the plastid-bearing species one or two plastids are present which have double thylakoids and a pyrenoid protruding on their inner side. An eyespot is present in some species. The plastids are surrounded by four membranes that are split into two pairs: a double plastid envelope, the periplastid membrane, which is the remnant of the plasma membrane of the ancestral symbiont, and a fold of the rough endoplasmic reticulum. The narrow space between the periplastid membrane and the chloroplast, known as periplastid space, is also the remnant of the secondary endosymbiont cytoplasm. This is the place for the accumulation of starch, which is different from other algae where the most important storage product is harboured either in the cytoplasm or in the chloroplast. In addition, the periplastid space hosts 80S ribosomes and the nucleomorph.

The nucleomorph shares four features with typical eukaryotic nuclei: a double membrane envelope with pores, DNA, self-replication, and a nucleolus where ribosomal RNA genes are transcribed (Gillott and Gibbs, 1980; Gibbs, 1981; Ludwig and Gibbs, 1985; Hansmann et al., 1986; Hansmann, 1988). The location of the nucleomorph varies considerably with respect to the pyrenoid and nucleus (Santore, 1982) and is considered an important taxonomic character with robust phylogenetic value (Novarino and Lucas, 1993; Clay et al., 1999; Hoef-Emden et al., 2002). Basically, two main states of this character are recognised, whereby the nucleomorph is either inside (Figure 11.1 [11]) or outside the pyrenoid. In the latter case it can assume different positions that characterise different genera, i.e. anterior to the pyrenoid, between the pyrenoid and the nucleus, or inside the nucleus (Clay et al., 1999).

Cryptomonads also stand apart from other groups of algae for their pigment composition. Chloroplasts can be blue, blue-green, reddish, red-brown, or yellow-brown, because the colour of the chlorophyll is masked by that of a variety of accessory photosynthetic pigments, which can occur in different proportions. Besides chlorophyll a and c_2, chloroplasts contain carotenoids, xanthophylls among which is alloxanthin, unique for the cryptomonads, and phycobiliproteins, Cr-phycocyanin and Cr-phycoerythrin. Unlike the red and blue-green algae, and glaucophytes, the only other algae possessing phycobiliproteins, the cryptomonad biliproteins are not contained in the phycobilisomes that are located on the external side of each thylakoids, but fill the thylakoid lumen (Gantt et al., 1971; Vesk et al., 1992; Spear-Bernstein and Miller, 1989). Moreover, each cryptomonad species contains either phycocyanin or phycoerythrin, but not both, and no allophycocyanin. Several distinct Cr-phycocyanin and Cr-phycoerythrin types exist which are designated according to their respective absorption maxima (Hill and Rowan, 1989) (Table 11.1). The phycobilin type is considered as a sound phylogenetic character, since each genus is normally characterized by one single type of either Cr-phycoerythrin or Cr-phycocyanin. The sole exception is the genus *Hemiselmis* Parke that contains both red and blue representatives. The red pigment Cr-phycoerythrin 555 in some *Hemiselmis* is in fact derived from phycocyanin (Marin et al., 1998).

LIFE CYCLE

As with many algal groups, where complex life cycles have been the origin of taxonomic and nomenclatural difficulties, cryptomonad classification is also running into problems in relation to the recent discoveries of heteromorphic life cycles (Hill and Wetherbee, 1986).

TABLE 11.1
Distribution of Different Kinds of Phycobiliproteins (Cr-PE: Phycoerythrin and Cr-PC: Phycocyanin) among Currently Recognized Cryptomonad Genera

Pigment	Colour	Absorbance Peak Wavelength (nm) [*]	Genera
Cr-PE I or 545	Red	540–550	*Geminigera*
			Guillardia
			Hanusia
			Plagioselmis
			Proteomonas
			Pyrenomonas/Rhodomonas
			Rhinomonas
			Storeatula
			Teleaulax
Cr-PE II or 555	Red	555	*Hemiselmis*
Cr-PE III or 566	Red	560–566	*Cryptomonas*
Cr-PC I or 569	Blue	569 and 630	*Falcomonas*
Cr-PC II or 615	Blue	615 and 577–585	*Hemiselmis*
Cr-PC III or 630	Blue	625–630 and 580–584	*Chroomonas*
Cr-PC IV or 645	Blue	640–650 and 580–585	*Chroomonas*
			Komma

[*] Range of absorbance. For Cr-PC the two absorbance peaks are given (the major one on the left, the minor one on the right).
Modified from Novarino, G., *Hydrobiologia*, 502: 225–270, 2003.

Cryptomonads reproduce asexually by mitosis and cytokinesis but have also been reported to undergo sexual reproduction. After the first undocumented reports of conjugation (Wawrik, 1969, 1971), fertilization was shown to occur in cultures of a cryptomonad tentatively assigned to *Chroomonas acuta* Utermöhl (Kugrens and Lee, 1988). Hill and Wetherbee (1986) described a species, *Proteomonas sulcata*, alternating between two morphs in its life cycle. Two kinds of flagellated cells differing in size, periplast structure, and flagellar apparatus were found in a single clonal culture of the species. Measurements of nuclear fluorescence after staining with a DNA-specific fluorescent dye demonstrated that one morph had twice the amount of DNA as compared to the other, thus providing indirect evidence for sexuality. Both forms, identified as haplo- and diplomorph, were able to reproduce vegetatively, but could also produce the other cell type. As discussed in the section "Molecular Phylogeny and Classification", additional evidence for the existence of dimorphism in the cryptomonad life cycle has been provided by a recent study combining morphological characters, nuclear ITS2, partial LSU rDNA, and nucleomorph SSU rDNA phylogenies in three freshwater genera: *Cryptomonas* Ehrenberg, *Campylomonas* Hill, and *Chilomonas* Ehrenberg (Hoef-Emden and Melkonian, 2003).

EVOLUTION

Like most algal groups, cryptomonads derived from a secondary endosymbiotic event (Greenwood et al., 1977; Douglas et al., 1991; Maier et al., 1991). However, they were the only group that retained part of the cytoplasm, including the nucleus, and the pigments of the endosymbiont.

Because cryptomonads have phycobiliproteins like red algae, and chlorophyll *c* like dinoflagellates, the possible ancestral endosymbiont could have been either a red alga or an intermediate between a red alga and a dinoflagellate. The discovery of the vestigial nucleus, the nucleomorph (Greenwood et al., 1977; Gillott and Gibbs, 1980; Gibbs, 1981; Ludwig and Gibbs, 1985, 1987), offered the chance for molecular studies that confirmed the hypothesis that cryptomonads are evolutionary chimaeras of two distinct eukaryotic cells and proved that the engulfed organism was a red alga (Douglas et al., 1991; Maier et al., 1991; McFadden, 1993). Other evidence using nuclear and nucleomorph 18S rRNA phylogenies (Cavalier-Smith et al., 1996), and also other genes (e.g. tubulin genes; Keeling et al., 1999), confirms the hypothesis that the origin of the cryptomonad nucleomorph was from a red alga. Further evidence is the presence of starch in the periplastid space, since starch is accumulated in the cytoplasm in red algae. As for the host component of the cryptomonad chimaera, no relatives have so far been found. It is currently hypothesised it was a heterotrophic organism, however results of some studies on glyceraldehyde-3-phosphate dehydrogenase (GAPDH) genes suggest that it could have been a phototrophic organism that lost its primary plastid after engulfing the red alga (Liaud et al., 1997).

Molecular analysis of nuclear and nucleomorph SSU rDNA (Marin et al., 1998; Hoef-Emden et al., 2002), immunochemical evidence (Guard-Friar et al., 1986), and phylogenetic analyses of protein sequences of the phycobiliprotein gene family (Sidler and Zubler, 1988; Sidler et al., 1990) have led to a hypothesis of phycobilin evolution in cryptomonads. Phycoerythrin is most probably the ancestral phycobilin, whereas the ancestral phycocyanin was lost after the endosymbiotic event. Cr-phycocyanin arose later in the blue-green cryptomonad genera *Hemiselmis* and *Chroomonas* Hansgirg (Marin et al., 1998; Hoef-Emden et al., 2002). The blue pigment eventually reverted to a phycoerythrin in some species of *Hemiselmis*.

An extremely interesting aspect from the biological and evolutionary point of view is the capacity of cryptomonads to survive ingestion by microheterotrophic predators. The colour of the marine planktonic ciliate *Myrionecta rubra* (Lohmann) Jankowski (formerly *Mesodinium rubrum*, a common red-tide-forming species, is due to the presence of cryptomonad endosymbionts (Barber et al., 1969; Hibberd, 1977). Certain dinoflagellates also regularly contain portions of cryptomonad cells, particularly plastids, e.g. *Gymnodinium acidotum* Nygaard (Wilcox and Wedemeyer, 1984) and some species of *Dinophysis* Ehrenberg (Hackett et al., 2003). This symbiosis was initially thought to be permanent (Hibberd, 1977), suggesting a third-level symbiogenesis between host and cryptomonad eventually leading to a new evolutionary step. More recent analyses, in fact, show that *M. rubra* does not maintain a permanent endosymbiont but rather needs to replace ageing chloroplasts periodically (Gustafson et al., 2000). Similarly, molecular analyses of both a 799 bp region of the *psbA* gene and 1221 bp region of the 16S rRNA gene from *Dinophysis* spp. and *Teleaulax amphioxeia* (Conrad) Hill, suggest that the plastid in *Dinophysis* is a so-called kleptoplastid (Takishita et al., 2002; Janson, 2004), i.e. may have been ingested and only temporarily retained. On the other hand, sequences of two nuclear-encoded GADPH genes from the dinoflagellate *Lingulodinium polyedrum* (Stein) Dodge (formerly *Gonyaulax polyedrum* Stein) were shown to form a monophyletic group with the plastid GADPH of cryptomonads, suggesting that a lateral exchange of genes occurred in an ancestral endosymbiotic event (Fagan et al., 1998). Therefore, whether the temporary associations observed nowadays are the first step of endosymbiotic events or other sources of genetic material and, hence, their possible role in the evolution of the hosts requires further elucidation.

MOLECULAR PHYLOGENY AND CLASSIFICATION

Prior to the advent of molecular analyses, the classification of cryptomonads as a group has undergone many changes. Following a first association with chrysomonads based on an apparent similarity in gullet and periplast structure (Pascher, 1911), a close relationship with the Dinophyceae was proposed (Pascher, 1914; Smith, 1933, 1938; Chapman, 1962; Bourrelly, 1970) and rejected several times (Fritsch, 1935; Pringsheim, 1944; Graham, 1951; Butcher, 1967), until

Cryptophyta, or Cryptista, were definitely separated as a subkingdom of the kingdom Chromista (Cavalier-Smith, 1986, 2003). Diagnostic characters for species, genera, and higher ranks have also changed significantly over time. After the first report of cryptomonad species, and the establishment of the first genus *Cryptomonas* (Ehrenberg, 1831), Karsten (1898) erected another genus, *Rhodomonas* Karsten, based on the red colour. Colour was hence extensively used as a diagnostic character taking advantage of the extreme chromatic diversity of cryptomonads. First classification schemes were based on features visible using light microscopy, such as the general organization of the thallus with motile and non-motile stages (Pascher, 1913, 1914; Oltmanns, 1922; West and Fritsch, 1932; Fritsch, 1935; Bourrelly, 1970; Christensen, 1980) and the structure of the furrow/gullet region (Pringsheim, 1944; Butcher, 1967). However, the discovery of new taxa and of new features highlighted the need for a reexamination of the characters suitable for establishing natural groups. The advent of ultrastructural investigations at the electron microscope provided new information on unique features and definitely proved the weakness of characters visible in light microscopy for species identification and classification (Santore, 1984; Klaveness, 1985; Novarino and Lucas, 1993).

Results of the first molecular analyses on a number of cryptomonads (Marin et al., 1998) were immediately used in combination with a set of characters (phycobiliprotein, furrow/gullet complex, inner and superficial periplast component, nucleomorph position, and rhizostyle) to redefine generic limits and suprageneric relationships (Clay et al., 1999) (Figure 11.2). With additional information on further strains and representatives of other genera (Deane et al., 2002; Hoef-Emden et al., 2002), Clay et al.'s (1999) classification was confirmed in part and refuted in others. For example, in agreement with Cavalier-Smith (1993) and Clay et al. (1999), but not with Novarino and Lucas (1993), a separate class is justified for the aplastidial genus *Goniomonas* Stein, whose relationship to plastid-containing cryptomonads is supported by many molecular studies (McFadden et al., 1994; Marin et al., 1998; Deane et al., 2002; Hoef-Emden et al., 2002). A more extensive recent sampling (von der Heyden et al., 2004) has also revealed a major goniomonad diversity and an ancient divergence between freshwater and marine species (Figure 11.3).

By contrast, a substantial change in classification introduced by recent molecular information is the evidence for a so far hidden dimorphism in the genus *Cryptomonas*, which implies that even two genera, *Campylomonas* and *Chilomonas*, and the family grouping them, the Campylomonadaceae, are superfluous (Hoef-Emden and Melkonian, 2003) (Figure 11.4). Profound morphological differences would separate these entities: *Cryptomonas* has an internal periplast component (IPC) consisting of round to oval plates, a furrow with a stoma and a fibrous furrow plate, a gullet, and a short, non-keeled rhizostyle. Conversely, *Campylomonas* has an IPC comprising a single sheet, a furrow with no stoma but with a scalariform furrow plate, a gullet, a long, keeled rhizostyle, and a unique structure termed "vestibular ligule". *Chilomonas* is secondarily colourless and possesses leucoplasts, but is otherwise identical to *Campylomonas*. Further molecular investigations using nuclear ribosomal DNA sequences, nucleomorph SSU rDNA (Hoef-Emden, 2005), and the plastid-encoded *rbc*L gene (Hoef-Emden et al., 2005) suggest that the loss of photosynthesis in the leucoplast-bearing members of *Cryptomonas* (formerly known as *Chilomonas*) has occurred independently more than once and has been accompanied by acceleration in the evolutionary rate. In some of these species that cluster in a single clade, only one of the two morphs is known, which has led to the hypothesis that they could have lost the sexual reproduction and, as a consequence, increased their evolutionary rate and lost photosynthesis (Hoef-Emden, 2005).

All nuclear SSU rDNA phylogenies (Marin et al., 1998; Deane et al., 2002; Hoef-Emden et al., 2002) basically recognise the same major lineages and identify nucleomorph position (internal or external to the pyrenoid) and phycobiliprotein type (phycoerythrin or phycocyanin) as the only characters present in each lineage in a single state (Figure 11.5). The presence of dimorphism is the most probable explanation to account for the close relationships between morphologically

Phylum Cryptophyta Cavalier-Smith emend. Clay, Kugrens and Lee

Class Goniomonadea Cavalier-Smith emend. Clay, Kugrens and Lee

Order Goniomonadida Novarino and Lucas emend. Clay, Kugrens and Lee
Family Goniomonadidae Hill
Genus *Goniomonas* Stein

Class Cryptophyceae Cavalier-Smith

Order Cryptomonadales emend. Clay, Kugrens and Lee
Family Cryptomonadaceae emend. Clay, Kugrens and Lee
Genus *Cryptomonas* Ehrenberg
Family Campylomonadaceae Clay, Kugrens and Lee
Genus *Campylomonas* Hill
Genus *Chilomonas* Ehrenberg

Order Pyrenomonadales emend. Clay, Kugrens and Lee
Family Pyrenomonadaceae Novarino and Lucas emend. Clay, Kugrens and Lee
Genus *Pyrenomonas* Santore
Genus *Rhinomonas* Hill and Wetherbee
Genus *Rhodomonas* Karsten emend. Hill and Wetherbee
Genus *Storeatula* Hill
Family Geminigeraceae Clay, Kurgens and Lee
Genus *Geminigera* Hill
Genus *Guillardia* Hill and Wetherbee
Genus *Hanusia* Deane, Hill, Brett and McFadden
Genus *Proteomonas* Hill and Wetherbee
Genus *Teleaulax* Hill
Family Chroomonadaceae Clay, Kugrens and Lee
Genus *Chroomonas* Hansgirg
Genus *Falcomonas* Hill
Genus *Komma* Hill
Family Hemiselmidaceae Butcher emend. Clay, Kugrens and Lee
Genus *Hemiselmis* Parke

FIGURE 11.2 Cryptomonad classification by Clay et al. (1999). The genera *Campylomonas* and *Chilomonas* are now considered synonyms of *Cryptomonas* (Hoef-Emden and Melkonian, 2003); therefore, the family Campylomonadaceae is superfluous.

distinct genera, and is probably widespread in cryptomonads, having also been suggested for *Rhinomonas* Hill and Wetherbee and *Storeatula* Hill species (Hoef-Emden et al., 2002).

A weakness shared by all phylogenies is the lack of support for the relationships among the major lineages and the poor resolution of the basal topologies (Marin et al., 1998; Deane et al., 2002; Hoef-Emden et al., 2002). In order to obtain a better resolved tree, the use of other genes that have evolved faster than nuclear 18S rDNA (e.g. the nucleomorph 18S rDNA) has been suggested (Deane et al., 2002). Outgroup rooting of the cryptomonad clade is difficult in this case, as it emerged from the first nucleomorph phylogeny (Cavalier-Smith et al., 1986), due to the large evolutionary distance between the cryptomonad endosymbiont and the red algae. However, when the outgroup taxa were excluded, the bootstrap frequencies for the internal nodes increased to significant levels (Hoef-Emden et al., 2002) (Figure 11.6).

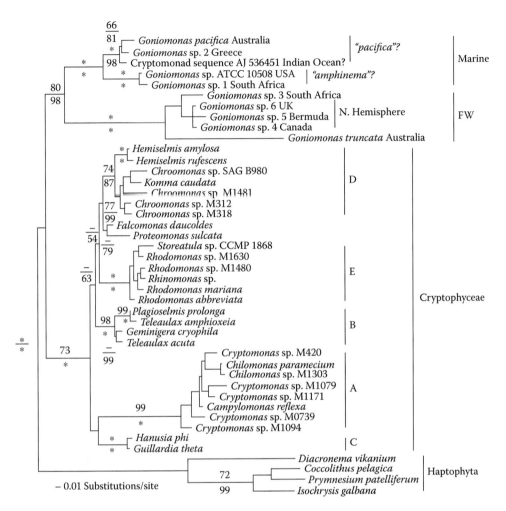

FIGURE 11.3 Goniomonad phylogeny based on SSU rDNA sequence comparisons inferred with the maximum likelihood method. Where bootstrap support was less than 50%, no values are shown (or -). Asterisks denote 100% bootstrap support. The clade comprising freshwater strains is labelled FW. (From *European Journal of Phycology*, 39: 343–350, 2004, Genetic diversity of goniomonads: an ancient divergence between marine and freshwater species, von der Heyden, S., Chao, E.E., and Cavalier-Smith, T., fig. 2, Taylor & Francis.)

CONCLUSIONS

Despite their acknowledged importance from the ecological and evolutionary point of view, the knowledge of cryptomonad diversity, phylogeny, and taxonomy is currently in an unsatisfactory state. Considerable progress in morphological studies has been achieved, yet the question of what the characters are that definitely circumscribe taxa at different ranks within cryptomonads is still unsolved. To this end, progress can be made combining molecular analyses with improved observation techniques. These should also include observation of live material (the original method to study cryptomonads) so as to exploit a whole range of such characters as colour, shape variability, and swimming behaviour that can only be studied by light microscopy. Considering results in other algal groups, it is also recommended that phylogenies are developed based on other genes which may give a better resolution of the intergeneric relationships as well as information on the evolution and phylogenetic value of phenotypic features.

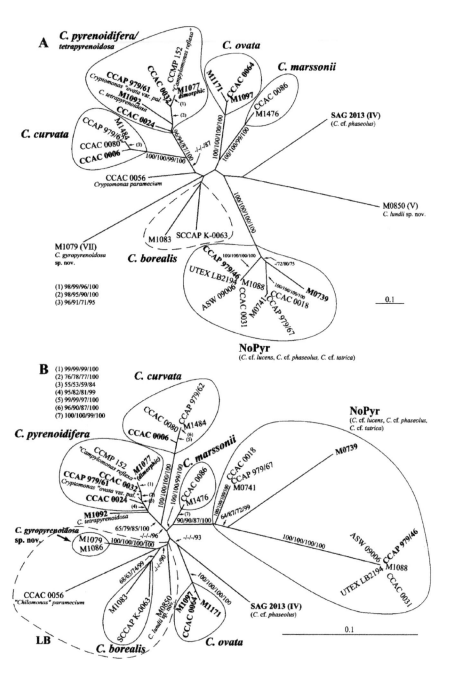

FIGURE 11.4 Unrooted maximum likelihood topologies of the nuclear ITS2 (A) and the partial nuclear LSU (B) rDNA sequences of *Cryptomonas* species. Boldface: strains with an inner periplast component made up of polygonal plates; regular: strains with a sheet-like inner periplast component; boldface and italic: dimorphic strains with both types of periplast in a clonal culture; in strains not part of a supported clade, the corresponding morphological group is added in parenthesis. Support values for the branches in order from left to right: distance analysis bootstrap/unweighted maximum parsimony bootstrap/maximum likelihood bootstrap/posterior probability × 100 (Bayesian analysis). Scale bars: substitutions per site. (Reprinted from *Protist*, 154: 371–409, 2003, Revision of the genus *Cryptomonas* (Cryptophyceae): a combination of molecular phylogeny and morphology provides insights into a long-hidden dimorphism, Hoef-Emden, K. and Melkonian, M., fig. 29A-B, with permission from Elsevier.)

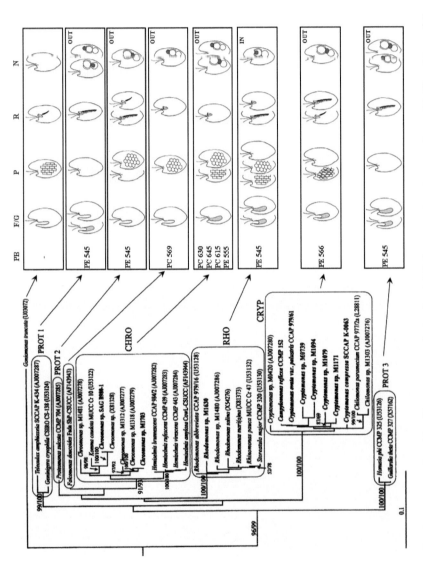

FIGURE 11.5 State of the main characters used to identify cryptomonads (to the right), in comparison to a maximum likelihood tree of the nuclear SSU rDNA data set. The tree is rooted with glaucocystophyte nuclear SSU rDNA sequences (not shown). Bootstrap values (1000 replicates): distance matrix calculated using the maximum-likelihood estimators of the suggested evolutionary model (hLRT: Tamura-Nei with G + I) and unweighted maximum parsimony (bootstrap values in order from left to right). – lnL = 8411.47946; scale bars, substitutions per site. (Maximum likelihood tree from *Journal of Molecular Evolution* 55: 161–179, 2002, Nuclear and nucleomorph SSU rDNA phylogeny in the Cryptophyta and the evolution of Cryptophyte diversity, Hoef-Emden, K., Marin, B. and Melkonian, M., fig. 4, with kind permission of Springer Science and Business Media.) F/G: organization of the furrow/gullet complex (presence of a gullet, in grey, a furrow or both); N: nucleomorph position (IN: inside the pyrenoid; OUT: outside the pyrenoid); P: periplast structure (polygonal scales or sheet-like); PB: phycobiliprotein type (PC: phycocyanin; PE: phycoerythrin); R: type of rhizostyle (absent or present, long and keeled or short and not keeled). Among the characters shown, the position of the nucleomorph (internal or external to the pyrenoid) and the phycobilin type are the only ones present in a single state in each lineage. Note that PE 555 of red *Hemiselmis* species is in fact a modified PC (Marin et al., 1998).

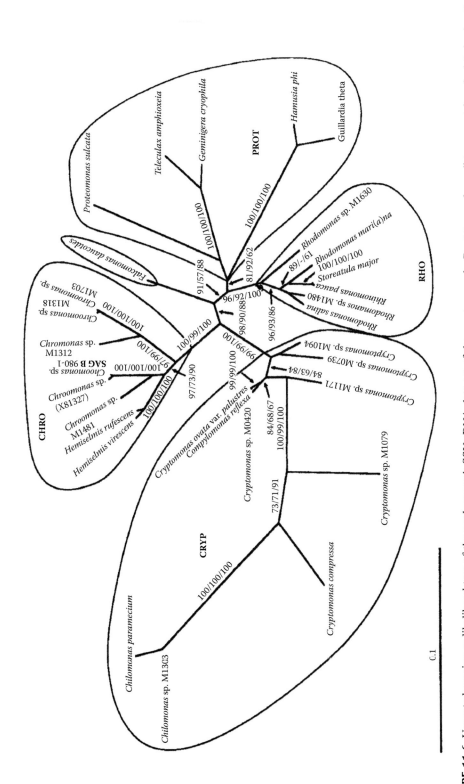

FIGURE 11.6 Unrooted maximum-likelihood tree of the nucleomorph SSU rDNA data set of the cryptomonads. Bootstrap values: distance matrix calculated using the maximum-likelihood estimators of the suggested evolutionary model (hLRT: 1000 replicates), unweighted maximum parsimony (1000 replicates), and maximum likelihood (1000 replicates; bootstrap values in order from left to right). Scale bar, substitutions per site. (From *Journal of Molecular Evolution* 55: 161–179, 2002, Nuclear and nucleomorph SSU rDNA phylogeny in the Cryptophyta and the evolution of Cryptophyte diversity, Hoef-Emden, K., Marin, B. and Melkonian, M., fig. 5A. With kind permission of Springer Science and Business Media.)

The addition of further molecular data from more species is also needed to improve the resolution of the phylogenetic analyses. More species than those that have been reexamined with modern techniques do exist, since most of the species and genera described in the age of morphological taxonomy have never been observed in the electron microscope, nor have they been cultured. In addition, the estimated total number of existing species ranges from 300 to 1200, one and a half to six times the number of currently known species (Novarino, 2003). Field investigations also demonstrate that a large part of cryptomonad diversity is still to be uncovered. Two recent studies in the Mediterranean Sea have revealed the presence of several taxa that do not fit the description of any described species (Novarino, 2005; Cerino and Zingone, 2006), while environmental DNA analyses at different places recovered a high number of sequences belonging to unknown cryptomonads (Massana et al., 2004; Romari and Vaulot, 2004).

Molecular studies have provided support to the existence of dimorphism in many taxa, yet sexuality is still to be demonstrated, since conjugation and meiosis have never been observed in relation to alternating morphs. Only in a limited number of strains in the same species has the alternation of morphs been witnessed, and dimorphism is in many cases only deduced from molecular similarity, whereby the alternate stage is still missing. Unfortunately, changes in morphology are not mentioned in the only study where conjugation was observed, where observations did not go beyond a four-flagellated planozygote (Kugrens and Lee, 1988). Clearly, many aspects of the life cycle of cryptomonads are still to be uncovered.

A considerable amount of information on cryptomonad genetic diversity is expected to stem from the use of molecular techniques, such as the development of genetic libraries, fingerprinting techniques, and specific probes. However, cultivation remains a vital tool to obtain material for morphological and pigment analyses as well as for life cycle, physiological, and molecular investigations. Only complementing these different approaches will it be possible to address the origin, evolution, and functional significance of genetic diversity.

ACKNOWLEDGMENTS

This work falls within the scope of the European Network of Excellence on Marine Biodiversity and Ecosystem Functioning in European Seas (MARBEF).

REFERENCES

Barber, R.T., White, A.W., and Siegelman, H.W. (1969) Evidence for cryptomonad symbiont in the ciliate, *Cyclotrichium meunieri*. *Journal of Phycology*, 5: 86–88.

Bourrelly, P. (1970) *Les algues d'eau douce. Initiation à la systématique. III Les algues bleues et rouge. Les Eugléniens, Peridiniens et Cryptomonadines*. N. Boubée e Cie, Paris.

Brett, S.J., Perasso, L., and Wetherbee, R. (1994) Structure and development of the cryptomonad periplast: a review. *Protoplasma*, 181: 106–122.

Brett, S.J. and Wetherbee, R. (1986) A comparative study of periplast structure in *Cryptomonas cryophila* and *C. ovata* (Cryptophyceae). *Protoplasma*, 131: 23–31.

Brett, S.J. and Wetherbee, R. (1996a) Periplast development in Cryptophyceae I. Changes in periplast arrangement throughout the cell cycle. *Protoplasma*, 192: 28–39.

Brett, S.J. and Wetherbee, R. (1996b) Periplast development in Cryptophyceae II. Development of the inner periplast component in *Rhinomonas pauca*, *Proteomonas sulcata* [haplomorph], *Rhodomonas baltica*, and *Cryptomonas ovata*. *Protoplasma*, 192: 40–48.

Brett, S.J. and Wetherbee, R. (1996c) Periplast development in Cryptophyceae III. Development of crystalline surface plates in *Falcomonas daucoides*, *Proteomonas sulcata* [haplomorph], and *Komma caudata*. *Protoplasma*, 192: 49–56.

Butcher, R.W. (1967) An introductory account of the smaller algae of the British coastal waters. Part IV: Cryptophyceae. *Fishery Investigations*, 4: 1–54.

Cavalier-Smith, T. (1986) The kingdom Chromista: origin and systematics. *Progress in Phycological Research*, 4: 309–347.

Cavalier-Smith, T. (1993) Kingdom Protozoa and its 18 phyla. *Microbiological Reviews*, 953–994.

Cavalier-Smith, T. (2002) Nucleomorphs: enslaved algal nuclei. *Current Opinion in Microbiology*, 5: 612–619.

Cavalier-Smith, T. (2003) Only six kingdoms of life. *Proceedings of the Royal Society of London B*, 271: 1251–1262.

Cavalier-Smith, T., Couch, J.A., Thorsteinsen, K.E., Gilson, P., Deane, J.A., Hill, D.R.A., and McFadden, G.I. (1996) Cryptomonad nuclear and nucleomorph 18S rRNA phylogeny. *European Journal of Phycology*, 31: 315–328.

Cerino, F. and Zingone, A. (2006) A survey of cryptomonad diversity and seasonality at a coastal Mediterranean site. *European Journal of Phycology*, 41: 363–378.

Chapman, V.J. (1962) *The Algae*. MacMillan, London.

Christensen, T. (1980) *Algae: A Taxonomic Survey*. Aio Tryk as, Odense.

Clay, B.L., Kugrens, P., and Lee, R.E. (1999) A revised classification of Cryptophyta. *Botanical Journal of the Linnean Society*, 131: 131–151.

Deane, J.A., Strachan, I.M., Saunders, G.W., Hill, D.R.A., and McFadden, G.I. (2002) Cryptomonad evolution: nuclear 18S rDNA phylogeny versus cell morphology and pigmentation. *Journal of Phycology*, 38: 1236–1244.

Douglas, S.E., Murphy, C.A., Spencer, D.F., and Gray, M.W. (1991) Cryptomonad algae are evolutionary chimaeras of two phylogenetically distinct unicellular eukaryotes. *Nature*, 350: 148–151.

Ehrenberg, C.G. (1831) *Über die Entwicklung und Lebensdauer der Infusionstiere; nebst ferneren Beiträgen zu einer Vergleichung ihrer organischen Systeme*. Akademie der Wissenschaften, Berlin.

Fagan, T., Woodland Hastings, J., and Morse, D. (1998) The phylogeny of glyceraldehyde-3-phosphate dehydrogenase indicates lateral gene transfer from cryptomonads to dinoflagellates. *Journal of Molecular Evolution*, 47: 633–639.

Faust, M.A. (1974) Structure of the periplast of *Cryptomonas ovata* var. *palustris*. *Journal of Phycology*, 10: 121–124.

Fritsch, F.E. (1935) *The Structure and Reproduction of the Algae*, Vol. 1. University Press, Cambridge.

Gantt, E. (1971) Micromorphology of the periplast of *Chroomonas* sp. (Cryptophyceae). *Journal of Phycology*, 7: 177–184.

Gantt, E., Edwards, M.R., and Provasoli, L. (1971) Chloroplast structure of the Cryptophyceae. Evidence for phycobiliproteins within intrathylakoidal spaces. *Journal of Cell Biology*, 48: 280–290.

Gibbs, S.P. (1981) The chloroplasts of some algal groups may have evolved from endosymbiotic eukaryotic algae. *Annals New York Academy of Sciences*, 361: 193–208.

Gillott, M.A. and Gibbs, S.P. (1980) The cryptomonad nucleomorph: its ultrastructure and evolutionary significance. *Journal of Phycology*, 16: 558–568.

Graham, H.W. (1951) Pyrrophyta. In *Manual of Phycology. An introduction to the Algae and Their Biology* (ed G.M. Smith), Chronica Botanica Company, Waltham, pp. 105–118.

Greenwood, A.D., Griffith, H.B., and Santore, U.J. (1977) Chloroplasts and cell compartments in Cryptophyceae. *British Phycological Journal*, 12: 119.

Guard-Friar, D., Eisenberg, B.L., Edwards, M.R., and MacColl, R. (1986) Immunochemistry on cryptomonad biliproteins. *Plant Physiology*, 80: 38–42.

Gustafson, D.E., Stoecker, D.K., Johnson, M.D., Van Heukelern, W.F., and Snelder, K. (2000) Cryptophyte algae are robbed of their organelles by the marine ciliate *Mesodinium rubrum*. *Nature*, 405: 1049–1052.

Hackett, J.D., Maranda, L., Su Yoon, H., and Bhattacharya, D. (2003) Phylogenetic evidence for the cryptophyte origin of the plastid of *Dinophysis* (Dinophysiales, Dinophyceae). *Journal of Phycology*, 39: 440–448.

Hansmann, P. (1988) Ultrastructural localization of RNA in cryptomonads. *Protoplasma*, 146: 81–88.

Hansmann, P., Falk, H., Scheer, U., and Sitte, P. (1986) Ultrastructural localization of DNA in two *Cryptomonas* species by use of a monoclonal DNA antibody. *European Journal of Cell Biology*, 42: 152–160.

Hibberd, D.J. (1977) Observations on the ultrastructure of the cryptomonad endosymbiont of the red-water ciliate *Mesodinium rubrum*. *Journal Marine Biology Association U.K.*, 57: 45–61.

Hibberd, D.J. (1979) The structure and phylogenetic significance of the flagellar transition region in the chlorophyll c-containing algae. *BioSystems*, 11: 243–261.

Hibberd, D.J., Greenwood, A.D., and Griffiths, H.B. (1971) Observations on the ultrastructure of the flagella and periplast in the Cryptophyceae. *British Phycological Journal*, 6: 61–72.

Hill, D.R.A. (1991) *Chroomonas* and other blue-green cryptomonads. *Journal of Phycology*, 27: 133–145.

Hill, D.R.A. and Rowan, K.S. (1989) The biliproteins of the Cryptophyceae. *Phycologia*, 28: 455–463.

Hill, D.R.A. and Wetherbee, R. (1986) *Proteomonas sulcata* gen. et sp. nov. (Cryptophyceae), a cryptomonad with two morphologically distinct and alternating forms. *Phycologia*, 25: 521–543.

Hill, D.R.A. and Wetherbee, R. (1989) A reappraisal of the genus *Rhodomonas* (Cryptophyceae). *Phycologia*, 28: 143–158.

Hoef-Emden, K. (2005) Multiple independent losses of photosynthesis and differing evolutionary rates in the genus *Cryptomonas* (Cryptophyceae): combined phylogenetic analyses of DNA sequences of the nuclear and the nucleomorph ribosomal operons. *Journal of Molecular Evolution*, 60: 183–195.

Hoef-Emden, K., Marin, B., and Melkonian, M. (2002) Nuclear and nucleomorph SSU rDNA phylogeny in the Cryptophyta and the evolution of Cryptophyte diversity. *Journal of Molecular Evolution*, 55: 161–179.

Hoef-Emden, K. and Melkonian, M. (2003) Revision of the genus *Cryptomonas* (Cryptophyceae): a combination of molecular phylogeny and morphology provides insights into a long-hidden dimorphism. *Protist*, 154: 371–409.

Hoef-Emden, K., Tran, H.-D., and Melkonian, M. (2005) Lineage-specific variations of congruent evolution among DNA sequences from three genomes, and relaxed selective constraints on *rbc*L in *Cryptomonas* (Cryptophyceae). *BMC Evolutionary Biology*, 5: 56–66.

Hovasse, R., Mignot, J.P., and Joyon, L. (1967) Nouvelles observations sur les trichocystes des Cryptomonadines et les "R bodies" des particules kappa de *Paramecium aurelia* Killer. *Protistologica*, 3: 241–255.

Janson, S. (2004) Molecular evidence that plastids in the toxin-producing dinoflagellate genus *Dinophysis* originate from the free-living cryptophyte *Teleaulax amphioxeia*. *Environmental Microbiology*, 6: 1102–1110.

Javornický, P. and Hindà, F. (1970) *Cryptomonas frigoris* spec. nova (Cryptophyceae), the new cyst-forming flagellate from the snow of the High Tatras. *Biològia (Bratislava)*, 25: 241–250.

Karsten, G. (1898) *Rhodomonas baltica* N. g. et sp. *Wissenschaftliches Meeresuntersuchungen, Abteilung Kiel, Neue Folge*, 3: 15–17.

Keeling, P.J., Deane, J.A., Hink-Schauer, C., Douglas, S.E., Maier, U.-G., and McFadden, G.I. (1999) The secondary endosymbiont of the cryptomonad *Guillardia theta* contains alpha-, beta-, and gamma-tubulin genes. *Molecular Biology and Evolution*, 16: 1308–1313.

Klaveness, D. (1985) Classical and modern criteria for determining species of Cryptophyceae. *Bulletin of Plankton Society of Japan*, 32: 111–123.

Klaveness, D. (1988) Ecology of the Cryptomonadida: a first review. In *Growth and Reproductive Strategies of Freshwater Phytoplankton* (ed. C.D. Sandgren), University Press, Cambridge, pp. 105–133.

Klaveness, D. (1988/1989) Biology and ecology of the Cryptophyceae: status and challenges. *Biological Oceanography*, 6: 257–270.

Kristiansen, J. and Kristiansen, A. (1999) A new species of *Chroomonas* (Cryptophyceae) living inside the submarine ikaite columns in the Ikkafjord, Southwest Greenland, with remarks on its ultrastructure and ecology. *Nordic Journal of Botany*, 19: 747–758.

Kugrens, P. and Lee, R.E. (1987) An ultrastructural survey of cryptomonad periplast using quick-freezing freeze fracture techniques. *Journal of Phycology*, 23: 365–376.

Kugrens, P. and Lee, R.E. (1988) Ultrastructure of fertilization in a cryptomonad. *Journal of Phycology*, 24: 385–393.

Kugrens, P., Lee, R.E., and Andersen, R.A. (1986) Cell form and surface patterns in *Chroomonas* and *Cryptomonas* cells (Cryptophyta) as revealed by scanning electron microscopy. *Journal of Phycology*, 22: 512–522.

Kugrens, P., Lee, R.E., and Andersen, R.A. (1987) Ultrastructural variations in cryptomonad flagella. *Journal of Phycology*, 23: 511–518.

Kugrens, P., Lee, R.E., and Corliss, J.O. (1994) Ultrastructure, biogenesis, and functions of extrusive organelles in selected non-ciliate protists. *Protoplasma*, 181: 164–190.

Lee, R.E. and Kugrens, P. (1986) The occurrence and structure of flagellar scales in some freshwater cryptophytes. *Journal of Phycology*, 22: 549–552.

Liaud, M.-F., Brandt, U., Scherzinger, M., and Cerff, R. (1997) Evolutionary origin of cryptomonad microalgae: two novel chloroplast/cytosol specific GAPDH genes as potential markers of ancestral endosymbiont and host cell components. *Journal of Molecular Evolution*, 44 (Supplement 1): S28–S37.

Ludwig, M. and Gibbs, S.P. (1985) DNA is present in the nucleomorph of Cryptomonads: further evidence that the chloroplast evolved from a eukaryotic endosymbiont. *Protoplasma*, 127: 9–20.

Ludwig, M. and Gibbs, S.P. (1987) Are the nucleomorphs of cryptomonads and *Chlorarachnion* the vestigial nuclei of eukaryotic endosymbionts? *Annals New York Academy of Sciences*, 503: 198–211.

Maier, U.-G., Hofmann, J.B., Eschbach, S., Wolters, J., and Igloi, G.L. (1991) Demonstration of nucleomorph-encoded eukaryotic small subunit ribosomal RNA in cryptomonads. *Molecular and General Genetics*, 230: 155–160.

Marin, B., Klingberg, M., and Melkonian, M. (1998) Phylogenetic relationships among the Cryptophyta: analyses of nuclear-encoded SSU rRNA sequences support the monophyly of extant plastid-containing lineages. *Protist*, 149: 265–276.

Massana, R., Balagué, V., Guillou, L., and Pedrós-Alió, C. (2004) Picoeukaryotic diversity in an oligotrophic coastal site studied by molecular and culturing approaches. *FEMS Microbiology Ecology*, 50: 231–243.

McFadden, G.I. (1993) Second-hand chloroplasts: evolution of cryptomonad algae. *Advances in Botanical Research*, 19: 139–230.

McFadden, G.I., Gilson, P.R., and Hill, D.R.A. (1994) *Goniomonas*: rRNA sequences indicate that this phagotrophic flagellate is a close relative of the host component of cryptomonads. *European Journal of Phycology*, 29: 29–32.

Mignot, J.-P., Joyon, L., and Pringsheim, E.G. (1968) Complements a l'etude cytologique des Crypto-monadines. *Protistologica*, 4: 493–506.

Moustaka-Gouni, M. (1995) Some aspects on the morphology and ecology of *Rhodomonas minuta* var. *nannoplanctica* and *R. lens* (Cryptophyceae) in two Greek lakes. *Nordic Journal of Botany*, 16: 335–343.

Novarino, G. (2003) A companion to the identification of cryptomonad flagellates (Cryptophyceae = Crypto-monadea). *Hydrobiologia*, 502: 225–270.

Novarino, G. (2005) Nanoplankton protists from the western Mediterranean Sea. II. Cryptomonads (Crypto-phyceae = Cryptomonadea). *Scientia Marina*, 69: 47–74.

Novarino, G. and Lucas, J.A.N. (1993) Some proposals for a new classification system of the Cryptophyceae. *Botanical Journal of the Linnean Society of London*, 111: 3–21.

Novarino, G., Warren, A., Kinner, N.E., and Harvey, R.W. (1994) Protists from a sewage contaminated aquifer on Cape Cod, Massachusetts. *Geomicrobiological Journal*, 12: 23–36.

Okamoto, N. and Inouye, I. (2005) The katablepharids are a distant sister group of the Cryptophyta: a proposal for Katablepharidophyta divisio nova/Kathablepharida phylum novum based on SSU rDNA and beta-tubulin phylogeny. *Protist*, 156: 163–179.

Oltmanns, F. (1922) *Morphologie und Biologie der Algen*. Verlag von Gustav Fischer, Jena.

Pascher, A. (1911) Über der Beziehungen der Cryptomonaden zu den Algen. *Berichte Deutsche Botanische Gesellschaft — Vereinigung für Angewandte Botanik*, 29: 193–203.

Pascher, A. (1913) Cryptomonadinae. In *Die Süsswasserflora von Deutschland, Österreichs und der Schweiz*. Verlag von Gustav Fischer, Jena, pp. 96–114.

Pascher, A. (1914) Über Flagellaten und Algen. *Ber. dt. Bot. Ges.*, 32: 136–160.

Paulsen, B.S., Vieira, A.A.H., and Klaveness, D. (1992) Structure of extracellular polysaccharides produced by a soil *Cryptomonas* sp. (Cryptophyceae). *Journal of Phycology*, 28: 61–63.

Pennick, N. (1981) Flagellar scales in *Hemiselmis brunnescens* Butcher and *H. virescens* Droop (Cryptophyceae). *Archiv für Protistenkunde*, 124: 267–270.

Perasso, L., Ludwig, M., and Wetherbee, R. (1997) The surface periplast component of the protist *Komma caudata* (Cryptophyceae) self-assembles from a secreted high-molecular-mass polypeptide. *Proto-plasma*, 200: 186–197.

Pringsheim, E.G. (1944) Some aspects of taxonomy in the Cryptophyceae. *New Phytologist*, 43: 143–150.

Roberts, K.R., Stewart, K.D., and Mattox, K.R. (1981) The flagellar apparatus of *Chilomonas paramecium* (Cryptophyceae) and its comparison with certain zooflagellates. *Journal of Phycology*, 17: 159–167.

Romari, K. and Vaulot, D. (2004) Composition and temporal variability of picoeukaryote communities at a coastal site of the English Channel from 18S rDNA sequences. *Limnology and Oceanography*, 49: 784–798.

Santore, U.J. (1977) Scanning electron microscopy and comparative micromorphology of the periplast of *Hemiselmis rufescens*, *Chroomonas* sp., *Chroomonas salina* and members of the genus *Cryptomonas* (Cryptophyceae). *British Phycological Journal*, 12: 255–270.

Santore, U.J. (1982) The distribution of the nuclemorph in the Cryptophyceae. *Cell Biology International Reports*, 6: 1055–1063.

Santore, U.J. (1983) Flagellar and body scales in the Cryptophyceae. *British Phycological Journal*, 18: 239–248.

Santore, U.J. (1984) Some aspects of taxonomy in the Cryptophyceae. *New Phytologist*, 98: 627–646.

Sidler, W., Nutt, H., Kumpf, B., Frank, G., Suter, F., Brenzel, A., Wehrmeyer, W., and Zubler, H. (1990) The complete amino acid sequence and the phylogenetic origin of phycocyanin-645 from the cryptophytan alga *Chroomonas* sp. *Biological Chemistry Hoppe-Seyler*, 371: 537–547.

Sidler, W. and Zubler, H. (1988) Structural and phylogenetic relantionship of phycoerythrins from cyanobacteria, red algae and Cryptophyceae. In *Photosynthetic Light-Harvesting Systems* (eds. H. Scheer and S. Schneider), Walter de Gruyter, Berlin, pp. 49–60.

Smith, G.M. (1933) *Freshwater Algae of the United States*. McGraw-Hill, New York.

Smith, G.M. (1938) *Cryptogamic Botany*. McGraw-Hill, New York.

Sommer, U. (1982) Vertical niche separation between two closely related planktonic flagellate species (*Rhodomonas lens* and *Rhodomonas minuta* v. *nannoplanctica*). *Journal of Plankton Research*, 4: 137–142.

Spear-Bernstein, L. and Miller, K.R. (1989) Unique location of the phycobiliprotein light-harvesting pigment in the Cryptophyceae. *Journal of Phycology*, 25: 412–419.

Takishita, K., Koike, K., Maruyama, T., and Ogata, T. (2002) Molecular evidence for plastid robbery (Kleptoplastidy) in *Dinophysis*, a dinoflagellate causing diarrhetic shellfish poisoning. *Protist*, 153: 293–302.

Vesk, M., Dwarte, D., Fowler, S., and Hiller, R.G. (1992) Freeze fracture immunocytochemistry of light-harvesting pigment complexes in a cryptophye. *Protoplasma*, 170: 166–176.

von der Heyden, S., Chao, E.E., and Cavalier-Smith, T. (2004) Genetic diversity of goniomonads: an ancient divergence between marine and freshwater species. *European Journal of Phycology*, 39: 343–350.

Wawrik, F. (1969) Sexualität bei *Cryptomonas* sp. und *Chlorogonium maximum*. *Nova Hedwigia*, 17: 283–292.

Wawrik, F. (1971) Zygoten und Cysten bei *Stenocalyx klarnetii* (Bourr.) Fott, *Stenocalyx inkonstans* Schmid und *Chroomonas acuta* Uterm. *Nova Hedwigia*, 21: 599–604.

West, G.S. and Fritsch, F.E. (1932) *A Treatise on the British Freshwater Algae*. University Press, Cambridge.

Wetherbee, R., Hill, D.R.A., and Brett, S.J. (1987) The structure of the periplast components and their association with the plasma membrane in a cryptomonad flagellate. *Canadian Journal of Botany*, 65: 1019–1026.

Wetherbee, R., Hill, D.R.A., and McFadden, G.I. (1986) Periplast structure of the Cryptomonad flagellate *Hemiselmis brunnescens*. *Protoplasma*, 131: 11–22.

Wilcox, L.W. and Wedemeyer, G.J. (1984) *Gymnodinium acidotum* Nygaard (Pyrrophyta), a dinoflagellate with an endosymbiotic cryptomonad. *Journal of Phycology*, 20: 236–242.

12 On dinoflagellate phylogeny and classification

Øjvind Moestrup and Niels Daugbjerg

CONTENTS

ABSTRACT

Cytologically, dinoflagellates constitute a very aberrant protist group. Molecular data indicate that the most primitive extant species are heterotrophic and belong to the Oxyrrhinales (free-living) and Syndiniales (parasitic). The remaining species usually form a single clade in the phylogenetic trees based on ribosomal DNA sequences, but their relative relationships are poorly resolved. Classification at generic, family, and order levels is presently undergoing major changes, and characters traditionally used to characterize genera (thecate or non-thecate, position of the cingulum, distance between the two ends of the cingulum, plate formula, etc.) are being replaced with features supported by both molecular and ultrastructural data (eyespot structure, path and construction of the "apical furrow" system, details of the flagellar apparatus, etc.). Dinoflagellate chloroplasts have arisen by at least eight independent symbioses, and some of these have resulted in stable symbioses, while others are kleptoplastidic.

INTRODUCTION

DEFINITION OF THE GROUP

The dinoflagellates is a group of unicellular or colony-forming protists comprising approximately 2000 extant and a similar number of extinct species. The great majority (perhaps 80%) are free-living, marine, planktonic, or benthic flagellates, while 20% are from similar habitats in freshwater.

A small number are parasitic (e.g. in copepods) or symbiotic (e.g. in corals). Estimation of species numbers is difficult as the taxonomy needs revision. Approximately half the described species are phototrophic ("algae"); the other half are heterotrophic (phagotrophic), free-living, or parasitic ("protozoa"). However, an increasing number of species is being found to be mixotrophic.

Motile cells of dinoflagellates are biflagellate, one flagellum typically inserted in a transverse furrow, the cingulum, which usually encircles the whole cell, and the other in a longitudinal furrow, the sulcus, restricted to the cell's antapical, ventral side. The nucleus contains large amounts of DNA, and the many chromosomes typically lack basic proteins (e.g. histones). Mitochondrial cristae are tubular. Many types of chloroplasts occur, believed to be the result of multiple symbioses between dinoflagellates and other groups of protists. An eyespot, if present, is located ventrally near the sulcus and connected within the cell to the flagellar apparatus by a conspicuous microtubular root, the R1. Trichocysts are common. A special, sometimes very conspicuous, organelle—the pusule—is believed to function in excretion. The cell periplast ("wall"), named the "amphiesma," comprises flat cisternae which in some species appear empty ("naked" or "athecate" dinoflagellates) and in other species contain cellulosic plates ("armoured" or "thecate" dinoflagellates). Cell division is typically by bipartition, each of the new cells in some cases inheriting half the parent amphiesma. Sexual reproduction has been described in a number of species but probably has been generally overlooked. Sexual reproduction is by isogamy or anisogamy and often (but not always) results in the formation of a resting spore, termed the hypnospore. The wall of the hypnospore contains extremely resistant sporopollenin-like material, called dinosporin. A number of marine species produce potent toxins that affect a variety of animals as well as humans. The toxic forms may cause fishkills (e.g. karlotoxins, brevetoxins), or the toxin may accumulate in crustaceans or molluscs (PSP, DSP, NSP, palytoxin) or in fish (ciguatera).

INTRODUCTION TO DINOFLAGELLATE TAXONOMY

Dinoflagellate research has a 225-year-old history, beginning with O.F. Müller in Copenhagen in the late 1700s. It received a major boost from Ehrenberg's (1838) work in Berlin, notably his book on the *Infusionsthierchen als vollkommende Organismen*, in which the family "Peridinaea" was erected, marking the beginning of dinoflagellate classification. During the latter half of the 1800s and the early part of the 1900s many people worked on dinoflagellates, prominent early names being Claparède and Lachmann in Switzerland and Germany, respectively, and von Stein in Prague. They were followed by Kofoid and Swezy in North America, and a long line of people in Central Europe (Lemmermann, Lindemann,Wołoszynská, etc.).The taxonomic work of this era was assembled by Johs Schiller in the famous *Kryptogamenflora von Deutschland, Österreich und der Schweiz* (Schiller, 1931–1937). This book, although printed in the 1930s, is still an important reference work for the identification of dinoflagellates. Termination of the premolecular age is marked by the appearance of another major book, *A Classification of Living and Fossil Dinoflagellates*, by Fensome et al. (1993). The title indicates the existence of fossil dinoflagellates, and the division or phylum Dinoflagellata is probably one of the oldest groups of eukaryotes. Thus, 1.5-billion-year old acritarch-like cells were recently discovered in sediments in Northern Territory, Australia (Javaux et al., 2001). Acritarchs are considered relatives of dinoflagellates, based on the presence of dinoflagellate-specific biological markers in the cell wall (Moldowan and Talyzina, 1998). This immense time span has given rise to a multitude of different types of dinoflagellates, some of which are cytologically highly unusual, probably more so than in any other group of protists. Special features include the sometimes very large amounts of DNA in the nuclear genome (up to 100 times as much as in the human genome), the lack of typical histones in the nucleus, the usually permanently condensed chromosomes, and the chloroplast minicircles amongst other features (see below).

The many findings of fossil dinoflagellate resting cysts led to the creation of a separate taxonomic system for these, resulting in two parallel nomenclatures for dinoflagellates. This has obviously been a source of confusion, and Fensome et al. (1993) is one of the first works to present a combined system for fossil and extant species.

The ongoing revolution in dinoflagellate taxonomy, classification, and phylogeny is the result of a combination of molecular studies and high-resolution electron microscopy, including reconstruction of the cell ultrastructure based on serial sectioning. The new data have turned ideas on dinoflagellate phylogeny upside down (compare, e.g. Saldarriaga et al., 2004, and Taylor, 1980). At the generic level, the genera of "naked" species have recently been examined in detail and the main genera have been redefined (Daugbjerg et al., 2000; Flø Jørgensen et al., 2004), and this is being followed by studies on the "thin-walled" species, the woloszynskioids ("*Glenodinium*") (Lindberg et al., 2005). An up-to-date taxonomy of the thecate dinoflagellates is still at its infancy. In this chapter, we provide an interim report of the present state of knowledge at three taxonomical levels: highest level (class/subclass), intermediate level (order/sometimes family), and generic level.

PHYLOGENY AND CLASSIFICATION AT CLASS OR SUBCLASS LEVEL, THE BASE OF THE PHYLOGENETIC TREE

In his much-reproduced phylogenetic tree from 1980, Max Taylor placed most heterotrophic genera of dinoflagellates at the top of the tree, whilst photosynthetic species occupied many of the basal branches. Molecular evidence (e.g. Saldarriaga et al., 2004) has turned the tree over, however, and the base of the tree is now occupied by heterotrophic species only, some free living and some parasitic.

Multiple protein phylogenies have shown that the heterotrophic marine species *Oxyrrhis marina* forms a sister group to the main group of dinoflagellates (Saldarriaga et al., 2003). Perkinsozoa and ciliates together form a sister group to this assemblage. This finding is in good agreement with the special type of mitosis found in *Oxyrrhis*, a type that resembles mitosis in many other eukaryotes but differs markedly from the typical dinoflagellate mitosis in being intranuclear (Triemer, 1982). Instead of typical histones (H1–H5), *Oxyrrhis* possesses one major type of basic protein, called Np23, associated with the chromosomes (Kato et al., 1997). *Oxyrrhis* is the coelacanth of the dinoflagellates, a phylogenetically isolated and (apparently) monotypic genus, which is undoubtedly dinoflagellate but has retained some ancient features. More work on genetic diversity and ultrastructure of *Oxyrrhis* is needed.

Using SSU data, Saldarriaga et al. (2001, 2004) also found *Noctiluca scintillans* at the base of the dinoflagellate tree, followed by the parasitic *Amoebophrya* of the Syndiniales (see also Skovgaard et al., 2005). The position of *Noctiluca* was not stable, however (Saldarriaga et al., 2004), and studies on additional noctilucoids are also required before their phylogenetic position can be firmly established. Mitotic data on the Syndiniales, however, agree with the separate and basal position of the order. Mitosis resembles the typical dinoflagellate mitosis in most respects but differs in the basal bodies forming the poles of the spindle, as in many other groups of flagellates (Hollande, 1974; Ris and Kubai, 1974). A positive reaction with Alkali Fast Green has been interpreted as showing the presence of histones (Cachon and Cachon, 1987), or a single basic protein as in *Oxyrrhis* (Rizzo, 1991). Based on the present evidence (mitosis and molecular studies), the division Dinoflagellata may therefore be divided into three major groups, the Oxyrrhinales, forming a sister group (perhaps a class) to the rest of the dinoflagellates, and, within the latter, the class Syndiniophyceae A.R. Loeblich III 1976, and the "core" dinoflagellates, the class Dinophyceae Fritsch in West and Fritsch (1927).

THE DINOFLAGELLATE CHLOROPLAST, ORIGINS

Chromalveolate protists are thought to share a common ancestor, a conclusion primarily based on molecular analyses of glyceraldehyde-3-phosphate dehydrogenase (Harper and Keeling, 2003), fructose-1,6-bisphosphate aldolate (Patron et al., 2004) and phosphoribulokinase (Petersen et al., 2006). Molecular data also indicate that the chromalveolate lineage acquired the chloroplast by uptake of a red alga in a single secondary endosymbiotic event (e.g. Cavalier-Smith, 1999, 2002;

Takishita and Uchida, 1999; Zhang et al., 1999). According to this hypothesis, ciliates secondarily lost the chloroplast. Among the chloroplast types known in dinoflagellates, the peridinin-containing type is thought to have evolved from the red algal chloroplast. As mentioned above, the branches at the base of the dinoflagellate phylogenetic tree are all occupied by heterotrophic species, indicating loss of chloroplasts or, perhaps, that chloroplasts were acquired by dinoflagellates after the emergence of *Oxyrrhis* and Syndiniophyceae (and perhaps Noctilucales). This is in conflict with the idea of a common origin of the chloroplasts in chromalveolate protists. The chloroplast situation in dinoflagellates is exceptionally complex, however, and a brief overview is given in Table 12.1.

As shown in Table 12.1, a symbiotic relationship between a dinoflagellate and another eukaryotic alga has become established at least eight times, but some of these are due to kleptoplastidy and are therefore of little or no phylogenetic significance. It has been speculated that the ancestral peridinin-containing chloroplast was replaced with chloroplasts from other algae, in other words, by additional (tertiary) endosymbiotic events (e.g. Chesnick et al., 1997; Tengs et al., 2000; Saldarriaga et al., 2001; Tamura et al., 2005) (Table 12.1). However, the symbioses have not all reached the same level of permanence. Thus, the endosymbiont in some cases is present in an almost unchanged form and was therefore probably established more recently (e.g. *Kryptoperidinium foliaceum*). In other species the symbiont has lost almost all organelles, and only the chloroplast remains (e.g. the peridinin-containing species), indicating an older, more well-established symbiosis. In some cases large variation in the chlorophyll fluorescence of the cells has been observed, indicating that the symbiosis is a case of kleptoplastidy, in which the prey is gradually digested (Koike et al., 2001).

Members of the Kareniaceae possess fucoxanthin as a major light-harvesting pigment in the chloroplasts (Bergholtz et al., 2006), as in heterokont and haptophyte algae. Analysis of chloroplast-encoded SSU rDNA gene sequences has confirmed that the peridinin-type chloroplast in this group has been superseded by a haptophyte-type chloroplast containing fucoxanthin and its derivatives, 19'-hexanoyloxy-fucoxanthin and 19'-butanoyloxy-fucoxanthin (Tengs et al., 2000). Ishida and Green (2002), based on PsbO sequence data (33-kDa oxygen-enhancer 1 protein), also confirmed that the peridinin-type chloroplast in *Karenia brevis* has been recently replaced with a haptophyte chloroplast. This study demonstrated, for the first time, replacement of a host nuclear gene with a gene originally located in the nucleus of the tertiary endosymbiont.

The dinoflagellate chloroplast: unique arrangement of genes in minicircles

Substantial morphological and biochemical differences exist between chloroplasts of extant red algae and peridinin-containing dinoflagellate chloroplasts, and the evolutionary origin of the peridinin-containing dinoflagellate chloroplast is therefore still debated. Recently, a very unusual and unique structure of the chloroplast DNA was discovered in the peridinin-containing chloroplasts (Zhang et al., 1999). Individual genes were located in plasmid-like minicircles with a size range of 2 to 3.7 kilobases (Zhang et al., 1999; Green, 2004). A few minicircles contained two to three genes, while some minicircles lacked recognizable encoding genes. This arrangement contrasts with most other plants (algae and land plants), in which the chloroplast genome comprises 100 to 200 kilobases and contains 100 to 250 genes (e.g. Turmel et al., 1999), while the genome of the peridinin-containing chloroplast contains only 15 known genes (Hackett et al., 2004). The remaining genes have either been transferred to the host nucleus or lost, or they have become modified to such an extent that they no longer can be recognized. The minimum number of genes required for keeping the photosynthetic apparatus working in the chloroplast is apparently less than 20. In *Ceratium horridum*, Laatsch et al. (2004), using *in situ* hybridization with minicircle-specific probes, convincingly demonstrated that the minicircles are located not in the chloroplasts but in the space between the chromosomes in the nucleus (i.e., nuclear, extrachromosomal). The physical

TABLE 12.1
Dinoflagellate Chloroplasts (or Kleptoplastids) Currently Recognized[1]

Type of Chloroplast	Description of Chloroplast Types
Type 1	The peridinin-containing species. The most common type, the endosymbiont is believed to be a red alga. Stable symbiosis, only the chloroplast remains (e.g. Ishida and Green, 2002; Petersen et al., 2006).
Type 2	The Kareniaceae. The chloroplast is a haptophyte. Stable symbiosis, only the chloroplast remains (e.g. Ishida and Green, 2002).
Type 3	The Dinophysiales Group I. *Dinophysis norvegica, D. acuta, D. tripos.* The chloroplast is a cryptophyte (Janson, 2004; Schnepf and Elbrächter, 1988). Only the symbiont chloroplast remains. Disputed whether it is a stable symbiosis or kleptoplastidy.
Type 4	The Dinophysiales Group II: *Dinophysis mitra* (and perhaps *D. rapa* (Hallegraeff and Lucas 1988)). The chloroplast is a haptophyte related to *Chrysochromulina* and *Prymnesium* (two slightly different symbionts known; Koike at al., 2005). Only the symbiont chloroplast remains. Probably kleptoplastidy.
Type 5	The genus *Lepidodinium*. The chloroplast is a green alga. Probably a stable symbiosis. Only the chloroplast remains. Two species studied in detail, *Lepidodinium viride* and *L. chlorophorum* (formerly *Gymnodinium chlorophorum*) (Elbrächter and Schnepf, 1996; Hansen and Moestrup, 2005; Hansen et al., 2007a; Watanabe et al., 1987). Additional species with green chloroplasts are known from light microscopy only.
Type 6	*Gymnodinium aeruginosum* and other (related?) forms with blue-green chloroplasts. The chloroplast is a cryptophyte. Some species have established a stable symbiosis with the cryptophyte symbiont and only the chloroplast remains, surrounded by three membranes (*Amphidinium wigrense*: Wilcox and Wedemayer, 1985). In other cases kleptoplastidy has been seen and the chloroplast must be renewed regularly (*Amphidinium poecilochroum*: Larsen, 1988). In others again (*Gymnodinium acidotum*: Wilcox and Wedemayer, 1984), an almost complete cryptophyte symbiont is present, which lacks any trace of the flagellar apparatus. Whether the group is monophyletic is presently unknown. The three species mentioned have retained the original generic names *Amphidinium* and *Gymnodinium*, pending further investigations. None of them belong here, as these genera have now been redefined (Daugbjerg et al., 2000; Flø Jørgensen et al., 2004).
Type 7	The *Durinskia/Krypthecodinium* group. The chloroplast is a diatom, either (7a) a pennate, or (7b) a centric diatom. The endosymbiont is relatively complete and retains, in addition to the chloroplasts, its nucleus, mitochondria, and so forth (e.g. Horiguchi and Pienaar, 1994). One may speculate that one diatom symbiont was exchanged for the other, many of the genes of the two symbionts being shared.
	This very unusual group includes *Durinskia, Kryptoperidinium, Galeidinium,* and *Dinothrix* and some species presently included in *Peridinium* (*P. quinquecorne*) and *Gymnodinium* (*G. quadrilobatum*) (Horiguchi, 2003; Tamura et al., 2005), but obviously belonging in other genera. Somewhat unexpectedly, the group appears to be monophyletic (SSU), a finding supported by the presence of the same, unique type of eyespot in the cells. The markedly diverse morphology of species belonging to this complex indicates a relatively old group, in which the original chloroplast has been modified into an eyespot, and new chloroplasts have been established more recently by uptake of diatoms. The group should be given at least family status, and the species referred to *Peridinium* and *Gymnodinium* must be transferred to other genera.
Type 8	*Podolampas*. The symbiont is apparently a heterokont alga (perhaps a dictyochophyte). Whole cells of the symbiont remain in the host cytoplassm, but no trace of the flagellar apparatus has been found (Schweikert and Elbrächter, 2004). Kleptoplastidy?

[1] *Polykrikos lebourae* Herdman is unusual within the otherwise heterotrophic genus *Polykrikos* in having yellow chloroplasts. The taxonomic identity of the chloroplast has not been established, but it may represent yet another example of kleptoplastidy (also *Pheopolykrikos*?).

placement of the minicircles discovered by Zhang et al. (1999) and Barbrook et al. (2001) was not examined, but it is known from *in situ* hybridization that at least some minicircles reside in the chloroplast genome in *Symbiodinum* (Takishita et al., 2003).

PHYLOGENY AND CLASSIFICATION AT THE ORDER LEVEL

While division of the dinoflagellates into the three major groups mentioned above is supported by morphological and molecular data, division into orders remains less clear. Fensome et al. (1993) divided the extant dinoflagellates into 14 orders (Table 12.2), but many of these are based on a single or very few morphological features. In several cases, molecular studies and detailed investigations of cell structure, including the flagellar apparatus, have shown the orders to be polyphyletic. The information is still very scattered, however. Phylogenies based on single genes residing in the dinoflagellate nucleus have predominantly used small and large subunit ribosomal sequences (e.g. Saunders et al., 1997; Saldarriaga et al., 2001; Daugbjerg et al., 2000). However, the relationships between the deep branches in the phylogenetic trees are not well supported in terms of bootstrap values and posterior probabilities from Bayesian analyses. Branch lengths are short for the divergent branches, indicating little phylogenetic signal to support this part of the tree, and perhaps a rapid radiation during a relatively short evolutionary time. In fact, some groups jump around depending on the algorithm applied for phylogenetic inference (Daugbjerg, personal observation). Murray et al. (2005) attempted concatenation of nuclear-encoded SSU and LSU rDNA sequences, but this approach did not provide higher levels of support for the divergent branches. Mitochondrial cytochrome *b* (*cob*) gene sequences from dinoflagellates have recently been published (Lin et al., 2002; Zhang et al., 2005; Zhang and Lin, 2005). Sequences of *cob* are still few (13 genera), so it is premature to evaluate the potential usefulness of this gene as a candidate for obtaining a better resolution of divergent branches. Since about 50% of the approximately 2000 described dinoflagellates are heterotrophic, chloroplast-encoded genes are of limited use for phylogenetic inference within the group.

The data presently available do not allow for a natural classification of the dinoflagellates into orders or families. However, some trends are appearing, and Figure 12.1 and Figure 12.2 illustrate our most recent phylogenetic tree, based on LSU sequences and including 19 new sequences.

TABLE 12.2
The Dinoflagellate Orders according to Fensome et al. (1993) but also Including Oxyrrhinales

Orders	Reference
Blastodiniales	Chatton (1906)
Desmocapsales	Pascher (1914)
Dinophysiales	Kofoid (1926)
Gonyaulacales	Taylor (1980)
Gymnodiniales	Apstein (1909)
Noctilucales	Haeckel (1894)
Oxyrrhinales	Sournia (1984)
Peridiniales	Haeckel (1894)
Phytodiniales	Christensen (1962 ex Loeblich, III, 1970)
Prorocentrales	Lemmermann (1910)
Ptychodiscales	Fensome et al. (1993)
Suessiales	Fensome et al. (1993)
Syndiniales	Loeblich III (1976)
Thoracosphaerales	Tangen (in Tangen et al., 1982)

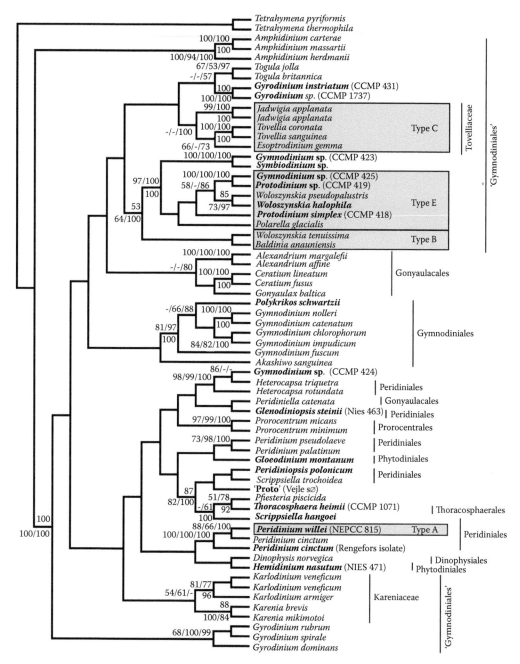

FIGURE 12.1 LSU rDNA phylogeny of 63 taxa of dinoflagellates inferred from neighbour-joining (NJ) analysis using maximum likelihood settings as suggested by Modeltest (ver. 3.7). Maxium parsimony analyses (1000 replications, not shown) produced eight equally parsimonious trees each 3337 steps (CI = 365, RI = 0.523). Bootstrap values or support from posterior probabilities of 50% or above are written to the left of internal nodes. The first numbers are from NJ analyses based on the maximum likelihood settings obtained using Modeltest (TrN+I+G model) and with 1000 replications. The second numbers are from unweighted maximum parsimony analyses (1000 replications), and the third numbers are posterior probabilities from Bayesian analyses and based on 9657 trees. Two ciliates (viz. *Tetrahymena pyriformis* and *T. thermophila*) comprised the outgroup. Species in boldface were determined in this study. The types of eyespot arrangements have been mapped on the tree. See Figure 12.3 for an explanation of these types. Division into orders follows Fensome et al. (1993).

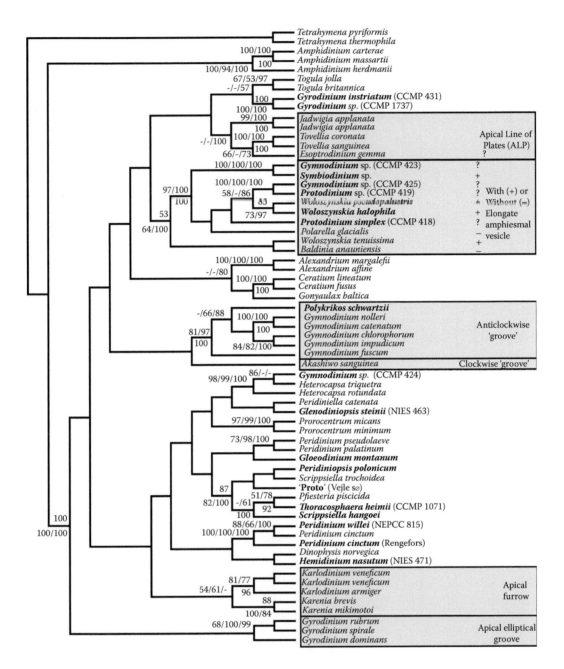

FIGURE 12.2 See Figure 12.1 for a tree description and information on tree statistics. Species in boldface were determined in this study. "Apical furrow" systems have been mapped on the tree.

The only large order supported frequently by SSU and LSU data is the **Gonyaulacales**. It is defined morphologically by its very asymmetrical plate tabulation. The other large order, **Gymnodiniales** is breaking up; thus, the molecular trees suggest that the Kareniaceae (*Karenia* and *Karlodinium* in Figure 12.1 and Figure 12.2) forms a separate, non-related order, and the family Tovelliaceae forms a family of unknown phylogenetic relationship. It may eventually be given order status. Isolates identified as *Gymnodinium* sp. fall into four separate groups in Figure 12.1 and Figure 12.2. The position of *Gyrodinium* (*G. rubrum*, *G. spirale*, *G. dominans*) is unresolved. The other taxa referred to *Gyrodinium*, *G. instriatum*, together with CCMP1737 form a separate genus of uncertain phylogeny.

The order **Peridiniales** also shows signs of falling apart in non-related groups, but the number of detailed studies is small. The order **Thoracosphaerales**, with the single genus *Thoracosphaera*, groups with other species forming calcified cysts such as *Scrippsiella*, and also with some of the peridinioids (though not the type species of *Peridinium*) (Figure 12.1 and Figure 12.2). There is no indication that it forms a separate order. The same was observed by Gottschling et al. (2005), using 5.8 S rDNA and conserved parts of ITS1 and ITS2. Inclusion of *Pfiesteria* in this clade needs to be verified (Figure 12.1 and Figure 12.2).

The phylogenetic relationship of the **Dinophysiales** and **Prorocentrales** is poorly understood, and their positions in Figure 12.1 and Figure 12.2 are not supported by high bootstrap values or posterior probabilities. The phylogenetic relationship between *Amphidinium* and other dinoflagellates is also poorly resolved. Among the groups not included in Figure 12.1 and Figure 12.2, we have found no molecular information on the **Ptychodiscales** and the **Desmocapsales**, the latter considered to be an early diverging group. Another putative early diverging species, *Haplodinium antjoliense* Klebs 1912, is undoubtedly identical to *Prorocentrum foveolata* Croome & Tyler, 1987.

Molecular trees indicate that the **Blastodiniales** is polyphyletic. The position of *Blastodinium* in the SSU trees varies depending on the algorithm used in the phylogenetic analysis, but it never groups with other parasitic species such as *Haplozoon* and *Amyloodinium* (Skovgaard et al., 2006). Skovgaard et al. (2006) found the motile cells of *Blastodinium* to be scrippsielloid rather than gymnodinioid (e.g. Taylor, 2004). Based on SSU rDNA sequence data another possible natural group of parasites appears to be emerging comprising *Amyloodinium*, *Pfiesteria*, *Pseudopfiesteria*, and *Paulsenella* (Kühn and Medlin, 2005). Thus, parasitism appears to have evolved independently at least three times within the dinoflagellates.

Available information on the **Noctilucales** is insufficient; only *Noctiluca* has been studied, and it often (but not invariably) appears close to the base of the SSU trees. The gametes were very recently shown to be biflagellate (Fukuda and Endoh, 2006), rather than uniflagellate as sometimes claimed (Zingmark, 1970).

Phytodiniales ("Dinococcales"), an order of coccoid genera, has received little attention so far, but it is almost certainly polyphyletic. Thus, *Gloeodinium viscosum* and *Halostylodinium arenarium* do not group together in the SSU trees. The contention that *Hemidinium nasutum* is the motile stage of the coccoid *Gloeodinium montanum* (Killian, 1924; Popovský, 1971) is not supported in our LSU trees (Figure 12.1): the two taxa do not occupy a sister position. This confirms the conclusions of Kelley and Pfiester (1990), based on culture studies, while the observations by Killian (1924) and Popovský (1971) were based on mixed samples. *Hemidinium* is thought to be thecate, while the zoospores and gametes of *Gloeodinium* appear to be naked.

The case of the order **Suessiales** will be reported separately below.

PHYLOGENY AND CLASSIFICATION AT THE GENERIC LEVEL

Classification into genera is based on selected morphological characters such as presence or absence of thecal plates, the number of thecal plates, the position of the flagella on the cell, the position of the cingulum, and the displacement of the two ends of the cingulum. The presence or absence of chloroplasts is not usually considered an important generic character. Many of the morphological characters are known to vary, however, and a stable taxonomic system reflecting phylogeny has been difficult to establish. With the new data from transmission electron microscopy (TEM), scanning electron microscopy (SEM), and molecular studies, the old system is slowly being replaced. Thecate and non-thecate species are found to group together, and there is a gradual transition between, on one side, distinctly thecate and, on the other, naked species, making it difficult to maintain these terms. Also, many chloroplast-containing species with a particular type of endosymbiont group together, indicating genuine phylogenetic relationship. Small differences in plate pattern, however, considered for a long time to be reliable generic characters, are not always supported as such by the molecular data. Morphological features supported by molecular data include the structure and path of the "apical furrow" system, the structure of the nuclear envelope, and certain details of the flagellar apparatus ultrastructure (Daugbjerg et al., 2000). The type of eyespot is one of the most reliable features (Figure 12.3), and in some cases this also applies to the type of chloroplast pigment present. Another promising feature is the morphology of the resting cyst (hypnozygote or hypnospore).

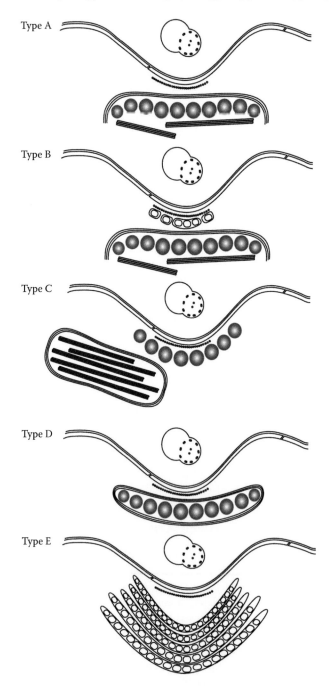

FIGURE 12.3 Five different types of eyespot arrangements in dinoflagellates. Type A is present in *Peridinium willei* and in many other groups of algae. Type B is present in *Woloszynskia tenuissima* and *Baldinia anauniensis*. Type C is characteristic of the Tovelliaceae. Type D is seen only in dinoflagellates with a diatom symbiont. Type E characterizes the main branch of the Suessiales, containing *Polarella* and related forms.

CASE STORY: THE SUESSIALES

The rapidly unfolding story of the order Suessiales gives an indication of the many changes that are taking place in our concept of dinoflagellate taxonomy and phylogeny. The order was erected only some 13 years ago (Fensome et al., 1993), and its first described members were fossils from the Mesozoic (the family Suessiaceae Fensome et al., 1993). However, Fensome et al. (1993) also included the family Symbiodiniaceae (symbionts in corals) in their new order because of the presence in the Symbiodiniaceae of seven latitudinal series of amphiesmal plates. Montresor et al. (1999) found the first truly free-living species, the polar flagellate *Polarella glacialis*, and the group has now rapidly grown to include also *Woloszynskia pseudopalustris*, *W. halophila*, *Protodinium simplex* Lohman (transferred, incorrectly it now appears, to *Gymnodinium* by Kofoid and Swezy, 1921), and *Gymnodinium bei* (e.g. Saldarriaga et al., 2001; Kremp et al., 2005; Lindberg et al., 2005). This will require emendation of the order as the number of latitudinal series in cells of *W. pseudopalustris* exceeds 10 (Moestrup, Lindberg, Daugbjerg, and Calado, in preparation). The order is characterized morphologically by a very special type of eyespot, unknown elsewhere in algae. Instead of eyespot globules, it comprises a series of cisternae with brick-like contents (Type E in Figure 12.3). Despite its unusual appearance this is undoubtedly a true eyespot and the same type occurs in *Gymnodinium natalense* Horiguchi and Pienaar and in the heterotroph *Prosoaulax lacustre* (J. Stein) Calado and Moestrup (Calado et al., 1998). These species almost certainly belong in the Suessiales (in Group II of Lindberg et al., 2005) but further studies are needed to clarify their phylogenetic position. A group related to the Suessiales has now appeared in the molecular trees, in which the cells are morphologically characterized by a more ordinary type of eyespot, with the carotenoid globules located within a chloroplast (Type B in Figure 12.3). The sister group (Group III in Lindberg et al., 2005) now also grows rapidly in a number of species. It presently comprises what is known as *Woloszynskia tenuissima*, and related species of the same genus, in addition to the new genus *Baldinia* ("the dominant green form" of Lake Tovel (Flaim et al., 2004; Hansen et al., 2007b) (Figure 12.1). The two sister groups comprise separate families, characterized in particular by the structure of the eyespot. As in *W. pseudopalustre*, cells of *W. tenuissima* are covered by more than ten latitudinal series of amphiesma plates (Crawford et al., 1970; Moestrup et al., submitted).

CONCLUSIONS

Applications of new methods are fundamentally changing ideas of dinoflagellate phylogeny and classification and additional, major, changes are likely, following the collapse of many of the traditionally used morphological features for classification. Thus, a substantial number of new genera will be described, and the old genera will be redefined. However, it will be some years before a natural classification scheme can be established. On the taxonomic side, identification of the many species (and sometimes genera) described with a few sketchy drawings from O.F. Müller onwards is hampering progress, and many species can never be identified with any degree of certainty. Future descriptions should be accompanied by DNA sequences to define the taxa more precisely. The wide geographical distribution of many (morpho)species is likely to be accompanied by genetic diversity, indicating that what is thought to be a widely distributed species is a species complex. This may apply in particular to freshwater species. Future molecular studies addressing the phylogenetic relationship at family and order levels should be based on concatenated multi-gene analyses, including nucleotide sequences from nuclear and mitochondrial genomes (i.e., host genes). The molecular data should be used in combination with information from detailed ultrastructural studies in a "total evidence" approach.

New LSU rDNA sequences

For this study, 19 new LSU rDNA sequences were determined from a wide range of dinoflagellates (Table 12.3). They were added to a data matrix comprising 44 other LSU rDNA dinoflagellate sequences available in GenBank. LSU rDNA from two ciliates formed the outgroup species in the phylogenetic analyses. Polymerase chain reaction (PCR) amplifications were based on extracted total genomic DNA of clonal cultures, on single cells (wild material), or on few-cells PCR (clonal cultures). The primers used were D1R-F (Scholin et al., 1994) in combination with "Dino-ND," a dinoflagellate-specific primer (Hansen and Daugbjerg, 2004). PCR amplifications, thermal cycle conditions, cleaning of PCR products, and cycle sequencing followed Hansen and Daugbjerg (2004). The phylogenetic analyses follow Calado et al. (2006). The tree topology shown in Figure 12.1 and Figure 12.2 resulted from neighbour-joining (NJ) analysis with maximum likelihood settings (Tamura and Nei, 1993, model) as suggested by Modeltest (ver. 3.7; Posada and Crandall, 1998). NJ bootstrap analyses included 1000 replications. Maximum parsimony bootstrap analyses also with 1000 replications and Bayesian analyses (500,000 generations) were also performed (Figure 12.1 and Figure 12.2).

TABLE 12.3
Novel Partial LSU rDNA Sequences Determined for this Study

Species	Strain Number	GenBank Accession Number
Glenodiniopsis steinii	NIES 463	EF205002
Gloeodinium montanum	Marburg isolate	EF205003
Gymnodinium sp.	CCMP 423	EF205004
Gymnodinium sp.	CCMP 424	EF205005
Gymnodinium sp.	CCMP 425	EF205006
Gyrodinium instriatum	CCMP 431	EF205007
Gyrodinium sp.	CCMP1737	EF205008
Hemidinium nasutum[7]	NIES 471	EF205009
Peridiniopsis polonicum[2]	—	EF205010
Peridinium cinctum[1,7]	—	EF205011
Peridinium willei	NEPCC 815	EF205012
Polykrikos schwartzii[3,7]	—	EF205013
"Proto" from Vejle sø[4,7]	—	—
Protodinium simplex	CCMP 418	EF205014
Protodinium sp.	CCMP 419	EF205015
Scrippsiella hangoei[5]	SHTV-2	EF205016
Symbiodinium sp.[6]	— (Australian isolate)	EF205017
Thoracosphaera heimii	CCMP 1071	EF205018
Woloszynskia halophila[5]	WHTV-C6	EF205019

[1] Isolates generously provided by Karin Rengefors.

[2] Cultures isolated by Niels Daugbjerg.

[3] Wild material isolated from Danish waters.

[4] Cultures isolated by Øjvind Moestrup.

[5] Isolates generously provided by Anke Kemp.

[6] Isolates generously provided by Karin Ulstrup.

[7] Sequence determination based on single-cell polymerase chain reaction (PCR).

Notes: CCMP strains from the Provasoli-Guillard National Center for Culture of Marine Phytoplankton and generously provided by Bob Andersen. NIES: culture from the National Institute for Environmental Studies, Microbial Culture Collection, Japan. NEPCC: culture from the Northeast Pacific Culture Collection, Canada. — = Information not available.

ACKNOWLEDGMENTS

We thank Charlotte Hansen ("Statens Naturhistoriske Museum") for running the cycle-sequences. Financial support was provided by the Danish Science Research Council (Grant no. 21-02-0539) and from the Villum-Kann Rasmussen Foundation. Karin Lindberg provided helpful comments on the text.

REFERENCES

Apstein, C. (1909) Die Pyrocysteen der Plankton-Expedition. In *Ergebnisse der Plankton-Expedition der Humboldt Stiftung*, 4, M, c: 1-27. Kiel and Leipzig, Lipsius & Tischer.

Barbrook, A.C., Symington, H., Nisbet, R.E.R., Larkum, A., and Howe, C.J. (2001) Organisation and expression of the plastid genome of the dinoflagellate *Amphidinium operculatum*. *Molecular Genetics and Genomics*, 266: 632–638.

Bergholtz, T., Daugbjerg, N., Moestrup, Ø., and Fernández-Tejedor, M. (2006) On the identity of *Karlodinium veneficum* and description of *Karlodinium armiger* sp. nov. (Dinophyceae), based on light- and electron microscopy, nuclear-encoded LSU rDNA and pigment composition. *Journal of Phycology*, 42: 170–193.

Cachon, J. and Cachon, M. (1987) Parasitic dinoflagellates. In *The Biology of Dinoflagellates* (ed. F.J.R. Taylor), Blackwells, Oxford, pp. 571–610.

Calado, A.J., Craveiro, S., and Moestrup, Ø. (1998) Taxonomy and ultrastructure of a freshwater, heterotrophic *Amphidinium* (Dinophyceae) that feeds on unicellular protists. *Journal of Phycology*, 34: 536–554.

Calado, A.J., Craveiro, S.C., Daugbjerg, N., and Moestrup, Ø. (2006) Ultrastructure and LSU rDNA-based phylogeny of *Esoptrodinium gemma* (Dinophyceae), with notes on feeding behavior and the description of the flagellar base area of a planozygote. *Journal of Phycology*, 42: 434–454.

Cavalier-Smith, T. (1999) Principles of protein and lipid targeting in secondary symbiogenesis: euglenoid, dinoflagellate, and sporozoan plastid origins and the eukaryote family tree. *Journal of Eukaryotic Microbiology*, 46: 347–366.

Cavalier-Smith, T. (2002) Chloroplast evolution: secondary symbiogenesis and multiple losses. *Current Biology*, 12: R62–R64.

Chatton, É. (1906) Les blastodinides, ordre nouveau de dinoflagellés parasites. *L'Ácademie des Sciences, Paris, Comptes rendus*, 143: 981–983.

Chesnick, J.M., Kooistra, W.H.C.F., Wellbrock, U., and Medlin, L.K. (1997) Ribosomal RNA analysis indicates a benthic pennate diatom ancestry for the endosymbionts of the dinoflagellates *Peridinium foliaceum* and *Peridinium balticum* (Pyrrhophyta). *Journal of Eukaryotic Microbiology*, 44: 314–320.

Christensen, T. (1962) Alger. In *Botanik* (eds T.W. Böcher, M. Lange and T. Sørensen), Munksgaard, Copenhagen, pp. 1–178.

Crawford, R.M., Dodge, J.D., and Happey, C.M. (1970) The dinoflagellate genus *Woloszynskia*. I. Fine structure and ecology of *W. tenuissima* from Abbott's Pool, Somerset. *Nova Hedwigia*, 19: 825–840.

Croome, R.L. and Tyler, P.A. (1987) *Prorocentrum playfairi* and *Prorocentrum foveolata*, two new dinoflagellates from Australian freshwaters. *British Phycological Journal*, 22: 67–75.

Daugbjerg, N., Hansen, G., Larsen, J., and Moestrup, Ø. (2000) Phylogeny of some of the major genera of dinoflagellates based on ultrastructure and partial LSU rDNA sequence data, including the erection of three new genera of unarmoured dinoflagellates. *Phycologia*, 39: 302–317.

Ehrenberg, C.G. (1838) *Die Infusionsthierchen als volkommende Organismen*, Voss, Berlin, Leipzig.

Elbrächter, M. and Schnepf, E. (1996) *Gymnodinium chlorophorum*, a new green, bloom-forming dinoflagellate (Gymnodiniales, Dinophyceae) with a vestigial prasinophyte endosymbiont. *Phycologia*, 35: 381–393.

Fensome, R.A., Taylor, F.J.R., Norris, G., Sarjeant, W.A.S., Wharton, D.I., and Williams, G.L. (1993) *A Classification of Living and Fossil Dinoflagellates*. Micropalaeontology, Special Publication, number 7, American Museum of Natural History.

Flaim, G., Hansen, G., Moestrup, Ø., Corradini, F., and Borghi, B. (2004) Reinterpretation of the dinoflagellate *Glenodinium sanguineum* in the reddening of Lake Tovel, Italian Alps. *Phycologia*, 43: 737–743.

Flø Jørgensen, M., Murray, S., and Daugbjerg, N. (2004) *Amphidinium* revisited. I. Redefinition of *Amphidinium* (Dinophyceae) based on cladistic and molecular phylogenetic analyses. *Journal of Phycology*, 40: 351–365.

Fukuda, Y. and Endoh, H. (2006) New details from the complete life cycle of the red-tide dinoflagellate *Noctiluca scintillans* (Ehrenberg) McCartney. *European Journal of Protistology*, 42: 209–219.

Gottschling, M., Keupp, H., Plötner, J., Knop, R., Willems, H., and Kirsch, M. (2005) Phylogeny of calcareous dinoflagellates as inferred from ITS and ribosomal sequence data. *Molecular Phylogenetics and Evolution*, 36: 444–455.

Green, B.R. (2004) The chloroplast genome of dinoflagellates—a reduced instruction set? *Protist*, 155: 23–31.

Hackett, J.D., Yoon, H.S., Soares, M.B., Bonaldo, M.F., Casavant, T.L., Scheetz, T.E., Nosenko, T., and Bhattacharya, D. (2004) Migration of the plastid genome to the nucleus in a peridinin dinoflagellate. *Current Biology*, 14: 213–218.

Haeckel, E. (1894) *Systematische Phylogenie. Entwurf eines natürlichen Systems der Organismen auf Grund ihrer Stammegeschichte, I. Systematische Phylogenie der Protisten und Pflanzen*, Reimer, Berlin.

Hallegraeff, G.M. and Lucas, I.A.N. (1988) The marine dinoflagellate genus *Dinophysis* (Dinophyceae): photosynthetic, neritic and non-photosynthetic, oceanic species. *Phycologia*, 27: 25–42.

Hansen, G., Botes, L., and de Salas, M. (2007a) Ultrastructure and LSU rDNA sequences of *Lepidodinium viride* reveal a close relationship to *Lepidodinium chlorophorum* comb. nov. (=*Gymnodinium chlorophorum*). *Phycological Research*, 55: 25–41.

Hansen, G. and Daugbjerg, N. (2004) Ultrastructure of *Gyrodinium spirale*, the type species of *Gyrodinium* (Dinophyceae), including a phylogeny of *G. dominans*, *G. rubrum* and *G. spirale* deduced from partial LSU rDNA sequences. *Protist*, 155: 271–294.

Hansen, G., Daugbjerg, N., and Henriksen, P. (2007b) *Baldinia anauniensis* gen. et sp. nov. a "new" dinoflagellate from Lake Tovel, Northern Italy. *Phycologia*, 46: 86–108.

Hansen, G. and Moestrup, Ø. (2005) Flagellar apparatus and nuclear chambers of the green dinoflagellate *Gymnodinium chlorophorum*. *Phycological Research*, 53: 169–181.

Harper, J. and Keeling, P. (2003) Nucleus-encoded, plastid targeted glyceraldehyde-3-phosphate dehydrogenase (GAPDH) indicates a single origin for chromalveolate plastids. *Molecular Biology and Evolution*, 20: 1730–1735.

Hollande, A. (1974) Étude comparée de la mitose syndinienne et de celle des Péridiniens libres et des Hypermastigines infrastructure et cycle évolutif des Syndinides parasites de Radiolaires. *Protistologica*, 10: 413–451.

Horiguchi, T. (2003) Diversity and phylogeny of dinoflagellates with a diatom endosymbiont and molecular phylogenetics of the genus *Protoperidinium*. Seventh International Conference on Modern and Fossil Dinoflagellates, Nagasaki 2003. Program, Abstracts & Participants, p. 53.

Horiguchi, T. and Pienaar, R.N. (1994) Ultrastructure of a new marine, sand-dwelling dinoflagellate, *Gymnodinium quadrilobatum* sp. nov. (Dinophyceae) with special reference to its endosymbiotic alga. *European Journal of Phycology*, 29: 237–245.

Ishida, K. and Green, B.R. (2002) Second- and third-hand chloroplasts in dinoflagellates: phylogeny of oxygen-evolving enhancer 1 (PsbO) protein reveals replacement of a nuclear-encoded plastid gene by that of a haptophyte tertiary endosymbiont. *Proceedings of the National Academy of Sciences, USA*, 99: 9294–9299.

Janson, S. (2004) Molecular evidence that plastids in the toxin-producing dinoflagellate genus *Dinophysis* originate from the free-living cryptophyte *Teleaulax amphioxeia*. *Environmental Microbiology*, 6: 1102–1106.

Javaux, E.J., Knoll, A.H., and Walter, M.R (2001) Morphological and ecological complexity in early eukaryotic ecosystems. *Nature*, 412: 66–69.

Kato, K.H., Moriyama, A., Huitorel, P., Cosson, J., Cachon, M., and Sato, H. (1997) Isolation of the major basic nuclear protein and its localization on chromosomes of the dinoflagellate, *Oxyrrhis marina*. *Biology of the Cell*, 89: 43–52.

Kelley, I. and Pfiester, L.A. (1990) Sexual reproduction in the freshwater dinoflagellate *Gloeodinium montanum*. *Journal of Phycology*, 26: 167–173.

Killian, C. (1924) Le cycle évolutif de *Gloeodinium montanum* (Klebs). *Archiv für Protistenkunde*, 50: 50–66.

Klebs, G. (1912) Über Flagellaten- und Algen-ähnlichen Peridineen. *Verhandlungen des Naturhistorisch-Medizinischen Vereines zu Heidelberg*, Neue Folge, 11: 369–451.

Kofoid, C.A. (1926) On *Oxyphysis oxytoxoides* gen. nov. and sp. nov. A dinophysoid dinoflagellate convergent toward the peridinioid type. *University of California Publications in Zoology*, 28: 203–216.

Kofoid, C.A. and Swezy, O. (1921) *The Free-Living Unarmoured Dinoflagellata*. Memoirs of the University of California, 5: 1–562.

Koike, K., Otobe, H., Takagi, M., Yoshida, T., Ogata, T., and Ishimaru, T. (2001) Recent occurrences of *Dinophysis fortii* (Dinophyceae) in the Okkirai Bay, Sanriku, Northern Japan, and related environmental factors. *Journal of Oceanography*, 57: 165–175.

Koike, K., Sekiguchi, H., Kobiyama, A., Takishita, K., Kawachi, M., Koike, K., and Ogata, T. (2005) A novel type of kleptoplastidy in *Dinophysis* (Dinophyceae): presence of haptophyte-type plastid in *Dinophysis mitra*. *Protist*, 156: 225–237.

Kremp, A., Elbrächter, M., Schweikert, M., Wolny, J.L., and Gottschling, M. (2005) *Woloszynskia halophila* (Biecheler) comb. nov.: a bloom-forming cold-water dinoflagellate co-occurring with *Scrippsiella hangoei* (Dinophyceae) in the Baltic Sea. *Journal of Phycology*, 41: 629–642.

Kühn, S.F. and Medlin, L.K. (2005) The systematic position of the parasitoid marine dinoflagellate *Paulsenella vonstoschii* (Dinophyceae) inferred from nuclear-encoded small subunit ribosomal DNA. *Protist*, 156: 393–398.

Laatsch, T., Zauner, S., Stoebe-Maier, B., Kowallik, K.V., and Maier, U.-G. (2004) Plastid-derived single gene minicircles of the dinoflagellate *Ceratium horridum* are localized in the nucleus. *Molecular Biology and Evolution*, 21: 1318–1322.

Larsen, J. (1988) An ultrastructural study of *Amphidinium poecilochroum* (Dinophyceae), a phagotrophic dinoflagellate feeding on small species of cryptophytes. *Phycologia*, 27: 366–377.

Lemmermann, E. (1910) III. Klasse. Peridiniales. In *Kryptogamenflora der Mark Brandenburg und angrenzender Gebiete, III Algen I (Schizophyceae, Flagellaten, Peridineen)*. Gebrüder Borntraeger, Leipzig, pp. 563–682.

Lin, S., Zhang, H., Spencer, D., Norman, J., and Gray, M. (2002) Widespread and extensive editing of mitochondrial mRNAs in dinoflagellates. *Journal of Molecular Biology*, 320: 727–739.

Lindberg, K., Moestrup, Ø., and Daugbjerg, N. (2005) Studies on woloszynskioid dinoflagellates I: *Woloszynskia coronata* re-examined using light and electron microscopy and partial LSU rDNA sequences, with description of *Tovellia* gen. nov and *Jadwigia* gen. nov. (Tovelliaceae fam. nov.). *Phycologia*, 44: 416–440.

Loeblich, A.R. III (1970) The amphiesma or dinoflagellate cell covering. In *North American Paleontological Convention, Chicago, September, 1969, Proceedings*, part G, 867–929.

Loeblich III, A.R. (1976) Dinoflagellate evolution: speculation and evidence. *Journal of Protozoology*, 23: 13–28.

Moestrup, Ø., Hansen, G., and Daugbjerg, N. Studies on woloszynskioid dinoflagellates III: On *Borghiella* gen. nov. and *B. dodgei* sp. nov., a cold-water species from Lake Tovel, N. Italy, and on *B. tenuissima* comb. nov. (syn. *Woloszynskia tenuissima*). *Phycologia* (submitted).

Moldowan, J.M. and Talyzina, N.M (1998) Biogeochemical evidence for dinoflagellate ancestors in the early Cambrian. *Science*, 281: 1168–1170.

Montresor, M., Procaccini, G., and Stoecker, D.K. (1999) *Polarella glacialis* gen. et sp. nov. (Dinophyceae): Suessiaceae are still alive! *Journal of Phycology*, 35: 186–197.

Murray, S., Flø Jørgensen, M., Ho, S.Y.W., Patterson, D.J., and Jermiin, L.S. (2005) Improving the analysis of dinoflagellate phylogeny based on rDNA. *Protist*, 156: 269–286.

Pascher, A. (1914) Über Flagellaten und Algen. *Berichte der deutschen botanischen Gesellschaft*, 32: 136–160.

Patron, N.J., Rogers, M.B., and Keeling, P.J. (2004) Gene replacement of fructose-1,6-bisphosphate aldolase supports the hypothesis of a single photosynthetic ancestor of chromalveolates. *Eukaryotic Cell*, 3: 1169–1175.

Petersen, J., Teich, R., Brinkmann, H., and Cerff, R. (2006) A "green" phosphoribulokinase in complex algae with red plastids: evidence for a single secondary endosymbiosis leading to haptophytes, cryptophytes, heterokonts and dinoflagellates. *Journal of Molecular Evolution*, 62: 143–157.

PopovskΔ, J. (1971) Some remarks on the life cycle of *Gloeodinium montanum* Klebs and *Hemidinium nasutum* Stein (Dinophyceae). *Archiv für Protistenkunde*, 113: 131–136.

Posada, D. and Crandall, K.A. (1998) MODELTEST: testing the model of DNA substitution. *Bioinformatics*, 14: 817–818.

Ris, H. and Kubai, D. (1974) An unusual mitotic mechanism in the parasitic protozoan *Syndinium* sp. *Journal of Cell Biology*, 60: 702–720.

Rizzo, P.J. (1991) The enigma of the dinoflagellate chromosome. *Journal of Protozoology*, 38: 246–252.

Saldarriaga, J.F., McEwan, M.L., Fast, N.M., Taylor, F.J.R., and Keeling, P.J. (2003) Multiple protein phylogenies show that *Oxyrrhis marina* and *Perkinsus marinus* are early branches of the dinoflagellate lineage. *International Journal of Systematic and Evolutionary Microbiology*, 53: 355–365.

Saldarriaga, J.F., Taylor, F.J.R. "Max", Cavalier-Smith, T., Menden-Deuer, S., and Keeling, P.J. (2004) Molecular data and the evolutionary history of dinoflagellates. *European Journal of Protistology*, 40: 85–111.

Saldarriaga, J.F., Taylor, F.J.R., Keeling, P.J., and Cavalier-Smith, T. (2001) Dinoflagellate nuclear SSU rRNA phylogeny suggests multiple plastid losses and replacements. *Journal of Molecular Evolution*, 53: 204–213.

Saunders, G.W., Hill, D.R.A., Sexton, J.P., and Andersen, R.A. (1997) Small-subunit ribosomal RNA sequences from selected dinoflagellates: testing classical evolutionary hypotheses with molecular systematic methods. In *Origin of Algae and Their Plastids* (ed. T. Bhattacharya), Springer, Wien, New York, pp. 237–259.

Schiller, J. (1931–1937) Dinoflagellatae (Peridineae) In *L. Rabenhorst's Kryptogamenflora von Deutschland, Österreich und der Schweiz*, Akademische Verlagsgesellschaft, Leipzig.

Schnepf, E. and Elbrächter, M. (1988) Cryptophycean-like double membrane-bound chloroplast in the dinoflagellate, *Dinophysis* Ehrenb.: evolutionary, phylogenetic and toxicological implications. *Botanica Acta*, 101: 196–203.

Scholin, C.A., Herzog, M., Sogin, M., and Anderson, D.M. (1994) Identification of group- and strain-specific genetic markers for globally distributed *Alexandrium* (Dinophyceae). II. Sequence analysis of a fragment of the LSU rDNA gene. *Journal of Phycology*, 30: 999–1011.

Schweikert, M. and Elbrächter, M. (2004) First ultrastructural investigations on the consortium between a phototrophic eukaryotic endosymbiont and *Podolampas bipes* (Dinophyceae). *Phycologia*, 43: 614–623.

Skovgaard, A., Massana, R., Balagué, V., and Saiz, E. (2005) Phylogenetic position of the copepod-infesting parasite *Syndinium turbo* (Dinoflagellata, Syndinea). *Protist*, 156: 413–423.

Skovgaard, A., Massana, R., and Saiz, E. (2006) Parasites of the genus *Blastodinium* (Blastodiniphyceae) are archetypical dinoflagellates according to SSU rDNA sequences and dinospore morphology. *Journal of Phycology*, 42 (in press).

Sournia, A. (1984) Classification et nomenclature de divers dinoflagellés marins (Dinophyceae). *Phycologia*, 23: 345–355.

Takishita, K., Ishikura, M., Koike, K., and Maruyama, T. (2003) Comparison of phylogenies based on nuclear-encoded SSU rDNA and plastid-encoded *psbA* in the symbiotic dinoflagellate genus *Symbiodinium*. *Phycologia*, 42: 285–291.

Takishita, K. and Uchida, A. (1999) Molecular cloning and nucleotide sequence analysis of psbA from dinoflagellates: origin of the dinoflagellate plastid. *Phycological Research*, 47: 207–216.

Tamura, K. and Nei, N. (1993) Estimation of the number of nucleotide substitutions in the control region of mitochondrial DNA in humans and chimpanzees. *Molecular Biology and Evolution*, 10: 512–526.

Tamura, M., Shimada, S., and Horiguchi, T. (2005) *Galeidinium rugatum* gen. et sp. nov. (Dinophyceae), a new coccoid dinoflagellate with a diatom endosymbiont. *Journal of Phycology*, 41: 658–671.

Tangen, K., Brand, L.E., Blackwelder, P.L., and Guillard, R.R.L. (1982) *Thoracosphaera heimii* (Lohmann) Kamptner is a dinophyte: observations on its morphology and life cycle. *Marine Micropalaeontology*, 7: 193–212.

Taylor, F.J.R. (1980) On dinoflagellate evolution. *BioSystems*, 13: 65–108.

Taylor, F.J.R. (2004) Illumination or confusion? Dinoflagellate molecular phylogenetic data viewed from a primarily morphological standpoint. *Phycological Research*, 52: 308–324.

Tengs, T., Dahlberg, O.J., Shalchian-Tabrizi, K., Klaveness, D., Rudi, K., Delwiche, C.F., and Jakobsen, K.S. (2000) Phylogenetic analyses indicate that the 19hexanoyloxy-fucoxanthin-containing dinoflagellates have tertiary plastids of haptophyte origin. *Molecular Biology and Evolution*, 17: 718–729.

Triemer, R.E. (1982) A unique mitotic variation in the marine dinoflagellate *Oxyrrhis marina* (Pyrrhophyta). *Journal of Phycology*, 18: 399–411.

Turmel, M., Otis, C., and Lemieux, C. (1999) The complete chloroplast DNA sequence of the green alga *Nephroselmis olivacea*: insights into the architecture of ancestral chloroplast genomes. *Proceedings of the National Academy of Sciences, USA*, 96: 10248–10253.

Watanabe, M.M., Takeda, Y., Sasa, T., Inouye, I., Suda, S., Sawaguchi, T., and Chihara, M. (1987) A green dinoflagellate with chlorophylls *a* and *b*: morphology, fine structure of the chloroplast and chlorophyll composition. *Journal of Phycology*, 23: 382–389.

West, G.S. and Fritsch, F.E. (1927) *A Treatise on the British Freshwater Algae*, Cambridge University Press, Cambridge.

Wilcox, L.W. and Wedemayer, G.J. (1984) *Gymnodinium acidotum* Nygaard (Pyrrhophyta), a dinoflagellate with an endosymbiotic cryptomonad. *Journal of Phycology*, 20: 236–242.

Wilcox, L.W. and Wedemayer, G.J. (1985) Dinoflagellate with blue-green chloroplasts derived from an endosymbiotic eukaryote. *Science*, 227: 192–194.

Zhang, H., Bhattacharya, D., and Lin, S. (2005) Phylogeny of dinoflagellates based on mitochondrial cytochrome b and nuclear small subunit rDNA sequence comparisons. *Journal of Phycology*, 41: 411–420.

Zhang, H. and Lin, S. (2005) Mitochondrial cytochrome b mRNA editing in dinoflagellates: possible ecological and evolutionary associations? *Journal of Eukaryotic Microbiology*, 52: 538–545.

Zhang, Z., Green, B.R., and Cavalier-Smith, T. (1999) Single gene circles in dinoflagellate chloroplast genomes. *Nature*, 400: 155–159.

Zingmark, R.G. (1970) Sexual reproduction in the dinoflagellate *Noctiluca miliaris* Suriray. *Journal of Phycology*, 6: 122–126.

13 Molecular genetics and the neglected art of diatomics

David G. Mann and Katharine M. Evans

CONTENTS

ABSTRACT

Molecular systematic methods have been applied to all levels of problem in the diatoms. A possible sister group to the diatoms has been identified, the primary evolutionary radiation has been shown to be among centric diatom lineages, and the systematic positions of some problematic genera have been established. Certain structural characteristics previously used to diagnose higher taxa have been shown to exhibit homoplasy, and examination of selected complexes of diatoms has shown that cryptic and semicryptic species are probably widespread. Population genetics show that some species are panmictic over large geographical areas, while others are strongly differentiated at the population level, either geographically or temporally. Ongoing research addresses the major biogeographical question of whether diatom species are ubiquitous. On the negative side, most gene trees are disappointing because of the low statistical support for basal nodes in group-wide analyses; independent data sources (cytology, fossils) also mostly fail to discriminate between competing hypotheses of relationships. Constraints on progress include the relative poverty of culture collections and lack of multi-gene sequence data. The worst problem is the poor state of "diatomics" (holistic information about diatom species): for most diatoms, low-grade information about cell wall morphology is all that is available. Improved phylogenies are therefore often unedifying.

INTRODUCTION

Diatoms are one of the least difficult groups of plants to recognize: (1) They are unicellular or colonial protists with a special type of cell wall (Figure 13.1 [1]), consisting of two overlapping halves (thecae). (2) The thecae consist of two larger elements at opposite ends of the cell (valves) and strips (girdle bands) or scales covering the region in between (Figure 13.1 [2] and Figure 13.2 [14]). (3) The cell wall is almost always silicified. (4) During the cell cycle, cells expand along one axis only (unidirectional growth, as in the fission yeast *Schizosaccharomyces*) (Figure 13.1 [3] and Figure 13.2 [9]). (5) Cell expansion is accommodated by addition of material to the edge of the inner of the two thecae, in the central, overlapping region of the wall. (6) Cells achieve cytokinesis by simple cleavage (Figure 13.1 [3]). (7) New valves are produced within the daughter cells after cytokinesis while they are still confined within the parent cell wall (Figure 13.1 [3]). (8) Therefore, in most (but not all) diatoms, the method of cell division leads to reduction in the mean cross-sectional area and mean overall size (and often the shape) of the cells during vegetative growth. (9) Size restitution takes place via a special cell, the auxospore (Figure 13.1 [6] and Figure 13.1 [8]), which is usually zygotic. (10) Sexual reproduction (Figure 13.1 [4] through Figure 13.1 [7]) is always associated with auxosporulation. (11) Vegetative cells are diploid and the life cycle is diplontic. All of these features are unusual in plants and most are unique. A detailed account of diatom characteristics is given by Round et al. (1990).

Structurally, there are two main types of diatoms. In centric diatoms, systems of ribs and pores radiate out from a central (occasionally eccentric) ring, the "annulus" (Figure 13.1 [10], Figure 13.1 [13], and Figure 13.3 [15]), though this is sometimes difficult to discern because of extra, superposed layers of silica (Figure 13.2 [14]). In pennate diatoms, on the other hand, the valve pattern is organized bilaterally, with systems of ribs and pores arranged about a longitudinal bar, called the "sternum" (Figure 13.3 [17] and Figure 13.3 [18]), which, like the annulus, is usually but not always central. Most pennate diatoms are elongate (Figure 13.1 [1], Figure 13.3 [17], and Figure 13.3 [18]). Centric diatoms are sometimes elongate (Figure 13.2 [11] and Figure 13.2 [12]), but generally less so than pennate diatoms, and most have valves that are circular or shortly elliptical, triangular, or polygonal. Although sexual reproduction is still undocumented in several major groups of diatoms, it appears that centric diatoms are primitively and usually oogamous, with uniflagellate sperm (Figure 13.1 [5] and Figure 13.1 [6]). Pennate diatoms, on the other hand, are usually isogamous (Figure 13.1 [7]), although the gametes may be differentiated biochemically and behaviourally (e.g. Chepurnov et al., 2004). Centric and pennate diatoms were recognized as separate orders or classes for most of the 20th century (e.g. Hustedt, 1927–1966; Fritsch, 1935).

Although diatoms were discovered at the beginning of the 18th century (Anonymous, 1703), they were not studied in detail until over a century later. Throughout the subsequent history of diatom systematics, technology has been an important driver. First there was the slow perfection of optical microscopes, essentially complete by 1900 and allowing some remarkable studies of living diatoms (Lauterborn, 1896) and their shells (e.g. Hustedt, 1928). Then there was a period of some stability from 1900 until World War II, when transmission electron microscopy (TEM) became generally available (e.g. Kolbe and Gölz, 1943; Desikachary, 1952). Scanning electron microscopy (SEM) started to make an important contribution in the late 1960s (Hasle, 1968; Round, 1970), and the more widespread availability of computers in the 1980s allowed implementation of hitherto impractical methods of classification and analysis (Stoermer and Ladewski, 1982; Williams, 1985; du Buf and Bayer, 2002). The pace of change quickened recently, with the establishment of novel means of communication via the Internet (e.g. see Mann et al., 2006) and the introduction (Medlin et al., 1988, 1991) and semi-automation of molecular genetic analysis.

Almost every previous technological advance has led to improvements in our ability to discern, interpret, and describe physical structures: we now see more than we did. By contrast, molecular genetic techniques, although they characterize parts of real molecules, have an almost metaphysical outworking: the data they produce are generally not valued by taxonomists for what they are in themselves, only

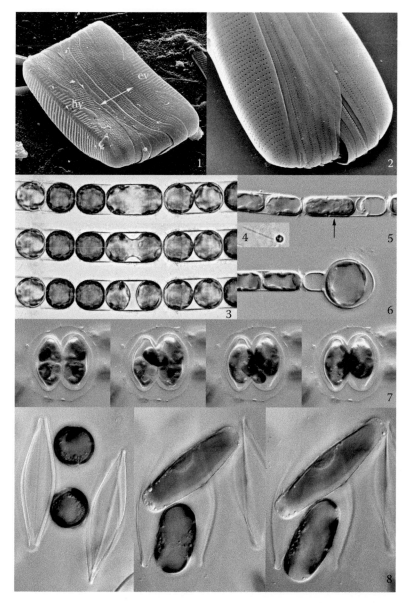

FIGURE 13.1 (1) Whole frustule of *Eunotia*, showing the two valves (hv and ev = hypovalve and epivalve) separated by sets of girdle bands (arrow). Note also the two short raphe slits on each valve (cf. 8). (2) Slightly disrupted frustule of *Parlibellus*, showing the overlap between the epitheca (left) and hypotheca, each of which consists of a valve and a set of girdle bands attached to it. (3) *Melosira*: a cell undergoing cytokinesis (after cell elongation and mitosis); the other cells in the chain are in interphase and are much shorter along the pervalvar axis. Note that pairs of cells from recent cell divisions are still contained within the parental cell wall, demonstrating the internalized cell wall formation present throughout the diatoms. (4) Uniflagellate sperm of *Actinocyclus*. (5) Differentiated auxospore mother-cell (arrow) of *Melosira* (such cells are usually oogonia in centric diatoms, but some develop into auxospores without fertilization), with an apochlorotic residual cell (at right). (6) Expanded auxospore of *Melosira*, still attached to small vegetative cells. (7) Plasmogamy in the morphologically isogamous pennate diatom *Placoneis*. (8) Development of auxospores in *Craticula*. At left, two spherical zygotes lie between the empty frustules of the gametangia. The zygotes differentiate into auxospores, which are constrained to expand along a single axis (right two photographs) through progressive formation of a perizonium of transverse silicified bands. The cells and frustules shown here and in the other plates are of moderate size for diatoms, i.e. 20–200 μm in maximum dimension.

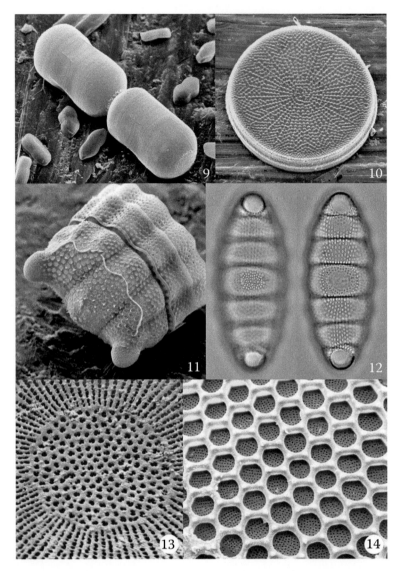

FIGURE 13.2 (9) Frustules of the radial centric diatom *Melosira*, in girdle view. (10) Frustule of the radial centric *Actinocyclus*, with radiating rows of pores. (11) The centric diatom *Biddulphia* as seen with SEM—a theca. (12) The centric diatom *Biddulphia* as seen with light microscopy—isolated valves. The valves are bipolar and lanceolate, but the striation is nevertheless radially organized. (Courtesy of Prof. F.E. Round.) (13) *Chrysanthemodiscus*: valve centre, showing ribs and lines of pores radiating from a ring (the annulus), within which pores are scattered ± evenly but irregularly. (Courtesy of Prof. F.E. Round.) (14) *Triceratium*: radiating rows of pores (an annulus can just be distinguished centrally) obscured by a superimposed system of hexagonal chambers. (Courtesy of Prof. F.E. Round.)

for what they imply about relationships between different organisms. A matrix of nucleotide or amino acid data is used to produce tree diagrams expressing our best estimates of evolution in selected genes. These "gene trees" are interpreted, with more or less care, as phylogenetic trees of organisms, and these in turn may be used as a basis for estimating the evolution and significance of morphological, cytological, reproductive, or other characters comprising the phenotype. It is rare that comparative studies of gene sequences by systematists are used to further our understanding of the genes and gene products themselves, rather than of the organisms that contain them; an exception is the analysis of

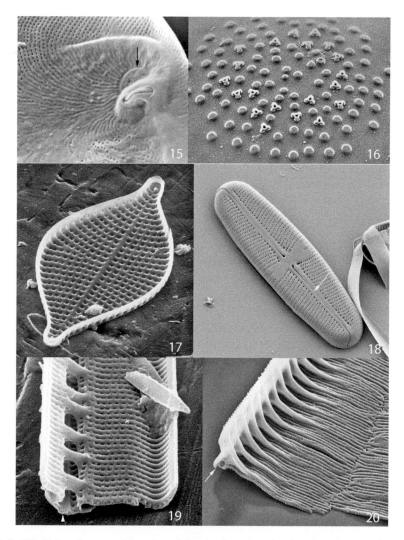

FIGURE 13.3 (15) *Guinardia*: part of the edge of the circular valve, showing the submarginal annulus (arrow). The slit-like structure is a rimoportula. (Courtesy of Prof. F.E. Round.) (16) Thalassiosirales valve: fultoportulae, each with a central pore and three satellite pores (scattered among radiating rows of valve pores, which are occluded by fine sieve-like membranes). (17) Interior of a valve of the araphid pennate diatom *Rhaphoneis*: transverse ribs and lines of pores extend out bilaterally from a principal longitudinal rib, the sternum. (18) *Sellaphora* valve, exterior, showing two raphe slits (e.g. arrow) incorporated within the sternum. (19) Interior of broken *Nitzschia* (Bacillariales) valve: a line of short, rib-like bridges of silica link the two sides of the valve together beneath the raphe (arrow). (20) Interior of broken *Cymatopleura* (Surirellales) valve, with flange-like fibulae subtending the raphe (arrow). Though similar in position, form, and presumably function, the fibulae of Bacillariales and Surirellales appear to have evolved independently.

seed-plant Rubisco by Kellogg and Juliano (1997). Thus, for the first time in the history of systematics, there is beginning to be a significant divergence between the kinds of characters we use to describe and recognize taxa and the characters that are most effective for determining relationships.

Relative to other groups of unicellular eukaryotes, diatoms have long had a sophisticated (≠ correct) taxonomy. The only groups of microalgae that can compete with them are the placoderm desmids and the armoured dinoflagellates. Diatom cell walls are easily preserved because of their silica content, and their shape, size, and patterns offer many characteristics that taxonomists can examine with the light microscope. These characteristics are remarkably constant within species and populations, once allowance has been made for the changes (which are generally highly predictable) that accompany size reduction during the life cycle; indeed, the huge industry of diatom palaeoecology (e.g. Stoermer and Smol, 1999) would otherwise have been impossible. By the 1970s, there were enough names of genera, species, varieties, and forms to fill an eight-volume, 4600-page catalogue (VanLandingham, 1967–1978). The introduction of SEM (an interim summary was provided by Round et al., 1990) and use of protoplast (e.g. Cox, 1987, 1996) and reproductive characters (e.g. Mann, 1989) led to refinements in both classification and identification. Evolutionary schemes were produced using both informal (e.g. Simonsen, 1979) and formal (e.g. Williams, 1985; Kociolek and Stoermer, 1988) methods.

MOLECULAR APPROACHES TO DIATOM SYSTEMATICS

Molecular approaches were introduced into diatom systematics by Medlin et al. (1988, 1991) during a study of *Skeletonema* that appeared to reveal a semicryptic species previously included within *S. costatum*. The gene used by Medlin et al. was the small-subunit (SSU = 18S) component of nuclear rDNA, and Medlin has continued to work mostly with this gene.

As will be seen from Table 13.1, molecular systematics of diatoms have been dominated (roughly 3:1) by analyses of nuclear rDNA, particularly the SSU gene and partial sequences of the large-subunit (LSU) gene. The SSU gene is generally regarded as a slowly evolving region and its use in diatoms reflects this: SSU studies have been focused on high-level relationships within the diatoms, or on the relationships of diatoms to other groups of eukaryotes (Table 13.1), although available data show that there are usually a few fixed differences even between closely related species (Behnke et al., 2004). *Rbc*L is another gene that evolves relatively slowly, and the primary inspiration for recent re-classifications of angiosperm families was the large *rbc*L matrix published by Chase et al. (1993). *Rbc*L has also been used by several authors (e.g. Daugbjerg and Guillou, 2001) to examine relationships among heterokonts. We therefore started to build a large data set of diatom sequences that includes representatives of many diatom orders but focuses on the raphid group (Mann et al., 2001). This remains largely unpublished but a few sequences were used by Jones et al. (2005). Edgar and Theriot (2004) and Amato et al. (2007) are among others to make extensive use of *rbc*L in diatoms. So far, there have been no multi-gene studies of evolution within the diatoms comparable to those in green algae (e.g. Nozaki et al., 2003).

Sequence-based methods have also been used to examine relationships between closely related species, particularly in the marine planktonic diatoms *Pseudo-nitzschia* and *Skeletonema* (Table 13.1), while Edgar and Theriot (2004) have made a combined analysis of molecular and morphological data to investigate evolution in the chain-forming centric diatom *Aulacoseira*. The original motive for studying *Pseudo-nitzschia* was economic: some *Pseudo-nitzschia* species produce the neurotoxin domoic acid (e.g. Fryxell and Hasle, 2003), which causes "amnesic shellfish poisoning." It was therefore important to determine the capacity of different species to produce domoic acid and to develop fast and reliable molecular methods to identify *Pseudo-nitzschia* species, which have few characters that can be scored reliably in the light microscope. Though the induction and mechanisms of toxin production remain important research topics, we doubt whether they are now the primary justification for molecular systematics research in *Pseudo-nitzschia*, especially given that domoic acid is now known to be produced by at least one species of the huge related genus *Nitzschia* (>500 spp) and reportedly also by a species of the phylogenetically distant genus *Amphora* (Fryxell and Hasle, 2003). *Pseudo-nitzschia* has simply become an excellent model system for examining microevolution in diatoms, because of the wealth of data about molecular systematic relationships (Table 13.1), distributions (Hasle et al., 1996; Hasle and Syvertsen, 1996; Hasle, 2002), the mating system (Davidovich and Bates, 1998; Chepurnov et al., 2005; Amato et al., 2007), ultrastructure (e.g. Hasle, 1965; Lundholm et al., 2002b), and ecology (e.g. Rines et al., 2002; Cerino et al., 2005).

Until recently, sequence-based studies of inter- and intra-specific variation in *Pseudo-nitzschia*, and more recently in *Skeletonema* (Sarno et al., 2005), have relied mostly on the hypervariable region of LSU rDNA. The ITS region, whose extensive use for phylogenetic inference at or below

TABLE 13.1

Molecular Systematic Studies of Diatoms

The list includes all those we have located that deal specifically with systematic relationships among diatoms, together with some key studies of larger groups (e.g. heterokonts) in which diatom sequences are a significant component and an eclectic choice of studies not focused on diatoms that could provide a good starting point for future phylogenetic studies. The availability of the nucleotide or amino acid alignment is recorded and whether the clones used (new clones only, for analyses using a mixture of new and previously published data) were available at the time in major culture collections (e.g. CCAP, CCMP, as opposed to university or private collections); these cultures may or may not still be available (several at least are not). We assessed specification of vouchers as best we could from the text or tables; remarks like "permanent slides were made" were not accepted as a specification. "Genbank" under cultures or voucher specification indicates that the study used previously published diatom sequences available in Genbank. D, P, L, and B refer to distance, parsimony, maximum-likelihood, and Bayesian methods of analysis. Subscripts e and b indicate exhaustive analysis (for parsimony) or bootstrapping, respectively. Nucl = nucleotide-based analysis; Prot = amino acids in alignment or analysis based on translated sequence of amino acids. In describing the level of the study, we generally disregarded the outgroups.

Gene	Location of Gene	Bases in Alignment	Parsimony Informative	Number of Diatom Taxa (Accessions)	Alignment Deposited	Cultures in Public Culture Collections	Voucher Specification	Methods of Analysis	Level of Study	Ref.
SSU rDNA	Nucleus	1798	7	2 (4)	No	Yes	No	**Nucl:** P_e	*Skeletonema*	Medlin et al. (1991)
SSU rDNA	Nucleus	1565/1720	275/327	11 (11)	No	Some	No	**Nucl:** D, P_b	Heterokonts	Medlin et al. (1993)
SSU rDNA	Nucleus	1767	?	10 (12)	Printed	No	No	**Nucl:** D, P_b	Diatoms	Douglas et al. (1994)
ITS1-5.8S–ITS2 rDNA	Nucleus	725	?	4 (7)	Printed	No	No	**Nucl:** P_e	Thalassiosirales	Zechman et al. (1994)
Partial LSU rDNA	Nucleus	?	?	5 (13)	No	No	No	**Nucl:** P_b	*Pseudo-nitzschia*	Scholin et al. (1994)
Partial LSU rDNA	Nucleus	275	105	5 (5)	Printed	No	No	**Nucl:** D, P_b	Eukaryotes	Philippe et al. (1994)
Partial LSU rDNA	Nucleus	400	?	8 (8)	Printed	Yes	No	**Nucl:** D_b, P_b	Diatoms	Sorhannus et al. (1995)
SSU rDNA	Nucleus	1597	?	6 (6)	No (supply on request)	GenBank	GenBank	**Nucl:** P_b comb mol/morph	Selected eukaryotes	Saunders et al. (1995)
*tuf*A	Plastid	?740	?	3	No (available from C. Delwiche)	No	No	**Prot:** P_b, **Nucl:** D_b, P_b, L_b	Life	Delwiche et al. (1995)
SSU rDNA	Nucleus	874	NA	5 (5)	No	GenBank	GenBank	**Nucl:** D_b	Selected eukaryotes	Sorhannus (1996)
SSU rDNA	Nucleus	?	?	11 (11)	?No longer available	No	No	**Nucl:** D_b	Chromalveolates	Van de Peer et al. (1996)

(Continued)

TABLE 13.1 (CONTINUED)
Molecular Systematic Studies of Diatoms

Gene	Location of Gene	Bases in Alignment	Parsimony Informative	Number of Diatom Taxa (Accessions)	Alignment Deposited	Cultures in Public Culture Collections	Voucher Specification	Methods of Analysis	Level of Study	Ref.
SSU rDNA	Nucleus	1711	?	29 (30)	No	GenBank	GenBank	**Nucl:** D, L	Heterokonts	Kooistra and Medlin (1996)
SSU rDNA	Nucleus	1739	528	29 (29)	No	Yes	Private collection	**Nucl:** D_b, P_b, L	Diatoms	Medlin et al. (1996a)
SSU rDNA	Nucleus	1739	528	29 (29)	No	Some	Private collection	**Nucl:** D_b, P_b, L	Diatoms	Medlin et al. (1996b)
rbcL	Plastid	? c. 1500	?	2 (2)	No	No	No	**Nucl:** P_b	Life	Chesnick et al. (1996)
rbcL–rbcS spacer	Plastid	43	?	2 (2)	Printed	No	No	**Nucl:** visual comparison	Eukaryotic plants	Chesnick et al. (1996)
rbcS	Plastid	? c. 417	?	2 (2)	No	No	No	**Nucl:** P_b	Life	Chesnick et al. (1996)
LHC[1]	Nucleus	**Prot:** 184	?	2 (2)[2]	Printed	GenBank	GenBank	**Nucl:** (D), F_b	Chromophytes	Caron et al. (1996)
Partial LSU rDNA	Nucleus	804	NA	7 (9)	Printed	?No	No	NA	NA	Miller and Scholin (1996)
SSU rDNA	Nucleus	?	?	25 (25)	No	No	No	**Nucl:** D_b	Heterokonts	Medlin et al. (1997a)
tufA	Plastid	?	?	3 (3)	No	No	No	**Nucl:** D_b	Life	Medlin et al. (1997a, 2000)
16S rDNA	Plastid	?	?	16 (16)	No	No	No	**Nucl:** D_b	Life	Medlin et al. (1997a)
rbcL	Plastid	?	?	3 (3)	No	No	No	**Nucl:** D_b	Life	Medlin et al. (1997a, 2000)
rbcL	Plastid	954[3]	?	4 (4)	No	Yes	No	**Nucl:** P_b, L	Heterokonts	Daugbjerg and Andersen (1997)
SSU rDNA	Nucleus	?	?	28 (30)	No	No	No	**Nucl:** D_b, L_b	Heterokonts	Medlin et al. (1997b)
SSU rDNA/LSU rDNA	Nucleus	Previously published	NA	11 (11)	Previously published	GenBank	GenBank	**Nucl:** D	Diatoms	Sorhannus (1997)
SSU rDNA	Nucleus	c. 1740?	?	9 (12)	No	No	No	**Nucl:** P	Thalassiosirales	Medlin and Simon (1998)

psbV–trnR–ORF–trnM– rpl19– petF	Plastid	1881	NA	2 (2)	Printed	Yes	No	**Nucl:** (*PetF*) D_b	Life	Gueneau et al. (1998); see also Gueneau et al. (1999)
SSU rDNA	Nucleus	507	?	8 (8)	Part printed	No	No	**Nucl:** L_b	*Aulacoseira*	Shcherbakova et al. (1998)
SSU rDNA	Nucleus	1601	?	7 (7)	No	GenBank	GenBank	**Nucl:** D_b, P_b	Heterokonts	Guillou et al. (1999)
FCP	Nucleus	**Prot:** 138	?	3 (3)	Printed	Yes	No	**Prot:** L_b	Eukaryotic plants	Eppard et al. (2000)
cox1	Mitochondrion	**Prot:** 353	?	9 (9)	No	Yes	No	**Prot:** D_b, P_b	Chromophytes	Ehara et al. (2000a, 2000b)
cox1 : intron reverse transcriptase/maturase	Mitochondrion	**Prot:** 146/98	NA	2 (2)	Printed	Yes	No	**Prot:** D_b	Eukaryotes	Ehara et al. (2000b)
SSU rDNA	Nucleus	1739	528	85 (85)	No	No	No	**Nucl:** D_b, P	Diatoms	Medlin et al. (2000)
16S rDNA	Plastid	?	?	12 (12)	No	No	No	**Nucl:** D	Life	Medlin et al. (2000)
TBl/Gap	Nucleus	**Prot:** ~250/~300	?	2 (2)	No	No	No	**Prot:** D_b, P_b	Life	Liaud et al. (2000)
rbcL	Plastid	1373	?	6 (6)	No	Some	No	**Nucl:** D_b, P_b, L_b	Heterokonts	Daugbjerg and Guillou (2001)
SSU rDNA	Nucleus	1702	236	22 (22)	No	Mostly no	No	**Nucl:** P_b, L_b	Diatoms	Beszteri et al. (2001)
5.8S + LSU/5.8S + LSU + SSU rDNA	Nucleus	? c. 3500/5300	NA	3 (3)	No	No	No	**Nucl:** D_b, (P, L)	Eukaryotes	Ben Ali et al. (2001)
SSU rDNA	Nucleus	1635	?	7 (7)	No	Yes	No	**Nucl:** D	*Thalassiosira*	Armbrust and Galindo (2001)
Partial -*tubulin*	Nucleus	671	?	1 (1)	Printed	Yes	No	**Nucl:** D	*Thalassiosira weissflogii*	Armbrust and Galindo (2001)
Partial *Sig1*	Nucleus	645	?	1 (7)[4]	Printed	Yes	No	**Nucl:** D	*Thalassiosira weissflogii*[5]	Armbrust and Galindo (2001)
SSU rDNA	Nucleus	1554	?	20 (20)	No	No	No	**Nucl:** P_b	Diatoms	Mayama and Kuriyama (2002)
Partial LSU rDNA	Nucleus	872	208	43 (56)	No	Some	No	**Nucl:** D_b, P_b, L_b	Bacillariales	Lundholm et al. (2002a)
Partial LSU rDNA	Nucleus	872	?	43 (60)	No	No	Yes (new taxa only)	**Nucl:** (D), P_b, L	Bacillariales	Lundholm et al. (2002b)
Partial LSU rDNA	Nucleus	856/570	~30/29	11 (23)	No	No	No	**Nucl:** D_b, P_b	*Pseudo-nitzschia*+	Orsini et al. (2002)

(Continued)

TABLE 13.1 (CONTINUED)
Molecular Systematic Studies of Diatoms

Gene	Location of Gene	Bases in Alignment	Parsimony Informative	Number of Diatom Taxa (Accessions)	Alignment Deposited	Cultures in Public Culture Collections	Voucher Specification	Methods cf Analysis	Level of Study	Ref.
Partial LSU rDNA	Nucleus	872	?	45 (59)	No	No	Yes (new taxa only)	**Nucl:** P_b, L	Bacillariales	Lundholm and Moestrup (2002)
SSU rDNA	Nucleus	~1400	?	15 (15)	On request	GenBank	GenBank	**Nucl:** D_b, P_b, L	Heterokonts	Kawachi et al. (2002)
rbcL	Plastid	~965	?	4 (4)	On request	GenBank	GenBank	**Nucl:** D_b, P_{t}, L	Heterokonts	Kawachi et al. (2002; omitted codon 3)
ITS2 rDNA	Nucleus	352	?	4 (8)	No	No	No (but images in paper)	**Nucl:** matrix of similarities only	*Stephanodiscus*	Wolf et al. (2002)
Partial SSU–ITS1–5.8S–ITS2–partial LSU rDNA	Nucleus	583	?	16 (24)	No	No	Yes (new taxa only)	**Nucl:** D_b, P_b, L_b	*Pseudo-nitzschia*	Lundholm et al. (2003)
SSU rDNA	Nucleus	?–1700	412	38 (38)	No	No	No	**Nucl:** L_b	Diatoms	Kooistra et al. (2003a)
SSU rDNA	Nucleus	?–1700	NA	96 (96)	No	No	No	**Nucl:** D_b	Diatoms	Kooistra et al. (2003b)
rpoA	Plastid	909	314	8 (8)	No	Yes	No	**Nucl:** D_b, P_{-}, L_b, B	Diatoms	Fox and Sorhannus (2003)
ITS2 rDNA	Nucleus	193	?	5 (7)	Printed	GenBank	GenBank	**Nucl:** D_b, P_t, L_b	*Stephanodiscus*	Wolf (2004)
SSU rDNA	Nucleus	1807	?	9 (9)	No[6]	GenBank	GenBank	**Nucl:** D_b, P_t, L	Eukaryotes	Kühn et al. (2004)
SSU rDNA	Nucleus	?–1800	?	?123 (123)[7]	No	Yes	No	**Nucl:** D	Diatoms	Sinninghe Damsté et al. (2004)
SSU rDNA	Nucleus	1764	764	109 (109)	No	Some	No	**Nucl:** D_b, P_t, B	Heterokonts or diatoms	Medlin and Kaczmarska (2004)
16S rDNA	Plastid	1422	526	16 (16)	No	Some	No	**Nucl:** D_b	Life	Medlin and Kaczmarska (2004)
SSU rDNA	Nucleus	1946	410	35	No	No	Yes (new taxa only)	**Nucl:** (P), L_b^8	Diatoms	Kooistra et al. (2004)

Gene region	Nucleus/plastid/ nucleus + plastid							Data	Taxon	Reference
SSU rDNA/rbcL/SSU rDNA + rbcL	nucleus + plastid	?/c. 1200/?	126/99/225	15 (23)/11 (18)/15 (23)	No	Some	Yes; supplementary material on Internet	**Nucl:** P_b	*Aulacoseira*	Edgar and Theriot (2004)
SSU rDNA	Nucleus	1698	215	26 (26)[9]	Yes	Some	Yes	**Nucl:** D_b, P_b, L_b	Pennate diatoms	Behnke et al. (2004)
ITS rDNA/5.8S rDNA	Nucleus	156/155	55/10	8 (13)[10]	Yes	No	Yes	**Nucl:** D_b, P_b, L_b	*Sellaphora*	Behnke et al. (2004)
ITS1-5.8S-ITS2 /LSU rDNA	Nucleus	988/?	157/?	1 (70)/2 (10)	No	No	No	**Nucl:** D_b, (P)	*Pseudo-nitzschia*	Orsini et al. (2004)
SSU rDNA	Nucleus	—[11]	?	126 (126)	Yes	GenBank	GenBank	**Nucl:** P	Diatoms	Sorhannus (2005)
rbcL	Plastid	?1382	?	10 (10)[12]	No	No	Some	**Nucl:** D_b, P_b, L_b	Chromalveolates	Tamura et al. (2005)
Partial *sit* (silicon transporters genes)	Nucleus	318	NA	4 (4)[13]	Na	No	No	**Nucl:** pairwise similarity matrix	Diatoms	Sherbakova et al. (2005)
Partial SIT (silicon transporters)	Nucleus	**Prot:** ?324	?	7 (7)[14]	No	No	No	**Prot:** P_b	Pennate and multipolar centric diatoms	Thamatrakoln and Hildebrand (2005)
SSU rDNA	Nucleus	1824	297	18 (36)	Yes	Some	Yes (but new taxa only)	**Nucl:** (P)[15], L_b	*Skeletonema*	Sarno et al. (2005)
Partial LSU rDNA	Nucleus	785	122	10 (35)	Yes	Minority; most at SZN	Yes (but new taxa only)	**Nucl:** (P), L_b[16]	*Skeletonema*	Sarno et al. (2005)
Partial LSU rDNA	Nucleus	868	55	16 (55)[17]	No	No	No	**Nucl:** D_b, P_b	Bacillariales	Cerino et al. (2005)
SSU rDNA	Nucleus	~ 1770	?	10(27)[18]	No	Yes	No	**Nucl:** P_b	*Skeletonema*	Alverson and Kolnick (2005)
ITS1–5.8S–ITS2	Nucleus	982	?	2 (13)[19]	No (on request)	No (on request)	No (on request)	**Nucl:** D_b, P_b	*Cyclotella*	Beszteri et al. (2005)
Partial LSU/SSU rDNA	Nucleus	?c. 600/?c. 1800	?/?	2[20](20)/2 (8)	No (on request)	No (on request)	No (on request)	**Nucl:** D_b, P_b/—	*Cyclotella*	Beszteri et al. (2005)
HSP90/actin + alpha-tubulin + beta-tubulin +HSP90	Nucleus	**Prot:** 516/1554	?	2 (2)	No (on request)	Yes	No	**Nucl:** D_b, P_b, L_b	Eukaryotes	Harper et al. (2005)
FBA[21]	Nucleus	**Prot:** c. 400	?	3 (3)[22]	No	Some	No	**Prot:** L_b	Life	Kroth et al. (2005)
rbcL	Plastid	1297	?	16 (16)	No	No	Yes	**Nucl:** P_b, L_b, B	Raphid diatoms	Jones et al. (2005)
Partial LSU rDNA	Nucleus	797	?	8 (56)	No	Some	No	**Nucl:** D, L_b	*Skeletonema*	Godhe et al. (2006)
ITS1/ITS2	Nucleus	~230/~300/~694	?	1 (24)/1 (24)	No	Some	No	**Nucl:** D	*Skeletonema marinoi*	Godhe et al. (2006)

(Continued)

TABLE 13.1 (CONTINUED)
Molecular Systematic Studies of Diatoms

Gene	Location of Gene	Bases in Alignment	Parsimony Informative	Number of Diatom Taxa (Accessions)	Alignment Deposited	Cultures in Public Culture Collections	Voucher Specification	Methods of Analysis	Level of Study	Ref.
SSU rDNA	Nucleus	?	?	156 (156)	No	No	No	**Nucl:** D_b	Diatoms	Kooistra et al. (2006)
SSU rDNA	Nucleus	1814	343	33 (43)	No	Mostly	No	**Nucl:** D_b, P_b, L, B	Thalassiosirales	Kaczmarska et al. (2006)
ITS1–5.8S–ITS2 rDNA	Nucleus	583	?	19 (38)	No	No	Some	**Nucl:** D_b, P_b, L_b, B	Pseudo-nitzschia	Lundholm et al. (2006)
SSU rDNA	Nucleus	?	?	? (222)	No[23]	No	No	**Nucl:** D_b, P_E, B	Diatoms	Sims et al. (2006)
Ice-binding proteins	?Nucleus	**Prot:** 277	NA	2 (2)	Printed	No	No	NA	Life	Janech et al. (2006)
SSU rDNA/rbcL	Nucleus	?/?	?/?	43 (46)/11 (11)	No	No	Image	**Nucl:** D_b, P_E, L_b	Diatoms and chromalveolates	Horiguchi and Takano (2006)
SIT[24]	Nucleus	**Prot:** 615	NA	10 (11)	Printed	Some	No	**Nucl:** L_b	Diatoms	Thamatrakoln et al. (2006)
SIG1	Nucleus	**Prot:** 197	NA	4 (64)	Yes	GenBank	GenBank	NA	Thalassiosira	Sorhannus and Kosakovsky Pond (2006)
SSU rDNA	Nucleus	1465	?	7 (7)	No	GenBank	GenBank	**Nucl:** L, B	Life	Berney and Pawlowski (2006)
SSU rDNA	Nucleus	1519	NA	32 (32)	No	GenBank	GenBank	**Nucl:** D_b	Heterokonts	Cavalier-Smith and Chao (2006)
SSU rDNA/rbcL	Nucleus/plastid	1345/1206	21/14	11 (11)	No	?Yes	Image	**Nucl:** D_b, P_m, L_b	Diatoms	Jung et al. (2006)
SSU rDNA	Nucleus	752	2034	c. 155 (189)	Yes	GenBank	GenBank	**Nucl:** P_b, L, B	Diatoms	Alverson et al. (2006)
ITS1–5.8S–ITS2	Nucleus	~720	?	1 (7)	No	No	± Yes	**Nucl:** D_b, P_m, L_b	Eunotia bilunaris	Vanormelingen et al. (2007)
Partial LSU rDNA	Nucleus	?c. 870	?	23 (25)	No	Some	Yes and images	**Nucl:** D_b, P, L_b	Bacillariaceae	Trobajo et al. (2006)
Partial LSU rDNA/ITS1–5.8S–ITS2 rDNA/rbcL	Nucleus and plastid	?/?/?	?	8 (92/61/78)	No	No	No	**Nucl:** L_b	Pseudo-nitzschia	Amato et al. (2007)

[1] Coding for fucoxanthin chlorophyll a/c binding proteins.

[2] Eleven intraclonal variants (9 in *Phaeodactylum*, 2 in *Odontella*).

[3] First and second codon positions only.

[4] Several copies of the *Sig1* gene with different sequences are present within clonal isolates of *T. weissflogii*; data for 40 intraclonal gene copies were given by Armbrust and Galindo (2001).

[5] Comparative data were tabulated for two other *Thalassiosira* species.

[6] Sequences from the ARB database.

[7] It is unclear whether divergent sequences said to be of the same taxon (e.g. *Rhizosolenia setigera*, *Cylindrotheca/Nitzschia closterium*) reflect inaccurate identification, faulty taxonomy, sequencing errors, or alignment inconsistencies.

[8] Parsimony data not shown.

[9] Demes have been treated as taxa.

[10] Behnke et al. (2004) sequenced a total of 23 clonal and intraclonal rDNA variants, but only 13 were used in their phylogenetic analysis.

[11] The sequences were derived from the aligned SSU rDNA data in the European ribosomal RNA database (www.psb.ugent.be/rRNA/index.html).

[12] Includes three diatom endosymbionts of dinoflagellates.

[13] Five different *sit* sequences included for *Cylindrotheca fusiformis*.

[14] Five different *sit* sequences included for *Cylindrotheca fusiformis* and three for *Thalassiosira pseudonana*.

[15] Parsimony data not shown.

[16] Parsimony data not shown.

[17] The named taxa correspond particularly poorly to the clades evident in the tree.

[18] Because of intraclonal variation, a total of 60 different SSU sequences were analysed.

[19] Forty-six different gene sequences recorded among and within diatom clones.

[20] Twelve different gene sequences recorded among and within diatom clones.

[21] Coding for fructose-1,6-bisphosphate aldolases.

[22] Several genes in *Thalassiosira pseudonana* and *Phaeodactylum tricornutum*, producing aldolases targeted to different cellular compartments.

[23] Sequences from the ARB database.

[24] Twenty-six different *sit* genes were isolated from ten species.

the genus level in angiosperms has been critically reviewed by Álvarez and Wendel (2003), was first used in diatoms as early as 1994 (Zechman et al., 1994), but relatively little for the next 10 years (Table 13.1). Behnke et al. (2004) used ITS sequences to examine relationships in the *Sellaphora pupula* species complex, demonstrating multiple divergent copies of ITS rDNA within single clones of *Sellaphora* species and extreme difficulty in aligning both ITS1 and ITS2 among different demes (we use "deme" to refer to a group of individuals, or a population, or a group of populations of a specified taxon that share particular phenotypic, genotypic, reproductive, ecological, or other characteristics, in a slight modification of the principles explained in detail by Gilmour and Heslop-Harrison, 1954). Some of the *Sellaphora* demes have subsequently been described as taxonomic species by Mann et al. (2004). Vanormelingen et al. (2007) have also detected intra-isolate and among-isolate ITS variation in *Eunotia bilunaris*. In *Sellaphora, Pseudo-nitzschia,* and *Eunotia,* ITS relationships have been compared not only with morphological variation but also with reproductive compatibility (Behnke et al., 2004; Amato et al., 2007; Vanormelingen et al., 2007). Amato et al. (2007) have evaluated how well four genetic markers (LSU, ITS1 and ITS2 rDNA, and *rbc*L) discriminate between species in the *Pseudo-nitzschia delicatissima/P. pseudodelicatissima* complex: ITS2 variation showed the best concordance with reproductive compatibility and morphology.

The impact of molecular methods on diatom systematics will be discussed in two sections, dealing in turn with supra- and infraspecific variation. Together with many others since the Modern Synthesis (e.g. Mayr, 1942) and in accordance with the arguments summarized recently by Coyne and Orr (2004), we regard species as real, not constructed more or less arbitrarily by the human mind. Their boundaries lie at the fuzzy interface between reticulate and hierarchical relationships (see also Mann, 1999), and the process by which new species are formed (speciation) is associated with restriction and loss of gene flow (reproductive isolation). According to this model, asexual organisms do not exist as species in the same way that sexual organisms do. However, available evidence indicates that diatoms are predominantly sexual organisms with the capacity to outbreed (e.g. Chepurnov et al., 2004).

SUPRA-SPECIFIC RELATIONSHIPS

THE ORIGINS OF DIATOMS

It is difficult to make an impartial assessment of the contribution molecular analyses have made to higher-level diatom systematics. Confronted with robust, well-supported conclusions from molecular studies, cynics will say that they knew it all anyway. If the results are even slightly equivocal, sceptics will say that they are totally unconvinced. Enthusiasts take hints of bootstrap support as a justification for extravagant scenario building. The ambitious may manipulate history to imply that systematics began with the invention of the polymerase chain reaction (PCR) machine. Of course, in this paper we will try to be scrupulously fair.

As a single entity recognized at the class or divisional level, the diatoms have emerged from almost two decades of molecular research as they entered it: as a single well-defined group. No organisms regarded previously as diatoms have been found not to be diatoms, and the only "unconventional" diatoms that have been added to the group are the shell-less endosymbionts of some dinoflagellates (e.g. Tamura et al., 2005; Horiguchi and Takano, 2006). Most of these endosymbiotic diatoms seem to have been derived from the Bacillariaceae, a group of fibulate raphid diatoms that arose relatively late in diatom evolution (e.g. Medlin et al., 2000) but existed by the upper Eocene, approximately 38 Mya (Schrader, 1969). However, *Peridinium quinquecorne* has apparently jettisoned its pennate symbiont in favour of a centric diatom (*Chaetoceros*) (Horiguchi and Takano, 2006). The diatoms have been shown to be a natural group, but given the combination of characters that almost all diatoms possess (our opening paragraph), this is not a surprise.

The most fundamental question remaining, therefore, is where did this remarkably well-defined group come from, and how did they acquire their trademark characteristics? Molecular systematics have not brought us the answer. Transmission electron microscopy of flagella (present in the sperm of centric diatoms) and chloroplast structure, carbohydrate storage (as β-1,3–linked glucans), and chloroplast pigments (possession of c-type chlorophylls, fucoxanthin, diatoxanthin, and diadinoxanthin) had shown long before 1990 that diatoms belong to the heterokonts, together with brown algae, chrysophytes, xanthophytes, eustigmatophytes, oomycetes, and some other classes of autotrophic and heterotrophic protists (van den Hoek et al., 1995). However, ultrastructure and pigmentation did not tell us how different heterokont groups are related to each other, or how they are related to other autotrophic protists with secondary endosymbionts derived from red algae, viz. the Haptophyta, Cryptophyta, and Dinophyta. These questions remain largely unanswered, despite the introduction of molecular methods (e.g. Goertzen and Theriot, 2003; Harper and Keeling, 2003; Harper et al., 2005) and the discovery of several previously unrecognized heterokont classes (e.g. Andersen et al., 1998; Kawachi et al., 2002). The sequencing of the complete nuclear, plastid, and mitochondrial genomes of *Thalassiosira pseudonana* (Armbrust et al., 2004) and the earlier sequencing of the *Odontella* plastid genome (Kowallik et al., 1995) provide almost limitless opportunities for using molecular systematics to examine relationships between diatoms and other organisms (e.g. Miyagishima et al., 2004; Vinogradov et al., 2005; Kroth et al., 2005; Foth et al., 2006), but there are currently few comparable data for other autotrophic heterokonts: there are no other plastid genomes besides *Odontella* and the only other nuclear genome is from the diatom *Phaeodactylum* (Scala et al., 2002; Maheshwari et al., 2005). However, the situation will soon change dramatically, with the sequencing of the genomes of *Aureococcus*, *Ectocarpus*, *Ochromonas*, and two other diatoms: *Pseudo-nitzschia* and *Fragilariopsis*.

There is just one well-supported new conclusion to be drawn from molecular data about the origins of diatoms: thanks to Guillou et al. (1999) and Daugbjerg and Guillou (2001), we know that one group of heterokonts, the Bolidophyceae, is particularly closely related to diatoms. Unfortunately, this discovery makes scarcely any contribution to understanding the evolution of diatom characteristics, because the known bolidophytes are tiny picoplanktonic flagellates that have no walls, are not known to produce any silicified structures or to metabolize silica, and seem to share no morphological or cytological characteristics with diatoms that are not also shared with other heterokonts. Their ploidy and life cycles are unknown (though it might be expected that such organisms would be haploid, given selection for extreme small size and low nutrient quotas). Mann and Marchant (1989) suggested that the closest relatives of the diatoms are the parmophytes, which are another group of picoplanktonic autotrophs that appear to be heterokonts and, unlike the bolidophytes, produce multipartite, patterned walls of silica that resemble diatom frustules in several ways. Unfortunately, no-one with access to parmophyte material has obtained any molecular sequence data (although attempts to culture parmophytes have failed, it would surely be possible to get sequences from environmental samples). Conceivably, parmophytes are resting stages in the life cycles of bolidophytes.

EVOLUTION WITHIN THE DIATOMS

Several papers have been published recently that review the contributions of molecular systematics to our understanding of relationships *within* the diatoms (Kooistra et al., 2003b; Medlin and Kaczmarska, 2004; Alverson and Theriot, 2005; Kooistra et al., 2006; Sims et al., 2006; Alverson et al., 2006), and we will not attempt to cover all the same ground. Instead, we will highlight some notable successes and then concentrate on what we do not know and the obstacles to further progress. So, first, what do we know and is it a surprise?

1. The "centric" group is paraphyletic, but there is a monophyletic group comprising most diatoms traditionally considered to be pennates (but see Kooistra et al., 2003a), specifically, those diatoms possessing a sternum. These features are evident in almost all trees

that contain representatives of several centric and pennate orders (an exception is the analysis by Van de Peer et al. [1996], but this contained only 11 diatoms), whatever the gene used—nuclear (e.g. Medlin and Kaczmarska, 2004), plastidial (e.g. Fox and Sorhannus, 2003), or mitochondrial (Ehara et al., 2000a). But did we not know this already? Well, no, we did not. Literature searches reveal remarkably few discussions of diatom evolution, and these few express even fewer strong opinions concerning the relationship between centrics and pennates. Fritsch noted that the diatoms "appear as a sharply circumscribed group of rather highly evolved forms which afford few points of vantage either for tracing of their phylogeny or of their affinities with other groups of Algae" (1935, p. 564), though he noted that fossil evidence suggested that the centrics are older than the pennates (p. 642). Round and Crawford acknowledged that fossil evidence "certainly points to a morphological sequence, centric → araphid → raphid" (1984), but considered that the fossil record was severely biased by selective preservation and loss of earlier material through erosion and re-working of deposits (Round and Crawford, 1981). They argued for a very early, Precambrian origin of diatoms and a rapid diversification of major groups, including both centrics and pennates, from a pool of ancestral diatoms (Round and Crawford, 1981; Round, 1981). So, an origin of pennate diatoms from centrics was not universally accepted before the advent of molecular systematics. Simonsen, however, was quite definite that pennate diatoms most likely developed "at some time of the early Tertiary or late Cretaceous, and they must have developed from the Centrales ... but from which ... we cannot tell. The Pennales were suddenly simply there" (1972). Simonsen (1979) later amplified and revised his views and summarized them in a tree that shows the early diversification of diatoms to be wholly among centric lineages; the pennate diatoms are shown as evolving from the Eupodiscaceae, a group of centric diatoms with elliptical, elongate, or multiangular valves. However, although this tree was undoubtedly based on great personal knowledge of diatom morphology and fossils, it is difficult to work out from Simonsen's descriptive text how he arrived at his conclusions.

2. The fossil record appears to be less misleading than Round and Crawford (1981) thought. Judging by the branching order in gene trees (poorly supported though many of these are), which agrees reasonably well with the first appearance of major groups in fossil deposits (Sims et al., 2006), and by molecular clock calculations (Kooistra and Medlin, 1996; Medlin et al., 1996a, 1997a, 1997b), the diatoms arose and diversified in the Mesozoic, as the fossils suggest, rather than in the Precambrian (see also Berney and Pawlowski, 2006). Nevertheless, the fossil record is certainly deficient. Using geochemical markers (highly branched C_{30} isoprenoid alkenes) that are apparently specific for *Rhizosolenia* among extant diatoms, Sinninghe Damsté et al. (2004) detected the rhizosolenid lineage in 91.5 Mya deposits, whereas the first preserved frustules of the group appear nearly 20 My later. Likewise, the earliest undoubted raphid diatoms recorded so far are from the Palaeocene and appear to be *Lyrella* species (Strel'nikova, 1992, where they are recorded as "*Navicula*"), but *rbc*L and SSU rDNA trees show that *Lyrella* is not a basal lineage within the raphid group (Jones et al., 2005).

3. Valve shape has been shown to have rather more significance than seemed to be the case 20 years ago, during the heyday of SEM studies. SEM revealed that groupings based on valve shape and symmetry often conflicted with those based on valve ultrastructure, and when extra data sets were introduced (e.g. cytological, reproductive), they tended to support ultrastructure-based classifications. For example, SEM studies of the asymmetrical diatom *Amphora* revealed several species that had valve and raphe structure like those of bilaterally symmetrical *Navicula*, and the chloroplasts and auxosporulation were also *Navicula*-like. These species were therefore reclassified into the Naviculaceae, as the genus *Seminavis* (Round et al., 1990; Danielidis and Mann, 2002; Chepurnov et al., 2002). Molecular data have confirmed that *Seminavis* is close to or within *Navicula* and

other shape-groups (e.g. *Gomphonema*) have also proved artificial (e.g. see the trees of Medlin and Kaczmarska, 2004, and Jones et al., 2005, based on SSU rDNA and *rbc*L, respectively). However, molecular approaches have provided evidence that some aspects of shape do not vary so capriciously and instead reflect major evolutionary events (Kaczmarska et al., 2001; see also Alverson et al., 2006). Thus, the major clades of centric diatoms revealed by molecular data are generally either "radial" or "multipolar" (Kooistra et al., 2003b; Medlin and Kaczmarska, 2004; Alverson and Theriot 2005), i.e. they either have circular valves (Figure 13.2 [9] and Figure 13.2 [10]), or valves that are elliptical, elongate, triangular, or multiangular (Figure 13.2 [11] and Figure 13.2 [12]). This seems to reflect a significant difference in the way the auxospores develop (von Stosch, 1982; Mann, 1994), although all generalizations are severely compromised by lack of data. In radial centrics, expansion is generally isodiametric (Figure 13.1 [6]), because the auxospore wall is homogeneous and either wholly organic or composed of an organic matrix in which small silica scales are embedded. In multipolar centric diatoms and pennate diatoms, on the other hand, auxospore expansion is accompanied by the formation of a system of silica bands (referred to as a properizonium in centric diatoms and as a perizonium in pennate diatoms, though they seem to be homologous structures), which constrain expansion to two, three, or more "soft spots" (Figure 13.1 [8]), producing bi- to multipolar shapes (Mann, 1994; Kaczmarska et al., 2001).

4. The Thalassiosirales, a group uniquely characterized by possession of special chitin-secreting organelles (fultoportulae = strutted processes: Figure 13.2 [16]), are not a basal group of centric diatoms, as previously thought. Thalassiosirales usually have circular valves, and Simonsen (1972, 1979) considered them to belong close to the Melosiraceae, a group of predominantly chain-forming "radial" centrics. Instead, they appear to have arisen from clades of elongate or multiangular ("multipolar") centric diatoms (Medlin et al., 1996a, 2000; Medlin and Kaczmarska, 2004). This implies that they have acquired circular valve morphologies secondarily, through loss of properizonium-associated, anisometric growth of the auxospore. Interestingly, a few Thalassiosirales do possess elliptical (e.g. McLaughlin, 1992) or polygonal valves (Economou-Amilli, 1979), but whether they possess a properizonium is unknown.

5. The raphid pennate diatoms (Figure 13.3 [18]) are monophyletic (Medlin et al., 2000; Kooistra et al., 2003b; Medlin and Kaczmarska, 2004). Consequently, it is most parsimonious to assume that the raphe system—a system of slits through the cell wall and specialized secretory and streaming areas of the underlying protoplast that together generate rapid surface-associated locomotion—has evolved only once. Among the raphid diatoms, it appears that the *Eunotia* group (Figure 13.1 [1]) is probably basal (Mayama and Kuriyama, 2002; Sims et al., 2006; Alverson et al., 2006), as required by the model of raphe evolution proposed by Mann (1984a).

6. Some revisions of diatom genera made using particularly extensive non-molecular data sets have received clear support from molecular data. For example, the marine genera *Lyrella* and *Petroneis* were separated from *Navicula* because of differences in valve and chloroplast type (Karayeva, 1978; Round et al., 1990; Mann and Stickle, 1993) and suggested to belong together in the same family because of apparent synapomorphies, for example, aspects of raphe structure (external central raphe endings opening into a tear-drop–shaped groove; crook-like central internal endings), chloroplasts appressed to the valves, and complex volate pore occlusions. *Rbc*L data confirm both that *Lyrella* and *Petroneis* should be separated from *Navicula*, and that it is reasonable to classify them together in the same family (Jones et al., 2005). Similar confirmation is available for other genera formerly included within *Navicula*, such as *Placoneis* (Jones et al., 2005, supporting Cox, 1987, Round et al., 1990, and Mann and Stickle, 1995) and *Sellaphora* (Behnke et al., 2004; Jones et al., 2005, supporting Mann, 1989, and Round et al., 1990).

7. Conversely, some structures that appeared similar in light and electron microscopy, and were interpreted as homologous, have been shown not to be. A good example is the fibula, defined as a bridge of silica subtending the raphe system (Figure 13.3 [19] and Figure 13.3 [20]). Fibulae are present in a dozen or more genera, which are always placed together in taxonomic treatments (e.g. Round et al., 1990), although differences in valve structure and the extent of the raphe system (whether it is only as long as the valve, running from one end to the other, or runs around most or all of both sides of the valve) lead to a grouping of the fibulate genera into three orders: the Bacillariales (Figure 13.3 [19]), Rhopalodiales, and Surirellales (Figure 13.3 [20]). In a cladistic analysis based wholly on morphological features, Ruck and Kociolek (2004) used two members of the Bacillariales (*Nitzschia scalaris* and *Simonsenia delognei*) as the out-groups for an analysis of the Surirellales. However, it has been shown by Medlin et al. (2000) that members of the Bacillariales belong to a different clade of raphid diatoms to the Surirellales and have almost certainly acquired fibulae independently. The nearest relatives of the Surirellales identified thus far are non-fibulate diatoms of the genus *Amphora* (Medlin et al., 2000; Medlin and Kaczmarska, 2004).

Another example is the lyre-shaped area adjacent to the raphe in *Lyrella* and *Fallacia*. Originally, both of these genera were classified together in *Navicula* sect. *Lyratae* (e.g. by Hustedt, 1927–1966), and when *Lyrella* was separated from *Navicula* by Karayeva (1978) it appears that all Lyratae were to be included in the new genus. On the basis of the ultrastructure of the lyre-shaped area (plain in *Lyrella*, with an overlying porous membrane in *Fallacia*) and the characteristics of the protoplast (chloroplast number and position, division of the nucleus always on the same side of the cell rather than on alternate sides with successive divisions), Round et al. (1990) separated the former Lyratae into two genera and suggested that one, *Fallacia*, belongs close to *Sellaphora*, which lacks lyre-shaped areas. The reasons for Round et al.'s (1990) classification have never been given in detail, but *rbc*L data (Evans et al., unpublished data) show that *Fallacia* and *Sellaphora* are indeed close relatives and that neither is close to *Lyrella* and *Petroneis*.

8. Although molecular data have not yet provided all the answers about phylogeny that we might want, it is increasingly clear that morphological (cell wall) data do not on their own give robust estimates of phylogeny. There is just too much homoplasy and too few characters. The unresolved polytomies and lack of bootstrap support in morphological analyses by Edgar and Theriot (2004) and Jones et al. (2005) demonstrate this well.

SPECIES AND INFRASPECIFIC RELATIONSHIPS

CRYPTIC SPECIES

In a review of the species concept in diatoms, Mann (1999) noted that no truly cryptic species had been found, only species that were very difficult to tell apart by eye. Recent work on *Pseudo-nitzschia* and *Skeletonema*, e.g. by Amato et al. (2007) and Sarno et al. (2005), likewise suggests that species initially distinguished on the basis of molecular or mating data will often subsequently be found to exhibit small morphological differences: they are "pseudocryptic." However, some of the differences are so slight that the species are effectively cryptic. We have recently attempted to clarify terminology, reserving "pseudocryptic" for species that are merely difficult to identify, "semicryptic" for species that can be told apart only when the observer has both morphological data and provenance information, and "cryptic" for species that cannot be separated morphologically under any circumstances. In each case, the criteria apply to identification of individuals, not populations. Semicryptic species exhibit partial overlap in the ranges of metric characters and/or the frequencies of alternative character states in qualitative characters. Thus, for example, two genetically distinct and reproductively isolated demes currently classified together in *Pseudo-nitzschia calliantha* differ in the mean number of sectors within each valve pore, but the ranges overlap for this character (Amato et al., 2007). In such cases, even though two species may differ significantly in the mean and dispersion of particular character states, identification of *all* individuals within a population is possible only when supplementary data are available (e.g. about variation within the whole population from which the individuals were derived, or about distributions in nature). Thus, in material from Blackford Pond, Edinburgh (which we have studied particularly intensively, e.g. Mann et al. 2004), all *Sellaphora* species can be told apart morphologically (though only with difficulty): *locally*, therefore, the *S. pupula* species complex is pseudocryptic. Elsewhere, however, genetically distinct demes can be found with morphologies that seem to overlap with the Blackford species, making purely morphology-based identification unsafe; on an international scale, therefore, variation in the *S. pupula* complex is semicryptic.

The best-studied species complexes are in the marine genera *Pseudo-nitzschia* (Orsini et al., 2004; Cerino et al., 2005; Lundholm et al., 2003, 2006; Amato et al., 2007) and *Skeletonema* (Kooistra et al., 2005; Sarno et al., 2005), and in the freshwater genus *Sellaphora* (Figure 13.3 [18]: Mann, 1989; Behnke et al., 2004). Together these studies demonstrate that assigning diatoms to individual taxa solely on the basis of morphological or physiological attributes may often be inadequate. One of the most pressing needs in micro-eukaryote taxonomy is the establishment of species definitions that are both meaningful and practical. Recently, a DNA barcoding system has been proposed, whereby all taxa would be "labelled" and subsequently identified according to the sequences of certain target genes. However, although a universal barcoding system has been supported strongly by its proponents (e.g. Hebert, et al., 2003; Blaxter, 2004), there are also many sceptics because of major unresolved issues. For example, it has not yet been determined how well molecular groupings correspond to biologically defined taxa (i.e. from morphology and breeding data), nor the range of molecular variation allowable within a barcoded unit, nor what genes are suitable in each major group. The species complexes listed above provide ideal systems in which to test rigorously the suitability of DNA barcodes in the designation and identification of diatom species (Evans et al., 2007).

In *Pseudo-nitzschia*, studies have concentrated on resolving taxonomic confusion as a result of morphological plasticity, variable toxin production, and reproductive isolation, particularly in the *P. pseudodelicatissima/P. cuspidata* (Lundholm et al., 2003), *P. delicatissima* (Orsini et al., 2004; Lundholm et al., 2006), and *P. galaxiae* (Cerino et al., 2005) species complexes. New species have been described, although insufficient sampling means that the biogeographies of these species are unknown, compared to the cosmopolitan nature of the morphospecies they replace (Hasle, 2002). Amato et al. (2007) have made a particularly detailed study of the *P. delicatissima* and *P. pseudodelicatissima* groups in the Bay of Naples and show that there is good concordance between groupings based on ITS2 gene sequences and reproductive compatibility, and that these groupings generally also show slight morphological separation. However, Coleman's (2005) study of the *Paramecium aurelia* complex shows, as might be expected, that reproductive isolation of sexual forms is not always accompanied by divergence in neutral regions of the genome. Thus, lack of significant ITS variation does not imply that speciation has not occurred.

Detailed studies have also been made of *Skeletonema* (Medlin et al., 1991; Kooistra et al., 2005; Sarno et al., 2005). As in *Pseudo-nitzschia*, morphological and molecular studies have uncovered diversity within what was previously assumed to be a single cosmopolitan species, *S. costatum*. Sarno et al. (2005) described four new species, and Kooistra et al. (2005) uncovered yet more diversity and found that the new *Skeletonema* taxa seem to be geographically confined; for example, *S. grethae* was detected only along the east coast of the United States.

In freshwater environments, intensive morphological, mating, and molecular studies have been conducted on the *Sellaphora pupula* species complex (Mann, 1984b; Mann, 1989; Mann and Droop, 1996;, Mann, 1999; Mann et al., 1999; Behnke et al., 2004; Mann et al., 2004). As in *Pseudo-nitzschia* and *Skeletonema*, there are many semicryptic and pseudocryptic species, which are morphologically similar but reproductively isolated, possess different mating systems, exhibit different degrees of genetic relationship to each other, and differ in sensitivity to parasites.

POPULATION GENETICS

Population genetics is the documentation of the distribution of genetic variation within and between populations of a species and the study of the evolutionary forces (mutation, migration, selection, and drift) that structure populations genetically, i.e. produce a non-random distribution of genetic variation (Hartl and Clark, 1997). Identifying populations (collections of individuals that live within sufficiently restricted areas that any member can potentially mate with any other member; Hartl and Clark, 1997), is important to our understanding of what constitutes a diatom species, since local populations are the evolving units of a species. Having a good understanding of diatom population genetics is vital if we are to understand the dispersal and, hence, biogeography and biodiversity of diatoms.

Diatoms vary in growth-form and habitat (epipelic, epilithic, epiphytic, planktonic, etc.) and breeding systems and all of these could have important impacts on the resultant population structure. Most research to date has focused on marine diatoms and all refers to planktonic forms. No data exist for benthic species, which by their very nature may be dispersed over shorter distances and, hence, mate with neighbours more frequently, leading to more pronounced population genetic structures.

Marine

Prior to Gallagher's (1980, 1982) pioneering research into the population genetics of *Skeletonema costatum* in Narragansett Bay (Rhode Island), the levels of genetic variation present within species of phytoplankton were unknown. On the whole, authors have tended to emphasize the likely clonal nature of diatom populations, produced by the long periods of mitotic division between rare sexual events (e.g. Richardson, 1995), rather than the obligatory nature and regularity of sex in most diatoms. Using allozymes to genotype 457 isolates, Gallagher demonstrated that *S. costatum* isolates were genetically variable (Gallagher, 1980) and subsequent work illustrated an even greater degree of physiological diversity, indicating that the relatively insensitive allozyme technique failed to detect all of the genetic diversity present (Gallagher, 1982). Despite this, two separate populations were identifiable, one associated with summer blooms and the other associated with winter blooms. Gallagher (1982) suggested that these two populations could represent individuals belonging to different species and recent evidence of cryptic speciation within *S. costatum* would appear to support this (Kooistra et al., 2005; Sarno et al., 2005).

The few studies conducted over the intervening years (e.g. Skov et al., 1997) also demonstrated genetic variability within marine planktonic diatoms, but these precluded detailed analyses of population structure because of small sample size and choice of molecular marker. It was not until 2000 that the first study to use microsatellites was published (Rynearson and Armbrust, 2000). Microsatellites are repetitive regions of DNA (e.g. CA or GA units) that are found in the genomes of every organism. They are useful to population geneticists because the length of the repeat region (i.e., the number of repeats) can vary between individuals of a species and so can act as part of a fingerprint to distinguish one individual from another. They are more variable than allozymes and more reliable than RAPDs (Tingey and del Tufo, 1993), and they exhibit co-dominant inheritance (i.e., both alleles at a locus can be detected and so it can be determined whether an individual is homozygous or heterozygous at each locus), which increases the information yielded (compared to AFLPs, for example). These features make microsatellites the markers of choice for population genetic analyses. The downside is that their initial development can be very time consuming. It has been noted previously that macroalgae have fewer and less polymorphic microsatellites than higher plants (Olsen, J. et al., 2002). Although isolating suitable microsatellite loci from diatoms can also be difficult, levels of polymorphism seem to be sufficient. Rynearson and Armbrust (2004) used only three loci, but the numbers of alleles per locus ranged from 10 to 78. In the only other microsatellite-based studies published to date, Evans and colleagues isolated nine microsatellite loci for *P. multiseries* (3 to 7 alleles per locus; Evans et al., 2004) and six for its closest known relative, *P. pungens* (6 to 24 alleles per locus; Evans et al. 2005). An additional potential problem is that diatom microsatellites appear to be more complex than those found in higher plants and animals, so that alleles can differ by just one base pair (bp) (Evans et al., 2005). Care and use of suitable positive controls are therefore required to ensure accurate genotyping of individuals, because a rise in temperature of 5°C has been shown to affect the allele length detected during genotyping by capillary electrophoresis by up to 0.7 bp (Davison and Chiba, 2003).

Rynearson and Armbrust (2000) used microsatellite markers to investigate genetic diversity within the planktonic centric marine diatom *Ditylum brightwellii* in the inland fjord of Puget Sound (Washington). They reported high levels of clonal and genetic diversity, but the small sample size (24 isolates) limited interpretation. Subsequently, hundreds of isolates were obtained from Puget Sound and an adjacent estuary, the Strait of Juan de Fuca (Rynearson and Armbrust, 2004; Rynearson et al., 2006). Concurrently, microsatellite-based work was published on *P. multiseries* (mostly from Canadian waters) and *P. pungens* (mostly from the North Sea: Evans and Hayes, 2004; Evans et al., 2004, 2005). The *P. pungens* work (Evans et al., 2005), in particular, provides a good complement to the *D. brightwellii* studies because oceanographic conditions differ significantly between the two areas.

Both studies demonstrate a large degree of clonal and genetic variation within planktonic marine diatoms. For example, 453 of the 464 North Sea *P. pungens* isolates genotyped were genetically distinct from each other and high levels of genetic variation were maintained even during bloom periods (Rynearson and Armbrust, 2005). However, despite the similarity of the spatial and temporal scales over which the isolates were obtained (approximately 100 km and 18 months in the Evans et al. [2005] and Rynearson and Armbrust [2004] studies), the structuring of the genetic variation differed markedly between the two species. These differences are probably best accounted for by the environments from which the isolates were obtained. The German North Sea is well mixed and so the genotyped isolates probably represent one population (significant F_{ST} values between isolates belonging to different groups, classified according to time or place of isolation, were at most 0.04, indicating weak genetic differentiation). In contrast, Puget Sound and the Strait of Juan de Fuca have only limited exchange of water and *D. brightwellii* isolates sampled from these waters belonged to four different populations (significant F_{ST} values between populations were up to 0.25, indicating a high degree of genetic differentiation; Rynearson and Armbrust, 2004; Rynearson et al., 2006). Despite this differentiation, it was thought likely that all isolates were members of the same species, because 18S and 5.8S rDNA sequences of selected isolates were identical and divergence of the less-conserved ITS region was at most 1.1% (Rynearson and Armbrust, 2004; Rynearson et al., 2006). Breeding experiments between isolates from each population should now be carried out to confirm these predictions. Until appropriate species concepts are established for diatoms and other microalgae, such an integrative approach is necessary if we are to truly understand population dynamics and, hence, speciation processes.

The fact that Rynearson and colleagues detected multiple populations (one of which has persisted for at least 7 years; Rynearson et al., 2006) within a relatively small area prompts a reassessment of our ideas of speciation in the marine environment, where barriers to dispersal are often not immediately apparent and where it is generally assumed that aquatic currents disperse organisms widely. But is *D. brightwellii* atypical? Evans et al.'s (2004) findings for *P. multiseries* suggest tentatively that its populations may be similarly structured, although a small sample size was involved (25 isolates). Here, a Russian isolate introduced 11 new alleles at six loci, relative to Canadian material (Evans et al., 2004). In contrast, work on *P. pungens* (Evans et al., 2005) showed that three Canadian isolates possessed only two alleles not found among the 464 German isolates, which is surprising given the considerable geographical separation and the conflicting results for its close relative *P. multiseries*. Castelyn et al. (2004) reported that ITS sequences from North Sea *P. pungens* isolates were identical and that all clones were sexually compatible; isolates collected from further afield, including New Zealand and the Pacific coast of the United States were also sexually compatible with the North Sea isolates, with no obvious loss of viability in the F1 generation (Chepurnov et al., 2005). These results support Hasle's (2002) view that *P. pungens* is a cosmopolitan species, which is presumably able to tolerate a wide range of environmental conditions. Before generalizations can be made, however, more studies need to be conducted, both over larger scales and in comparable environments to those from which the *D. brightwellii* isolates were obtained (Rynearson and Armbrust, 2004).

Freshwater

Work in freshwater environments has lagged behind that in the marine environment, despite the fact that it is easier to envisage potential barriers to dispersal and therefore to test hypotheses relating to the dispersal and biogeography of diatoms. Currently, little is known about how much gene flow could occur between populations of freshwater microalgae. The little information available suggests that terrestrial or subaerial algae are more easily spread than lotic or lentic species (e.g. during colonization of Surtsey: Behre and Schwabe, 1970). There are a few observations relevant to dispersal of freshwater phytoplankton (e.g. Parsons et al., 1966; Atkinson, 1971), but none that apply to freshwater benthic microalgae.

All freshwater diatom population genetics studies have focused on planktonic species and all have used molecular markers with well-known associated drawbacks (e.g. isozymes) and/or small sample sizes (*Asterionella formosa*, Soudek and Robinson, 1983, and De Bruin et al., 2004; *Stephanodiscus*, Zechman et al., 1994; and *Fragilaria capucina*, Lewis et al., 1997). The results differ but suggest overall that freshwater diatoms have limited dispersal capabilities and that geographic patterns exist, although, at least for Lewis et al. (1997), samples were collected along a broad temperature gradient and so genetic distinctiveness between populations could in part be due to thermal ecotypes.

What don't we know and what impedes us making progress?

1. Despite the addition of more taxa to molecular analyses, we do not know the branching order of the major lineages within the centric diatoms. SSU phylogenetic trees that have *Bolidomonas* as the outgroup and include several tens or >100 species representing many of the families and orders of centric and pennate diatoms show the multipolar centrics (Fox and Sorhannus, 2003; Sinninghe Damsté et al., 2004) or both the radial and the multipolar centrics (Medlin et al., 2000; Medlin and Kaczmarska, 2004, Sorhannus, 2005) as paraphyletic (see also Alverson et al., 2006; Cavalier-Smith and Chao, 2006). Even where analyses do show the radial centrics or the multipolar centrics + pennates as monophyletic, there is generally little statistical support. Thus, in a recently published SSU analysis (Sims et al., 2006, figure 1), based on thousands of sequences in the ARB database (www.arb-home.de), there is good support only for monophyly of the pennate diatoms (Bacillariophyceae), not of the radial centrics or bi/multipolar diatoms. The Bayesian analysis in Sims et al.'s (2006) figure 2 gives apparently strong support to the idea of a basal dichotomy between the radial and the [multipolar centric + pennate] clades, but this is an exception (see also Kooistra et al., 2003b; Horiguchi and Takano, 2006). Despite the paraphyly of the radial centrics and multipolar centrics in some of their analyses, Medlin and Kaczmarska (2004) decided to recognize both as classes.

    However, let us assume for the sake of argument that both the radial centrics and the multipolar centrics are indeed monophyletic. What would that mean for our understanding of diatom evolution? (1) The shape and symmetry of the ancestral diatoms would probably be unknown, unless we discover a new sister group for extant diatoms or make remarkable new discoveries in the fossil record. This is because the bolidophytes have few or no characteristics that can be used to polarize morphological character state transitions in the diatoms. The first diverse flora of diatoms (from a Lower Cretaceous marine deposit in Antarctica; Gersonde and Harwood, 1990) includes both circular and bipolar centric diatoms (discussed by Sims et al., 2006), presumably indicating that the "radial" and "multipolar" groups had already diverged from each other. (2) All extant multipolar centric diatoms would have to be descended from the diatom that also gave rise to all extant pennate diatoms, so that any morphological features shared by pennate diatoms and any of the multipolar diatoms would either have to be homoplasies or plesiomorphic for the whole pennate–multipolar centric clade. In addition, the origins of any autapomorphies of pennate diatoms, such as the sternum or isogamy, would probably remain unrecoverable.

2. Within each of the araphid and raphid pennate diatoms, although some groupings receive strong support, the overall course of evolution is not obvious and the origins of many groups are obscure.

3. The inability of currently used genetic sequence analyses to generate robust hypotheses about relationships and evolutionary trends implies one or more of the following: (a) the gene sequences used are not long enough (see below, point 4); (b) their evolution proceeds too quickly (producing saturation and homoplasy) or too slowly (some nodes will not be resolved); (c) taxon sampling is inadequate; (d) morphological evolution is poorly coupled to DNA sequence evolution; (e) different loci or parts of loci are giving conflicting evolutionary signals; and (f) there really were bursts of cladogenesis, perhaps during and immediately after environmental crises, e.g. at the Cretaceous–Tertiary (KT) boundary or during the less famous Triassic–Jurassic extinction event (Olsen, P. et al., 2002). Unfortunately, we do not have enough information at the moment to determine what combination of these or other factors is to blame. In relation to point (f), the fossil record does not show a massive turnover of taxa at the KT boundary (Sims et al., 2006), but it is important to remember that the fossil record is heavily biased toward planktonic diatoms, whereas the greatest diatom diversity today—and presumably therefore many key evolutionary transitions in the past—is in the benthos.

4. Most diatom phylogenies have been constructed from less than 2000 aligned nucleotides, whereas Wortley et al. (2005) suggest that approximately 10,000 may be required for "difficult" cases. Multi-gene phylogenies are surely the way forward, if we really want to know the origins of diatom diversity. Genes like *rbc*L and the more slowly evolving elements of rDNA are not unsuitable for phylogenetic analysis at most levels of the taxonomic hierarchy in diatoms; we just need more like them.

At first sight, developing a multi-gene phylogeny looks straightforward—a simple though tedious and expensive extension to the work represented in Table 13.1. Unfortunately, it will be difficult to build on past work, because in many cases the material used for SSU rDNA or other previous studies is no longer available. This is because of the special difficulties of maintaining diatoms in culture. Many taxa have never been successfully cultivated *in vitro*—a problem that is not unique to diatoms—but even in those that can be grown, the life span of clones is generally limited by obligate size reduction (Mann and Chepurnov, 2004; Chepurnov et al., 2004). For a culture to persist, the diatom must undergo auxosporulation to restore maximum cell dimensions and allow further vegetative growth (it will then usually be genetically heterogeneous). Auxosporulation often fails, however, because the mating system may prohibit sexual reproduction within a clone (pennate diatoms are frequently heterothallic), or conditions may not permit gametogenesis, or inbred progeny may not be viable (Chepurnov et al., 2004). Thus, most of the *Sellaphora* clones used for the ITS–SSU rDNA study by Behnke et al. (2004, originally from our collections) are now dead, as are many of the clones used by Medlin and co-workers for their earlier analyses (we have searched the CCAP and CCMP collections for the strains specified). Those diatoms that survive long term in culture must lack (or bypass) a sexual cycle or must be able to tolerate inbreeding; they are the atypical weeds of the diatom world (cf. *Phaeodactylum*). Hence, making a multi-gene phylogeny of diatoms will in many cases involve starting from scratch with new cultures.

5. The uncertainties in species identification in diatoms (not only because of the existence of cryptic species but also because of fuzzy concepts of species, accidental error, or simple incompetence) mean that it is dangerous to assume that sequences recorded for a particular species do in fact represent that taxon, or that the taxon itself is a meaningful entity. For example, Sinninghe Damsté et al. (2004) used several cultures labelled *Rhizo-solenia setigera*, but these appear in three places in their phylogenetic tree, on the ends of long branches. Such oddities can be studied and explained if cultures are still available. Otherwise, it ought to be possible to check the identity of the material studied, from preserved material or slides. Table 13.1 shows, however, that voucher material is rarely deposited safely in a recognized herbarium (examples where this was done are Edgar and Theriot, 2004; Behnke et al., 2004). Furthermore, to meet the highest standards of scientific reproducibility, alignments should be deposited in accessible data banks, especially alignments of rDNA (for some protein-encoding sequences, the alignment may be unambiguous). Table 13.1 shows that this ideal, too, is rarely met.

6. Many genera, families, and orders have not been sampled or are represented by one or a few species. Of the 22 orders of centric diatoms recognized by Round et al. (1990), eight remain totally unsampled, and several of the families of raphid diatoms are also missing. Furthermore, as Alverson and Theriot (2005) have noted, Round et al.'s (1990) conceptual framework was more phenetic than phylogenetic, so that their groupings may often be based on symplesiomorphies, not synapomorphies. Consequently, designing molecular sampling strategies on the basis of Round et al.'s classification will greatly underestimate the work needed to produce even a skeletal phylogeny. There is also the problem of poor coverage of taxa in culture collections, for the reasons listed earlier. However, with care, sequence data (even for two or more genes coded for in different genomes) can be recovered from a single cell whose morphology has been recorded (Sherbakova et al., 2000; Takano and Horiguchi, 2005). So, providing that we can determine in advance what set of genes is appropriate, it may not be necessary to culture "difficult" taxa.

7. Examining the relative pace and constancy of molecular and morphological evolution requires more data than we currently have, and matched data are essential. There is a considerable danger that, after molecular data have been used to group isolates into species, only one or two isolates will be studied in depth for their morphology and cytology, on the unjustifiable assumption that they are typical of their clade. Of all the studies listed in Table 13.1, only Lundholm et al. (2003) and Edgar and Theriot (2004) seem to have made a sustained attempt to obtain and analyse morphological and molecular data for the same accessions; Edgar and Theriot (2004) also included morphological data for related fossil diatoms. In a study of some raphid diatoms, Jones et al. (2005) constructed a tree from a matrix of morphological, cytological, and reproductive characters for comparison with an *rbc*L gene tree. However, the taxa included were only a small subset of those that would have been needed to examine evolution of all raphid diatoms and were selected to test only a few, very restricted hypotheses of relationships.

8. The system of families, genera, and species that we probe with molecular tools is for the most part a system that has been built up from comparative morphology of just one part of the diatom cell—the valve (for most diatomists, it almost seems as though the valve is the diatom). Mann and Cox spent several years attempting to increase the information content of classifications of raphid diatoms through studies of chloroplast morphology and sexual reproduction (e.g. Cox, 1987; Mann, 1989; Mann and Stickle, 1993, 1995), although it must be admitted that the approach taken was often as flawed as that described under point 7 above, being based on the use of "exemplar" species that it was hoped were representative of groupings based on cell wall ultrastructure. Where it has been possible to check them using molecular data, the taxonomic realignments

suggested by Cox and Mann have proved on the whole to be just (although it is also becoming evident that several new paraphyletic groups, e.g. *Nitzschia sensu stricto*, *Navicula sensu stricto*, were created as a by-product: Lundholm et al., 2002a; Simpson and Mann, unpublished). By contrast, genera based wholly on wall characters—whether the genera are long established or recently described—often prove to be poly- or paraphyletic. Examples are the raphid genus *Eolimna* (Schiller and Lange-Bertalot, 1997), which is polyphyletic (Beszteri et al., 2001; Behnke et al., 2004), and the centric *Thalassiosira*, which is apparently paraphyletic (Kaczmarska et al., 2006), despite being revised rigorously by Hasle and others using SEM data (producing the system summarized by Hasle and Syvertsen, 1996).

9. However, the principal deficiency is that, even when robust phylogenies are available, they are often profoundly unsatisfying. The information content of the existing classification is low, having been built almost exclusively from valve data (although we acknowledge that it is nevertheless richer than in many other protists). Hence, molecular phylogenies of diatoms easily become little more than exercises in linking names together. Sometimes there have been attempts to plot information about morphological or cytological characters onto gene trees, but this is generally done "by eye" (e.g. Medlin and Kaczmarska, 2004), rather than by formal reconciliation using some specified criterion such as parsimony, and the information is often so sparse and inconsistent as to make generalization dangerous: Sims et al. (2006) are notably more cautious about correlations between molecular phylogenies and non-molecular data, compared to Medlin et al. (2000).

   What is lacking is a science of "diatomics": a concerted, consistent attempt to survey, record, and categorize the morphological, cytological, nuclear, growth, reproductive, and other characteristics of diatom species. Gathering such information can be easy (plastid morphology, gross nuclear structure) or difficult (determining protoplast ultrastructure from thin sections) or time consuming (examining auxospore formation and development), but without it, molecular phylogenies will often be trees of name tags.

10. We turn now to infraspecific variation. We are just beginning to understand gene flow between microbes in marine environments, and it is becoming increasingly apparent that population dynamics are much more complex than had been assumed. Barriers to gene flow exist in seemingly open aquatic environments, as do locally adapted populations, which helps to explain how diversity in marine systems has arisen and why it is higher than has been concluded from surveys based on morphospecies concepts. Now that microsatellite markers exist for a few species, progress in our understanding of phytoplankton population dynamics should be much more rapid. *Pseudo-nitzschia pungens* is an apparently cosmopolitan species and has been the focus of detailed studies of its life history (Chepurnov et al., 2005); it therefore serves as an ideal model. Understanding the bloom dynamics of its closest relative, *P. multiseries,* a diatom often connected with outbreaks of amnesic shellfish poisoning, should be a priority in view of the differences that seem to exist in genetic structure between *P. multiseries* and *P. pungens.* There is some evidence for cross-amplifiability of microsatellites between these two species (Evans and Hayes, 2004; Evans et al., 2004). The increasing amount of information available from the *Thalassiosira pseudonana* and *Phaeodactylum tricornutum* sequencing projects (projects to sequence the *Fragilariopsis cylindrus* and *Pseudo-nitzschia multiseries* genomes are in progress: information on these is available at www.jgi.doe.gov/sequencing/index.html) will aid the development of markers (both neutral and those potentially under selection) in these species and their closest relatives, though the absence of any good information about the distribution and ecology of *Phaeodactylum* will be a severe hindrance to interpretation. Studies that assess population differentiation and gene flow need to be carried out on a global scale to determine the extent of human-mediated dispersal (e.g. due to transportation of cells in ships' ballast water) versus dispersal via oceanic currents.

11. No explanation of diatom evolution can be adequate if it does not address the diversification of benthic diatoms, which outnumber planktonic species by an order of magnitude or more. No population genetic studies have yet been made of benthic diatoms, and we have therefore begun work on the population genetics and biogeography of the *Sellaphora pupula* species complex. This work should also allow the first detailed comparisons between marine and freshwater: does the division of the habitat into lakes or rivers ("water islands and isthmuses") constitute more effective barriers to gene flow than are ever present in the sea?

12. Other questions that need to be answered to aid our understanding of population dynamics include the fate of cells between bloom periods, especially for diatoms that (apparently) lack resting stages. Also, methods to measure the incidence of sexual reproduction in field populations need to be developed, because direct observation is impractical and it is therefore hard to assess the importance of sex in the production and maintenance of the high levels of genetic diversity that have been detected (Evans et al., 2005). In North Sea *P. pungens* populations, sexual reproduction appears to occur frequently because all six microsatellite loci were in Hardy–Weinberg equilibrium (Maynard Smith, 1989). It is only through conducting such research that a true understanding of the population genetics, dispersal, biogeography, and biodiversity of microalgae will be gained. Until then, the general (and possibly highly misleading) consensus (cf. Finlay and Fenchel, 2004) will continue to grow, that microorganisms do not possess biogeographies and are relatively species poor.

ACKNOWLEDGMENTS

We thank Professor F.E. Round for permission to use his archive of his scanning electron micrographs (now at the Royal Botanic Garden Edinburgh). Alberto Amato, Dr Victor Chepurnov, Dr Eileen Cox, Dr Richard Crawford, Dr Wiebe Kooistra, Dr Linda Medlin, Dr Marina Montresor, Prof. Aloisie Poulíčková, Prof. Frank Round, Prof. Koen Sabbe, Dr Rosa Trobajo, Dr Pieter Vanormelingen, Prof. Wim Vyverman, and Dr Adriana Zingone are appreciated for discussions over many years of diatom evolution and speciation. Participation of Dr Katharine M. Evans was supported by a Natural Environment Research Council Fellowship (NE/C518373/1). David Mann thanks the Royal Society for an equipment grant enabling purchase of a Reichert Polyvar photomicroscope; Frieda Christie (Royal Botanic Garden Edinburgh) for help with scanning electron microscopy; and Dr Gillian Simpson and Carolyn Guihal for assistance with molecular systematic research.

REFERENCES

Alvarez, I. and Wendel, J.F. (2003) Ribosomal ITS sequences and plant phylogenetic inference. *Molecular Phylogenetics and Evolution*, 29: 417–434.

Alverson, A.J., Cannone, J.J., Gutell, R.R., and Theriot, E.C. (2006) The evolution of elongate shape in diatoms. *Journal of Phycology*, 42: 655–668.

Alverson, A.J. and Kolnick, L. (2005) Intragenomic nucleotide polymorphism among small subunit (18S) rDNA paralogs in the diatom genus *Skeletonema* (Bacillariophyta). *Journal of Phycology*, 41: 1248–1257.

Alverson, A.J. and Theriot, E.C. (2005) Comments on recent progress toward reconstructing the diatom phylogeny. *Journal of Nanoscience and Nanotechnology*, 5: 57–62.

Amato, A., Kooistra, W.H.C.F., Levialdi Ghiron, J.H., Mann, D.G., Pröschold, T., and Montresor, M. (2007) Reproductive isolation among sympatric cryptic species in marine diatoms. *Protist*, 158: 193–207.

Andersen, R.A., Potter, D., Bidigare, R.R., Latasa, M., Rowan, K., and O'Kelly, C.J. (1998) Characterization and phylogenetic position of the enigmatic golden alga *Phaeothamnion confervicola*: ultrastructure, pigment composition and partial SSU rDNA sequence. *Journal of Phycology*, 34: 286–298.

Anonymous (1703) Two letters from a Gentleman in the Country, relating to Mr. Leeuwenhoek's Letter in Transaction, no. 283. *Philosophical Transactions of the Royal Society of London*, 23 (288): 1494–1498 and 1498–1501.

Armbrust, E.V., Berges, J.A., Bowler, C., Green, B.R., Martinez, D., Putnam, N.H., Zhou, S., Allen, A.E., Apt, K.E., Bechner, M., Brzezinski, M.A., Chaal, B.K., Chiovitti, A., Davis, A.K., Demarest, M.S., Detter, J.C., Glavina, T., Goodstein, D., Hadi, M.Z., Hellsten, U., Hildebrand, M., Jenkins, B.D., Jurka, J., Kapitonov, V.V., Kröger, N., Lau, W.W., Lane, T.W., Larimer, F.W., Lippmeier, J.C., Lucas, S., Medina, M., Montsant, A., Obornik, M., Parker, M.S., Palenik, B., Pazour, G.J., Richardson, P.M., Rynearson, T.A., Saito, M.A., Schwartz, D.C., Thamatrakoln, K., Valentin, K., Vardi, A., Wilkerson, F.P., and Rokhsar, D.S. (2004) The genome of the diatom *Thalassiosira pseudonana*: ecology, evolution, and metabolism. *Science*, 306: 79–86.

Armbrust, E.V. and Galindo, H.M. (2001) Rapid evolution of a sexual reproduction gene in centric diatoms of the genus *Thalassiosira*. *Applied and Environmental Microbiology*, 67: 3501–3513.

Atkinson, K.M. (1971) Further experiments in dispersal of phytoplankton by birds. *Wildfowl*, 22: 98–99.

Behnke, A., Friedl, T., Chepurnov, V.A., and Mann, D.G. (2004) Reproductive compatibility and rDNA sequence analyses in the *Sellaphora pupula* species complex (Bacillariophyta). *Journal of Phycology*, 40: 193–208.

Behre, K. and Schwabe, G.H. (1970) Auf Surtsey/Island im Sommer 1968 nachgewiesene nicht marine Algen. (Über die natürliche Frühbesiedlung postvulkanischer Substrate oberhalb des Litorals). *Schriften der Naturwissenschaftlichen Vereins für Schleswig-Holstein*, Supplement: 31–100.

Ben Ali, A., De Baere, R., Van der Auwera, G., De Wachter, R., and Van de Peer, Y. (2001) Phylogenetic relationships among algae based on complete large-subunit rRNA sequences. *International Journal of Systematic and Evolutionary Microbiology*, 51: 737–749.

Berney, C. and Pawlowski, J. (2006) A molecular time-scale for eukaryote evolution recalibrated with the continuous microfossil record. *Philosophical Transactions of the Royal Society of London, B*, 273: 1867–1872.

Beszteri, B., Ács, E., Makk, J., Kovács, G., Márialideti, K., and Kiss, K.T. (2001) Phylogeny of six naviculoid diatoms based on 18S rDNA sequences. *International Journal of Systematic and Evolutionary Microbiology*, 51: 1581–1586.

Beszteri, B., Ács, E., and Medlin, L.K. (2005) Ribosomal DNA sequence variation among sympatric stains of the *Cyclotella meneghiniana* complex (Bacillariophyceae) reveals cryptic diversity. *Protist*, 156: 317–333.

Blaxter, M.L. (2004) The promise of a DNA taxonomy. *Philosophical Transactions of the Royal Society of London, B*, 359: 669–679.

Caron, L., Douady, D., Quinet-Szely, M., de Goër, S., and Berkaloff, C. (1996) Gene structure of a chlorophyll a/c-binding protein from a brown alga: presence of an intron and phylogenetic implications. *Journal of Molecular Evolution*, 43: 270–280.

Castelyn, G., Chepurnov, V.A., Sabbe, K., Mann, D.G., Vannerum, K., and Vyverman, W. (2004) Species structure of *Pseudo-nitzschia pungens* (Grunow ex Cleve) Hasle: molecular, morphological and reproductive data. In *Abstracts, 18th International Diatom Symposium*, Midzyzdroje, Poland, 2–7 September 2004 (eds A. Witkowski, T. Radziejewska, B. Wawrzyniak-Wydrowska, D. Daniszewska-Kowalczyk, and M. Bak), p. 98.

Cavalier-Smith, T. and Chao, E.E.-Y. (2006) Phylogeny and megasystematics of phagotrophic heterokonts (kingdom Chromista). *Journal of Molecular Evolution*, 62: 388–420.

Cerino, F., Orsini, L., Sarno, D., Dell'Aversano, C., Tartaglione, L., and Zingone, A. (2005) The alternation of different morphotypes in the seasonal cycle of the toxic diatom *Pseudo-nitzschia galaxiae*. *Harmful Algae*, 4: 33–48.

Chase, M.W., Soltis, D.E., Olmstead, R.G., Morgan, D., Les, D.H., Mishler, B.D., Duvall, M., Price, R.A., Hills, H.G., Qiu, Y.-L., Kron, K.A., Rettig, J.H., Conti, E., Palmer, J.D., Manhart, J.R., Sytsma, K.J., Michaels, H.J., Kress, W.J., Karol, K.G., Clark, W.D., Hedren, M., Gaut, B.S., Jansen, R.K., Kim, K.-J., Wimpee, C.F., Smith, J.F., Furnier, G.R., Strauss, S.H., Xiang, Q.-Y., Plunkett, G.M., Soltis, P.S., Swensen, S.M., Williams, S.E., Gadek, P.A., Quinn, C.J., Eguiarte, L.E., Golenberg, E., Learn, G.H. Jr., Graham, S.W., Barrett, S.C.H., Dayanandan, S., and Albert, V.A. (1993) Phylogenetics of seed plants: an analysis of nucleotide sequences from the plastid gene *rbc*L. *Annals of the Missouri Botanical Garden*, 80: 528–580.

Chepurnov, V.A., Mann, D.G., Sabbe, K., Vannerum, K., Castelyn, G., Verleyen, E., Peperzak, L., and Vyverman, W. (2005) Sexual reproduction, mating system, chloroplast dynamics and abrupt cell size reduction in *Pseudo-nitzschia pungens* from the North Sea (Bacillariophyta). *European Journal of Phycology*, 40: 379–395.

Chepurnov, V.A., Mann, D.G., Sabbe, K., and Vyverman, W. (2004) Experimental studies on sexual reproduction in diatoms. *International Review of Cytology*, 237: 91–154.

Chepurnov, V.A., Mann, D.G., Vyverman, W., Sabbe, K., and Danielidis, D.B. (2002) Sexual reproduction, mating system and protoplast dynamics of *Seminavis* (Bacillariophyta). *Journal of Phycology*, 38: 1004–1019.

Chesnick, J.M., Morden, C.W., and Schmieg, A.M. (1996) Identity of the endosymbiont of *Peridinium foliaceum* (Pyrrophyta): analysis of the *rbc*LS operon. *Journal of Phycology*, 32: 850–857.

Coleman, A.W. (2005) *Paramecium aurelia* revisited. *Journal of Eukaryotic Microbiology*, 52: 68–77.

Cox, E.J. (1987) *Placoneis* Mereschkowsky: the re-evaluation of a diatom genus originally characterized by its chloroplast type. *Diatom Research*, 2: 145–157.

Cox, E.J. (1996) *Identification of Freshwater Diatoms from Live Material*. Chapman & Hall, London.

Coyne, J.A. and Orr, H.A. (2004) *Speciation*. Sinauer, Sunderland, MA.

Danielidis, D.B. and Mann, D.G. (2002) The systematics of *Seminavis* (Bacillariophyta): the lost identities of *Amphora angusta*, *A. ventricosa* and *A. macilenta*. *European Journal of Phycology*, 37: 429–448.

Daugbjerg, N. and Andersen, R.A. (1997) A molecular phylogeny of the heterokont algae based on analyses of chloroplast-encoded *rbc*L sequence data. *Journal of Phycology*, 33: 1031–1041.

Daugbjerg, N. and Guillou, L. (2001) Phylogenetic analyses of Bolidophyceae (Heterokontophyta) using *rbc*L gene sequences support their sister group relationship to diatoms. *Phycologia*, 40: 153–161.

Davidovich, N.A. and Bates, S.S. (1998) Sexual reproduction in the pennate diatoms *Pseudo-nitzschia multiseries* and *P. pseudodelicatissima* (Bacillariophyceae). *Journal of Phycology*, 34: 126–137.

Davison, A. and Chiba, S. (2003) Laboratory temperature variation is a previously unrecognised source of genotyping error during capillary electrophoresis. *Molecular Ecology Notes*, 3: 321–323.

De Bruin, A., Ibelings, B.W., Rijkeboer, M., Brehm, M., and van Donk, E. (2004) Genetic variation in *Asterionella formosa* (Bacillariophyceae): is it linked to frequent epidemics of host-specific parasitic fungi? *Journal of Phycology*, 40: 823–830.

Delwiche, C.F., Kuhsel, M., and Palmer, J.D. (1995) Phylogenetic analysis of *tuf*A sequences indicates a cyanobacterial origin of all plastids. *Molecular Phylogenetics and Evolution*, 4: 110–128.

Desikachary, T.V. (1952) Electron microscopic study of the diatom-wall structure. *Journal of Scientific and Industrial Research*, 11B: 491–500.

Douglas, D.J., Landry, D., and Douglas, S.E. (1994) Genetic relatedness of toxic and nontoxic isolates of the marine pennate diatom *Pseudonitzschia* (Bacillariophyceae): phylogenetic analysis of 18S rRNA sequences. *Natural Toxins*, 2: 166–174.

Du Buf, J.M.H. and Bayer, M.M. (eds.) (2002) *Automatic Diatom Identification*. Series in Machine Perception and Artificial Intelligence, vol. 51. World Scientific Publishing, Singapore.

Economou-Amilli, A. (1979) Two new taxa of *Cyclotella* Kütz. from Lake Trichonis, Greece. *Nova Hedwigia*, 31: 467–477.

Edgar, S.M. and Theriot, E.C. (2004) Phylogeny of *Aulacoseira* (Bacillariophyta) based on molecules and morphology. *Journal of Phycology*, 40: 771–788.

Ehara, M., Inagaki, Y., Watanabe, K.I., and Ohama, T. (2000a) Phylogenetic analysis of diatom cox1 genes and implications of a fluctuating GC content on mitochondrial genetic code evolution. *Current Genetics*, 37: 29–33.

Ehara, M., Watanabe, K.I., and Ohama, T. (2000b) Distribution of cognates of group II introns detected in mitochondrial *cox*1 genes of a diatom and a haptophyte. *Gene*, 256: 157–167.

Eppard, M., Krumbein, W.E., von Haeseler, A., and Rhiel, E. (2000) Characterization of *fcp* 4 and *fcp*12, two additional genes encoding light harvesting proteins of *Cyclotella cryptica* (Bacillariophyceae) and phylogenetic analysis of this complex gene family. *Plant Biology*, 2: 283–289.

Evans, K.M., Bates, S.S., Medlin, L.K., and Hayes, P.K. (2004) Microsatellite marker development and genetic variation in the toxic marine diatom *Pseudo-nitzschia multiseries* (Bacillariophyceae). *Journal of Phycology*, 40: 911–920.

Evans, K.M. and Hayes, P.K. (2004) Microsatellite markers for the cosmopolitan marine diatom *Pseudo-nitzschia pungens*. *Molecular Ecology Notes*, 4: 125–126.

Evans, K.M., Kühn, S.F., and Hayes, P.K. (2005) High levels of genetic diversity and low levels of genetic differentiation in North Sea *Pseudo-nitzschia pungens* (Bacillariophyceae) populations. *Journal of Phycology*, 41: 506–514.

Evans, K.M., Wortley, A.H., and Mann, D.G. (2007) An assessment of potential "barcode" genes (cox1, rbcL, 18s and ITS rDNA) and their effectiveness in determining relationships in Sellaphora (Bacillariophyta). *Protist*, 158.

Finlay, B.J. and Fenchel, T. (2004) Cosmopolitan metapopulations of free-living microbial eukaryotes. *Protist*, 155: 237–244.

Foth, B.J., Goedecke, M.C., and Soldati, D. (2006) New insights into myosin evolution and classification. *Proceedings of the National Academy of Sciences of the USA*, 103: 3681–3686.

Fox, M.G. and Sorhannus, U.M. (2003) *RpoA*: a useful gene for phylogenetic analysis in diatoms. *Journal of Eukaryotic Microbiology*, 50: 471–475.

Fritsch, F.E. (1935) *The Structure and Reproduction of the Algae*, vol. 1. Cambridge University Press, Cambridge.

Fryxell, G.A. and Hasle, G.R. (2003) Taxonomy of harmful diatoms. In *Manuals on Harmful Marine Microalgae* (eds G.J. Hallegraeff, D.M. Anderson, and A.D. Cembella), UNESCO, Paris, pp. 465–509.

Gallagher, J.C. (1980) Population genetics of *Skeletonema costatum* (Bacillariophyceae) in Narragansett Bay. *Journal of Phycology*, 16: 464–474.

Gallagher, J.C. (1982) Physiological variation and electrophoretic banding patterns of genetically different seasonal populations of *Skeletonema costatum* (Bacillariophyceae). *Journal of Phycology*, 18: 148–162.

Gersonde, R. and Harwood, D.M. (1990) Lower Cretaceous diatoms from ODP Leg 113 Site 693 (Weddell Sea). Part 1: Vegetative cells. *Proceedings of the Ocean Drilling Program, Scientific Results*, 113: 365–402.

Gilmour, J.S.L. and Heslop-Harrison, J.S. (1954) The deme terminology and the units of micro-evolutionary change. *Genetica*, 27: 147–161.

Godhe, A., McQuoid, M.R., Karunasagar, I., Karunasagar, I., and Rehnstam-Holm, A.S. (2006) Comparison of three common molecular tools for distinguishing among geographically separated clones of the diatom *Skeletonema marinoi* Sarno et Zingone (Bacillariophyceae). *Journal of Phycology*, 42: 280–291.

Goertzen, L.R. and Theriot, E.C. (2003) Effect of taxon sampling, character weighting, and combined data on the interpretation of relationships among the heterokont algae. *Journal of Phycology*, 39: 423–439.

Gueneau, P., Loiseaux-de Goër, S., and Williams, K.P. (1999) The GC-rich region and TC trn arm found in the *petF* region of the *Thalassiosira weissflogii* plastid genome encode a tmRNA. *European Journal of Phycology*, 34: 533–535.

Gueneau, P., Morel, F., Laroche, J., and Erdner, D. (1998) The *petF* region of the chloroplast genome from the diatom *Thalassiosira weissflogii*: sequence, organization and phylogeny. *European Journal of Phycology*, 33: 203–211.

Guillou, L., Chrétiennot-Dinet, M.-J., Medlin, L.K., Claustre, H., Loiseaux-de Goër, S., and Vaulot, D. (1999) *Bolidomonas*: a new genus with two species belonging to a new algal class, the Bolidophyceae (Heterokonta). *Journal of Phycology*, 35: 368–381.

Harper, J.T. and Keeling, P.J. (2003) Nucleus-encoded, plastid-targeted glyceraldehyde-3-phosphate dehydrogenase (GAPDH) indicates a single origin for chromalveolate plastids. *Molecular Biology and Evolution*, 20: 1730–1735.

Harper, J.T., Waanders, E., and Keeling, P.J. (2005) On the monophyly of chromalveolates using a six-protein phylogeny of eukaryotes. *International Journal of Systematic and Evolutionary Microbiology*, 55: 487–496.

Hartl, D.L. and Clark, A.G. (1997) *Principles of Population Genetics*. Sinauer, Sunderland, MA.

Hasle, G.R. (1965) *Nitzschia* and *Fragilariopsis* species studied in the light and electron microscopes. II. The genus *Pseudo-nitzschia*. *Skrifte utgitt av det Norske Videnskaps-Akademi i Oslo, I. Mat.-Naturvidensk. Klasse, Ny Serie*, 18: 1–49.

Hasle, G.R. (1968) The valve processes of the centric diatom genus *Thalassiosira*. *Nytt Magasin for Botanik*, 15: 193–201.

Hasle, G.R. (2002) Are most of the domoic acid-producing species of the diatom genus *Pseudo-nitzschia* cosmopolites? *Harmful Algae*, 1: 137–146.

Hasle, G.R., Lange, C.B., and Syvertsen, E.E. (1996) A review of *Pseudo-nitzschia*, with special reference to the Skagerrak, North Atlantic, and adjacent waters. *Helgoländer wissenschaftlicher Meeresuntersuchungen*, 50: 131–175.

Hasle, G.R. and Syvertsen, E.E. (1996) Marine diatoms. In *Identifying Marine Diatoms and Dinoflagellates* (ed. C. Tomas), Academic Press, San Diego, pp. 5–385.

Hebert, P.D.N, Cywinska, A., Ball, S.L., and deWaard, J.R. (2003) Biological identifications through DNA barcodes. *Proceedings of the Royal Society of London, B*, 270: 313–321.

Horiguchi, T. and Takano, Y. (2006) Serial replacement of a diatom endosymbiont in the marine dinoflagellate *Peridinium quinquecorne* (Peridiniales, Dinophyceae). *Phycological Research*, 54: 193–200.

Hustedt, F. (1927–1966) Die Kieselalgen Deutschlands, Österreichs und der Schweiz unter Berücksichtigung der übrigen Länder Europas sowie der angrenzenden Meeresgebiete. In *Dr. L. Rabenhorsts Kryptogamenflora von Deutschland, Österreich und der Schweiz*, vol. 7. Akademische Verlagsgesellschaft, Leipzig.

Hustedt, F. (1928) Untersuchungen über den Bau der Diatomeen. IV. Zur Morphologie und Systematik der Gattungen *Denticula* und *Epithemia*. V. Über den Bau der Raphe bei der Gattung *Hantzschia*. VI. Zur Kenntnis der Gattung *Cylindrotheca*. *Berichte der Deutschen Botanischen Gesellschaft*, 46: 148–164.

Janech, M.G., Krell, A., Mock, T., Kang, J.S., and Raymond, J.A. (2006) Ice-binding proteins from sea ice diatoms (Bacillariophyceae). *Journal of Phycology*, 42: 410–416.

Jones, H.M., Simpson, G.E., Stickle, A.J., and Mann, D.G. (2005) Life history and systematics of *Petroneis* (Bacillariophyta), with special reference to British waters. *European Journal of Phycology*, 40: 43–71.

Jung, W., Joo, H.M., Hong, S.S., Kang, J.-S., Choi, H.-G., and Kang, S.-H. (2006) Morphology and molecular data for antarctic cryophilic microalga, *Porosira pseudodenticulata*. *Algae*, 21: 169–174.

Kaczmarska, I., Beaton, M., Benoit, A.C., and Medlin, L.K. (2006) Molecular phylogeny of selected members of the order Thalassiosirales (Bacillariophyta) and evolution of the fultoportula. *Journal of Phycology*, 42: 121–138.

Kaczmarska, I., Ehrman, J.M., and Bates, S.S. (2001) A review of auxospore structure, ontogeny, and diatom phylogeny. In *Proceedings of the 16th International Diatom Symposium* (ed. A. Economou-Amilli), University of Athens Press, Athens, Greece, pp. 153–168.

Karayeva, N.I. (1978) Novye rod iz semejstva Naviculaceae West. *Botanicheskii Zhurnal*, 63: 1593–1596.

Kawachi, M., Inouye, I., Honda, D., O'Kelly, C-J., Bailey, J.C., Bidigare, R.R., and Andersen, R.A. (2002) The Pinguiophyceae classis nova, a new class of photosynthetic stramenopiles whose members produce large amounts of omega-3 fatty acids. *Phycological Research*, 50: 31–48.

Kellogg, E.A. and Juliano, N.D. (1997) The structure and function of RuBisCo and their implications for systematic studies. *American Journal of Botany*, 84: 413–428.

Kociolek, J.P. and Stoermer, E.F. (1988) A preliminary investigation of the phylogenetic relationships of the freshwater, apical pore field-bearing cymbelloid and gomphonemoid diatoms (Bacillariophyceae). *Journal of Phycology*, 24: 377–385.

Kolbe, R.W. and Gölz, E. (1943) Elektronenmikroskopische Diatomeenstudien. *Berichte der Deutschen Botanischen Gesellschaft*, 61: 91–98.

Kooistra, W.C.H.F., Chepurnov, V., Medlin, L.K., De Stefano, M., Sabbe, K., and Mann, D.G. (2006) Evolution of the diatoms. In *Plant Genome: Biodiversity and Evolution. Vol. 2B: Lower Groups* (eds A.K. Sharma and A. Sharma). Oxford and IBH Publishing, New Delhi, India, and Science Publishers, Enfield, NH, pp. 117–178.

Kooistra, W.H.C.F., De Stefano, M., Mann, D.G., Salma, N., and Medlin, L.K. (2003a) Phylogenetic position of *Toxarium*, a pennate-like lineage within centric diatoms (Bacillariophyceae). *Journal of Phycology*, 39: 185–197.

Kooistra, W.H.C.F., De Stefano, M., Mann, D.G., and Medlin, L.K. (2003b) The phylogeny of the diatoms. *Progress in Molecular and Subcellular Biology*, 33: 59–97.

Kooistra, W.H.C.F., Forlani, G., Sterrenburg, F.A.S., and De Stefano, M. (2004) Molecular phylogeny and morphology of the marine diatom *Talaroneis posidoniae* gen. et sp. nov. (Bacillariophyta) advocate the return of the Plagiogrammaceae to the pennate diatoms. *Phycologia*, 43: 58–67.

Kooistra, W.H.C.F., Hargraves, P., Anderson, R.A., Balzano, S., Zingone, A., and Sarno, D. (2005) Pseudo-cryptic diversity and phylogeography in the centric diatom *Skeletonema*. *Phycologia* 40 (4, Supplement): 56.

Kooistra, W.H.C.F. and Medlin, L.K. (1996) Evolution of the diatoms (Bacillariophyta) IV. A reconstruction of their age from small subunit rRNA coding regions and the fossil record. *Molecular Phylogenetics and Evolution*, 6: 391–407.

Kowallik, K.V., Stoebe, B., Schaffran, I., Kroth-Pancic, P., and Freier, U. (1995) The chloroplast genome of a chlorophyll a+c containing alga, *Odontella sinensis*. *Plant Molecular Biology Reporter*, 13: 336–342.

Kroth, P.G., Schroers, Y., and Kilian, O. (2005) The peculiar distribution of class I and class II aldolases in diatoms and in red algae. *Current Genetics*, 48: 389–400.

Kühn, S., Medlin, L., and Eller, G. (2004) Phylogenetic position of the parasitoid nanoflagellate *Pirsonia* inferred from nuclear-encoded small subunit ribosomal DNA and a description of *Pseudopirsonia* n. gen. and *Pseudopirsonia mucosa* (Drebes) comb. nov. *Protist*, 155: 143–156.

Lauterborn, R. (1896) *Untersuchungen über Bau, Kernteilung und Bewegung der Diatomeen*. W. Engelmann, Leipzig.

Lewis, R.J., Jensen, S.I., DeNicola, D.M., Miller, V.I., Hoagland, K.D., and Ernst, S.G. (1997) Genetic variation in the diatom *Fragilaria capucina* (Fragilariaceae) along a latitudinal gradient across North America. *Plant Systematics and Evolution*, 204: 99–108.

Liaud, M.-F., Lichtlé, C., Apt, K., Martin, W., and Cerff, R. (2000) Compartment-specific isoforms of TPI and GAPDH are imported into diatom mitochondria as a fusion protein: evidence in favor of a mitochondrial origin of the eukaryotic glyolytic pathway. *Molecular Biology and Evolution*, 17: 213–223.

Lundholm, N., Daugbjerg, N., and Moestrup, Ø. (2002a) Phylogeny of the Bacillariaceae with emphasis on the genus *Pseudo-nitzschia* (Bacillariophyceae) based on partial LSU rDNA. *European Journal of Phycology*, 37: 115–134.

Lundholm, N., Hasle, G.R., Fryxell, G.A., and Hargraves, P.E. (2002b) Morphology, phylogeny and taxonomy of species within the *Pseudo-nitzschia americana* complex (Bacillariophyceae) with descriptions of two new species, *Pseudo-nitzschia brasiliana* and *Pseudo-nitzschia linea*. *Phycologia*, 41: 480–497.

Lundholm, N. and Moestrup, Ø. (2002) The marine diatom *Pseudo-nitzschia galaxiae* sp. nov. (Bacillariophyceae): morphology and phylogenetic relationships. *Phycologia*, 41: 594–605.

Lundholm, N., Moestrup, Ø., Hasle, G.R., and Hoef-Emden, K. (2003) A study of the *Pseudo-nitzschia pseudodelicatissima/cuspidata* complex (Bacillariophyceae): what is *P. pseudodelicatissima*? *Journal of Phycology*, 39: 797–813.

Lundholm, N., Moestrup, Ø., Kotaki, Y., Hoef-Emden, K., Scholin, C., and Miller, P. (2006) Inter- and intraspecific variation of the *Pseudo-nitzschia delicatissima* complex (Bacillariophyceae) illustrated by rRNA probes, morphological data and phylogenetic analyses. *Journal of Phycology*, 42: 464–481.

Maheshwari, U., Montsant, A., Goll, J., Krishnasamy, S., Rajyashri, K.R., Patell, V.M. and Bowler, C. (2005) The Diatom EST database. *Nucleic Acids Research*, 33: D344–D347.

Mann, D.G. (1984a) An ontogenetic approach to diatom systematics. In *Proceedings of the 7th International Diatom Symposium* (ed. D.G. Mann), O. Koeltz, Koenigstein, pp. 113–144.

Mann, D.G. (1984b) Observations on copulation in *Navicula pupula* and *Amphora ovalis* in relation to the nature of diatom species. *Annals of Botany*, 54: 429–438.

Mann, D.G. (1989) The diatom genus *Sellaphora*: separation from *Navicula*. *British Phycological Journal*, 24: 1–20.

Mann, D.G. (1994) The origins of shape and form in diatoms: the interplay between morphogenetic studies and systematics. In *Shape and Form in Plants and Fungi* (eds. D.S. Ingram and A.J. Hudson), Academic Press, London, pp. 17–38.

Mann, D.G. (1999) The species concept in diatoms (Phycological Reviews 18). *Phycologia*, 38: 437–495.

Mann, D.G. and Chepurnov, V.A. (2004) What have the Romans ever done for us? The past and future contribution of culture studies to diatom systematics. *Nova Hedwigia*, 79: 237–291.

Mann, D.G., Chepurnov, V.A., and Droop, S.J.M. (1999) Sexuality, incompatibility, size variation and preferential polyandry in natural populations and clones of *Sellaphora pupula* (Bacillariophyceae). *Journal of Phycology*, 35: 152–170.

Mann, D.G. and Droop, S.J.M. (1996) Biodiversity, biogeography and conservation of diatoms. *Hydrobiologia*, 336: 19–32.

Mann, D.G., Droop, S.J.M., Hicks, Y.A., Marshall, A.D., Martin, R.R., and Rosin, P.L. (2006) Alleviating the taxonomic impediment in diatoms: prospects for automatic and web-based identification systems. In *Proceedings of the 18th International Diatom Symposium* (ed. A. Witkowski), Biopress, Bristol, pp. 265–285.

Mann, D.G. and Marchant, H. (1989) The origins of the diatom and its life cycle. In *The Chromophyte Algae: Problems and Perspectives* (eds J.C. Green, B.S.C. Leadbeater, and W.L. Diver) (Systematics Association Special Volume 38), Clarendon Press, Oxford, pp. 305–321.

Mann, D.G., McDonald, S.M., Bayer, M.M., Droop, S.J.M., Chepurnov, V.A., Loke, R.E., Ciobanu, A., and du Buf, J.M.H. (2004) Morphometric analysis, ultrastructure and mating data provide evidence for five new species of *Sellaphora* (Bacillariophyceae). *Phycologia*, 43: 459–482.

Mann, D.G., Simpson, G.E., Sluiman, H.J., and Möller, M. (2001) *rbc*L gene tree of diatoms: a second large data-set for phylogenetic reconstruction. *Phycologia*, 40 (4, Supplement): 1–2.

Mann, D.G. and Stickle, A.J. (1993) Life history and systematics of *Lyrella*. *Nova Hedwigia, Beiheft*, 106: 43–70.

Mann, D.G. and Stickle, A.J. (1995) Sexual reproduction and systematics of *Placoneis* (Bacillariophyta). *Phycologia*, 34: 74–86.

Mayama, S. and Kuriyama, A. (2002) Diversity of mineral cell coverings and their formation processes: a review focused on the siliceous cell coverings. *Journal of Plant Research*, 115: 289–295.

Maynard Smith, J. (1989) *Evolutionary Genetics*. Oxford University Press, Oxford, 325 pp.

Mayr, E. (1942) *Systematics and the Origin of Species*. Columbia University Press, New York, 334 pp.

McLaughlin, R.B. (1992) *Cyclotella jonesii*, a new diatom species from Pliocene deposits at Chiloquin, Oregon, U.S.A. *Diatom Research*, 7: 95–101.

Medlin, L.K., Elwood, H.J., Stickel, S., and Sogin, M.L. (1988) The characterization of enzymatically amplified eukaryotic 16S-like rRNA-coding regions. *Gene*, 71: 491–499.

Medlin, L.K., Elwood, H.J., Stickel, S., and Sogin, M.L. (1991) Morphological and genetic variation within the diatom *Skeletonema costatum* (Bacillariophyta): evidence for a new species, *Skeletonema pseudocostatum*. *Journal of Phycology*, 27: 514–524.

Medlin, L.K. and Kaczmarska, I. (2004) Evolution of the diatoms: V. Morphological and cytological support for the major clades and a taxonomic revision. *Phycologia*, 43: 245–270.

Medlin, L.K., Kooistra, W.H.C.F., Gersonde, R., Sims, P.A., and Wellbrock, U. (1997b) Is the origin of the diatoms related to the end-Permian mass extinction? *Nova Hedwigia*, 65: 1–11.

Medlin, L.K., Kooistra, W.H.C.F., Gersonde, R., and Wellbrock, U. (1996a) Evolution of the diatoms (Bacillariophyta). II. Nuclear-encoded small-subunit rRNA sequence comparisons confirm a paraphyletic origin for the centric diatoms. *Molecular Biology and Evolution*, 13: 67–75.

Medlin, L.K., Kooistra, W.H.C.F., Gersonde, R., and Wellbrock, U. (1996b) Evolution of the diatoms (Bacillariophyta): III. Molecular evidence for the origin of the Thalassiosirales. *Nova Hedwigia, Beiheft*, 112: 221–234.

Medlin, L.K., Kooistra, W.H.C.F., Potter, D., Saunders, G.W., and Andersen, R.A. (1997a) Phylogenetic relationships of the "golden algae" (haptophytes, heterokont chromophytes) and their plastids. *Plant Systematics and Evolution*, 11 (Supplement): 187–219.

Medlin, L.K., Kooistra, W.H.C.F., and Schmid, A.-M.M. (2000) A review of the evolution of the diatoms—a total approach using molecules, morphology and geology. In *The Origin and Early Evolution of the Diatoms: Fossil, Molecular and Biogeographical Approaches* (eds A. Witkowski and J. Siemiska), W. Szafer Institute of Botany, Polish Academy of Sciences, Cracow, Poland, pp. 13–35.

Medlin, L.K. and Simon, N. (1998) Phylogenetic analysis of marine phytoplankton. In *Molecular Approaches to the Study of the Ocean* (ed. K.E. Cooksey), Chapman & Hall, London, pp. 161–186.

Medlin, L.K., Williams, D.M., and Sims, P.A. (1993) The evolution of the diatoms (Bacillariophyta). I. Origin of the group and assessment of the monophyly of its major divisions. *European Journal of Phycology*, 28: 261–275.

Miller, P.E. and Scholin, C.A. (1996) Identification of cultured *Pseudo-nitzschia* (Bacillariophyceae) using species-specific LSU rRNA-targeted fluorescent probes. *Journal of Phycology*, 32: 646–655.

Miyagishima, S., Nozaki, H., Nishida, Keishin, Nishida, Keiji, Matsuzaki, M., and Kuroiwa, T. (2004) Two types of FtsZ proteins in mitochondria and red-lineage chloroplasts: the duplication of FtsZ is implicated in endosymbiosis. *Journal of Molecular Evolution*, 58: 291–303.

Nozaki, H., Misumi, O., and Kuroiwa, T. (2003) Phylogeny of the quadriflagellate Volvocales (Chlorophyceae) based on chloroplast multigene sequences. *Molecular Phylogenetics and Evolution*, 29: 58–66.

Olsen, J.L., Sadowski, G., Stam, W.T., Veldsink, J.H., and Jones, K. (2002) Characterization of microsatellite loci in the marine seaweed *Ascophyllum nodosum* (Phaeophyceae; Fucales). *Molecular Ecology Notes*, 2: 33–34.

Olsen, P.E., Kent, D.V., Sues, H.D., Koeberl, C., Huber, H., Montanari, A., Rainforth, E.C., Fowell, S.J., Szajna, M.J., and Hartline, B.W. (2002) Ascent of dinosaurs linked to an iridium anomaly at the Triassic–Jurassic boundary. *Science*, 296: 1305–1307.

Orsini, L., Procaccini, G., Sarno, D., and Montresor, M. (2004) Multiple rDNA ITS-types within the diatom *Pseudo-nitzschia delicatissima* (Bacillariophyceae) and their relative abundances across a spring bloom in the Gulf of Naples. *Marine Ecology Progress Series*, 271: 87–98.

Orsini, L., Sarno, D., Procaccini, G., Poletti, R., Dahlmann, J., and Montresor, M. (2002) Toxic *Pseudo-nitzschia multistriata* (Bacillariophyceae) from the Gulf of Naples: morphology, toxin analysis and phylogenetic relationships with other *Pseudo-nitzschia* species. *European Journal of Phycology*, 37: 247–257.

Parsons, W.M., Schlichting, H.E., and Stewart, K.W. (1966) In-flight transport of algae and protozoa by selected Odonata. *Transactions of the American Microscopical Society*, 85: 520–527.

Philippe, H., Sörhannus, U., Baroin, A., Perasso, R., Gasse, F., and Adoutte, A. (1994) Comparison of molecular and paleontological data in diatoms suggests a major gap in the fossil record. *Journal of Evolutionary Biology*, 7: 247–265.

Richardson, J.L. (1995) Dominance of asexuality in diatom life cycles: evolutionary, ecological and taxonomic implications. In *Proceedings of the 13th International Diatom Symposium* (eds. D. Marino and M. Montresor), Biopress, Bristol, pp. 129–137.

Rines, J.E.B., Donaghay, P.L., Dekshenieks, M.N.M., Sullivan, J.M., and Twardowski, M.S. (2002) Thin layers and camouflage: hidden *Pseudo-nitzschia* spp. (Bacillariophyceae) populations in a fjord in the San Juan Islands, Washington, USA. *Marine Ecology Progress Series*, 225: 123–137.

Round, F.E. (1970) The genus *Hantzschia* with particular reference to *H. virgata* v. *intermedia* (Grun.) comb. nov. *Annals of Botany*, 34: 75–91.

Round, F.E. (1981) Some aspects of the origin of diatoms and their subsequent evolution. *BioSystems*, 14: 483–486.

Round, F.E. and Crawford, R.M. (1981) The lines of evolution of the Bacillariophyta. I. Origin. *Proceedings of the Royal Society of London, B*, 211: 237–260.

Round, F.E. and Crawford, R.M. (1984) The lines of evolution of the Bacillariophyta. II. The centric series. *Proceedings of the Royal Society of London, B*, 221: 169–188.

Round, F.E., Crawford, R.M., and Mann, D.G. (1990) *The Diatoms. Biology and Morphology of the Genera.* Cambridge University Press, Cambridge, 747 pp.

Ruck, E.C. and Kociolek, J.P. (2004) Preliminary phylogeny of the family Surirellaceae (Bacillariophyta). *Bibliotheca diatomologica*, 50: 1–236.

Rynearson, T.A. and Armbrust, E.V. (2000) DNA fingerprinting reveals extensive genetic diversity in a field population of the centric diatom *Ditylum brightwellii*. *Limnology and Oceanography*, 45: 1329–1340.

Rynearson, T.A. and Armbrust, E.V. (2004) Genetic differentiation among populations of the planktonic marine diatom *Ditylum brightwellii* (Bacillariophyceae). *Journal of Phycology*, 40: 34–43.

Rynearson, T.A. and Armbrust, E.V. (2005) Maintenance of clonal diversity during a spring bloom of the centric diatom *Ditylum brightwellii*. *Molecular Ecology*, 14: 1631–1640.

Rynearson, T.A., Newton, J.A., and Armbrust, E.V. (2006) Spring bloom development, genetic variation and population succession in the planktonic diatom *Ditylum brightwellii*. *Limnology and Oceanography*, 51: 1249–1261.

Sarno, D., Kooistra, W.H.C.F., Medlin, L.K., Percopo, I., and Zingone, A. (2005) Diversity in the genus *Skeletonema* (Bacillariophyceae). II. An assessment of the taxonomy of *S. costatum*-like species with the description of four new species. *Journal of Phycology*, 41: 151–176.

Saunders, G.W., Potter, D., Paskind, M.P., and Andersen, R.A. (1995) Cladistic analyses of combined traditional and molecular data sets reveal an algal lineage. *Proceedings of the National Academy of Sciences of the USA*, 92: 244–248.

Scala, S., Carels, N., Falciatore, A., Chiusano, M.L., and Bowler, C. (2002) Genome properties of the diatom *Phaeodactylum tricornutum*. *Plant Physiology*, 129: 1–10.

Schiller, W. and Lange-Bertalot, H. (1997) *Eolimna martinii* n. gen., n. sp. (Bacillariophyceae) aus dem Unter-Oligozän von Sieblos/Rhön im Vergleich mit ähnlichen rezenten Taxa. *Paläontologische Zeitschrift*, 71: 163–172.

Scholin, M.A., Villac, M.C., Buck, K.R., Powers, J.M., Fryxell, G.A., and Chavez, F.P. (1994) Ribosomal DNA sequences discriminate among toxic and non-toxic *Pseudonitzschia* species. *Natural Toxins*, 2: 152–165.

Schrader, H.-J. (1969) Die pennaten Diatomeen aus dem Obereozän von Oamaru, Neuseeland. *Nova Hedwigia, Beiheft*, 28: 1–124.

Sherbakova, T.A., Kiril'chik, S.V., Likhoshvaj, E.V., and Grachev, M.A. (1998) Filogeneticheskoe polozhenie diatomovykh vodoroslej roda *Aulacoseira* iz ozera Bajkal po rezul'tatam sravneniya nukleotidnykhposledovatel'nostej uchastka gena 18S rRNK. *Molekulyarnaya Biologiya*, 32: 735–740.

Sherbakova, T.A., Masyukova, Yu.A., Safonova, T.A., Petrova, D.P., Vereshagin, A.L., Minaeva, T.V., Adelshin, R.V., Triboy, T.I., Stonik, I.V., Aizdaitcher, N.A., Kozlov, M.V., Likhoshway, Ye.V., and Grachev, M.A. (2005) Conserved motif CMLD in silicic acid transport proteins of diatoms. *Molekulyarnaya Biologiya*, 39: 303–316.

Sherbakova, T.A., Rubtsov, N.B., Likhoshway, Ye.V., and Grachev, M.A. (2000) Combined SEM ultrastructure studies and PCR with individual diatom cells. *Diatom Research*, 15: 349–354.

Simonsen, R. (1972) Ideas for a more natural system of the centric diatoms. *Nova Hedwigia, Beiheft*, 39: 37–53.

Simonsen, R. (1979) The diatom system: ideas on phylogeny. *Bacillaria*, 2: 9–71.

Sims, P.A., Mann, D.G., and Medlin, L.K. (2006) Evolution of the diatoms: insights from fossil, biological and molecular data. *Phycologia*, 45: 361–402.

Sinninghe Damsté, J.S., Muyzer, G., Abbas, B., Rampen, S.W., Masse, G., Allard, W.G., Belt, S.T., Robert, J.-M., Rowland, S.J., Moldowan, J.M., Barbanti, S.M., Fago, F.J., Denisevich, P., Dahl, J., Trindade, L.A.F., and Schouten, S. (2004) The rise of the rhizosolenid diatoms. *Science*, 304: 584–587.

Skov, J., Lundholm, N., Pocklington, R., Rosendahl, S., and Moestrup, Ø. (1997) Studies on the marine planktonic diatom *Pseudo-nitzschia*. 1. Isozyme variation among isolates of *P. pseudodelicatissima* during a bloom in Danish coastal waters. *Phycologia*, 36: 374–380.

Sorhannus, U. (1996) Higher ribosomal RNA substitution rates in Bacillariophyceae and Dasycladales than in Mollusca, Echinodermata, and Actinistia–Tetrapoda. *Molecular Biology and Evolution*, 13: 1032–1038.

Sorhannus, U. (1997) The origination time of diatoms: an analysis based on ribosomal RNA data. *Micropaleontology*, 43: 215–218.

Sorhannus, U. (2005) Diatom phylogenetics inferred based on direct optimization of nuclear-encoded SSU rRNA sequences. *Cladistics*, 20: 487–497.

Sorhannus, U., Gasse, F., Perasso, R., and Baroin Tourancheau, A. (1995) A preliminary phylogeny of diatoms based on 28S ribosomal RNA sequence data. *Phycologia*, 34: 65–73.

Sorhannus, U. and Kosakovsky Pond, S.L. (2006) Evidence for positive selection on a sexual reproduction gene in the diatom genus *Thalassiosira* (Bacillariophyta). *Journal of Molecular Evolution*, 63: 231–239.

Soudek, D. and Robinson, G.G.C. (1983) Electrophoretic analysis of the species and population structure of the diatom *Asterionella formosa*. *Canadian Journal of Botany*, 61: 418–433.

Stoermer, E.F. and Ladewski, T.B. (1982) Quantitative analysis of shape variation in type and modern populations of *Gomphoneis herculeana*. *Nova Hedwigia, Beiheft*, 73: 347–386.

Stoermer, E.F. and Smol, J.P. (eds.) (1999) *The Diatoms: Applications for the Environmental and Earth Sciences*. Cambridge University Press, Cambridge.

Stosch, H.A. von (1982) On auxospore envelopes in diatoms. *Bacillaria*, 5: 127–156.

Strel'nikova, N.I. (1992) *Paleogenovye diatomovye vodorosli*. Izdatel'stvo S.-Peterburgskogo Universiteta, St. Petersburg.

Takano, Y. and Horiguchi, T. (2005) Acquiring scanning electron microscopical, light microscopical and multiple gene sequence data from a single dinoflagellate cell. *Journal of Phycology*, 42: 251–256.

Tamura, M., Shimada, S., and Horiguchi, T. (2005) *Galeidinium rugatum* gen. et sp. nov. (Dinophyceae), a new coccoid dinoflagellate with a diatom endosymbiont. *Journal of Phycology*, 41: 658–671.

Thamatrakoln, K., Alverson, A.J., and Hildebrand, M. (2006) Comparative sequence analysis of diatom silicon transporters: toward a mechanistic model of silicon transport. *Journal of Phycology*, 42: 822–834.

Thamatrakoln, K. and Hildebrand, M. (2005) Approaches for functional characterization of diatom silicic acid transporters. *Journal of Nanoscience and Nanotechnology*, 5: 158–166.

Tingey, S.V. and del Tufo, J.P. (1993) Genetic analysis with random amplified polymorphic DNA markers. *Plant Physiology*, 101: 349–352.

Trobajo, R., Mann, D.G., Chepurnov, V.A., Clavero, E., and Cox, E.J. (2006) Auxosporulation and size reduction pattern in *Nitzschia fonticola* (Bacillariophyta). *Journal of Phycology*, 42: 1353–1372.

van den Hoek, C., Mann, D.G., and Jahns, H.M. (1995) *Algae. An Introduction to Phycology*. Cambridge University Press, Cambridge.

Van de Peer, Y., Van der Auwera, G., and De Wachter, T. (1996) The evolution of stramenopiles and alveolates as derived by "substitution rate calibration" of small ribosomal subunit RNA. *Journal of Molecular Evolution*, 42: 201–210.

VanLandingham, S.L. (1967–1978) *Catalogue of the Fossil and Recent Genera and Species of Diatoms and Their Synonyms*, 8 vols. J. Cramer, Lehre, Germany.

Vanormelingen, P., Chepurnov, V.A., Mann, D.G., Cousin, S., Sabbe, K., and Vyverman, W. (2007) Congruence of morphological, reproductive and ITS rDNA sequence data in some Australasian *Eunotia bilunaris* (Bacillariophyta). *European Journal of Phycology*, 42: 61–79.

Vinogradov, S.N., Hoogewijs, D., Bailly, X., Arredondo-Peter, R., Guertin, M., Gough, J., Dewilde, S., Moens, L., and Vanfleteren, J.R. (2005) Three globin lineages belonging to two structural classes in genomes from the three kingdoms of life. *Proceedings of the National Academy of Sciences of the USA*, 102: 11386–11389.

Williams, D.M. (1985) Morphology, taxonomy and inter-relationships of the ribbed araphid diatoms from the genera *Diatoma* and *Meridion* (Diatomaceae: Bacillariophyta). *Bibliotheca diatomologica*, 8: 1–228.

Wolf, M. (2004) The secondary structure of the ITS2 transcript in *Cyclotella* and *Stephanodiscus* (Thalassiosiraceae, Bacillariophyta). *Diatom Research*, 19: 135–142.

Wolf, M., Scheffler, W., and Nicklisch, A. (2002) *Stephanodiscus neoastraea* and *Stephanodiscus heterostylus* (Bacillariophyta) are one and the same species. *Diatom Research*, 17: 445–451.

Wortley, A.H., Rudall, P.J., Harris, D.J., and Scotland, R.W. (2005) How much data are needed to resolve a difficult phylogeny? Case study in Lamiales. *Systematic Biology*, 54: 697–709.

Zechman, F.W., Zimmer, E.A., and Theriot, E.C. (1994) Use of ribosomal DNA internal transcribed spacers for phylogenetic studies in diatoms. *Journal of Phycology*, 30: 507–512.

14 Classification of the Phaeophyceae from past to present and current challenges

Bruno de Reviers, Florence Rousseau, and Stefano G.A. Draisma

CONTENTS

ABSTRACT

For a long time, it has been impossible to unravel phylogenetic relationships within the brown algae. Various systems of classification of the Phaeophyceae on the basis of morphology have been proposed. One of the main disputes concerning brown algal classification was whether to recognize a narrower or a wider circumscription for the Ectocarpales and whether or not some or all of the diplobiontic brown algae with conceptacles (i.e., Durvillaeales, Notheiales, and Ascoseirales) belong or, at least, are closely related to the Fucales. Our knowledge has greatly improved since phylogenies can be reconstructed based on DNA sequence data. This has led to a completely new phylogenetic hypothesis (Ectocarpales is not basal and Fucales is not sister to all other brown algae), and many taxa have moved to other or newly created orders and families. This upheaval is summarized: previous and current hypotheses are reviewed and an updated classification is provided.

INTRODUCTION

The brown algae are gathered in a class named either Melanophyceae (Rabenhorst, 1863), Fucophyceae (Warming, 1884), or Phaeophyceae (Kjellman 1891 [1891–1896]). All these names are in accordance with the ICBN (Greuter et al., 2000), see Silva (1980). The name the most in use at the present time is Phaeophyceae.

In the tree of life, brown algae belong to the subregnum Heterokonta (Cavalier-Smith 1986 = Stramenopiles Patterson 1989), defined by heterokont flagellated cells (either monads or, as in the Phaeophyceae, reproductive cells), i.e. cells possessing one flagellum bearing tubular tripartite structures named mastigonemes and a second one either naked or bearing different flagellar structures or reduced to the basal body or even lost. The phototrophic heterokont organisms (the golden-brown algae) are gathered in the phylum Ochrophyta (Cavalier-Smith, 1995a, 1995b; Cavalier-Smith and Chao, 1996). These algae are defined by their plastids of red algal secondary endosymbiontic origin, with

267

lamellae made of three stacked thylakoids, including a girdle lamella that runs continuously beneath the plastid four membranes. The phylum Ochrophyta is subdivided in about 14 classes (Reviers, 2003), depending on authors, three of which being more closely related to the brown algae: the (predominantly) freshwater Xanthophyceae (= Tribophyceae) and Phaeothamniophyceae, and the marine Schizocladiophyceae, the last (currently containing a single species) being considered sister to the Phaeophyceae (Kawai et al., 2003). Two autapomorphies (own derived characters, evolutionary innovations) are distinctive for brown algae: sexual reproduction involving peculiar uni- and plurilocular zoidangia, and plasmodesmata (protoplasmic connections between adjacent cells). The plasmodesmata of brown algae are not homologous to those of the green lineage of which they differ in lacking the desmotubule (Kawai et al., 2003).

Besides these autapomorphies, the Phaeophyceae are multicellular photoautotrophic organisms, currently classified in about 285 genera (Reviers, personal data) and some 1792 (www.algaebase.org) or 2000 (Van den Hoek et al., 1995) species, a number difficult to assess precisely. Their size ranges from a few millimetres to 40 metres and even more in the largest seaweed known: *Macrocystis pyrifera* (Linnaeus) C. Agardh (Womersley, 1954). The golden colour of their plastids is due to carotenoid pigments, in particular, fucoxanthin and violaxanthin, that they possess in addition to chlorophylls a, c_1, and c_2 (Jeffrey, 1989). Their food storage carbohydrate, laminarin, is a hydrosoluble β_{1-3} glucan, some chains of which are terminated by a mannitol unit (Craigie, 1974). Cell walls of brown algae contain cellulose, alginic acid, an acidic polyuronid widely used in diverse industries, and fucoidans (fucose-rich sulphated polysaccharides) (Kloareg and Quatrano, 1988). Brown algal cells contain physodes, vesicular bodies containing phloroglucinol polymers (phlorotannins) that Schoenwaelder and Clayton (1998, 1999) have shown to be excreted as a component of cell walls.

The brown algae are almost exclusively marine; only a few species belonging to eight genera (i.e., *Bodanella*, *Pseudobodanella*, *Heribaudiella*, *Lithoderma*, *Sphacelaria*, *Pleurocladia*, *Porterinema*, and *Ectocarpus*) occur in freshwater (Bourrelly, 1981; West and Kraft, 1996). The freshwater representatives of the latter three genera are euryhaline. All freshwater species have been reported from the northern hemisphere, except for the cosmopolitan marine species *Ectocarpus siliculosus* (Dillwyn) Lyngbye that has been found in a fully freshwater habitat in Australia (West and Kraft, 1996). Brown algae are a typical component of the marine littoral flora, from the subpolar areas to the equator. However, their greatest diversity is observed in cold and temperate waters. Rockweeds (Fucaceae) are typical of the tidal zone in north-Atlantic temperate areas. Kelps (a term usually referring to members of the Laminariales, but sometimes also used in a broader sense for other large brown algae) form forests in the sublittoral zone of cold and temperate areas, except in the Antarctic. Dominant brown algae encountered in tropical waters are mainly species assigned to Sargassaceae and Dictyotales. Brown algae are benthic organisms, although two pelagic species of *Sargassum* are scattered in an area of the Atlantic Ocean known as the Sargasso Sea, and *Pylaiella littoralis* (Linnaeus) Kjellman can also become pelagic and form brown tides in the state of Massachusetts (Wilce et al., 1982). One single lichenic symbiosis, involving the brown alga *Petroderma*, has been confirmed (Peters and Moe, 2001; Sanders et al., 2004, 2005).

Our knowledge of the brown algae has greatly benefited from DNA-sequence data. The consequence of their use has led to an overthrow of their classification and also, since the year 2001, to a completely new phylogenetic hypothesis implying a new concept of their evolution. The story of this upheaval is summarized below.

HISTORICAL SUBDIVISIONS OF THE BROWN ALGAE AND ORDER DELINEATION

Although his main criterion is thallus organisation, Lamouroux (1813a, 1813b) was the first author to define plant groups using colour, i.e. considering their pigments. However, he did not determine any relationship between colour and the concept of accessory pigments and it was not until Kützing (1843) that this connection was made. In subsequent systems of the early 19th century, colour had no priority over morphology and some red algae or even the lichen *Lichina* were classified in the brown algae. Until Areschoug (1847), the brown algal genera were scattered either in the algae as a whole or in families not always considered as closely related. Some improvements slowly appeared from the year 1840 (Areschoug, 1847; Agardh, 1848).

During the 19th century, the dictyotalean algae often had a separate placement. While C. Agardh (1824) and J. Agardh (1848) considered the Dictyotaceae as brown algae, Cohn (1865) considered them as close relatives of the Florideae because of the non-swimming spores producing tetrasporangia of *Dictyota*. Hauck (1885 [1883–1885]) tentatively placed them in the Phaeophyceae and this placement was followed by De Toni (1895), Oltmanns (1904 [1904–1905]), and subsequent authors. However, Kjellman (1891 [1891–1896]) still considered the Phaeophyceae and the Dictyotales as separate groups, although very close relatives.

A more modern concept of the brown algae appeared with Chodat (1909) who distinguished the "Phéophycées" from the "Chlorophycées" on the basis of absence of starch, presence of "*paramylum*" (i.e., laminarin), accessory pigments obscuring the chlorophyll, plastids without pyrenoids or with pyrenoids different from that known in green algae, and zoids "*asymétriques*" or with "*cils* [flagella] *asymétriques*" in the former. However, Chodat included in this group heterokont algae, euglenoids, and dinoflagellates.

At present, the definition of the brown algae includes numerous features and two autapomorphies (see Introduction). Three main historical periods can be distinguished with respect to the subdivision of the brown algae (Reviers and Rousseau, 1999). From Areschoug (1847) to Oltmanns (1904 [1904–1905]), the classification is greatly influenced by the progressive discovery of the sexuality of algae (including the browns), e.g., Areschoug distinguished algae with (Cyclosporae) and without (Episporae) conceptacles. Important studies of this period include those of Areschoug (1847), Thuret (1850–1851, 1854–1855), Hauck (1885 [1883–1885]), Kjellman (1891 [1891–1896]), De Toni (1895), and Oltmanns (1904 [1904–1905]).

From Kylin (1917) to Setchell and Gardner (1925), including Oltmanns (1922) and Taylor (1922), a consistent division of the class in orders appeared (based on the knowledge of that period). Camille Sauvageau, a pioneer in the field of life histories, discovered the life history of *Saccorhiza polyschides* (Lightfoot) Batters and the Laminariales (Sauvageau 1915 and following works). The classification is then based on the type of life history (similar or dissimilar generations), the type of gamy (iso-, aniso-, oo-), and the type of spore (motile or not).

The third period can be dated from Kylin (1933) to 2001. It is also from that time that phylogenetic hypotheses were more or less clearly formulated, but care has to be taken not to apply our current cladistic concepts to pre-Hennig (1950, 1966) studies. In 1933, Kylin proposed a system (Figure 14.1) founded to a large extent upon life-histories (at the class level) and upon vegetative

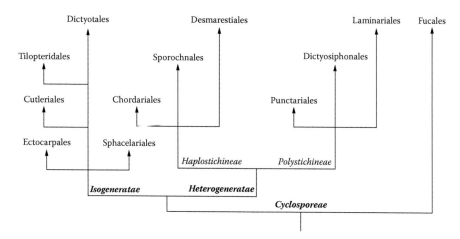

FIGURE 14.1 Diagram of Kylin's (1933, p. 93) system. (From Kylin, H., *Lunds Universitets Årsskrift, Ny Földj*, Avd. 2, Band 29 (Nr 7), 1933, p. 93. With permission.)

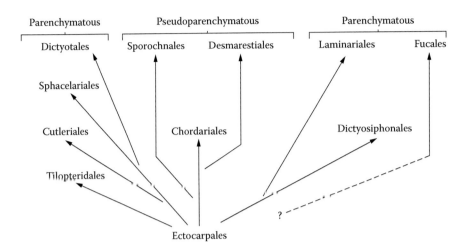

FIGURE 14.2 Diagram produced by Papenfuss. (From Papenfuss, G.F., Svensk Botanisk Tidskrift utgivfen af svenska botaniska föreningen, Stockholm, 45, 8, 1951b; and Papenfuss, G.F., In Proceedings of the Seventh International Botanical Congress, Stockholm, 1950 (1953), p. 822. With permission.)

structure (at the subclass level). This system appears as a logical result of the observations made since the end of the 19th century. Kylin (1933) placed the orders of brown algae in three classes: Isogeneratae, Heterogeneratae, and Cyclosporeae. The class Heterogeneratae was further divided in two subclasses: Haplostichineae and Polystichinae. Fritsch (1945) and Papenfuss (1951b, 1953, 1955) rejected Kylin's classes and placed all orders in one class Phaeophyceae (Figure 14.2). Fritsch also pointed out that that Kylin's grouping obscures the fact that polystichous thalli also occur in the Cyclosporeae (Fucales) and the Isogeneratae (Dictyotales, Sphacelariales). Both Kylin and Papenfuss systems received wide recognition. In more recent studies (Christensen, 1980; Wynne and Kraft, 1981; Wynne, 1982; Van den Hoek et al., 1995; Womersley, 1987) subdivision of the brown algae in classes and subclasses was generally no longer maintained and only one class was recognized. However, there have been a few exceptions that always aimed to give a separate position to the Fucales. Kylin's Cyclosporeae was reinstated as the subclass Cyclosporidae by Scagel (1966; Figure 14.3) and followed by Wynne and Loiseaux (1976; Figure 14.4), on the basis of the different flagellation in the Fucales (and probably Durvillaeales), the other orders of brown algae being placed in the subclass Phaeophycidae.

The orders, as currently conceived, appeared in Oltmanns (1922). Since that time, ordinal delineation in the Phaeophyceae traditionally has been based on the type of life cycle, type of gamy, mode of growth, and the filamentous vs parenchymatous construction of the thallus (e.g. Oltmanns, 1922; Kylin, 1933; Fritsch, 1945; Papenfuss, 1951a, b; Scagel, 1966; Nakamura, 1972; Wynne and Loiseaux, 1976; Van den Hoek and Jahns, 1978; Bold and Wynne, 1985; Christensen, 1980; Clayton, 1981; Womersley, 1987; Van den Hoek et al., 1995). Various systems of classification of the Phaeophyceae based on these morphological grounds have been proposed (See Table 5 in Reviers and Rousseau, 1999). One of the main disputes concerning brown algal classification has been whether to recognize narrower (*sensu stricto*) or wider (*sensu lato*) circumscription for the Ectocarpales; this topic has been reviewed in Rousseau and Reviers (1999a). Another matter of debate was whether or not some or all the diplobiontic brown algae with conceptacles (Durvillaeales, Notheiales, and Ascoseirales) belonged or, at least, were closely related to the Fucales (reviewed in detail in Rousseau and Reviers, 1999b). A detailed treatment of history and concepts of classification and phylogeny within the Phaeophyceae is reported in Reviers and Rousseau (1999).

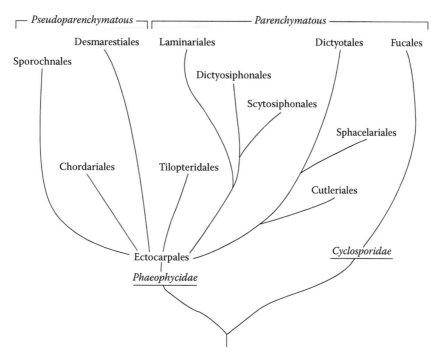

FIGURE 14.3 Diagram produced by Scagel. (From Scagel, R.F., *Oceanography and Marine Biology, Annual Review*, 4, 130, 1966. With permission.)

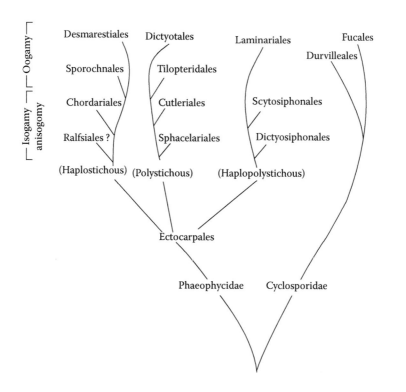

FIGURE 14.4 Diagram produced by Wynne and Loiseaux (1976, p. 448).

MOLECULAR PHYLOGENIES

With the possibility to generate DNA sequence data and steadily improving computational power for cladistic analyses, insights into the evolution of brown algae have improved drastically.

No cladistic analysis of morphological characters has ever been performed in order to reconstruct brown algal evolution. Ancestral and derived status of morphological characters was assigned speculatively, because different authors attributed varying importance to those characters. Actually, knowledge about morphology and biochemistry (e.g. pheromones) is unequally distributed within the brown algae, and the number of these characters is furthermore limited. In addition, hypotheses of primary homologies are difficult to assess. For instance, are the sori of unilocular sporangia intermixed with paraphyses homologous in *Laminaria, Saccorhiza,* and *Adenocystis*? Are the conceptacles of *Durvillaea, Splachnidium,* and *Fucus* homologous? The topology of the reconstructed phylogenetic tree will depend on the answer to these questions when coding the characters. DNA sequences make a set of characters independent from morphological and biochemical ones. Our understanding of the classification and phylogeny of brown algae has thus undergone marked changes since the early 1990s because of the contribution of molecular phylogenies that can be confronted with morphological and biochemical data.

The year 1993 saw a pioneer study by Tan and Druehl, using SSU rDNA (Tan and Druehl, 1993). The successive main steps of the contribution of molecular phylogenies of the Phaeophyceae are summarized in Table 14.1; some matter of debate with phylogenetic consequences is outlined below.

As soon as 1993, molecular results have suggested that an order Ectocarpales *sensu lato* made sense. Indeed, a broad concept of Ectocarpales, including the Chordariales, Dictyosiphonales, and Scytosiphonales, leads to a monophyletic group well supported in the trees constructed in this study, the Chordariales and Dictyosiphonales being polyphyletic. Later, some Ralfsiales (brown crusts), the Tilopteridaceae, and algae with stellate plastids were shown not to belong to the Ectocarpales (Tan and Druehl, 1994; Siemer et al., 1998; Peters and Clayton, 1998; Rousseau and Reviers, 1999a) and Rousseau and Reviers (1999a) proposed a new delineation for Ectocarpales *sensu lato*, including Ectocarpales *sensu stricto*, Chordariales, Dictyosiphonales, Punctariales, and Scytosiphonales, all of which have one or several plastids, each with one or several stalked pyrenoids, and excluding Tilopteridaceae, Ralfsiales *sensu* Nakamura (1972), Scytothamnales, *Asteronema, Bachelotia,* and *Asterocladon* which have either plastids without pyrenoids or non-pedunculate pyrenoids. Rousseau and Reviers (1999a) considered stalked pyrenoids to be a synapomorphy (shared derived character) of the Ectocarpales and a diagnostic character of the order. Multiple plastids without a pyrenoid appear to be the plesiomorphic condition in the brown algae. This concept of the Ectocarpales has been confirmed by more recent studies (Cho et al., 2004; Kawai and Sasaki, 2004; Kawai et al., 2005). The pyrenoid and, more generally, plastid ultrastructure, has turned out to be a very important phylogenetic marker in the brown algae (on this topic, see Reviers and Rousseau, 1999, and references quoted herein). Peters and Ramírez (2001), using partial LSU and SSU rDNA, *rbc*L and *rbc*L-*rbc*S spacer sequence data, proposed a new concept for families in the Ectocarpales in order to reconcile nomenclature and phylogeny: five families (including the Adenocystaceae erected by Rousseau et al., 2000) instead of the numerous ones considered previously (23 listed in Reviers and Rousseau, 1999). As more ectocarpalean DNA sequences are being determined, more monophyletic clades deserving family status may become apparent. The delineation of the currently recognized families is far from being clear; e.g.: Cho and Boo (2006) have generated sequences of *Petrospongium* that appeared closely related to the Ectocarpaceae but may deserve creation of a new family.

Early studies (Tan and Druehl, 1993, 1994, 1996; Saunders and Kraft, 1995) showed the limited informativeness of SSU rDNA alone to infer phylogenetic relationships within the brown algae. Later on, LSU nuclear rDNA (from Rousseau et al., 1997), as well as plastid-encoded *rbc*L (from Draisma et al., 2001) and *psa*A (from Cho et al., 2004) have proved to be much more informative at the class level.

Rousseau and Reviers (1999b) used combined partial SSU and LSU rDNA to include *Notheia* (anisogamous and of simple construction, previously in the Notheiales a result already suggested by Saunders and Kraft in 1995) and *Durvillaea* (of filamentous construction and with diffuse growth, previously in the Durvillaeales) within the Fucales, an order considered until then, strictly oogamous

TABLE 14.1

Main Molecular Studies Having Led Either to Modification of the Classification at the Family or Ordinal Level, or to New Phylogenetic Hypotheses

Year and Authors	Molecular Marker(s) Used	Main Result at Family and/or Ordinal Level
1993 Tan and Druehl	Nuclear 18S rDNA (complete)	The Chordariales and Dictyosiphonales are not monophyletic groups; the concept of Ectocarpales *s.l.* is supported; 18S rDNA is poorly informative
1994 Tan and Druehl	Nuclear 18S rDNA (partial)	*Ralfsia* and *Analipus* appear not to belong to the Ectocarpales *s.l.*
1995 Saunders and Kraft	Nuclear 18S rDNA (partial)	*Notheia* belongs to the Fucales
1996 Tan and Druehl	Nuclear 18S rDNA (partial)	The Laminariales as conceived at that time appear polyphyletic
1997 Rousseau et al.	C'1-D2 of the 26S nuclear rDNA	Utility of 26S nrDNA
1998 Peters	Partial 18S nuclear rDNA + ITS 1 and 2	The genus *Halosiphon* (Halosiphonaceae) is reinstated for *Chorda tomentosa* (Chordaceae)
1998 Peters and Clayton	Partial 18S nuclear rDNA and partial 26S nuclear rDNA + ITS 1 and 2	Erection of the order Scytothamnales for brown algae with a single central stellate plastid with one central pyrenoid immersed in the stroma for three genera previously placed in the Ectocarpales *s.l.*
1998 Siemer et al.	Plastid encoded *rbc*L + *rbc*LS spacer	Ectocarpales *s.s.*, Chordariales, and Dictyosiphonales are not monophyletic groups; the concept of Ectocarpales *s.l.* is supported; the Tilopteridaceae are excluded from the Ectocarpales although this last result is not clearly established
1999 Boo et al.	Complete 18S nuclear rDNA	Erection of the family Pseudochordaceae
1999 (a) Rousseau and Reviers	Partial 18S + 26S nuclear rDNAs	The Ectocarpales are defined by their plastid(s), each with one or several stalked pyrenoids
1999 (b) Rousseau and Reviers	Partial 18S + 26S nuclear rDNAs	The Notheiales and Durvillaeales are merged in the Fucales, definition of which is consequently emended; the paraphyletic Cystoseiraceae are merged in the Sargassaceae *sensu* De Toni
2000 Rousseau et al.	Partial 18S + 26S nuclear rDNAs	Creation of the family Adenocystaceae within the Ectocarpales
2001 Peters and Ramírez	Partial 18S + 26S nuclear rDNAs + plastid encoded *rbc*L + *rbc*LS spacer	The number of families within the Ectocarpales is restricted to only five: Acinetosporaceae, Adenocystaceae, Chordariaceae s.l., Ectocarpaceae, and Scytosiphonaceae
2000 Kawai and Sasaki	Plastid-encoded *rbc*L, 18S nuclear rDNA, and ITS 1 and 2	Erection of the families Akkesiphycaceae and Halosiphonaceae
2001 Sasaki et al.	Plastid-encoded *rbc*L + *rbc*LS spacer, 5.8S nuclear rDNA + ITS 2 + partial 26S nuclear rDNA	Phyllariaceae and Halosiphonaceae are shown to be related to the Tilopteridaceae and far from the Laminariales
2001 Rousseau et al.	Partial 18S + partial 26S nuclear rDNAs and partial 18S + complete 26S nuclear rDNA on a subdataset	First comprehensive phylogeny; the Fucales are not sister to the rest of the brown algae; Strasburger's hypothesis is confirmed; the Ectocarpales do not make an early divergence but the Dictyotales do; apical growth is

(Continued)

TABLE 14.1 (CONTINUED)
Main Molecular Studies Having Led Either to Modification of the Classification at the Family or Ordinal Level, or to New Phylogenetic Hypotheses

Year and Authors	Molecular Marker(s) Used	Main Result at Family and/or Ordinal Level
		ancestral, not derived as previously thought; *Nemoderma* appears to be sister to the Fucales and *Phaeosiphoniella* does not belong to Tilopteridaceae
2001 Draisma et al.	Plastid-encoded *rbc*L + partial 18S and 26S nuclear rDNA	Same main results as Rousseau et al. (2001) and additional results: *Choristocarpus* is sister to the rest of the brown algae, *Onslowia* and *Verosphacela* do not belong to the Choristocarpaceae but are related to the Sphacelariales; the Cladostephaceae should be merged within the Sphacelariaceae
2001 Draisma and Prud'homme van Reine	Based on the molecular results of Draisma et al. (2001), see above	Creation of the family Onslowiaceae
2003 Draisma et al.	Plastid-encoded *rbc*L	Most complete phylogeny constructed with *rbc*L
2003 Burrowes et al.	*rbc*L + partial 26S nrDNA and *rbc*L + complete 26S nrDNA on a subdataset	Same main results as Draisma et al. (2001), inclusion of *Microzonia* into the Syringodermatales
2004 Cho et al.	Plastid-encoded *rbc*L + *psa*A + *psb*A	Creation of the order Ishigeales for the genus *Ishige* (previously placed in the Ectocarpales) having plastids lacking a pyrenoid; *Ishige* is very distantly related to the Ectocarpales
2004 Kawai and Sasaki	Plastid-encoded *rbc*L and 5.8S + partial 26S nuclear rDNAs + ITS 2	Creation of the family Stschapoviaceae related to the Tilopteridaceae
2005 Kawai et al.	Plastid-encoded *rbc*L and partial 18S + 26S nuclear rDNAs	Creation of the family Phaeostrophionaceae for the genus *Phaeostrophion* (previously placed in the Ectocarpales) having plastids lacking a pyrenoid; *Phaeostrophion* is rather related to the Sphacelariales
2006 Cho and Boo	Plastid-encoded *rbc*L + *psa*A	*Petrospongium* is closely related to the Ectocarpaceae
2006 Cho et al.	Plastid-encoded *psa*A	Creation of the Xiphophoraceae and Bifurcariopsidaceae within the Fucales
2006 Lane et al.	Regions in the nuclear, chloroplast, and mitochondrial genomes	Creation of the Costariaceae in the ALL group and emendation of the three other families within this group
2007 Kawai et al.	*rbc*L and partial 18S rDNA	Reinstatement of the order Discosporangiales for *Discosporangium* and *Choristocarpus* and of the family Discosporangiaceae for the former genus
Submitted Parente et al.	*rbc*L and the C1-D2 domain of the 26S nuclear rDNA	Exclusion of several taxa from the Ralfsiales
Submitted Phillips et al.	Complete 26S nuclear rDNA	Proposal for erecting the Nemodermatales, Onslowiales, and Phaeosiphoniellaceae

and parenchymatous, suggesting a new definition of this order. A further result of this study is that the genus *Splachnidium* is not closely related to the Fucales. Conceptacles in *Splachnidium* and the Fucales are thus not homologous despite their similar ontogeny.

In 1996, Tan and Druehl, still with SSU rDNA, suggested (unfortunately it was poorly supported) that the Laminariales, as conceived at that time, might be polyphyletic (Tan and Druehl, 1996). There is indeed a large range of morphological diversity known within these algae: their growth is either by intercalary meristem or not, paraphyses are unicellular or multicellular, the pheromone is often lamoxirene but sometimes another compound (multifidene or yet undetermined but known not to be lamoxirene [Maier, 1995]), zoospores have an eyespot and a flagellar swelling in some and not in others, spermatozoids do or do not have an eyespot, spermatangia are unilocular or plurilocular, there is either oogamy or anisogamy (see Table 5 in Reviers and Rousseau, 1999). Polyphyly of the Laminariales is thus a result consistent with morphological and biochemical knowledge. A review of the types of swimming cells by Kawai (1992a, 1992b) might suggest such a result: e.g., zoospores of *Akkesiphycus*, *Chorda*, *Halosiphon*, the Phyllariaceae, and *Pseudochorda* have an eyespot and a flagellar swelling, whereas zoospores of Alariaceae, Lessoniaceae, and Laminariaceae (three families later named the ALL group) lack an eyespot and a flagellar swelling. Furthermore, pheromone in the ALL group is consistently lamoxirene, whereas other pheromones occur in the other taxa (when known). Paraphyses are unicellular with a thickened top in the ALL group, whereas paraphyses are either multicellular or unicellular without a thickened top in the other taxa. However, this diversity may be quite possible in a monophyletic clade. A first scan of molecular techniques (rDNA sequence data) on this topic was carried out by Peters (1998) who showed that *Chorda filum* (Linnaeus) Stackhouse was not sister of "*Chorda*" *tomentosa* and that the name *Halosiphon tomentosus* (Lyngbye) Jaasund should be reinstated for the latter. Kawai and Sasaki (2001) erected the family Halosiphonaceae for this species. Boo et al. (1999), using complete SSU rDNA sequences, created the family Pseudochordaceae for the two species of the genus *Pseudochorda*, within the Laminariales. In their study, the Laminariales appeared once more polyphyletic but the tree was poorly supported because the SSU lacks enough variation. Later on, Sasaki et al. (2001), using sequences of the plastid RuBisCO cistron and the nuclear ribosomal cistron, clearly demonstrated that the Laminariales are polyphyletic, because Phyllariaceae and Halosiphonaceae are more closely related to the Tilopteridaceae than to other Laminariales. Furthermore, Kawai and Sasaki (2001) created the family Akkesiphycaceae for the genus *Akkesiphycus*, thought for long to belong to the Dictyosiphonales. More recently, Lane et al. (2006) reorganized the ALL group based on a phylogenetic analysis of over 6000 bp from regions in the nuclear, chloroplast, and mitochondrial genomes. They maintained the families Alariaceae, Laminariaceae, and Lessoniaceae, but with vastly different compositions, as well as proposed the new family Costariaceae for the genera *Agarum*, *Costaria*, *Dictyoneurum*, and *Thalassiophyllum*.

All these studies focussed on a limited number of taxa. Rousseau et al. (2001) produced the first comprehensive phylogeny of the Phaeophyceae using partial SSU and LSU rDNA sequences. This result was immediately and independently confirmed by Draisma et al. (2001) using the same rDNA regions plus *rbc*L sequences. The *rbc*L phylogeny was later updated with additional taxa (Draisma et al., 2003). Their results showed that, contrary to the previous phylogenetic hypotheses (see references in previous section), the Ectocarpales do not make an early divergence and the Fucales are not the sister group of the rest of the brown algae as sometimes suggested (Kylin, 1933; Papenfuss, 1951b; Scagel, 1966; Wynne and Loiseaux, 1976). The Dictyotales make an early divergence: consistently, Clayton (1984) suggested the Dictyotales being an independent lineage not closely related to other brown algae, based on unique sperm with a posterior reduced flagellum and meiosporangia differing from the typical brown algal unilocular organs; moreover, in early classifications, Dictyotales were often not considered to be brown algae. Conceptacles seem to have evolved independently in *Splachnidium*, the Fucales, and in *Ascoseira*. *Nemoderma tingitanum* Schousboe *in* Bornet, an anisogamous Northeast Atlantic and Mediterranean crustose brown alga, previously placed in the Ralfsiales but possessing several plastids (instead of one as in other Ralfsiales) without pyrenoids, may be sister to the Fucales, but the Rousseau et al. (2001) study showed no support and the Draisma et al. study (2001) did not include it. Draisma et al. (2001) showed that the genus *Choristocarpus* (not included in the study by Rousseau et al., 2001) makes the very first divergence and this was later confirmed by Burrowes et al. (2003). A new paradigm of brown algal phylogeny was thus born in 2001: Isomorphic life cycle and apical growth are ancestral, contrary to all previous phylogenetic hypotheses. Strasburger's (1906) hypothesis was confirmed: the gametophytic phase of the Fucales is extremely reduced and included in the sporophyte. A further result of the study by Draisma et al. (2001, 2002) was to modify some generic and familial concepts in the Sphacelariales, in particular, the monotypic family Cladostephaceae has to be merged in the Sphacelariaceae. Draisma and Prud'homme van Reine (2001) created a new family Onslowiaceae for the genera *Onslowia* and *Verosphacela*, previously placed in the Choristocarpaceae, and which position, although close to the Sphacelariales, remained somewhat unclear. Kawai et al. (2007) recently reported on the phylogenetic placement of the monotypic genus *Discosporangium* based on *rbc*L and partial 18S rDNA sequences. Together with *Choristocarpus*, *Discosporangium* forms a monophyletic clade that is sister to all other brown algae, and Kawai et al. (2007) proposed to reinstate the order Discosporangiales Schmidt 1937 to accommodate them.

Cho et al. (2004) combined three plastid genes: *rbc*L + *psa*A + *psb*A (the latter two for the first time). They showed that the two species of the genus *Ishige* do not belong to Ectocarpales but make an early divergence (after *Choristocarpus*). They created the new order Ishigeales to accommodate them. *Ishige* species do not have any pyrenoids, which is consistent with the new delineation of the Ectocarpales according to Rousseau and Reviers (1999a). This study also confirms independently that apical growth and an isomorphic life cycle are plesiomorphic within the Phaeophyceae. The next clade to diverge after the Ishigeales includes the Onslowiaceae (*rbc*L data only), Sphacelariales, Syringodermatales, and Dictyotales, but there is no branch support. This phylogenetic hypothesis, however, is notable because these algae share a polystichous construction, whereas other multiseriate brown algae (with the exception of Tilopteridaceae and Cutleriaceae) are parenchymatous.

Kawai and Sasaki (2004), using *rbc*L and rDNA, showed that the poorly known genus *Stschapovia* deserves the creation of a new family sharing numerous morphological similarities with the Tilopteridaceae. They placed their new family Stschapoviaceae in the order Tilopteridales with a new broad concept, including Tilopteridaceae, Halosiphonaceae, and Phyllariaceae. The study did not include any representative of the Cutleriaceae, a family shown to be closely related to Tilopteridaceae (Rousseau et al., 2001; Cho et al., 2004).

Kawai et al. (2005) showed that the genus *Phaeostrophion* does not belong to the Ectocarpales and, consistently, does not possess any pyrenoids. It is closely related to the Sphacelariales and the new family Phaeostrophiaceae was created to accommodate this genus. The family was left *incertae sedis* and not incorporated in the Sphacelariales.

STATE OF THE SCIENCE IN 2006

Our knowledge of the Phaeophyceae has tremendously improved, thanks to molecular techniques. However, the puzzle is far from being completed. New orders and families have just been published or are under review for publication: monotypic Bifurcariopsidaceae and monogeneric (two species) Xiphophoraceae within the Fucales (Cho et al., 2006) (based on new *psa*A and previous published rDNA sequence data), Onslowiales (one family: Onslowiaceae), monotypic Nemodermatales, and monotypic Phaeosiphoniellaceae (Philips et al., submitted; based on complete LSU nrDNA sequence data and *rbc*L). Other expected issues are possible new families in the Ectocarpales (Cho and Boo, 2006, using *psa*A, Racault et al. unpublished results presented at the 54th Annual Meeting of the British Phycological Society in January 2006, using *rbc*L and *rbc*L/S spacer) and exclusion of many taxa from the Ralfsiales (*Petroderma*, *Heribaudiella*, *Pseudolithoderma* spp.; Parente et al., unpublished results presented at the Eighth International Phycological Congress, in Durban, South Africa, in August 2005, using *rbc*L and the C1-D2 domain of the LSU). Our present knowledge of brown algal classification is summarized in Table 14.2 and Figure 14.5.

TABLE 14.2
Update of Brown Algal Classification

Discosporangiales
  Choristocarpaceae
  Discosporangiaceae
Ishigeales
  Ishigeaceae
Petrodermatales (*submitted*)
  Petrodermataceae (*submitted*)
Dictyotales
  Dictyotaceae
  Dictyotopsidaceae
Sphacelariales
  Sphacelariaceae
  Stypocaulaceae
  ? Phaeostrophiaceae
  ? Lithodermataceae
  ? *Bodanella*
Onslowiales (*submitted*)
  Onslowiaceae
Syringodermatales
  Syringodermataceae
Ascoseirales
  Ascoseiraceae
Nemodermatales (*submitted*)
  Nemodermataceae
Fucales
  Seiroccoceae
  Notheiaceae
  Sargassaceae (incl. Cystoseiraceae)
  Durvillaeaceae
  Bifurcariopsidaceae
  Himanthaliaceae
  Hormosiraceae
  Xiphophoraceae
  Fucaceae
Tilopteridales (incl. Cutleriales)
  Tilopteridaceae
  Cutleriaceae
  Phyllariaceae
  ? Halosiphonaceae
  ? Stschapoviaceae
Related to Laminariales
  Phaeosiphoniellaceae (*submitted*)
Laminariales (*Kelp*)
  Chordaceae
  Akkesiphycaceae
  Pseudochordaceae
  « ALL group »
  Alariaceae
  Costariaceae
  Lessoniaceae
  Laminariaceae

(*Continued*)

TABLE 14.2 (CONTINUED)
Update of Brown Algal Classification

Ectocarpales
  Chordariaceae
  Ectocarpaceae
  Petrospongium (sister of Ectocarpaceae)
  Acinetosporaceae
  Scytosiphonaceae
  Adenocystaceae
  Further families expected
Order and family names have to be created for the genus *Asterocladon* which is related to the Ectocarpales
Scytothamnales
  Asteronema
  Splachnidiaceae
  Scytothamnaceae
  Stereocladon
Desmarestiales
  Arthrocladiaceae
  Desmarestiaceae
Sporochnales
  Sporochnaceae
Ralfsiales
  Ralfsiaceae
Incertae sedis
Bachelotia (forms its own branch in the brown algal crown or is related to the Scytothamnales ?), *Herpodiscus* (related
  to Sphacelariales ?), *Pseudobodanella* (very close to *Porterinema* according Bourrelly, 1981), *Porterinema* (related to
  Ishigeales and *Petroderma*?), *Zosterocarpus*

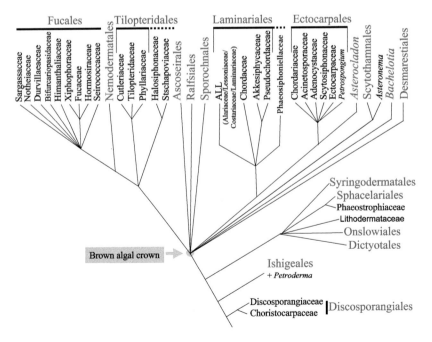

FIGURE 14.5 Synthetic tree presenting an update of brown algal phylogeny in 2006.

FUTURE RESEARCH

On a small scale, several subgroups such as the Cystoseiraceae (nowadays merged with Sargassaceae, Fucales) and the Ectocarpales need further phylogenetic investigations. Highly variable markers and, possibly, population genetics methods may turn out to be necessary for some taxa such as, e.g., *Cystoseira* spp. Molecular markers will be developed for barcoding and are expected to become routine tools, extremely useful for notoriously difficult taxa such as *Sargassum* spp., *Cystoseira* spp., some Dictyotales, and some Ectocarpales. Several studies reveal distinct clades within a (putative) morphological species indicating that it actually might comprise several cryptic species (Kawai et al., 2001; Kim and Kawai, 2002; Kim et al., 2003; Kawai et al., 2007).

On a large scale, the main challenge will be the resolution of the brown algal crown, an obligate step for the understanding of brown algal evolution. Another interesting topic will be to state whether or not Dictyotales, Sphacelariales, Onslowiales, and Syringodermatales (DSOS) make a clade sister to the brown algal crown, or if they make successive early divergent lineages (i.e., a grade), with the Syringodermatales sister of the crown as suggested by some phylogenetic studies. In the first hypothesis, the DSOS clade will be composed of polystichous taxa (haplostichous construction of *Sphacella*, and oligostichous contruction of *Onslowia* and *Verosphacela* being interpreted as secondarily derived), while the other brown algae are either haplostichous, oligostichous, or parenchymatous, with the exception of the Tilopteridaceae and Cutleriaceae. The DSOS clade could thus become the

---

In a few cases DNA sequence data have been published under wrong taxonomic names because of misidentification of the studied specimen or because DNA of another organism (contamination) was amplified in the PCR. These data can cause confusion and therefore we try to give an overview of sequence data that must be interpreted with caution. However, we cannot guarantee that this overview is complete. In Tan and Druehl (1993), a putative *Desmarestia* sp. gametophyte was included and this specimen grouped with *Syringoderma phinneyi* with 99% bootstrap support in a partial SSU rDNA neighbor-joining tree. It is now known that Desmarestiales and Syringodermatales are not sister orders and that this specimen was misidentified and it is not included in their 1994 publication. Lee and King (1996) published a phylogeny of five Dictyotaceae genera based on ITS1-5.8S-ITS2 rDNA sequences. However, a nucleotide-nucleotide BLAST search reveals that these sequences belong to green algae, in spite of the fact that the authors examined their dictyotacean field collected plants carefully for epiphytes and washed them twice with filtered seawater. The *rbc*L sequence of the laminarialean *Thallasiophyllum clathrus*, firstly reported in Kawai and Sasaki (2001) but with collection data in Kawai et al. (2001), clearly is a chimera of *T. clathrus* and an ectocarpalean "epiphyte"; nucleotides 285–982 in DDBJ submission AB05793 (the section between the primer pair rbc-F2 and rbc-R3) are ectocarpalean (Lane et al., 2006). The *Phaeosiphoniella cryophila* specimen in Sasaki et al. (2001) (DDBJ accession numbers AB045259 and AB045275) was misidentified and is actually *Haplospora* sp. The studies by Rousseau et al. (2001) and Philips et al. (submitted) include another specimen from the type locality identified as *Phaeosiphoniella cryophila* and which does not group with *Haplospora* and *Tilopteris* (Tilopteridales). Burrowes et al. (2003) determined a LSU rDNA sequence (AY157701) for *Choristocarpus tenellus* that differs substantially from the partial LSU rDNA sequence (AJ287442) that was determined by Draisma et al. (2001). Since these two studies used a different culture (respectively from Villefranche [France] and Ischia [Italy]), one may think that *C. tenellus* represents two cryptic species. However, Burrowes et al. also determined a partial *rbc*L sequence for their French specimen (not submitted) and this sequence differed only by three nucleotides from the Italian specimen in a stretch of 795 nt of *rbc*L. Since Burrowes determined the LSU rDNA sequence directly from the PCR product and Draisma from a cloned fragment, the former may be considered the most plausible. Moreover, the phylogenetic position of *C. tenellus* in the LSU-rDNA-based tree by Burrowes is congruent with the *rbc*L tree (*C. tenellus* is the most basal brown alga), whereas *C. tenellus* groups with Sphacelariales *s.s.* in the rDNA tree by Draisma. Maybe different paralogous rRNA genes exist in *C. tenellus*, but further investigation is needed to provide a decisive answer.

BOX 1 Erroneously published DNA sequences.

"Eu-polystichidae." If this DSOS clade exists, the inter-ordinal relationships between the Dictyotales, Sphacelariales, Onslowiales, and Syringodermatales remain to be resolved.

International collaboration involving sequences of numerous markers may resolve these large-scale problems as is beginning to be the case for eukaryotes (Arisue et al., 2005). The complete genome of *Ectocarpus siliculosus* (Peters et al., 2004) will be available soon and used to identify new markers and design further probes. Comparative genome organisation may turn out not to be highly phylogenetically informative, but inserted sequences like the intron observed in the mito-chondrial genome of some brown algae seem to be a promising tool (Oudot-Le Secq et al., 2006). Furthermore, other introns, transposable elements families, repeated elements, LINEs (long inter-spersed nuclear elements), SINEs (short interspersed nuclear elements) (Friesen et al., 2001), or any rare genetic event may be detected when comparing complete genomes and used for phylogenetic reconstruction.

REFERENCES

Agardh, C.A. (1824) *Systema algarum*. Lundae [Lund], XXXVIII + 312 p.

Agardh, J.G. (1848) Vol. I. *Species genera et ordines Fucoidearum seu descriptiones succintae specierum, generum et ordinum, quibus Fucoidearum classis constituitur. Species genera et ordines algarum seu descriptiones succintae specierum, generum et ordinum, quibus algarum regnum constituitur.* Lundae [Lund]: apud C.W.K. Gleerup; Lipsiae [Leipzig]: apud T.O. Weigel; Parisiis [Paris]: apud Masson; Londini [London]: apud W. Pamplin.

Areschoug, J.E. (1847 [1846]) Phyceae scandinaviceae marinae. *Nova acta regiae societatis scientiarum Upsaliensis*, 13: 223–382, pls 1–9. [Also appeared in book form, Upsala, 1850].

Arisue, N., Hasegawa, M., and Hashimoto, T. (2005) Root of the Eukaryota tree as inferred from combined maximum likelihood analyses of multiple molecular sequence data. *Molecular Biology and Evolution*, 22: 409–420.

Bold, H.C. and Wynne, M.J. (1985) *Introduction to the Algae, Structure and Reproduction*, 2nd edition. Prentice Hall, Englewood Cliffs, NJ.

Boo, S.M., Lee, W.J., Yoon, H.S., Kato, A., and Kawai, H. (1999) Molecular phylogeny of Laminariales (Phaeophyceae) inferred from small subunit ribosomal DNA sequences. *Phycological Research*, 47: 109–114.

Bourrelly, P. (1981) *Les Algues d'eau douce. Initiation à la systématique. II. Les Algues jaunes et brunes: Chrysophycées, Phéophycées, Xanthophycées et Diatomées.* Soc. N. Ed. Boubée, Paris, 1–517.

Burrowes, R., Rousseau, F., Müller, D.G., and Reviers, B. de (2003) Taxonomic placement of *Microzonia* (Phaeophyceae) in Syringodermatales based on *rbc*L and 28S nrDNA sequences. *Cryptogamie, Algologie*, 24: 63–73.

Cavalier-Smith, T. (1986) The kingdom Chromista: origin and systematics. *Progress in Phycological Research* (eds. F.E. Round and D.J. Chapman), 4: 309–347.

Cavalier-Smith, T. (1995a) Membrane heredity, symbiogenesis, and the polyphyletic origins of algae. In *Biodiversity and Evolution* (eds. R. Arai, M. Kato, and Y. Doi). National Science Museum Foundation, Tokyo, pp. 75–114.

Cavalier-Smith, T. (1995b) The phylogeny and classification of zooflagellates. *Cytology*, 37: 1010–1029.

Cavalier-Smith, T. and Chao, E.E. (1996) 18S rRNA sequence of *Heterosigma carterae* (Raphidophyceae), and the phylogeny of heterokont algae (Ochrophyta). *Phycologia*, 35: 500–510.

Cho, G.Y. and Boo, S.M. (2006) Phylogenetic position of *Petrospongium rugosum* (Ectocarpales, Phaeo-phyceae): insights from the protein-coding plastid *rbc*L and *psa*A gene sequences. *Cryptogamie, Algologie*, 27: 3–15.

Cho, G.Y., Lee, S.H., and Boo, S.M. (2004) A new brown algal order, Ishigeales (Phaeophyceae), established on the basis of plastid protein-coding *rbc*L, *psa*A, and *psb*A region comparisons. *Journal of Phycology*, 40: 921–936.

Cho, G.Y., Rousseau, F., Reviers, B. de, and Boo, S.M. (2006) Phylogenetic relationships within the Fucales (Phaeophyceae) assessed by the photosystem I coding *psa*A sequences. *Phycologia*, 45: 512–519.

Chodat, R. (1909) *Etude critique et expérimentale sur le polymorphisme des algues*. Genève.

Christensen, T. (1980–1994) *Algae. A Taxonomic Survey*. AiO Print Ltd., Odense, Part 1, 1980; Part 2, 1994.

Clayton, M.N. (1981) Phaeophyta. In *Marine Botany an Australian Perspective* (eds. M.N. Clayton and R.J. Kings). Longman Cheshire, Melbourne, pp. 104–137.

Clayton, M.N. (1984) Evolution of the Phaeophyta with particular reference to the Fucales. *Progress in Phycological Research* (eds. F.E. Round and D.J. Chapman), 3: 12–46.

Cohn, F. (1865) *Über einige Algen von Helgoland.* Beiträge zur näheren Kenntnis und Verbreitung der Algen, herausgegeben von Dr L. Rabenhorst, Heft 2, Leipzig, E. Kummer, pp. 17–40, 3 pls.

Craigie, J.S. (1974) Storage products. In *Algal Physiology and Biochemistry* (ed. W.P.D. Stewart). University of California Press, Berkeley, pp. 206–235.

De Toni, G.B. (1895) *Sylloge algarum omnium hucusque cognitarum. Digessit Doct. J. Bapt. De Toni,[...]* Vol. III Fucoideae. Patavii: sumptibus auctoris, typis seminarii. Praemonitus.

Draisma, S.G.A., Olsen, J.L., Stam, W.T., and Prud'homme van Reine, W.F. (2002) Phylogenetic relationships within the Sphacelariales (Phaeophyceae): *rbc*L, RUBISCO spacer and morphology. *European Journal of Phycology*, 37: 385–401.

Draisma, S.G.A., Peters, A.F., and Fletcher, R.L. (2003) Evolution and taxonomy in the Phaeophyceae: effects of the molecular age on brown algal systematics. In *Out of the Past* (ed. T.A. Norton), pp. 87–102.

Draisma, S.G.A. and Prud'homme van Reine, W.F. (2001) Onslowiaceae fam. nov. (Phaeophyceae). *Journal of Phycology*, 37: 647–649.

Draisma, S.G.A., Prud'homme van Reine, W.F., Stam W.T., and Olsen, J.A. (2001) A reassessment of phylogenetic relationships within the Phaeophyceae based on RUBISCO large subunit and ribosomal DNA sequences. *Journal of Phycology*, 37: 586–603.

Friesen, N., Brandes, A., and Heslop-Harrison, J.S. (2001) Diversity, origin, and distribution of retrotransposons (gypsy and copia) in conifers. *Molecular Biology and Evolution*, 18: 1176–1188.

Fritsch, F.E. (1945) *The Structure and Reproduction of Algae*, Vol. II. Cambridge University Press, Cambridge.

Greuter, W. (Chairman), McNeill, J. (Vice-Chairman), Barrie, F.R., Burdet, H.M., Demoulin, V., Filgueiras, T.S., Nicolson, D.H., Silva, P.C., Skog, J.E., Trehane, P., and Turland, N.J. (Members) (2000) *International Code of Botanical Nomenclature* (Saint-Louis Code) adopted by the 16th International Botanical Congress St. Louis, MO, July–August 1999/Königstein, prepared and edited by Koeltz Scientific Books, 2000.

Hauck, F. (1883–1885) *Die Meeresalgen Deutschland und Oesterreichs.* In Dr L. Rabenhorst's *Kryptogamen-Flora von Deutschland, Oesterreich und der Schweiz.* 2. Auflage. 2. Band. Leipzig, 1–320 (1883), 321–512 (1884), 513–575 (1885).

Hennig, W. (1950) *Grundzünge einer Theorie der phylogeneteischen Systematic.* Deutscher Zentralverlag, Berlin.

Hennig, W. (1966) *Phylogenetic Systematics.* University of Illinois Press, Urbana.

Jeffrey, S.W. (1989) Chlorophyll *c* pigments and their distribution in the Chromophyte algae. In *The Chromophyte Algae, Problems and Perspectives* (eds. J.C. Green, B.S.C. Leadbeater, and W.L. Diver). The Systematics Association special volume 38. Published for the Systematics Association by Clarendon Press, Oxford, pp. 13–36.

Kawai, H. (1992a) A summary of chloroplasts and flagellated cells in the Phaeophyceae. *The Korean Journal of Phycology*, 7: 33–43.

Kawai, H. (1992b) Green flagellar autofluorescence in brown algal swarmers and their phototactic responses. *Botanical Magazine, Tokyo*, 105: 171–184.

Kawai, H., Hanyuda, T., Draisma, S.G.A., and Müller, D.G. (2007) Molecular phylogeny of *Discosporangium mesarthrocarpum* (Phaeophyceae) with a reinstatement of the order Discosporangiales. *Journal of Phycology* 43: 186–194.

Kawai, K., Maeba, S., Sasaki, H., Okuda, K., and Henry, E.C. (2003) *Schizocladia ischiensis*: A new filamentous marine chromophyte belonging to a new class, Schizocladiophyceae. *Protist*, 154: 211–228.

Kawai, H. and Sasaki, H. (2001) [2000] Molecular phylogeny of the brown algal genera *Akkesiphycus* and *Halosiphon* (Laminariales), resulting in the circumscription of the new families Akkesiphycaceae and Halosiphonaceae. *Phycologia*, 39: 416–428. [Note: Effective publication date on cover page 4: February 7, 2001.]

Kawai, H. and Sasaki, H. (2004) Morphology, life history, and molecular phylogeny of *Stschapovia flagellaris* (Tilopteridales, Phaeophyceae) and the erection of the Stschapoviaceae fam. nov. *Journal of Phycology*, 40: 1156–1169.

Kawai, H., Sasaki, H., Maeda, Y., and Arai, S. (2001) Morphology, life history, and molecular phylogeny of *Chorda rigida*, sp. nov. (Laminariales, Phaeophyceae) from the Sea of Japan and the genetic diversity of *Chorda filum. Journal of Phycology*, 37: 130–142.

Kawai, H., Sasaki, H., Maeba, S. and Henry, E. (2005) Morphology and molecular phylogeny of *Phaeostrophion irregulare* (Phaeophyceae) with a proposal for Phaeostrophionaceae *fam. nov.*, and a review of Ishigeaceae. *Phycologia*, 44: 169–182.

Kim, S.-H. and Kawai, H. (2002) Taxonomic revision of *Chordaria flagelliformis* (Chordariales, Phaeophyceae) including novel use of the intragenic spacer region of rDNA for phylogenetic analysis. *Phycologia*, 41: 328–339.

Kim, S.-H., Peters, A.F., and Kawai, H. (2003) Taxonomic revision of *Sphaerotrichia divaricata* (Ectocarpales, Phaeophyceae), with a reappraisal of *S. firma* from the north-west Pacific. *Phycologia*, 42: 183–192.

Kjellman, F.R. (1891–1896) Phaeophyceae (Fucoideae). In *Die natürlichen Pflanzenfamilien...* (eds. A. Engler and K. Prantl). I. Teil, 2. Abteilung, Leipzig, pp. 176–192 (1891), 193–288 (1893), 289–297 (1896).

Kloareg, B. and Quatrano, R.S. (1988) Structure of the cell walls of marine algae and ecophysiological functions of the matrix polysaccharides. *Oceanography and Marine Biology Annual Review*, 26: 259–315.

Kützing, F. (1843) *Phycologia generalis oder Anatomie, Physiologie und Systemkunde der Tange*. F.A. Brockhaus, Leipzig.

Kylin, H. (1917) Über die Entwicklungsgeschiste und die systematische Stellung der Tilopterideen. *Sonderabdruck aus den Berichten der Deutschen Botanischen Gesellschaft*, Band 35 (Heft 3): 298–310.

Kylin, H. (1933) Über die Entwicklungsgeschiste der Phaeophyceen. *Lunds Universitets Årsskrift, Ny Földj*, Avd. 2, Band 29 (Nr 7).

Lamouroux, J.V.F. (1813a) Essai sur les genres de la famille des thalassiophytes non articulées. *Annales du Muséum d'Histoire Naturelle*, 20: 21–47,115–139, 267–293, pls 7–13. [Note: Also issued as an independently paginated book (Lamouroux, 1813b) with an index and list of errata.]

Lamouroux, J.V.F. (1813b) *Essai sur les Genres de la Famille des Thalassiophytes non articulées*. C. Dufour, Paris. [Note: Repaginated version of Lamouroux (1813a) with an index and list of errata added.]

Lane, C.E., Mayes, C., Druehl, L.D., and Saunders, G.W. (2006) A multi-gene molecular investigation of the kelp (Laminariales, Phaeophyceae) supports substantial taxonomic re-organization. *Journal of Phycology*, 42: 493–512 (corrigendum on p. 962).

Lee, W.J. and King, R.J. (1996) The molecular characteristics of five genera of Dictyotaceae (Phaeophyta) from Australia: based on DNA sequences of nuclear rDNA Internal Transcribed Spacer (ITS) and 5.8S. *Algae*, 11: 381–388.

Maier, I. (1995) Brown algal pheromones. *Progress in Phycological Research* (eds F.E. Round and D.J. Chapman), 11: 45–102.

Nakamura, Y. (1972) A proposal on the classification of the Phaeophyta. In *Contributions to the Systematics of Benthic Marine Algae of the North Pacific* (eds. I.A. Abbott and M. Kurogi,). Japanese Society of Phycology, Kobe, Japan, pp. 147–155.

Oltmanns, F. (1904–1905) *Morphologie und Biologie der Algen*, 1st ed., Gustav Fischer, Jena. Vol. I. Spezieller Teil, 1904; Vol. II. Allgemeiner Teil, 1905.

Oltmanns, F. (1922) *Morphologie und Biologie der Algen. Vol. II. Phaeophyceae-Rhodophyceae*, 2nd ed., Gustav Fischer, Jena.

Oudot-Le Secq, M.-P., Loiseaux-de Goër, S., Stam, W.T., and Olsen, J.L. (2006) Complete mitochondrial genomes of the three brown algae (Heterokonta: Phaeophyceae) *Dictyota dichotoma*, *Fucus vesiculosus* and *Desmarestia viridis*. *Current Genetics*, 49: 47–58.

Papenfuss, G.F. (1951a) Phaeophyta. In *Manual of Phycology—An Introduction to the Algae and Their Biology* (ed. G.M. Smith). A New Series of Plant Science Books, Vol. 27. The Chronica Botanica Co., Waltham, MA, pp. 119–158.

Papenfuss, G.F. (1951b) Problems in the classification of the marine algae. *Svensk Botanisk Tidskrift utgivfen af svenska botaniska föreningen*, Stockholm, 45: 4–11.

Papenfuss, G.F. (1953) Problems in the Classification of the Marine Algae of the tropical and southern pacific. *Proceedings of the Seventh International Botanical Congress*, Stockholm, 1950, pp. 822–823.

Papenfuss, G.F. (1955) Classification of the algae. In *A Century of Progress in the Natural Sciences, 1853–1953*. California Academy of Science, San Francisco, pp. 115–224.

Patterson, D.J. (1989) Stramenopiles: chromophytes from a protistan perspective. In *The Chromophyte Algae, Problems and Perspectives* (eds. J.C. Green, B.S.C. Leadbeater, and W.L. Diver). The Systematics Association special volume 38. Published for the Systematics Association by Clarendon Press, Oxford, pp. 357–379.

Peters, A. (1998) Ribosomal DNA sequences support taxonomic separation of the two species of *Chorda*: reinstatement of *Halosiphon tomentosus* (Lyngbye) Jaasund (Phaeophyceae, Laminariales). *European Journal of Phycology*, 33: 65–71.

Peters, A.F. and Clayton, M.N. (1998) Molecular and morphological investigation of three brown algal genera with stellate plastids: evidence for Scytothamnales ord. nov. (Phaeophyceae). *Phycologia*, 37: 106–113.

Peters, A.F., Marie, D., Scornet, D., Kloareg, B., and Cock, J.M. (2004) Proposal of *Ectocarpus siliculosus* (Ectocarpales, Phaeophyceae) as a model organism for brown algal genetics and genomics. *Journal of Phycology*, 40: 1079–1088.

Peters, A.F. and Moe, R.L. (2001) DNA sequences confirm that *Petroderma maculiforme* (Phaeophyceae) is the brown algal phycobiont of the marine lichen *Verrucaria tavaresiae* (Verrucariales, Ascomycota) from central California. *Bulletin of the California Lichen Society*, 8: 41–43.

Peters, A.F. and Ramírez, M.E. (2001) Molecular phylogeny of small brown algae, with special reference to the systematic position of *Caepidium antarcticum* (Adenocystaceae, Ectocarpales). *Cryptogamie algologie*, 22: 187–200.

Rabenhorst, L. (1863) *Kryptogamen-Flora von Sachsen...* Leipzig.

Reviers, B. de (2003) *Biologie et phylogénie des algues. Tome 2.* Belin, Paris.

Reviers, B. de and Rousseau, F. (1999) Towards a new classification of the brown algae. *Progress in Phycological Research* (eds. F.E. Round and D.J. Chapman), 13: 107–201.

Rousseau, F., Burrowes, R., Peters, A.F., Kuhlenkamp, R., and Reviers, B. de (2001) A comprehensive phylogeny of the Phaeophyceae based on nrDNA sequences resolves the earliest divergences. *Compte Rendus de l'Académie des Sciences*, 324: 305–319.

Rousseau, F., Leclerc, M.-C., and Reviers, B. de (1997) Molecular phylogeny of European Fucales (Phaeophyceae) based on large subunit rDNA sequence comparisons. *Phycologia*, 36: 438–446.

Rousseau, F. and Reviers, B. de (1999a) Circumscription of the order Ectocarpales (Phaeophyceae): bibliographical synthesis and molecular evidence. *Cryptogamie, Algologie*, 20: 5–18.

Rousseau, F. and Reviers, B. de (1999b) Phylogenetic relationships within the Fucales (Phaeophyceae) based on combined partial SSU + LSU rDNA sequence data. *European Journal of Phycology*, 34: 53–64.

Rousseau, F., Reviers, B. de, Leclerc, M.-C., Asensi, A., and Delépine, R. (2000) Adenocystaceae fam. nov. (Phaeophyceae) based on morphological and molecular evidence. *European Journal of Phycology*, 35: 35–43.

Sanders, W.B., Moe, R.L., and Ascaso, C. (2004) The intertidal marine lichen formed by the pyrenomycete fungus *Verrucaria tavaresiae* (Ascomycotina) and the brown alga *Petroderma maculiforme* (Phaeophyceae): thallus organization and symbiont interaction. *American Journal of Botany*, 91: 511–522.

Sanders, W.B., Moe, R.L., and Ascaso, C. (2005) Ultrastructural study of the brown alga *Petroderma maculiforme* (Phaeophyceae) in the free-living state and in lichen symbiosis with the intertidal marine fungus *Verrucaria tavaresiae* (Ascomycotina). *European Journal of Phycology*, 40: 353–361.

Sasaki, H., Flores-Moya, A., Henry, E.C., Müller, D.G., and Kawai, H. (2001) Molecular phylogeny of Phyllariaceae, Halosiphonaceae and Tilopteridales (Phaeophyceae). *Phycologia*, 40: 123–134.

Saunders, G.W. and Kraft, G.T. (1995) The phylogenetic affinities of *Notheia anomala* (Fucales, Phaeophyceae) as determined from partial small-subunit rRNA sequences. *Phycologia*, 34: 383–389.

Sauvageau, C. (1915) Sur la sexualité hétérogamique d'une Laminaire (*Saccorhiza bulbosa*). *Comptes Rendus de l'Académie des Sciences,* 161: 796–799.

Scagel, R.F. (1966) The Phaeophyceae in perspective. *Oceanography and Marine Biology, Annual Review*, 4: 123–194.

Schmidt, O.C. (1937) Choristocarpaceen und Discosporangiaceen. *Hedwigia*, 77: 1–4.

Schoenwaelder, M.E.A. and Clayton, M.N. (1998) Secretion of phenolic substances into the zygote wall and cell plate in embryos of *Hormosira* and *Acrocarpia* (Fucales, Phaeophyceae). *Journal of Phycology*, 34: 969–980.

Schoenwaelder, M.E.A. and Clayton, M.N. (1999) The presence of phenolic compounds in isolated cell walls of brown algae. *Phycologia*, 38: 161–166.

Setchell, W.A. and Gardner, N.L. (1925) The marine algae of the Pacific coast of North America. III Melanophyceae. *University of California Publications in Botany*, 8: 383–898 incl. pls 34–107.

Siemer, B.L., Stam, W.T., Olsen, J.L., and Pedersen, P.M. (1998) Phylogenetic relationships of the brown algal orders Ectocarpales, Chordariales, Dictyosiphonales, and Tilopteridales (Phaeophyceae) based on Rubisco large subunit and spacer sequences. *Journal of Phycology*, 34: 1038–1048.

Silva, P.C. (1980) Names of classes and families of living algae. *Regnum Vegetabile*, 103: 156 p.

Strasburger, E. (1906) Zur Frage eines Generationwechsels bei Phaeophyceen. *Botanische Zeitung,* 64: 1–7.

Tan, I.H. and Druehl, L.D. (1993) Phylogeny of the Northeast Pacific brown algal (Phaeophycean) orders as inferred from 18S rRNA gene sequences. In *14th International Seaweed Symposium* (eds. A.R.O.

Chapman, M.T. Brown, and M. Lahaye), Brest, France, 16–21 August 1992, *Hydrobiologia*, 260–261: 699–704.

Tan, I.H. and Druehl, L.D. (1994) A molecular analysis of *Analipus* and *Ralfsia* (Phaeophyceae) suggests the order Ectocarpales is polyphyletic. *Journal of Phycology*, 30: 721–729.

Tan, I.H. and Druehl, L.D. (1996) A ribosomal DNA phylogeny supports the close evolutionary relationships among the Sporochnales, Desmarestiales, and Laminariales (Phaeophyceae). *Journal of Phycology*, 32: 112–118.

Taylor, R. (1922) Recent studies of Phaeophyceae and their bearing classification. *The Botanical Gazette*, 74: 431–441.

Thuret, G. (1850–1851) Recherche sur les zoospores et les anthéridies des Cryptogames. *Annales des Sciences naturelles, Botanique*, série 3. First part (1850), Zoospores des algues 14: 214–256, pls 16–31; second part (1851), Anthéridies des Cryptogames 16: 5–39, pls 1–15. Also exists as a consolidated reprint, Paris, L. Martinet, [1]–93, pls 16–31, 1–15 (1851).

Thuret, G. (1854–1855) Recherches sur la fécondation des Fucacées et les anthéridies des algues. *Annales des Sciences naturelles, Botanique*, série 4. First part (1854), 2: 197–214, pls 12–15; second part (1855), 3: 5–28, pls 2–4. Also exists as a consolidated reprint, Victor Masson, Paris, [1]–22, pls 12–15, [23]–46, pls 2–4 (1855).

Van den Hoek, C. and Jahns, H.M. (1978) *Algen. Einführung in die Phykologie.* Thieme, Stuttgart.

Van den Hoek, C., Mann, D., and Jahns, H.M. (1995) *Algae. An Introduction to Phycology.* Cambridge University Press, Cambridge.

Warming, E. (1884) *Haanbog i den systematiske botanik. Anden gjennemsete udgave.* Kjøbenhavn.

West, J.A. and Kraft, G.T. (1996) *Ectocarpus siliculosus* (Dillwyn) Lyngbye from Hopkins River Falls, Victoria—The first record of a freshwater brown alga in Australia. *Muelleria*, 9: 29–33.

Wilce, R.T., Schneider, C.W., Quinlan, A.V., and Van den Bosch, K. (1982) The life history and morphology of free-living *Pilayella littoralis* (L.) Kjellm. (Ectocarpaceae, Ectocarpales) in Nahant Bay, Massachusetts. *Phycologia*, 21: 336–354.

Womersley, H.B.S. (1954) The species of *Macrocystis* with special reference to those on southern Australian coasts. *University of Californy Publications in Botany*, 27: 109–132.

Womersley, H.B.S. (1987) *The Marine Benthic Flora of Southern Australia*. Part II. Adelaide, South Australian Government Printing Division.

Wynne, M.J. (1982) Phaeophyceae. In *Synopsis and Classification of Living Organisms*, Vol. I (ed. S.P. Parker), McGraw-Hill, New York, pp. 115–125.

Wynne, M.J. and Kraft, G.T. (1981) In *The Biology of Seaweeds* (eds C.S. Lobban and M.J. Wynne). Botanical Monographs, Vol. 17, Oxford, London, Edinburgh, Boston, Melbourne: Blackwell Scientific Publications, pp. 743–750.

Wynne, M.J. and Loiseaux, S. (1976) Recent advances in life history studies of the Phaeophyta. *Phycologia*, 15: 435–452.

15 Molecular systematics of the Chrysophyceae and Synurophyceae

Robert A. Andersen

CONTENTS

ABSTRACT

The definition of the chrysophytes has changed over time, especially after electron microscopic and molecular systematic techniques were applied. Historically, the group was classified based upon light microscopic observations that emphasized flagellar number, life forms, and other morphological features; these were summarized using three keystone publications. Electron microscopic observations caused taxonomic revisions, especially at the class level, and as groups were removed from the Chrysophyceae *sensu stricto*, a number of new classes were formed. Molecular phylogenetic analysis brought to question the separation of the class Synurophyceae and the classification of the Chrysophyceae based upon morphology. This chapter provides more critical examination of some taxa examined previously, and it analyses new SSU rRNA and *rbc*L gene sequences that provide greater taxon sampling. The separation of the two classes was weakly supported in a combined gene analysis, but the data were not sufficiently robust to conclusively resolve the question. The use of capsoid, coccoid, and flagellate life forms for higher-level classification of the Chrysophyceae was incongruent with the clades recovered in the molecular phylogenetic analyses. As with previous recent publications, the number of visible flagella (using light microscopy) did not correspond with recovered clades in the molecular analyses. However, smaller clades of exclusively uniflagellate or biflagellate taxa were recovered. The incongruence between morphological classifications and supported clades on gene trees suggested two points. First, the morphological features and the gene trees agreed for small terminal clades, and therefore the incongruence between the two data sets occurs only for higher-level classifications. Second, the polyphyly of simple flagellates (e.g., *Chromulina, Ochromonas*) in molecular phylogenetic analyses suggested that flagellate forms evolved more than once, perhaps by reductive evolution. This, in turn, suggested that the ancestral chrysophyte was not a flagellate.

INTRODUCTION

The chrysophytes are a group of predominately freshwater algae that are classified in two closely related classes, the Chrysophyceae and Synurophyceae (Pascher, 1914; Andersen, 1987). Common genera include *Chromulina, Dinobryon, Mallomonas, Ochromonas*, and *Synura*. Although there are a number of differences between the two classes (Table 15.1), both share one unique morphological character, a siliceous cyst that has a special pore that is sealed with a plug (Hibberd, 1977b; Sandgren, 1980). The Chrysophyceae and Synurophyceae also share a number of features with other heterokont algae (= stramenochromes) (Andersen, 2004b). For example, chloroplasts have lamellae formed by three adpressed thylakoids, which in turn are surrounded by a sac-like girdle lamella (Dodge, 1973). Chloroplast pigments usually include chlorophyll *a* and *c* as well as a wide range of carotenes and xanthophylls (Andersen and Mulkey, 1983; Jeffrey and Vesk, 1997). The typical cellular storage product is a beta-linked glucan (laminarin, chrysolaminarin) (Quillet, 1955; Archibald et al., 1963). Swimming cells usually have a hairy immature flagellum and a second, smooth flagellum (Leadbeater, 1989).

The chrysophytes are almost exclusively microscopic organisms. The cells may be naked (sometimes embedded in a gelatinous matrix) or covered with organic scales, silica scales, organic loricae (sometimes mineralized), organic thecae, or rarely cell walls. The two classes differ in many other morphological and biochemical respects (Table 15.1), and these differences remain the basis for their separation (Andersen, 1987, 1989; Andersen and Preisig, 2002; Preisig and Andersen, 2002).

TABLE 15.1
Similarities and Differences between the Chrysophyceae and Synurophyceae

Character	Chrysophyceae	Synurophyceae
Chlorophylls	c_1 and c_2	c_1 only
Basal body orientation	~ 90° angle	~ parallel
Photoreceptor type	Short flagellum-eyespot	2 flagellar swellings
Eyespot	Often present	Never present
Nuclear envelope	Confluent with chloroplast ER	Not confluent with plastid ER
Silica scale symmetry	Biradially symmetrical	Bilaterally symmetrical
Stomatocyst	Present	Present
Carotenoids	Photosynthetically active fucoxanthin; violaxanthin cycle	Photosynthetically active fucoxanthin; violaxanthin cycle
Flagellar hairs	Present; short and long lateral filaments	Present; short and long lateral filaments
Transitional helix	5–6 gyres	6–9 gyres
Phagotrophy	Common	Absent
Silica scale formation	In cytoplasm	Associated with plastid

Detailed molecular phylogenetic analyses for the two classes have been based primarily on SSU rRNA and *rbc*L genes, although the LSU rRNA gene has been examined for select species (Ben Ali et al., 2002). Results show either weak support for separation of the two classes (e.g., Andersen et al., 1999) or a paraphyletic Chrysophyceae (e.g., Lavau et al., 1997; Caron et al., 1999). Consequently, the class level taxonomy for the chrysophytes remains unresolved in terms of molecular phylogenetics. Similarly, morphologically based orders, families, and some genera are not recovered in these molecular phylogenies. Thus, traditional classifications based upon light microscopic morphology are incongruent with molecular phylogenetic analyses. Simply stated, the classification of the chrysophytes is chaotic when molecular data are used.

SEVERAL KEYSTONE PUBLICATIONS

The history of the chrysophytes has been reviewed recently (Andersen, 2004a) and only keystone works will be mentioned as a historical perspective. The first keystone publication was that of Pascher (1913), and it drew upon earlier works, especially Stein (1878), Klebs (1892a, 1892b), and Senn (1900). Pascher classified the Chrysophyceae into three subclasses based upon the predominant vegetative life stage (flagellate, capsoid, amoeboid) (Table 15.2). Flagellates were divided

TABLE 15.2
The Class Chrysophyceae Classified by Pascher (1913)*

Subclass	Order	Family	Subfamily	Genera
Euchrysomonadinae	Chromulinales	Chrysapidaceae		*Chrysapsis*
		Euchromulinaceae	Chromulinae	*Chromulina, Pyramidochrysis*
			Sphaleromantideae	*Sphaleromantis*
			Kytochromulineae	*Chrysococcus*
			Lepochromulineae	*Kephyrion, Lepochromulina, Chrysopyxis*
			Cyrotophorae	*Pedinella, Cyrtophora, Palatinella*
		Mallomonadaceae	Mallomonadineae	*Mallomonas, Microglena*
			"Colonial"	*Chrysosphaerella*
	Isochrysidales	Isochrysidaceae	Isochrysidineae	*Syncrypta*
			Lepisochrysidineae	*Deripyxis, Stylochrysallis*
		Euhymenomonadaceae	Hymenomonadeae	*Chlorodesmus, Hymenomonas, Synura*
	Ochromonadales	Euochromonadaceae	Euochromonadinae	*Cyclonexis, Ochromonas, Uroglena, Uroglenopsis*
			Lepochromonadineae	*Dinobryon, Halobryon, Kephyriopsis, Pseudokephyrion, Poterioochromonas*
Chrysocapsinae		Chrysocapsaceae		*Chrysocapsa, Phaeosphaera*
		Hydruraceae		*Hydrurus*
Rhizochrysidinae		Rhizochryidaceae		*Chrysidiastrum, Chrysostephanosphaera, Lagynion, Rhizochrysis, Stylococccus*

* Pascher used the predominant life stage (flagellate, capsoid, amoeboid) to assign subclasses and flagellar type (one flagellum, two equal flagella, two unequal flagella) to assign orders within the subclass Euchrysomonadinae.

TABLE 15.3
The Class Chrysophyceae Classified by Bourrelly (1957)*

Subclass	Superorder	Order	Genera
With cell walls	No flagellate stage	Phaeoplacales	*Phaeoplaca*
		Stichogloeales	*Stichogloea, Phaeodactylum, Phaeoschizochlamys*
	Biflagellate stage	Phaeothamniales	*Phaeothamnion, Apistonema, Chrysonephos*
		Chrysapionales	*Chrysapion, Koinopodion, Sarcinochrysis*
	Uniflagellate stage	Thallochrysidales	*Thallachrysis, Phaeodermatium, Chrysomeris*
		Chrysosphaerales	*Chrysosphaera, Epicystis*
No cell walls	No flagellate stage	Chrysosaccales	*Chrysosaccus Phaeosphaera, Chalkopyxis*
		Rhizochrysidales	*Rhizochrysis, Chrysarachnion, Stylococcus, Lagynion*
	Biflagellate stage	Ochromonadales	*Ochromonas, Dinobryon, Synura, Phaeocystis, Naegelliella*
		Isochrysidales	*Isochrysis, Prymnesium, Deripyxis, Tessellaria*
	Uniflagellate stage	Silicoflagellales	*(Not addressed in detail)*
		Craspedomonadales	*Monosiga, Salpingoeca, Phalansterium*
		Chromulinales	*Chromulina, Chrysamoeba, Chrysochaete, Hydrurus*

* Bourelly used cell walls (presence/absence) to assign subclasses and flagella (absent, one flagellum, two flagella) to assign some orders.

into three orders based upon the flagellar morphology as viewed in the light microscope. Families and subfamilies were constructed using other morphological characters (e.g., cell walls, colonies). The classification was artificial, designed more for categorizing taxa than for showing evolutionary trends; phylogenetic remarks were rare (e.g., p. 10 suggests that capsoid organisms may be linked to gel-forming flagellates such as *Chromulina nebulosa*).

The second keystone publication was that of Bourrelly (1957); it summarized the many new taxa that were described since 1913, and it provided an intuitive phylogenetic classification (Table 15.3). Bourrelly divided the class into two groups (= subclasses) based upon the presence or absence of cell walls. Within each of the two subclasses, he divided the taxa based upon flagellar condition (no flagella, one flagellum, two flagella). Finally, within orders, predominant vegetative life stages (e.g., capsoid, amoeboid), presence of loricas or scales, presence of colonies, and so forth were used to define families. Therefore, Bourrelly (1957) more or less inverted the scheme of Pascher (1913) when he gave weight to characters that define family to subclass groups.

The third keystone publication was by Bourrelly (1968) whose new classification disregarded the presence/absence of cell walls and defined subclasses based upon flagellar condition (Table 15.4). He acknowledged presence of two flagella, one visible only by electron microscopy, in the Chromulinales. He combined the Chromulinales and Ochromonadales in his subclass Heterochrysophycidées. He established a subclass for those where no flagellate stage was known, and he continued to recognize the haptophytes and choanoflagellates as subclasses despite growing evidence against their inclusion (e.g., Christensen, 1962). Starmach (1985), in a revision of the Süsswasserflora, closely followed Bourrelly's (1968) classification, although Starmach recognized the haptophytes as a separate class. However, Starmach continued to classify *Deripyxis* as a haptophyte because he considered the pseudopodia to be two equal flagella.

All three historic keystone papers relied entirely on morphological features and life history stages that were evident from light microscopic observation. Electron microscopy brought forward a wealth of new data that caused major changes in the classification of golden algae.

TABLE 15.4
The Class Chrysophyceae Classified by Bourrelly (1968), Illustrating His Use of Flagella (None, Two Unequal, Two Equal, One) for Determining Subclasses and Dominant Vegetative State for Determining Orders, Suborders, and Families

Subclass	Order	Suborder	Family	Genera
Acontochrysophycidées	Phaeoplacales		Sphaeridiothicacées	*Sphaeridothrix*
			Phaeoplacacées	*Phaeoplaca*
	Stichogloeales		Stichogloaeacées	*Stichogloea*
	Chrysosaccales		Chrysosaccacées	*Chrysosaccus, Phaeosphaera*
	Rhizochrysidales		Rhizochrysidacées	*Chrysarachnion, Rhizochrysis*
			Stylococcacées	*Bitrichia, Lagynion*
Heterochrysophycidées	Chromulinales	Thallochrysidinées	Thallochrysidacées	*Phaeodermatium*
			Chrysoméridacées	*Chrysoclonium*
		Chrysosphaerinées	Chrysophaeracées	*Chrysosphaera, Epicystis*
		Chromulinées	Chromulinacées	*Chromulina*
			Chrysococcacées	*Chrysococcus, Kephryion*
			Pédinellacées	*Pedinella*
			Chrysocapsacées	*Chrysocapsa, Chrysonebula, Kremastochrysopsis*
			Chrysochaetacées	*Chrysochaete*
			Hydruracées	*Hydrurus*
			Chrysamoebacées	*Chrysamoeba*
			Kybotionacées	*Kybotion*
			Myxochrysidacées	*Myxochrysis*
	Ochromonadales	Ochromonadinées	Ochromonadacées	*Ochromonas, Spumella, Uroglena*
			Dinobryacées	*Dinobryon, Epipyxis*
			Synuracées	*Mallomonas, Paraphysomonas, Synura*
			Rutnéracées	*Kremastochrysis, Ruttnera*
			Naegeliellacées	*Naegeliella*
		Phaeothamniinées	Phaeothamniacées	*Phaeothamnion*
		Chrysapioninées	Chrysapionacées	*Chrysapion*
Isochrysophycidées	Isochrysidales		Isochrysidacées	*Tesselaria*
			Dérépyxidacées	*Derepyxis*
			Diacronématacées	*Diacronema*
	Prymnésiales		Prymnésiacées	*Chrysochromulina*
			Coccolithophoracées	*Hymenomonas*
Craspédomonadophycidées	Monosigales		Monosigacées	*Codosiga, Monosiga*
			Salpingoecacées	*Salpingoeca*
			Phalanstériacées	*Phalansterium*

MAJOR CLASSIFICATION CHANGES BASED ON ULTRASTRUCTURAL
AND MOLECULAR DATA

During the past few decades, systematic studies have resulted in the removal of many taxa that were considered chrysophytes by earlier workers. Choanoflagellates were brought into the Chrysophyceae by Bourrelly (1957) and retained as a subclass by Starmach (1985), although Leadbeater (1972), Hibberd (1975) and others showed that choanoflagellates had no relationship to the Chrysophyceae. Choanoflagellates are currently classified in the opisthokonts (Adl et al., 2005).

TABLE 15.5
A List of Species Studied, including Their Culture Strain Number, Collection Site, and GenBank Accession Numbers for the 18S rRNA and *rbc*L Genes

Species	Strain Number	Collection Site	18S rRNA	*rbc*L
Chromophyton cf. *rosanoffii*	CCMP2751	Small cave (fourth) near Tokyo, Japan	EF165106	EF165164
Chromophyton cf. *rosanoffii*	CCMP2753	Small cave (second) near Tokyo, Japan	EF165107	EF165165
Chromulina cf. *nebulosa*	CCMP2719	34.7500S 150.4800E Pool, Barren Grounds, Lyrebird Creek, Natural Stone Bridge, New South Wales, Australia	EF165101	EF165180
Chromulina nebulosa	CCMP263	Bog, Minnesota, USA	AF123285	AF155076
"*Chromulina*" sp.	SAG17.97	Bog pool, Obergurgl/Oetztal, Tyrol, Austria	EF165103	EF165151
Chrysamoeba mikrokonta	CCMP1857	47.2275N 88.1636W plankton, roadside ditch, Gay, Michigan, USA	AF123287	EF165182
Chrysamoeba pyrenoidifera	CCMP1663	40.3666S 175.6000E Greenhouse, Massey University, Palmerston North, New Zealand	AF123286	
Chrysamoeba tenera	UTCC273	Pine Lake, Ontario, Canada	EF165102	EF165181
Chrysocapsa paludosa	CCMP380	37.5700S 145.6600E St. Ronan's well, along Maroondah Highway, near Narbethong, Victoria, Australia	EF165145	EF165149
Chrysocapsa vernalis	CCMP277	42.3400N 88.2900W Volo Bog Nature Preserve, McHenry County, Illinois, USA	AF123283	AF155877
Chrysocapsa vernalis	CCMP278	42.3400N 88.2900W Volo Bog Nature Preserve, McHenry County, Illinois, USA	EF165105	EF165148
"*Chrysocapsa*" sp.	UTCC280	Dickie Lake, Ontario, Canada	EF165130	EF165153
Chrysolepidomonas dendrolepidota	CCMP293	47.4374N 87.9611W Lake Medora, Keeweenaw County, Michigan, USA	AF123297	AF015570
Chrysonebula flava	ACOI-647	33.4333N 16.73333W Madeira, Azores	EF165104	EF165150
Chrysonephele palustris	Field colony	Central Tasmania, Australia	U1196	
Chrysosaccus sp.	CCMP295	36.1000N 94.1000W southeast bridge pond, Washington County, Arkansas, USA	AF044845	
Chrysosaccus sp.	CCMP1156	Lake Michigan, EPA Station 41, 20 m depth, USA	EF165120	EF165167
Chrysosaccus sp.	CCMP368	47.2600N 88.1161W road ditch, 2.9 miles north of Burnette Park, Keeweenaw County, Michigan, USA	EF165121	EF165166
Chrysosphaerella sp.	Field colony	Lake Nickerbocker, Boothbay, Maine, USA	EF185316	EF185313
Chrysoxys sp.	CCMP591	48.5200N 123.1580W West San Juan Island, Washington, USA	AF123302	EF165176
Cyclonexis annularis	CCMP1858	43.8581N 69.6467W West Harbor Pond, Maine, USA	AF123292	
Dinobryon cylindricum	SWD3	51.0833N 114.0833W Glenmore Reservoir, Calgary, Alberta, Canada	EF165140	EF165157
Dinobryon sertularia	CCMP1859	51.0000N 114.0500W Glenmore Reservoir, Calgary, Alberta, Canada	AF123289	

TABLE 15.5 (CONTINUED)
A List of Species Studied, including Their Culture Strain Number, Collection Site, and GenBank Accession Numbers for the 18S rRNA and *rbc*L Genes

Species	Strain Number	Collection Site	18S rRNA	*rbc*L
Dinobryon cf. *sociale*	UTCC392	Pond, Sippewissett, Woods Hole, Massachusetts, USA	EF165141	EF165158
Dinobryon sociale var. *americana*	CCMP1860	43.8470N 69.6470W adjacent to Route 27, lower end of West Harbor Pond, West Boothbay Harbor, Maine, USA	AF123291	EF165156
Epipyxis aureus	CCMP385	36.0400N 94.0500W Sequoyah Pond, near Sequoyah Lake, Fayetteville, Arkansas, USA	AF123301	EF165155
Epipyxis pulchra	CCMP382	36.0400N 94.0500W Sequoyah Pond, near Sequoyah Lake, Fayetteville, Arkansas, USA	AF123298	AF015571
Hibberdia magna	CCMP453	52.1333N 0.0475W pond, Wimpole, Cambridgeshire, England, United Kingdom	M7331	AF015572
Kremastochrysis sp.	CCMP260	Massachusetts, USA	AF123282	EF165152
Lagynion cf. *ampullaceum*	CCMP2727	Unknown collection site	EF165146	EF165161
Lagynion scherffelii	CCMP465	36.1000N 94.1000W Sequoyah Pond, Washington County, Arkansas, USA	AF123288	EF165162
Mallomonas adamas	MUCC287	Australia	U73225	
Mallomonas akrokomos	MUCC288	Australia	U73229	
Mallomonas annulata	CCMP474	42.3400N 88.2900W Bluff Lake, Lake County, Illinois, USA	EF165127	EF165193
Mallomonas annulata	MUCC289	Australia	U73230	
Mallomonas asmundae	CCMP1658	42.3400N 88.2900W Volo Bog Nature Preserve, McHenry County, Illinois, USA	M87333	AF015585
Mallomonas caudata	FW644	San Juan Island, Washington, USA	U73228	
Mallomonas insignis	CCMP2549	43.8581N 69.6467W West Harbor Pond, Maine, USA	EF165118	EF165198
Mallomonas matvienkoae	MUCC290	Australia	U73227	
Mallomonas rasilis	CCMP479	47.2520N 88.4680W Chrysopyxis pond, Calumet, Michigan, USA	M55285	EF165195
Mallomonas rasilis	MUCC292	Australia	U73231	
Mallomonas splendens	CCMP1782	37.4770S 144.5600E from freshwater reservoir, Mount Macedon, Victoria, Australia	EF165147	
Mallomonas striata var. *serrata*	MUCC295	Australia	U73232	
Mallomonas striata var. *serrata*	CCMP2059	Australia		EF165194
Naegeliella flagellifera	CCMP280	42.3400N 88.2900W Lake Defiance, McHenry County, Illinois, USA	AF123284	EF165154
Ochromonas danica		Bog-pool, Everdrup, Denmark	M32704	
Ochromonas danica	CCMP588	36.1000N 94.1000W Lake Fayetteville, Fayetteville, Arkansas, USA	EF165108	
Ochromonas cf. *gloeopara*	CCMP2060	17.0168N 88.0690W Southwest shore of Man-of-War Cay, Belize	EF165113	EF165171

(*Continued*)

TABLE 15.5 (CONTINUED)
A List of Species Studied, including Their Culture Strain Number, Collection Site, and GenBank Accession Numbers for the 18S rRNA and *rbc*L Genes

Species	Strain Number	Collection Site	18S rRNA	*rbc*L
Ochromonas cf. *gloeopara*	CCMP2718	38.9800S 146.2800E Darby River, Wilson's Promontory, Victoria, Australia	EF165112	EF165170
Ochromonas cf. *perlata*	CCMP2732	47.4374N 87.9611W Lake Medora, Keeweenaw County, Michigan, USA	EF165143	EF165187
Ochromonas sphaerocystis	CCMP586	36.1350N 92.6400W Rush Creek, northwest Arkansas, USA	AF123294	EF165185
Ochromonas cf. *sphaerocystis*	CCMP2061	47.6500N 116.7700W USGS station 4, Couer D'Alene Lake (north), Kootenai County, Idaho, USA.	EF165123	EF165186
Ochromonas tuberculata	CCMP1861	42.3400N 88.2900W Volo Bog Nature Preserve, McHenry County, Illinois, USA	AF123293	EF185315
Ochromonas vasocystis	CCMP2741	Portugal	EF165111	EF165184
Ochromonas sp.	AC22	Aquarium, Luc-sur-Mer Marine Station, Basse-Normandie, France	EF165138	EF165203
Ochromonas sp.	AC24	Sallanelles, Basse-Normandie, France	EF165124	EF165175
Ochromonas sp.	AC25	Odet estuary, Brittany, France	EF165136	EF165177
Ochromonas sp.	AC514	Blainville, Basse-Normandie, France	EF165125	EF165204
Ochromonas sp.	ACOI-1258	Portugal	EF165115	EF165173
Ochromonas sp.	CCMP1147	48.5440N 123.0100W limestone quarry, San Juan Island WA, USA	EF165137	EF165205
Ochromonas sp.	CCMP1149	38.7020N 72.3667W Oceanus cruise 83 station II	EF165139	EF165202
Ochromonas sp.	CCMP1278	37.8600S 144.9300E Port Phillip Bay, Hobsons Bay, Melbourne, Victoria, Australia	U42382	EF165174
Ochromonas sp.	CCMP1393	38.7020N 72.3667W Oceanus cruise 83 station II	EF165142	EF185314
Ochromonas sp.	CCMP1899	77.8333S 163.0000E McMurdo Sound, Antarctica	EF165133	EF165159
Ochromonas sp.	CCMP2740	Rhode Island, USA	EF165144	
Ochromonas sp.	CCMP584	34.0000N 65.0000W Oceanus Cruise 83, Station I	U42381	
Ochromonas sp.	CCMP592	39.7550N 70.6666W Oceanus cruise 100	EF165135	EF165201
Ochromonas sp.	CCMP2767	Russia	EF165110	EF165183
Ochromonas sp.	MBIC10896	Pacific Ocean		EF165168
Ochromonas sp.	SAG933.10	Pond, Solling, Germany	EF165109	
Ochromonas sp.	CCMP2761	51.4833N 13.6333 acidic mining lake 111, eastern Germany	EF165126	EF165200
Oikomonas mutabilis		Nitobe Botanical Garden, University of British Columbia, Vancouver, British Columbia, Canada	U42454	
Paraphysomonas bandaiensis	Hflag	Sargasso Sea, Atlantic Ocean	AF109322	
Paraphysomonas butcheri	DB4	Patuxant River, Maryland, USA	AF109326	
Paraphysomonas foraminifera	SOTON A	United Kingdom?	Z38025	

TABLE 15.5 (CONTINUED)
A List of Species Studied, including Their Culture Strain Number, Collection Site, and GenBank Accession Numbers for the 18S rRNA and *rbc*L Genes

Species	Strain Number	Collection Site	18S rRNA	*rbc*L
Paraphysomonas foraminifera	TPC2	Twin Peaks Chimney, hydrothermal vent, 2550 meters depth, Pacific Ocean	AF174376	
Paraphysomonas imperforata	SR3	Sakonnet River, Rhode Island, USA	AF109324	
Paraphysomonas imperforata	VS1	Vineyard Sound, Massachusetts, USA	AF109323	
Paraphysomonas sp.	HD	Baltic Sea, Germany	AJ236863	
Phaeoplaca thallosa	CCMP634	Uncertain; possibily from Busse Resevoir, Arlington Heights, Illinois, USA	AF123296	EF165160
Picophagus flagellatus	CCMP1953	11.1333S 150.0000W OLIPAC cruise, Equatorial Pacific	AF185051	
Poterioochromonas malhamensis	MBIC-HT2	Japan	AB023070	
Poterioochromonas malhamensis	SAG933.1c	Mountain pond, Yorkshire, United Kingdom	EF165114	EF165169
Poterioochromonas stipitata	CCMP1862	47.2275N 88.1636W submerged vegetation, roadside ditch, Gay, Michigan USA	AF123295	EF165172
Spumella obliqua		Lake Constance, Baden-Württemberg, Germany	AJ236860	
Spumella sp.	15G	Lake Behler See, Schleswig-Holstein, Germany	AJ236857	
Spumella sp.	37G	Pond near Plön, Schleswig-Holstein, Germany	AJ236858	
Spumella sp.	SpiG	Lake Plußsee, Schleswig-Holstein, Germany	AJ236862	
Synura curtispina	CCMP847	47.2917N 88.0721W Winter's Creek, Keeweenaw County, Michigan, USA	EF165128	EF165196
Synura mammillosa	MUCC298	Australia	U73220	
Synura petersenii	CCMP851	Unknown	U73224	EF165188
Synura petersenii	CCMP854	42.2400N 88.0500W pool below spillway, Island Lake ,Island Lake, Illinois, USA	EF165116	EF165189
Synura petersenii	CCMP857	Unknown	EF165117	EF165190
Synura petersenii	MUCC300	Australia	U73223	
Synura petersenii	SAG24.86	Botanical Garden, Goettingen, Germany	EF165129	EF165191
Synura sphagnicola	CCMP1705	Keeweenaw County, Michigan, USA	U73221	EF165197
Synura spinosa	A471	Myra Pond, Washington County, Arkansas, USA	M87336	
Synura uvella	CCMP870	42.3400N 88.2900W Volo Bog Nature Preserve, McHenry County, Illinois, USA		AF015586
Synura uvella	CCMP871	47.2325N 88.1700W Tobacco River, Keeweenaw County, Michigan, USA	U73222	EF165192
Tessellaria volvocina	CCMP1781	34.2667S 142.2167E Billabong, Yarra River, Melbourne, Victoria, Australia	EF165119	EF165199
Tessellaria volvocina	MUCC302	Australia	U73219	

(Continued)

TABLE 15.5 (CONTINUED)
A List of Species Studied, including Their Culture Strain Number, Collection Site, and GenBank Accession Numbers for the 18S rRNA and *rbc*L Genes

Species	Strain Number	Collection Site	18S rRNA	*rbc*L
Uroglena sp.	SW2C	51.0833N 114.0833W Glenmore Reservoir, Calgary, Alberta, Canada	EF165132	
Uroglenopsis americana	CCMP1863	51.0833N 114.0833W Glenmore Reservoir, Calgary, Alberta, Canada	AF123290	EF165179
Uroglenopsis americana	SW1A7	51.0833N 114.0833W Glenmore Reservoir, Calgary, Alberta, Canada	EF165131	EF165178
Unidentified capsoid sp.	CCMP349	47.3000N 88.0500W roadside ditch, near Burnette Park, Keeweenaw County, Michigan, USA	EF165122	
Unidentified capsoid sp.	CCMP1161	47.3000N 88.0500W roadside ditch, near Burnette Park, Keeweenaw County, Michigan, USA		EF165163
Unidentified sp.	CCAP909/9	48.0700N 121.1500W snow alga, Pugh Mountain, Washington, USA	M87332	
Unidentified sp.	CCMP261	48.0700N 121.1500W snow alga, Pugh Mountain, Washington, USA	EF185317	
Unidentified sp.	CCCM41	Vancouver, British Columbia, Canada	EF165134	

Many of the Isochrysidales were removed and placed in a separate class, the Haptophyceae, (Christensen, 1962; Hibberd, 1976). Today, the group is recognized as the division Haptophyta with two classes, Pavlovophyceae and Prymnesiophyceae (Edvardsen et al., 2000). Pascher's (1913) original Isochrysidales included *Synura* (but not *Mallomonas*). Bourrelly (1957, 1968) brought the silica-scaled algae together in his family Synuraceae even though it contradicted his uniflagellate and biflagellate classification. The Synuraceae *sensu* Bourrelly were subsequently placed in a separate class by Andersen (1987) based upon ultrastructural and biochemical characters (Table 15.1).

The uniflagellates placed in Pascher's (1913) subfamily Cyrtophoreae (= Family Pedinellaceae *sensu* Bourrelly, 1957; Starmach, 1985) have been removed from the chrysophytes (e.g., Saunders et al., 1995; Daugbjerg and Andersen, 1997a, 1997b) and placed in Silva's (1980) class Dictyochophyceae. This change was based upon ultrastructural observations, chloroplast pigmentation, and molecular data. Many of the larger marine chrysophytes were transferred to the Chrysomeridales (Gayral and Billard, 1977). This order was based primarily on flagellar hair structure and life histories. Approximately 20 years later, some genera were moved to the Pelagophyceae (Saunders et al., 1995) based upon ultrastructural and molecular data. The remaining Chrysomeridales *sensu* Gayral and Billard were placed without explanation in a new Chrysomerophyceae (Cavalier-Smith et al., 1995); it is assumed that the authors were separating this group from the Chrysophyceae for reasons previously outlined by others (e.g., O'Kelly, 1989; Andersen, 1991).

A number of the remaining Chrysophyceae *sensu* Bourrelly, often with cell walls, were placed in the class Phaeothamniophyceae (Bailey et al., 1998). This class was erected using ultrastructural, biochemical, and molecular data. Finally, one species (*Polypodochrysis teissieri*) was placed into the recently described Pinguiophyceae (Kawachi et al., 2002a, 2002b). Again, the new class was based upon ultrastructural, biochemical, and molecular data.

UNCERTAIN CHRYSOPHYTE TAXA

A number of taxa have been tentatively classified in the Chrysophyceae, but compelling evidence for their inclusion is absent. The Parmales, a group of silica-scaled oceanic organisms, has been tentatively assigned to the Chrysophyceae (Booth and Marchant, 1987, 1988). Electron microscopic evidence shows a chloroplast with a girdle lamella, but this feature is common amongst most heterokont algae and cannot be used to classify an organism into the Chrysophyceae. Photosynthetic pigments, flagellar structure, and gene sequences are good characters for classifying heterokont algae, and these features are completely unknown for the Parmales. Therefore, the Parmales must remain as an enigmatic group of algae.

Myxochrysis was described by Pascher (1916) as a plasmodial alga, but it has not been reported again. *Ruttneria* was described by Geitler (1942) as a subaerial alga, and it has not been investigated since then. *Sphaleromantis* was described by Pascher (1910) as a strongly flattened uniflagellate cell with two chloroplasts. Two scaled taxa were assigned to the genus (Harris, 1963; Manton and Harris, 1966; Pienaar, 1976) but later these were transferred to a separate genus, *Chrysolepidomonas* (Peters and Andersen, 1993a, 1993b), that is distinctly different from *Spheromantis sensu* Pascher.

Picophagus is a marine, colourless flagellate that consumes bacteria. *Picophagus* has been investigated using electron microscopy, and its phylogeny was examined using the 18S rRNA sequence (Guillou et al., 1999). This organism is closely related to the Chrysophyceae, based upon molecular phylogenetic analysis, but it appears as the most basal taxon if forced in the ingroup. The lack of a chloroplast and photosynthetic pigments in *Picophagus* are a hindrance when classifying the flagellate. A similar taxon is *Chlamydomyxa*, which was described by Archer (1875), and it had been classified as a xanthophyte (e.g., Bourrelly, 1968). A recent investigation by Wenderoth et al. (1999) suggested that *Chlamydomyxa* was related to the Chrysophyceae, but like *Picophagus*, it was a very basal taxon in the analysis. When the sequences of *Picophagus* and *Chlamydomyxa* are divided into two parts and blasted to GenBank, the highly matched sequences vary. For *Chlamydomyxa*, the front half had a high match with a number of colourless chrysophytes such as *Spumella* and the rear half matched brown algae with high scores. For *Picophagus*, the front half had high match scores with some aquatic fungi, while the back half had high scores for colourless chrysophytes. Therefore, the sequences as well as other features for these two organisms are unusual.

CURRENT STATUS FOR THE CHRYSOPHYCEAE AND SYNUROPHYCEAE

Kristiansen and Preisig (2001) provided a modern scheme classifying chrysophytes, including the questionable taxa mentioned previously. Their classification does not incorporate molecular studies, and like Bourrelly, they based ordinal classification on vegetative life stage morphology (flagellate, coccoid, etc.). Kristiansen and Preisig, like Bourrelly (1968), argued that the uniflagellate organisms (viz. light microscopy) have a short second flagellum when examined with an electron microscope (e.g., Fauré-Frémiet and Rouiller, 1957; Hibberd, 1971, 1976, 1977a; Andersen, 1986). However, unlike Bourrelly, they combined all flagellates into a single order, and they relegated (apparent) flagellar number to a family level character.

This chapter reports new observations on the Chrysophyceae using light microscopy and cultures, and these observations suggest that in some cases, several life stages (e.g., flagellate, capsoid, coccoid) can occur for a single species. Also, new molecular data (18S rRNA and *rbc*L genes) are provided for the Chrysophyceae and Synurophyceae. The conflicts between molecular and morphological data are re-examined in light of the microscopic observations and the new DNA sequences. The chapter also re-examines the molecular phylogenetic relationship between the Chrysophyceae and Synurophyceae using the greater taxon sampling from these new sequences.

TAXONOMIC OBSERVATIONS

During the course of this study, cultures were examined repeatedly in an effort to provide the best possible identifications, and some cultures previously identified with other names have been renamed in this paper. There were several reasons why identity problems existed. The vegetative states of some Chrysophyceae were variable under culture conditions (i.e., flagellate, capsoid, amoeboid, etc.). Cysts were rarely produced in culture, especially morphologically distinctive cysts. Finally, flagellar number was difficult to determine for those organisms with a very short mature flagellum.

1. *Chrysocapsa paludosa* (Korsh.) nov. comb. Andersen

Basonym: *Phaeocapsa paludosa* Korsh. Korshikov, A.A. 1924, Protistologiceskije zametki. *Arch. Russ. Protistol.* 3: 113–127.

Culture strain CCMP380 was collected from St. Ronan's well, Victoria, Australia, where cells grew attached individually to the stem of a vascular plant. The morphological condition in the field was identical to that of *Phaeocapsa paludosa* Korsh. [= *Epichrysis paludosa* (Korsh.) Pascher 1925], and the name *Epichrysis paludosa* was used earlier (Andersen et al., 1999). Descriptions for both *Phaeocapsa* and *Epichrysis* emphasized their epiphytic habit. Starmach (1973, 1975) described an organism that may be identical (*Chrysocapsa flavescens* Starmach), and it was epiphytic on the red alga *Batrachospermum*. Cells had a single parietal chloroplast that lacked an eyespot, and there were 1–2 contractile vacuoles in each cell (Figure 15.1a, Figure 15.1b, and Figure 15.1e). Gelatinous

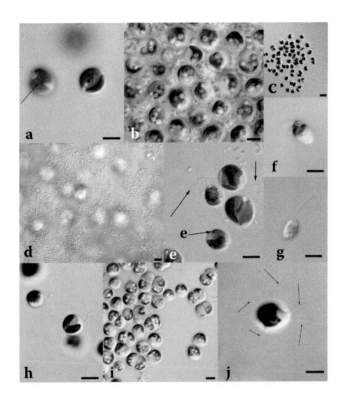

FIGURE 15.1 (Please see colour insert following page 76) (a and b) *Chrysocapsa paludosa*—CCMP380: (a) Two cells in opposing view, showing the connection (arrow) between the two deeply divided chloroplast lobes. Scale bar = 5 μm. (b) Densely packed cells from an old colony showing individual gels around each cell, a chloroplast that fills only about half of each cell, and numerous globules inside and outside the cells. Scale bar = 5 μm. (c) *Naegeliella flagellifera*—CCMP280. Typical colony in culture with gelatinous hairs. Scale bar = 20 μm. (d and e) *Kremastochrysis* sp.—CCMP260: (d) Surface film of bacteria with hyponeustonic cells out of focus below the surface. (e) Flagellate cells showing the long immature flagellum (arrows) and the eyespot (e). Scale bar = 5 μm. (f) *Chrysosaccus* sp.—CCMP2717. Zoospore with an obovate shape, a small chloroplast, a long immature flagellum, a short mature flagellum, and an eyespot. Scale bar = 5 μm. (g) Unidentified chrysophyte—CCMP1161. Zoospore with oval shape, a small chloroplast, and a long immature flagellum. Scale bar = 5 μm. (h and i) *Chrysosaccus* sp.: (h) CCMP295. Vegetative cells with spherical shape, a large chloroplast that fills the cell, and an immature flagellum that beats very slowly. Scale bar = 5 μm. (i) CCMP296. Vegetative cells with spherical shape organized into a capsoid colony, cells not flagellated, several cells in division stages. (j) Unidentified chrysophyte—CCMP1161. Amoeboid cell with several fine pseudopods (arrows). Scale bar = 5 μm.

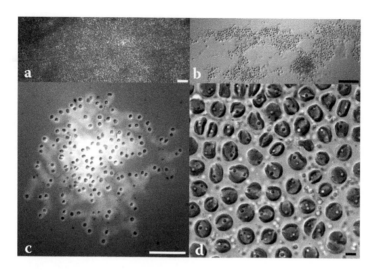

FIGURE 15.2 (a) *Kremastochrysis* sp.—CCMP260. Low magnification image of an old hyponeustonic sheet of cells growing just below the cell surface; sheet removed by touching a coverglass to the water surface. The cell layer is mostly monostromatic, but may become 2–3 cells thick in older regions. Scale bar = 250 μm. (b through d) *Chrysocapsa paludosa*—CCMP380: (b) Low maginification image of an old, reticulate colony; colony removed by Pasteur pipette. Scale bar = 100 μm. (c) Younger colony surrounded by India ink to demonstrate the presence of a colonial gel. Scale bar = 50 μm. (d) Vegetative cells in a medium-aged colony showing a large chloroplast that nearly fills the cell, fewer globules inside the cells, and cell division in a few cells. Scale bar = 5 μm.

colonies were either present or absent under differing field conditions (Korshikov, Starmach gelatinous colony in field samples; Korshikov, Pascher, Andersen—no gelatinous colony in field samples). However, in culture, CCMP380 produced gelatinous colonies that were at times identical to those described by Korshikov (1924) and Starmach (1973, 1975) (Figure 15.1c). In old cultures, very large, reticulate colonies were produced (Figure 15.2b). CCMP380 lacked a cell wall, consistent with observations by Korshikov (1924) and Pascher (1925).

Bourrelly (1957) emended the generic description of *Chrysosphaera* and transferred Korshikov's alga to *Chrysosphaera* [*Chrysosphaera paludosa* (Korsh.) Bourrelly]. However, a fundamental character for *Chrysosphaera* is vegetative reproduction by autospores or zoospores. For CCMP380, vegetative cell division is not by autospores or zoospores; cells divide vegetatively as is characteristic for capsoid organisms (Figure 15.1e). Cell division may not be obvious in field material; however, in culture the mode of vegetative cell division is easily observed. Based upon morphological observations and SSU rRNA and *rbc*L gene sequences (see below), I conclude that CCMP380 should belong to the genus *Chrysocapsa*, and a new combination is proposed.

2. *KREMASTOCHRYSIS* SP.

Culture strain CCMP260 was collected in Massachusetts by Ralph Lewin and brought into culture. The strain was named *Chromophyton rosanoffii* Woronin because of its neustonic habit of growth. Recently, two additional strains were collected from Japan that appear to more accurately represent *Chromophyton* (CCMP2751, CCMP2753). Therefore, strain CCMP260 was re-examined. By carefully picking up cells from the water surface onto a microscope coverglass, it was discovered that CCMP260 grew hyponeustonically, i.e., the cells extended downward into the water. The surface film had a layer of bacteria, often associated above an algal cell that was below the water surface (Figure 15.1d). Conversely, *Chromophyton* cells grew epineustonically, i.e., the cells extended above the air–water interface (Woronin, 1880; Lund, 1942; Vischer, 1943; Petersen and Hansen, 1958).

Krematochrysis (Pascher, 1942) and *Krematochrysopsis* (Bourrelly, 1957) differ only in the number of flagella visible in the light microscope. Pascher did not consider flagellar number to be significant because his original paper included one species with two flagella and a second species with only one flagellum. Because Bourrelly (1957) placed strong emphasis on flagellar number (Table 15.3), he created a new genus for the second species. Cells of strain CCMP260 were 4 to 12 μm in diameter, and each cell had a single visible flagellum, and a single chloroplast (with eyespot) that nearly filled the cell (Figure 15.1e). In old cultures, the entire water surface became covered by a subsurface layer of cells (Figure 15.2a). The layer was usually one cell layer thick in young cultures and became two to three cells thick in areas of older cultures.

Kremastochrysis pendens is very similar to CCMP260; however, it has two visible flagella, there is no eyespot, and the cells are slightly larger. *Kremastochrysis ocellata* Pascher differs in that the cells are up to 25 μm in diameter and there are two chloroplasts (one with eyespot) that are narrow and bandlike. Therefore, strain CCMP260 may represent an undescribed species of *Kremastochrysis*. *Kremastochrysis* sp. (CCMP260) also has distinctly different gene sequences from *Chromophyton* cf. *rosanoffii* (see below).

3. *Naegeliella flagellifera* Correns

Strain CCMP280 was previously named *Chrysochaete britannica* (Goodward) Rosenberg (e.g., Andersen et al., 1997, 1999), but after further study, it is called *Naegeliella flagellifera* Correns. This alga—by any name—is so distinct that it cannot be mistaken (Correns, 1892). The minor differences observed by others do not warrant, in my opinion, separate generic or species distinction (Scherffel, 1927; Godward, 1933; Rosenberg 1941; Petersen and Hansen, 1960; Dop, 1978).

Correns (1892) provided the original diagnosis as part of a thorough study. He believed that he observed zoospores with two laterally inserted flagella, but this was not conclusive; he did state that he found no protoplasmic strand inside the gelationous hairs and he observed basal sheaths around gelatinous hairs. The species *N. natans* Scherffel was based upon Scherffel's (1927) observations of protoplasmic strands in the gel stalks, the lack of a basal sheath, and the description of a uniflagellate zoospore. Godward (1933) provided additional observations and distinguished a new species, *Naegeliella britannica* Godward, based upon gel characteristics around the cell. Rosenberg (1941) transferred *N. britannica* to a new genus, *Chrysochaete*, based upon the presence of uniflagellate zoospores. That is, she emphasized flagellar number when distinguishing *Naegeliella* and *Chrysochaete*. Petersen and Hansen (1960) re-classified *N. natans* as a variety [*Naegeliella flagellifera* var. *natans* (Scherffel) Petersen et Hansen]. They considered the tentative association of the biflagellate zoospore given by Correns as questionable and not a good basis for distinguishing two genera. Tschermak-Woess (1970) classified the alga as *Chysochaete* and she discovered the presence of a cryptic pyrenoid. Dop (1978) described the ultrastructure and he confirmed the presence of a pyrenoid near the nucleus. Dop also conducted studies on the gelatinous hair production under varying concentrations of nitrate (hair structure increased as nitrate concentration decreased).

Strain CCMP280 expressed various morphologies in culture. The cells had a single, deeply lobed and often V-shaped chloroplast with a pyrenoid (Figure 15.3a). The typical colony had a cluster of cells and gelatinous hairs extended out and away from the cells (Figure 15.1c and Figure 15.3b). The gelatinous material around the cells was often laminated (Figure 15.3a), sometimes a basal sheath was present (Figure 15.3c), and a variety of odd morphologies could be found (e.g., Figure 15.3f). Some cells had gelatinous strands with so-called protoplasmic extensions (Figure 15.3d and Figure 15.3e). The gelatinous hairs connected to the cell itself, but this was evident only after straining (e.g., methylene blue in Figure 15.3d). The so-called protoplasmic extensions could be observed only after various harsh treatments, e.g., after treatment with dilute NaOH and Lugol's Iodine solution (Figure 15.3e). The protoplasmic extensions were not always present, even on gelatinous hairs from the same colony. Zoospores were rarely observed, but they had a single visible anterior flagellum (not shown).

CCMP280 was highly variable in culture, and except for the dubious biflagellate zoospore, it was comparable to the descriptions prepared by others (see above). Petersen and Hansen (1958) considered

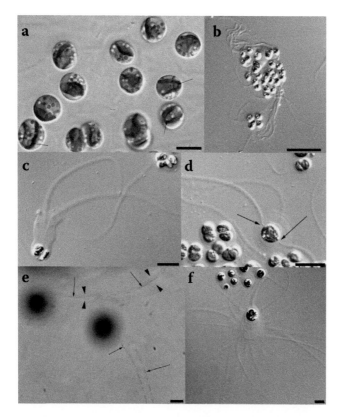

FIGURE 15.3 (a through f) *Naegeliella flagellifera* – CCMP280: (a) High magnification of a mature colony with laminated gelatinous material around the cells. Note the deep V-shaped chloroplast in most cells, which is characteristic for this species. Scale bar = 10 μm. (b) Low magnification of a typical colony showing the cells in the center of the colony and gelatinous hairs extending away from the cells. Scale bar = 50 μm. (c) A few basal cells with a gelatinous sheath enclosing two gelatinous hairs. Scale bar = 20 μm. (d) A cell stained with methylene blue to show the attachment of the gelatinous hairs to the cell body (arrows). Scale bar = 20 μm. (e) Gelatinous hairs treated with dilute NaOH and stained with Lugol's Iodine to show the delicate central thread broken into short, dark segments (arrows) inside the gelatinous wall (arrowheads). Scale bar = 5 μm. (f) An unusual cell with a towering, laminate gelatinous column with numerous gelatinous hairs extending from the end opposite the cell. Scale bar = 10 μm.

that the biflagellate cell was mistakenly associated with *Naegeliella* by Correns (1892), and Correns himself was not certain about the association of the biflagellate cell as a zoospore. I agreed with their decision. I demonstrated all the other morphological features that have been associated with the alga under its various names, and in conclusion, I used the name *Naegeliella flagellifera*.

4. CHRYSOSACCUS SP.

Organisms tentatively named *Chrysosaccus* sp. were isolated from four different locations in the United States, and multiple isolates were obtained for three of the four locations. CCMP295, CCMP296, CCMP1162, and CCMP2717 were isolated from samples collected on March 6, 1978, from a small pond, colloquially known as Southeast Bridge Pond, near Fayetteville, Arkansas. CCMP 1163 was collected on March 28, 1981, from Volo Bog, Illinois. CCMP368 and CCMP2716 were isolated from samples collected on August 11, 1984, from a small pool near Burnette Park, Keeweenaw County, Michigan. CCMP1156 and CCMP1157 were isolated from a Lake Michican sample collected in April 1987 by Dr. Norm Andresen.

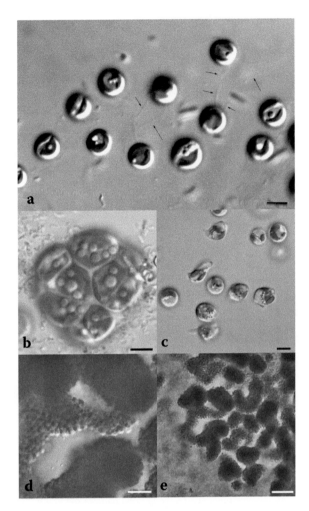

FIGURE 15.4 (a and b) *Chrysosaccus* sp.—CCMP1156: (a) Fine connections (arrows) between colonial cells; colony with gelatineous matrix. Scale bar = 5 µm. (b) Small spheroidal packet with tightly packed cells. Scale bar = 5 µm. (c through e) Unidentified chrysophyte—CCMP1161: (c) Amoeba-like cells with thick pseudopod extending from the posterior end of the cell intermixed with typical spherical cells. Scale bar = 5 µm. (d and e) High and low magnifications of an old, reticulate colony. Note that the colony has nodules of densely packed cells up to 25 cells thick that are held together with regions only 1–5 cells thick. Scale bar = 5 µm.

The morphological features of all isolates were nearly identical. Cells were almost always naked, and usually forming colonies held together by a thin, gelatinous substance (Figure 15.1i). Although single cells were isolated into culture, the cells formed gelatinous colonies in culture. Cell walls were rarely observed, and they were very thin when present (not shown). One culture (CCMP1156) produced small packets of abutted cells that lay inside the larger and typical amorphous colony (Figure 15.4b).

Cells were typically spherical, with a single, parietal chloroplast that filled most of the cell. The chloroplast lacked an eyespot, and cells had 1–2 contractile vacuoles. Nonflagellate colonial cells had the gross appearance of *Chrysocapsa*, but the cells sometimes had fine intercellular connections that are unreported for *Chrysocapsa* (Figure 15.4a). Sometimes a culture had many vegetative cells with flagella that beat very slowly (Figure 15.1h), i.e., the flagellate cells were still embedded in the colonial gel and they did not swim around. True zoospores were produced and they escaped from the colonial gel; the zoospores were more elongate and swam rapidly. Zoospores were biflagellate and they sometimes had an eyespot (Figure 15.1f). Cysts were simple and unornamented (not shown). Vegetative cell division was the common form of growth (Figure 15.1i).

The type diagnosis for *Chrysosaccus incompletus* Pascher described cells in groups of four, cells that abutted each other, and he described the presence of rings in the gelatinous material (Pascher, 1925). Bourrelly (1957) described a second species, *C. sphaericus* Bourrelly, and Starmach described a third species, *C. epilithicus* Starmach, (Starmach, 1985). *Phaeosphaera gelatinosa* West and West superficially resembled these strains (West and West, 1903).

Of these, the culture strains in this study most closely resembled *C. epilithicus*. Except for rare occasions in CCMP1156, cells never grouped together or formed laminate rings in the gel matrix, and even under these conditions, they did not resemble *Chrysosaccus incompletus*. The presence of flagellate cells, both slowly beating vegetative cells and rapidly swimming zoospores, are not reported for any *Chrysosaccus* species. However, Pascher acknowledged that his description was incomplete, and the remaining descriptions are from field samples.

Phaeosphaera was described as forming cylindrical, branching colonies. Colony shape may be attributable to a flowing water habitat, and therefore it is possible that a different colonial morphology occurs in test tube culture. However, the presence of cells with a slowly beating flagellum does not conform to previous descriptions, and it seems unlikely that this feature was overlooked.

The precise identity of the *Chrysosaccus* strains remains unknown, and even the placement in *Chrysosaccus* must be considered tentative. Sequence data for CCMP368 and CCMP1156 are presented in Figure 15.1 through Figure 15.3; sequences for all strains are essentially identical.

5. UNIDENTIFIED CAPSOID CHRYSOPHYTE SPECIES

CCMP349 and CCMP1161 were from the same collection site in northern Michigan and represent a single taxon. CCMP349 was previously identified as *Chrysosphaera parvula* (Pascher) Bourrelly (Andersen et al., 1997, 1999). However, further study cast doubt on that identification. Specifically, *Chrysosphaera* was described as reproducing via autospores or zoospores, yet these strains showed clear evidence of vegetative cell division similar to that found in capsoid organisms (not shown).

In old cultures, CCMP349 and CCMP1161 produced large gelatinous colonies (ca. 1–3 cm) that superficially resembled *Dermatochrysis* (Figure 15.4d and Figure 15.4e). However, *Dermatochysis* has a monostromatic thallus (Entwisle and Andersen, 1990). The thallus of CCMP349 and 1161 had vast regions that were several (2–6) cells thick, and much thicker balls of cells (ca. 25 cells thick) developed in certain areas of the thallus. The thallus was never perforate in the thin regions, but open areas developed where the balls of cells occurred (Figure 15.4d and Figure 15.4e). Vegetative cells were naked, usually spherical or oblong in shape, with a single chloroplast that lacked an eyespot. Two types of amoeboid cells were also observed. One type had thin pseudopods radiating from around the cell (Figure 15.1j), and the other type had a broad pseudopodial "tail" (Figure 15.4c). Zoospores were elongate, not spherical, and they had two flagella; the mature flagellum was short and not always visible (Figure 15.1g). Zoospores escaped from the colonial gel matrix and swam rapidly. Cysts were smooth and without ornamentation (not shown). We have sequenced genes from ACOI-514, a strain identified as *Dermatochrysis*; the sequences of CCMP349 and CCMP1161 differed significantly from those of ACOI-514 (data not shown).

The earlier identification was based upon the similarity to *Chrysosphaera parvula* cells (see *Chrysobotrys parvula* Pascher 1925), not to colony shape and development. The vegetative colony of CCMP349 and CCMP1161 showed some resemblance to capsoid organisms such as *Chrysonebula*, *Celloniella*, and *Hydrurus* in that the cells were occasionally elongated and possessed a cytoplasmic "tail." Gene sequences from *Chrysonebula flava* Starmach were included in this study, and *C. flava* was not closely related to CCMP349 and CCMP1161. *Celloniella* and *Hydrurus* were considered to be restricted to very cold streams, and they have very distinctive wings or flanges on their cysts; neither habit nor cyst applied to our strains. Because I was unable to provide a clear identification, I referred to this organism as an unidentified capsoid chrysophyte species.

6. OTHER TAXONOMIC NOTES

Organisms with a morphological appearance consistent with traditional morphological genera but with a molecular phylogenetic relationship distinct from the type species (or well-established species) were given provisional status (e.g., "*Chromulina*" or "*Chrysocapsa*"). This problem was most acute with regard to *Ochromonas*. The type species, *O. triangulata*, has never been observed since its original description (Vysotskii, 1887).

Poterioochromonas also presents a confusing situation. The type species, *P. stipitata* Scherffel, has a distinct lorica morphology, and CCMP1862 is an excellent representative for this species. *Poterioochromonas malhamensis* (Prings.) Peterfi is a more complicated situation. The lorica is sometimes present and sometimes absent, even in the authentic culture (SAG933.1). Furthermore, many "*Ochromonas*" species will produce a conical "tail," and these tails have never been carefully investigated to see if a chitinous lorica is present.

Some of the *Synura petersenii* strains could have been identified as *S. glabra* but were not. Unfortunately, CCMP851 and CCMP857 were once strains of *S. echinulata* and *S. mollispina*, respectively (based upon transmission electron microscopy [TEM] observation of scales) (see Andersen et al., 1997). However, apparently due to mistakes in labeling the cultures, the cultures became occupied by *S. petersenii*.

Stein (1963) described *Chromulina chionophila* from a Canadian snow sample. The alga was described as a flattened flagellate. One culture strain has been named *Chromulina chionophila* (CCAP909/9; subsequently CCMP261). A culture by this name was collected from snow and isolated by Ronald Hoham, but apparently it has been contaminated or misidentified because both culture collection strains have smaller, spherical cells with no eyespot. The identity of CCAP909/9 and CCMP261 remains unclear. Morphologically, the alga resembles the *Chrysosaccus* strains (see above) and its gene sequences are very similar to those from the *Chrysosaccus* strains. Furthermore, the strain has been erroneously named *Chromulina chromophila* (Bhattacharya et al., 1992), causing further confusion.

MOLECULAR PHYLOGENETIC ANALYSIS

Almost all DNA samples were obtained from culture strains, although the *Chrysosphaerella* sequences were produced from single-colony polymerase chain reaction (PCR) amplifications (repeated five times). Cultures were obtained from several public culture collections and some personal cultures from colleagues. Additional sequences were obtained from GenBank.

DNA was extracted from cultures using a CTAB extraction method (Doyle and Doyle, 1987). PCR was performed using standard primers (see Andersen et al., 1999; Daugbjerg and Andersen, 1997b), and the sequencing reactions were run on an ABI310 DNA sequencer. Sequence data were edited in Sequencher then aligned using ClustalX. Evolutionary models were evaluated using Modeltest 3.06 (Posada and Crandall, 1998). A general time reversible (GTR) model with gamma correction for among-site rate variation and invariant sites was recommended for both the *rbc*L and 18S rDNA data sets. Bayesian analyses were run in MrBayes (Huelsenbeck, 2000) on the *rbc*L alone, 18S rDNA alone, and combined data with the GTR model (nst = 6) with rates set to invgamma and nucleotides set to equal. Bayesian analyses were also run using specified parameters from Modeltest. All analyses were performed using four Markov chains, with 2,000,000 generations per chain and had burn-ins of 1000 trees. Phylogenetic estimates based on Modeltest parameters entered into MrBayes were virtually identical to those in which MrBayes estimated the parameters (lset nst = 6 rates = invgamma settings), and therefore only the latter are shown.

The *rbc*L phylogenetic tree (Figure 15.5) shared ten clades with the 18S rDNA data (Figure 15.6), and the combined analysis (Figure 15.7) also shared 10 clades with the *rbc*L tree. The basal relationships among these clades varied in the analyses; higher posterior probabilities occurred in the *rbc*L analysis (Figure 15.5) and the combined data analysis (Figure 15.7). The 18S rDNA analysis had a basal polychotomy with seven unresolved branches (Figure 15.6). The combined analysis had weak posterior probability support for separating the Chrysophyceae and Synurophyceae (Figure 15.7); however, neither gene by itself was able to resolve monophyletic relationships for these two classes.

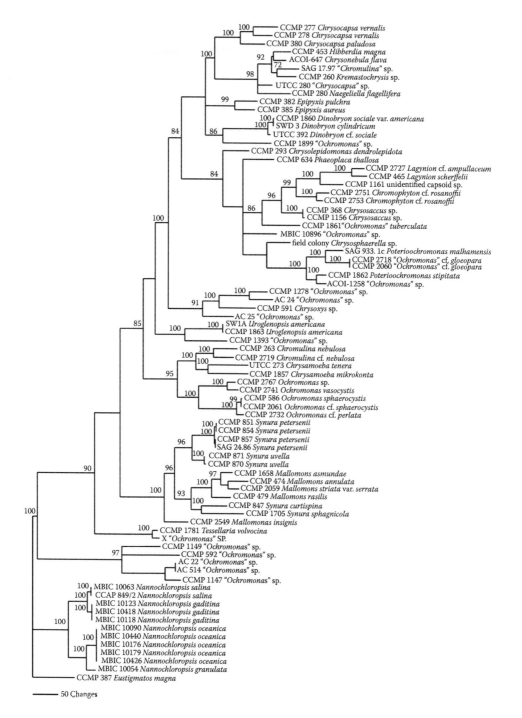

FIGURE 15.5 Phylogenetic tree based upon *rbc*L gene sequences (635 parsimony informative characters, L = 4824, CI = 0.21, RI = 0.79). Posterior probabilities >70% are shown.

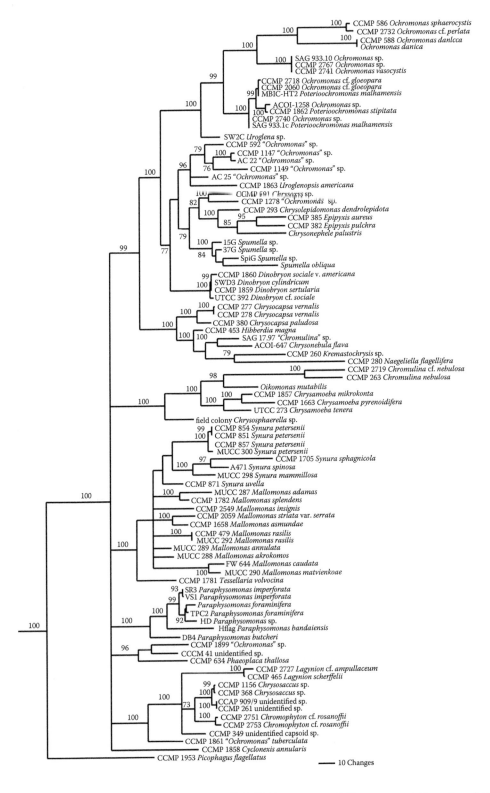

FIGURE 15.6 Phylogenetic tree based upon SSU rRNA sequences (442 parsimony informative characters, L = 3381, CI = 0.27, RI = 0.75). Eustigmatophyceae outgroup (not shown) is the same as in Figure 15.5 and Figure 15.7. Posterior probabilities >70% are shown.

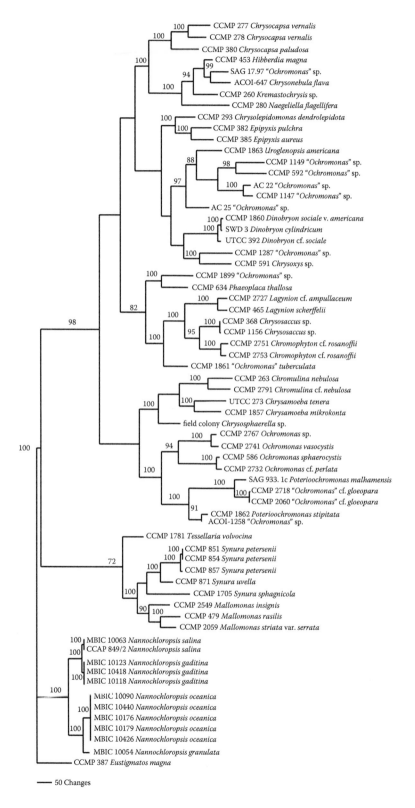

FIGURE 15.7 Phylogenetic tree based upon the combined *rbc*L and SSU rRNA sequences (1077 parsimony informative characters, L = 6599, CI = 0.27, RI = 0.73). Posterior probabilities >70% are shown.

DISCUSSION

The morphological classification of chrysophytes, as summarized in three keystone papers (Table 15.2 through Table 15.4), is incongruent with past molecular phylogenetic analyses. In an effort to resolve the incongruence, this study attempted (a) to carefully evaluate the identity of some problematic taxa using cultures and light microscopy, and (b) to add a significant number of new taxa to the molecular phylogenetic analyses. Unfortunately, these attempts failed, incongruence still exists, and the classification of chrysophytes remains chaotic. From the class and order levels to the genus and species levels, it is often unclear how to identify organisms, how to resolve evolutionary conflicts in morphological and molecular data sets, and how to describe and classify organisms in a meaningful way. The following discussion attempts to look at high- and low-level classification problems, and it attempts to offer some explanation for the seemingly bewildering differences between morphology and molecular data sets.

DISTINGUISHING CLASSES

Some earlier molecular phylogenetic studies found a paraphyletic relationship for the Chrysophyceae and Synurophyceae (e.g., Lavau et al., 1997; Caron et al., 1999), however, the combined analysis in our study (Figure 15.7) suggests that there may be molecular support, albeit weak support, to corroborate morphological and biochemical characters used to separate the two classes (Table 15.1) (Andersen, 1987, 1989). While the overall assemblage of morphological, biochemical, and molecular data provide support for the continued separation of the Chrysophyceae and Synurophyceae, the issue is by no means resolved. Further study on all fronts may be required to achieve compelling evidence for the higher-level classification of these two groups. It will be particularly important to use non-plastid genes in future studies so that the colourless, silica-scaled *Paraphysomonas* can be included in multigene analyses.

BROAD RELATIONSHIPS WITHIN THE SYNUROPHYCEAE

Pipes et al. (1991) and Pipes and Leedale (1992) studied the enigmatic synurophyte *Tessellaria volvocina*, and they suggest *Tessellaria* has features that are intermediate between the typical Chrysophyceae and typical Synurophyceae taxa. The molecular phylogenetic study presented here supports their opinion because *Tessellaria* was basal to the other Synurophyceae taxa. The combined gene analysis suggests that *Synura* and *Mallomonas* are monophyletic genera (Figure 15.7), but the *rbc*L gene alone did not support monophyletic genera (Figure 15.5). Interestingly, Siver and Wolfe (2005) found synurophyte silica scales from Eocene deposits approximately 47 million years old, and the scales are essentially identical to those produced by today's living taxa (i.e., *Mallomonas insignis* and *Synura uvella*). Was stasis of scale morphology over millions of years accompanied by stasis of genes? Does this reveal anything about the evolution of nonmineralized organisms that do not fossilize? Unfortunately, very little is known about the fossil history of chrysophytes and therefore there is no wealth of fossils to help us resolve conflicts between morphology and molecules.

BROAD RELATIONSHIPS WITHIN THE CHRYSOPHYCEAE

The historical separation of the Chrysophyceae into "uniflagellate" and "biflagellate" groups (see Introduction, Table 15.2 through Table 15.4) was not supported in any of the three molecular phylogenetic analyses that are presented in this chapter. Even though there was weak support for the basal clades, the subdivision of the Chrysophyceae into monophyletic Chromulinales and Ochromonadales clades is not possible. As it has become evident during the past 50 years, all flagellate cells of the Chrysophyceae appear to have two flagella (or at least two basal bodies). Therefore, classification by flagellar number is an artifact due to the resolving power of the light microscope. Both morphology (i.e., electron microscopy) and molecular phylogenetic analysis agree on this point. Therefore, no classification of the Chrysophyceae should be based on flagellar number.

How to proceed? The Chrysophyceae literature contains several examples of very similar organisms that are classified in different genera (and different orders) based upon the number of visible flagella. For example, *Kephyrion* species have one visible flagellum, whereas *Pseudokephyrion* species have two visible flagella. Other examples are *Kremastochysis—Kremastochrysopsis* and *Naegeliella—Chrysochaete*. It was argued above that *Naegeliella* should be used rather than *Chrysochaete*, and it was implicit that *Kremastochrysis* should not be split to recognize *Kremastochrysopsis*. Nevertheless, while it could be argued for complete abandonment of flagellar number, a careful examination of organismal activity and phylogenetic trees (Figure 15.5 through Figure 15.7) suggests that flagellar number may have some taxonomic value. Andersen et al. (1999) argued that a visible and obvious second flagellum is correlated with phagotrophy. The uppermost large clade in Figure 15.6 (*Ochromonas sphaerocystis* to *Dinobryon* cf. *sociale*) is supported with a 100% posterior probability and contains only organisms that feed phagotrophically with aid of the relatively long second flagellum. However, *Paraphysomonas* also uses a second flagellum for phagotrophy, and the *Paraphysomonas* clade is not included within the aforementioned clade. Nonphagotrophic organisms with only one visible flagellum also form well-supported clades (e.g., *Chromulina nebulosa-Chrysamoeba* spp. clade; *Chrysocapsa* spp.—*Hibberdia*, etc.). Therefore, while flagellar number is not a good taxonomic character, phagotrophy based upon a relatively long and easily visible second flagellum appears to measure an evolutionary signal that is consistent for both morphology and molecules.

Molecular phylogenetic study with several species of *Chromulina* and *Ochromonas*, often considered the most typical genera for the Chrysophyceae, showed that the genera were polyphyletic. However, of interest given the discussion above, no single clade had *Chromulina* and *Ochromonas* species as sister taxa. The polyphyletic relationships in the molecular trees are incongruent with the morphology used to define the genera, but there appears to be some congruence (evolutionary signal?) related to flagella because species of the two genera are kept separate using both morphology and molecules. Is this still related to phagotrophy or something more? Presumably, naked single-cell flagellates have evolved independently several times from other life forms by reductive evolution (e.g., capsoid colonies, flagellate colonies, loricate cells). Perhaps the polyphyly comes from repeated events of reductive evolution. If this is the case, then we must invert our ideas about chrysophyte evolution. That is, rather than following the Blackman (1900) model of a single flagellate giving rise to progressively more complex organisms, perhaps the ancestral chrysophyte was a nonflagellate organism with many cells (e.g., *Phaeoplaca*)?

Cells and cysts that agree with the type description of *Chromulina nebulosa* (Cienkowski, 1870) identify the clade that must be called *Chromulina*, and presumably, with further study, older names such as *Chrysomonas*, *Pseudochromulina*, and *Chromulinella* may be applied in some cases. In the case of *Ochromonas*, the type species has never been reported since its original description (Vysotskii, 1887). The original description was remarkably thorough, including many drawings of vegetative cells, mitotic cells, and (remarkably) flagellar transformation. We (Dr. Olga Gravrilova and I) attempted to find the type species in samples collected from the type locality but *Ochromonas triangulata* was never observed. Failing its rediscovery, and considering that DNA is unavailable from the type material (ink drawings), an epitype might be established to resolve the taxonomic ambiguity. However, is it possible to establish an epitype in this instance? The *Ochromonas triangulata* is thoroughly described, and there is no ambiguity surrounding its description. In addition to identity of the type species, all *Ochromonas* species are described largely on cell morphology and this causes problems for culture work (usually cultures are necessary to obtain DNA for molecular systematics) because the cells express many morphological forms in culture. For those species with distinctive cysts, identification is more reliable (e.g., *O. sphaerocystis*, *O. tuberculata*). However, many species are described without cysts or with simple, unornamented cysts. Consequently, the trees have many taxa listed as "*Ochromonas*" sp. and distinctive taxa, such as *O. tuberculata*, are distantly related to *O. danica* and its relatives. Further work is required to resolve these problems.

Lorica formation did not define a monophyletic group in the phylogenetic trees, but it has been used (with flagellar number) to define families (Table 15.2 through Table 15.4). A number of common loricate species always formed monophytic groups for their genus (i.e., *Dinobryon, Epipyxis, Lagynion*), although admittedly, the number of species for each genus was limited. However, for example, the family Dinobryaceae *sensu* Bourrelly (Table 15.4) was not monophyletic in these trees. Furthermore, the loricate *Poterioochromonas* had non-loricate *Ochromonas*-type taxa interspersed, and *Poterioochromonas* as a genus did not form a monophyletic clade. *Dinobryon* and the smooth-scaled loricate species of *Epipyxis* had, at one time, been placed into *Dinobryon*, while the scallop-scaled loricate species of *Epipyxis* were placed in *Halobryon* (e.g., Huber-Pestalozzi, 1941). More recent treatments combined the two *Epipyxis* loricate types (e.g., Hilliard and Asmund, 1963), and our analysis of *E. aureus* (smooth) and *E. pulchra* (scalloped) supported this view. The fundamental difference between *Dinobryon* and *Epipyxis* loricas is that the former are spun with cellulose fibres (Franke and Herth, 1973; Herth, 1979; Herth and Zugenmaier, 1979), while the latter are assembled from organic scales (Hilliard and Asmund, 1963). Interestingly, *Chrysolepidomonas* produces ornate organic scales that coat the cell but are not assembled into a lorica, and *Chrysolepidomonas* has 100% support at the base of the *Epipyxis* clade. This suggests that *Epipyxis* evolved from a scaled algal intermediate, assuming that both were evolutionarily derived from a naked flagellate. In summary, gross morphology of the lorica is misleading, but the detailed morphology and development of loricas provides phylogenetic signal such that the morphology and molecules are congruent.

Many capsoid taxa (i.e., *Chrysocapsa, Chrysonebula, Hibberdia, Naegeliella*) formed a clade with other taxa that produce gelatinous colonies ("*Chromulina*" and *Kremastochrysis*) in the phylogenetic trees, however, other capsoid taxa (e.g., *Chrysosaccus* strains and the unidentified chrysophyte strains, CCMP349 and CCMP1161) were distantly related. Therefore, the capsoid habit, or the morphological presence of a gelatinous matrix, does not define a monophyletic clade in the molecular phylogenetic trees. Interestingly, the type species of *Chromulina* (*C. nebulosa*) produces large gelatinous colonies where cells swim slowly inside the gel sac (Cienkowski, 1870; Andersen, 1986), a life form or habit named the pseudopalmelloid condition (Andersen, 1986). Nevertheless, in all three molecular phylogenetic analyses (Figure 15.5 through Figure 15.7), *C. nebulosa* was most closely related to free-living species of *Chrysamoeba*. The organisms tentatively called *Chrysosaccus* express a life form somewhat intermediate between the more thickly gelatinous colonies of nonflagellate cells (*Chrysocapsa, Naegeliella*) and the thin, watery gels of the swimming *Chromulina nebulosa*. That is, *Chrysosaccus* cells sometimes have vegetative cells with a slowly beating flagellum even though the cell does not usually swim around inside the gelatinous matrix. Is this an evolutionary trend? Molecular studies say no. If gelatinous material provides evolutionary signal, it must be in the relationship of organisms with a very specific type rather than as a general trait amongst all.

Flagellate colony formation is a morphological character for separating genera (e.g., *Uroglenopsis* from *Ochromonas, Synura* from *Mallomonas*). The molecular phylogenetic analyses generally support the monophyly of the colonial genera. However, culture studies suggest caution. *Uroglenopsis*, when in culture, forms beautiful colonies during the early growth period, but as the culture ages, the colony completely disperses and individual *Ochromonas*-like cells continue to live and grow in cell number. Similarly, at least for *Synura petersenii*, colonies of *Synura* are large and robust when cultures have been recently subcultured, however, in old cultures there are mostly single cells or colonies of 1–3 cells. If *Uroglenopsis* responds similarly in nature, forming large numbers of uniflagellate cells, it would be very difficult to identify an individual swimming cell from *Ochromonas* spp. In the case of *Synura*, it is easy to distinguish single cells from *Mallomonas* because scales are distinctive; if scales were lacking from both genera, the problem would resemble that for *Uroglenopsis*. Finally, molecular evidence for separating *Uroglena* and *Uroglenopsis* is consistent with morphological distinctions. The separation or combination of the two genera has been long debated (e.g., Huber-Pestalozzi, 1941; Bourrelly, 1957, Wujek, 1976). *Uroglena* has cells that are united at the center of the colony by elongate cytoplasmic strains, whereas *Uroglenopsis* has cells united by bifurcating gel stalks. The SSU rRNA data show the two genera in two different and highly supported clades (Figure 15.6). We have not completed the *rbc*L sequence for *Uroglena*, but partial sequence data suggest the *rbc*L will support this finding.

OTHER PROBLEMS

At least some, and perhaps many, species of the Chrysophyceae express morphological plasticity in response to environmental changes (e.g., temperature, water movement, nutrient concentration, predation). For example, CCMP380 was isolated as a single cell that grew attached to a vascular plant and it appeared like *Epichrysis* in the field. However, in culture, it developed capsoid colonies like *Chrysocapsa*. Similarly, the *Chrysosaccus* strains were isolated as single cells, and as was previously known, *Naegeliella* produced a range of morphologies in culture. Therefore, it is likely that the same species has been described more than once, i.e., one form has one species name and another form has a second species name. In part, the problem can be addressed by careful culture work, although interpretation of culture variants should be cautious. Alternatively, the problem can now be attacked using single cells and gene sequences amplified by PCR. That is, if a rare species (or rare form of a common species) is found in nature, a single cell can be isolated and placed into a PCR tube for processing (realistically, repeated several times for confirmation). If the gene sequence proves to be identical with that of another morphologically distinguished species, then it will become more convincing that the environment is influencing morphology.

The topic of sexual reproduction is rarely raised for the chrysophytes (Wawrik, 1960, 1970, 1972; Sandgren and Flanagin, 1986). Therefore, haploid and diploid stages of the same organism or male and female dimorphism may also cause problems. It is generally believed that sexual reproduction is necessary to produce cysts (Sandgren, 1980, 1989; Sandgren and Flanagin, 1986). Culture isolates of *Ochromonas* prepared with more than one cell will often produce cysts, which then aid in species identification. When single cell isolations are made, cysts (or at least ornate cysts) are not produced. Consequently, the identification of culture strains from single cell isolations becomes problematic because of morphological plasticity of the vegetative cells and the lack of ornate cysts. Sexual crossing studies using culture strains, so prevalent for volvocalean algae, may offer insights when trying to understand naked flagellates such as *Ochromonas*.

SUMMARY

The morphological and molecular data sets are incongruent when examined at the broadest taxonomic levels (e.g., subclasses, orders). That is, when considering all chrysophytes, the use of light microscopic flagellar number and the use of life forms provides conflicting results. However, when the same morphological features are used at lower taxonomic levels (e.g., family, genus), then there may be agreement between the two data sets. It remains possible that the molecular data are not providing good phylogenetic signal, and this can be tested using more or different genes. The eventual resolution of conflicts between morphology and molecules will also require careful reexamination of organisms, their various life forms, and their sexual reproduction to more precisely define the morphological features. For example, the production of vegetatively dividing cells held within a colonial gel can be broadly termed capsoid, but that definition does not carry phylogenetic significance for the Chrysophyceae because there are several types of gelatinous colonies. To recover phylogenetically useful morphological characters, the morphological features must be critically examined. After critical examination, it should be possible to find substantial congruence between morphological and molecular data sets.

ACKNOWLEDGMENTS

I acknowledge the receipt of cultures from Dr Olga Gravrilova and Dr Susan Watson as well as from the ACOI, AlgaBank, CCAP, CCMP, MBIC, SAG, UTCC, and UTEX culture collections.

I thank Dr Norman Andresen for the Lake Michigan sample. This research was supported by U.S. National Science Foundation grants 02-12138, 02-06590, 04-44418, and 06-29564.

REFERENCES

Adl, S.M., Simpson, A.G.B., Farmer, M.A., Andersen, R.A., Anderson, O.R., Barta, J.R., Bowser, S.S., Brugerolle, G., Fensome, R.A., Fredericq, S., James, T.Y., Karpov, S., Kugrens, P., Krug, J., Lane, C.E., Lewis, L.A., Lodge, J., Lynn, D.H., Mann, D.G., McCourt, R.M., Mendoza, L., Moestrup, Ø., Mozley-Standridge, S.E., Nerad, T.A., Shearer, C.A., Smirnov, A.V., Spiegel, F.W., and Taylor, M.F.J.R. (2005) The new higher level classification of eukaryotes with emphasis on the taxonomy of protists. *Journal of Eukaryotic Microbiology*, 52: 399–451.

Andersen, R.A. (1986) Some new observations on *Saccochrysis piriformis* Korsh. emend. Andersen (Chrysophyceae). In *Chrysophytes: Aspects and Problems* (eds J. Kristiansen and R.A. Andersen), Cambridge University Press, Cambridge, pp. 107–118.

Andersen, R.A. (1987) The Synurophyceae *classis nov.*, a new class of algae. *American Journal of Botany*, 74: 337–353.

Andersen, R.A. (1989) The Synurophyceae and their relationship to other golden algae. *Beih. Nova Hedwigia*, 95: 1–26.

Andersen, R.A. (1991) The cytoskeleton of chromophyte algae. *Protoplasma*, 164: 143–159.

Andersen, R.A. (2004a) Heterokont phylogeny and origin(s) of plastids. *The Japanese Journal of Phycology*, 52 (Supplement): 153–162.

Andersen, R.A. (2004b) Biology and systematics of heterokont and haptophyte algae. *American Journal of Botany*, 91: 1508–1522.

Andersen, R.A., Morton, S.L., and Sexton, J.P. (1997) CCMP—Provasoli-Guillard National Center for Culture of Marine Phytoplankton—List of Strains. *Journal of Phycology*, 33 (Supplement): 1–75.

Andersen, R.A. and Mulkey, T.J. (1983) The occurrence of chlorophylls c_1 and c_2 in the Chrysophyceae. *Journal of Phycology*, 19: 289–294.

Andersen, R.A. and Preisig, H. (2002) Synurophyceae. In *An Illustrated Guide to the Protozoa*, 2nd ed., Vol. 2. (eds. J.J. Lee, G.F. Leedale, and P.C. Bradbury), Society of Protozoologists, Lawrence, KS, pp. 759–775.

Andersen, R.A., Van de Peer, Y., Potter, D., Sexton, J.P., Kawachi, M., and LaJeunesse, T. (1999) Phylogenetic analysis of the SSU rRNA from members of the Chrysophyceae. *Protist*, 150: 71–84.

Archer, W. (1875) *Chlamydomyxa labyrinthuloides* nov. gen. et nov. sp. *Quarterly Journal of Microscopic Science, New Series*, 15: 107–130.

Archibald, A.R., Cunningham, W.L., Manners, D.J., and Stark, J.R. (1963) Studies on the metabolism of Protozoa. 10. The molecular structure of reserve polysaccharides from *Ochromonas malhamensis* and *Peranema trichophorum*. *Biochemical Journal*, 88: 444–451.

Bailey, J.C., Bidigare, R.R., Christensen, S.J., and Andersen, R.A. (1998) Phaeothamniophyceae *classis nova*: a new lineage of chromophytes based upon photosynthetic pigments, *rbc*L sequence analysis and ultrastructure. *Protist*, 149: 245–263.

Ben Ali, A., DeBaere, R., Van der Auwere, G., DeWachter, R., and Van de Peer, Y. (2002) Evolutionary relationships among heterokont algae (the autotrophic stramenopiles) based upon combined analyses of small and large subunit ribosomal RNA. *Protist*, 153: 123–132.

Bhattacharya, D., Medlin, L., Wainright, P.O., Ariztia, E.V., Bibeau, C., Stickle, S.K., and Sogin, M.L. (1992) Algae containing chlorophylls $a + c$ are paraphyletic: molecular evolutionary analysis of the chromophyta. *Evolution*, 46: 1801–1817.

Blackman, F.F. (1900) The primitive algae and the flagellata. An account of modern work bearing on the evolution of the algae. *Annals of Botany*, 14: 647–688.

Booth, B.C. and Marchant, H.J. (1987) Parmales, a new order of marine Chrysophyceae, with the descriptions of three new genera and seven new species. *Journal of Phycology*, 23: 245–260.

Booth, B.C. and Marchant, H.J. (1988) Triparmaceae, a substitute name for a family in the Order Parmales (Chrysophyceae). *Journal of Phycology*, 24: 124.

Bourrelly, P. (1957) Recherches sur les Chrysophycées: morphologie, phylogénie, systématique. *Revue Algologique, mémoire hors-séries*, 1: 1–412.

Bourrelly, P. (1968) *Les Algues d'eau douce. II. Algues jaunes et brunes*. Boubée, Paris.

Caron, D.A., Lim, E.L., Dennett, M.R., Gast, R.J., Kosman, C., and DeLong, E.F. (1999) Molecular phylogenetic analysis of the hetrotrophic chrysophyte genus *Paraphysomonas* (Chrysophyceae), and the design of rRNA-targeted oligonucleotide probes for two species. *Journal of Phycology*, 35: 824–837.

Cavalier-Smith, T., Chao, E.E., and Allsopp, M.T.E.P. (1995) Ribosomal RNA evidence for chloroplast loss within Heterokonta: pedinellid relationships and a revised classification of ochristan algae. *Archiv für Protistenkunde*, 145: 209–220.

Christensen, T. (1962) *Alger*. In *Botanik*, Bd. 2. Systematisk Botanik Nr. 2. (eds T.W. Böcher, M.C. Lange, and T. Sørensen), Munksgaard, Copenhagen, pp. 1–178.

Cienkowski, L. (1870) Ueber Palmellaceen und einige Flagellaten. *Archiv für Mikroskopische Anatomie*, 7: 421–438.

Correns, C. (1892) Über eine neue braune Süsswasseralge *"Naegeliella flagellifera"* n. g. et sp. *Berichte Deutsche Botanische Gesellschaft*, 10: 629–636.

Daugbjerg, N. and Andersen, R.A. (1997a) Phylogenetic analyses of the *rbc*L sequences from haptophytes and heterokont algae suggest their chloroplasts are unrelated. *Molecular Biology and Evolution*, 14: 1242–1251.

Daugbjerg, N. and Andersen, R.A. (1997b) A molecular phylogeny of the heterokont algae based on analyses of chloroplast-encoded *rbc*L sequence data. *Journal of Phycology*, 33: 1031–1041.

Dodge, J.D. (1973) *The Fine Structure of Algal Cells*. Academic Press, London.

Dop, A.J. (1978) Systematics and morphology of *Chrysochaete brittanica* (Godward) Rosenberg and *Phaeoplaca thallosa* Chodat (Chrysophyceae). *Acta Botanica Neerlandica*, 27: 35–60.

Doyle, J.J. and Doyle, J.L. (1987) A rapid DNA isolation procedure for small quantities of fresh leaf tissue. *Phytochemical Bulletin*, 19: 11–15.

Edvardsen, B., Eikrem, W., Green, J.C., Andersen, R.A., Moon-van der Staay, S.Y., and Medlin, L.K. (2000) Phylogenetic reconstructions of the Haptophyta inferred from 18S ribosomal DNA sequences and available morphological data. *Phycologia*, 39: 19–35.

Entwisle, T.J. and Andersen, R.A. (1990) A re-examination of *Tetrasporopsis* (Chrysophyceae) and the description of *Dermatochrysis* gen. nov. (Chrysophyceae): a monostromatic alga lacking cell walls. *Phycologia*, 29: 263–274.

Fauré-Frémiet, E. and Rouiller, C. (1957) Le flagelle interne d'une Chrysomonadale: *Chromulina psammobia*. *Comptes Rendus de l'Academie des Sciences, Paris*, 244: 2655–2657.

Franke, W.W. and Herth, W. (1973) Cell and lorica fine structure of the chrysomonad alga, *Dinobryon sertularia* Ehr. (Chrysophyceae). *Archiv für Mikrobiologie*, 91: 323–344.

Gayral, P. and Billard, C. (1977) Synopsis du nouvel ordre des Sarcinochrysidales (Chrysophyceae). *Taxon*, 26: 241–245.

Geitler, L. (1942–1943) Eine neue atmophytische Chrysophyceae *Ruttneria spectabilis* n.g. n.sp. *Internationale Revue der Gesamten Hydrobiologie*, 43: 100–109.

Godward, M.B. (1933) Contributions to our knowledge of British algae. III. The genus *Naegeliella* in Britain. IV. On a form of *Phaeothamnion*. *Journal of Botany*, 71: 33–44.

Guillou, L., Chrétiennot-Dinet, M.-J., Moon-van der Staay, S.Y., Boulben, S., and Vaulot, D. (1999) *Symbiomonas scintillans* gen. and sp. nov. and *Picophagus flagellatus* gen. et sp. nov. (Heterekonta): two new heterotrophic flagellates with picoplanktonic size. *Protist*, 150: 383–398.

Harris, K. (1963) Observations on *Sphaleromantis tetragona*. *Journal of General Microbiology*, 33: 345–348.

Herth, W. (1979) Behaviour of the chrysoflagellate alga, *Dinobryon divergens*, during lorica formation. *Protoplasma*, 100: 345–351.

Herth, W. and Zugenmaier, P. (1979) The lorica of *Dinobryon*. *Journal of Ultrastructural Research*, 69: 262–272.

Hibberd, D.J. (1971) Observations on the cytology and ultrastructure of *Chrysamoeba radians* Klebs (Chrysophyceae). *British Phycological Journal*, 6: 207–223.

Hibberd, D.J. (1975) Observations on the ultrastructure of the choanoflagellate *Codosiga botrytis* (Ehr.) Saville-Kent, with special reference to the flagellar apparatus. *Journal of Cell Science*. 17: 191–219.

Hibberd, D.J. (1976) The ultrastructure and taxonomy of the Chrysophyceae and Prymnesiophyceae (Haptophyceae): a survey with some new observations on the ultrastructure of the Chrysophyceae. *Botanical Journal of the Linnean Society*, 72: 55–80.

Hibberd, D.J. (1977a) The cytology and ultrastructure of *Chrysonebula holmesii* Lund (Chrysophyceae), with special reference to the flagellar apparatus. *British Phycological Journal*, 12: 369–383.

Hibberd, D.J. (1977b) Ultrastructure of cyst formation in *Ochromonas tuberculata* (Chrysophyceae). *Journal of Phycology*, 13: 309–320.

Hilliard, D.K. and Asmund, B. (1963) Studies on Chrysophyceae from some ponds and lakes in Alaska. II. Notes on the genera *Dinobryon*, *Hyalobryon* and *Epipyxis* with decriptions of new species. *Hydrobiologia*, 22: 331–397.

Huber-Pestalozzi, G. (1941) *Das Phytoplankton des Süsswassers. Systematik und Biologie, Teil 2 (1): Chrys-ophyceen, farblose Flagellaten, Heterokonten.* In *Die Binnengewässer,* Band 16, Teil 2(1) (ed. A. Thienemann), Schweizerbart, Stuttgart, pp. 1–365.

Huelsenbeck, J. (2000) MRBAYES: Bayesian inference of phylogeny. Distributed by author. Department of Biology, University of Rochester, New York.

Jeffrey, S.W. and Vesk, M. (1997) Introduction to marine phytoplankton and their pigment signatures. In *Phytoplankton Pigments in Oceanography* (eds. S.W. Jeffrey, R.F.C. Mantoura, and S.W. Wright), UNESCO Publishing, Paris, France, pp. 37–84.

Kawachi, M., Inouye, I., Honda, D., O'Kelly, C.J., Bailey, J.C., Bidigare, R.R., and Andersen, R.A. (2002a) The Pinguiophyceae *classis nova,* a new class of chromophyte algae whose members produce large amounts of omega-3 fatty acids. *Phycological Research,* 50: 31–47.

Kawachi, M., Noël, M.-H., and Andersen, R.A. (2002b) Re-examination of the marine "chrysophyte" *Poly-podochrysis teissieri* (Pinguiophyceae). *Phycological Research,* 50: 91–100.

Klebs, G. (1892a) Flagellatenstudien. I. *Zeitschrift für Wissenschaftliche Zoologie,* 55: 265–351.

Klebs, G. (1892b) Flagellatenstudien. II. *Zeitschrift für Wissenschaftliche Zoologie,* 55: 353–445.

Korshikov, A.A. (1924) Protistologiceskije zametki. *Arch. Russ. Protistol.* 3: 113–127. (In Russian)

Kristiansen, J. and Preisig, H. (eds.) (2001) *Encyclopedia of Chrysophyte Genera.* Bibliotheca Phycologica Band 110, J. Cramer, Berlin.

Lavau, S., Saunders, G.W., and Wetherbee, R. (1997) A phylogenetic analysis of the Synurophyceae using molecular data and scale case morphology. *Journal of Phycology,* 33: 135–151.

Leadbeater, B.S.C. (1972) Fine-structural observations on some marine choanoflagellates from the coast of Norway. *Journal of the Marine Biological Association of the United Kingdom,* 52: 67–79.

Leadbeater, B.S.C. (1989) The phylogenetic significance of flagellar hairs in the Chromophyta. In *The Chromophyte Algae: Aspects and Problems* (eds. J.C. Green, B.S.C. Leadbeater, and W.L. Diver), Systematics Association Special Vol. 38, Clarendon Press, Oxford, pp. 145–165.

Lund, J.W.G. (1942) Contributions to our knowledge of British Chrysophyceae. *The New Phytologist,* 41: 274–292.

Manton, I. and Harris, K. (1966) Observations on the microanatomy of the brown flagellate *Sphaleromantis tetragona* Skuja with special reference to the flagellar apparatus and scales. *Botanical Journal of the Linnean Society,* 59: 397–403.

O'Kelly, C.J. (1989) The evolutionary origin of the brown algae: information from studies of motile cell ultrastructure. In *The Chromophyte Algae: Problems and Perspectives* (eds J.C. Green, B.S.C. Leadbeater, and W.L. Diver), Systematics Association Special Volume 38, Clarendon Press, Oxford, pp. 255–289.

Pascher, A. (1910) Der Grossteich bei Hirschberg in Nord-Böhmen. *Monographien und Abhandlungen zur Internationale Revue der gesamten Hydrobiologie und Hydrographie,* 1: 1–66.

Pascher, A. (1913) Chrysomonadinae. In *Die Süsswasser-Flora Deutschlands, Oesterreichs und der Schweiz* (ed. A. Pascher), G. Fischer, Jena, 2: 7–95.

Pascher, A. (1914) Über Flagellaten und Algen. *Berichte der Deutschen botanischen Gesellschaft,* 32: 136–160.

Pascher, A. (1916) Rhizopodialnetze als Fangsvorrichtung bei einer plamodialen Chrysomonade. *Archiv für Protistenkunde,* 37: 15–30.

Pascher, A. (1925) Die braune Algenreihe der Chrysophyceen. *Archiv für Protistenkunde,* 52: 489–564.

Pascher, A. (1942) Über einige mit Schwimmschirmen versehene Organismen der Wasseroberfläche. *Beiheft Botanisches Centralblatt,* 61: 462–487.

Peters, M.C. and Andersen, R.A. (1993a) The fine structure and scale formation of *Chrysolepidomonas dendrolepidota* gen. et sp. nov. (Chrysolepidomonadaceae fam. nov., Chrysophyceae). *Journal of Phycology,* 29: 469–475.

Peters, M.C. and Andersen, R.A. (1993b) The flagellar apparatus of *Chrysolepidomonas dendrolepidota* (Chrysophyceae), a single-celled monad covered with organic scales. *Journal of Phycology,* 29: 476–485.

Petersen, J.B. and Hansen, J.B. (1958) On some neuston organisms. I. *Saertryk af Botanisk Tidsskrift,* 54: 93–110.

Petersen, J.B. and Hansen, J.B. (1960) On some neuston organisms. II. *Saertryk af Botanisk Tidsskrift,* 56: 197–234.

Pienaar, R.N. (1976) The microanatomy of *Sphaleromantis marina* sp. nov. (Chrysophyceae). *British Phyco-logical Journal,* 11: 83–92.

Pipes, L.D. and Leedale, G.F. (1992) Scale formation in *Tessellaria volvocina* (Synurophyceae). *British Phycological Journal,* 27: 1–29.

Pipes, L.D., Leedale, G.F., and Tyler, P.A. (1991) The ultrastructure of *Tessellaria volvocina* (Synurophyceae). *British Phycological Journal,* 26: 259–278.

Posada, D. and Crandall, K.A. (1998) Modeltest: testing the model of DNA substitution. *Bioinformatics,* 14, 817–818.

Preisig, H. and Andersen, R.A. (2002) Chrysophyceae. In *An Illustrated Guide to the Protozoa,* 2nd ed., Vol. 2 (eds J.J. Lee, G.F. Leedale, and P.C. Bradbury), Society of Protozoologists, Lawrence, KS, pp. 693–730.

Quillet, M. (1955) Sur Ia nature chimique de la leucosine, polysaccharide de réserve caractéristique des Chrysophycees, extraite d'*Hydrurus foetidus. Comptes Rendus Hebdomadaires de Séances, de l'Academie des Sciences, Paris,* 240: 1001–1003.

Rosenberg, M. (1941) *Chrysochaete,* a new genus of Chrysophyceae allied to *Naegeliella. New Phytologist,* 40: 304–315.

Sandgren, C.D. (1980) An ultrastructural investigation of resting cyst formation in *Dinobryon cylindricum* Imhof (Chrysophyceae, Chrysophycota). *Protistologica,* 16: 259–276.

Sandgren, C.D. (1989) SEM investigations of statospore (stomatocyst) development in diverse members of the Chrysophyceae and Synurophyceae. *Nova Hedwigia Beiheft,* 95: 45–69.

Sandgren, C.D. and Flanagin, J. (1986) Heterothallic sexuality and density dependent encystment in the chrysophycean alga *Synura petersenii* Korsh. *Journal of Phycology,* 22: 206–216.

Saunders, G.W., Potter, D., Paskind, M.P., and Andersen, R.A. (1995) Cladistic analyses of combined traditional and molecular data sets reveal an algal lineage. *Proceedings of the National Academy of Sciences, USA,* 92: 244–248.

Scherffel, A. (1927) Beitrag zur Kenntnis der Chrysomonadineen. II. *Archiv für Protistenkunde,* 57: 331–361.

Senn, G. (1900) Flagellata. In *Die natürlichen Pflanzenfamilien nebst ihren Gattungen und wichtigeren Arten* [...] (eds. A. Engler and K. Prantl), Teil I, Abt 1a. W. Engelmann, Leipzig, pp. 93–188, figs. 63–140.

Silva, P.C. (1980) Names of classes and families of living algae. *Regnum Vegetabile,* 103: 1–156.

Siver, P.A. and Wolfe, A.P. (2005) Eocene scaled chrysophytes with pronounced modern affinities. *International Journal of Plant Sciences,* 166: 533–536.

Starmach, K. (1973) *Chrysocapsa epiphytica* sp. n. (Chrysophyceae). *Bulletin de l'Academie Polonaise des Sciences. Serie II. Sciences Biologiques,* 21: 611–613.

Starmach, K. (1975) New name for the species *Chrysocapsa epiphytica* (= *Ch. flavescens*) Starm. *Acta Hydrobiologica,* 17: 309.

Starmach, K. (1985) Chrysophyceae und Haptophyceae. In *Süsswasserflora von Mitteleuropa,* Band 1, (eds. H. Ettl, J. Gerloff, H. Heynig, and D. Mollenhauer), *Süsswasserflora von Mitteleuropa,* Gustav Fischer Verlag, Stuttgart, Germany, Band 1, 1–515.

Stein, J.R. (1963) A *Chromulina* (Chrysophyceae) from snow. *Canadian Journal of Botany,* 41: 1367–1370.

Stein, F. von (1878) *Der Organismus der Infusionsthiere,* Abt. 3 (1). Engelmann, Leipzig.

Tschermak-Woess, E. 1970. Die wechselnde Sichtbarkeit des Pyrenoids und die Zoosporen von *Chrysochaete britannica. Oesterreichische Botanische Zeitschrift,* 119: 514–520.

Vischer, W. (1943) Über die Goldalge *Chromophyton rosanoffii* Woronin. *Berichte der Schweizer Botanische Gesellschaft,* 53: 91–101.

Vysotskii, A.V. (1887) Mastigophora i Rhizopoda, najdennyja v Vejsovom i Repnom ozerach. *Trudy Obshchestvo Ispytatelei Prirody pri Imperatorskom Kharkovskom Universitete, Kharkov,* 21: 119–140.

Wawrik, F. (1960) Sexualität bei *Mallomonas fastigata* var. *kriegerii. Archiv für Protistenkunde,* 104: 541–544.

Wawrik, F. (1970) Isogamie bei *Synura petersenii* Korschikov. *Archiv für Protistenkunde,* 112: 259–261.

Wawrik, F. (1972) Isogame Hologamie in der Gattung *Mallomonas* Perty. *Nova Hedwigia,* 23: 353–362.

Wenderoth, K., Marquaardt, J., Fraunholz, M., van de Peer, Y., Wastl, J., and Maier, U.G. (1999) The taxonomic position of *Chlamydomyxa labyrinthuloides. European Journal of Phycology,* 34: 97–108.

West, W. and West, G.S. (1903) Notes on freshwater algae. III. *Journal of Botany,* 41: 33–40.

Woronin, M. (1880) *Chromophyton Rosanoffii. Bot. Zeitung.,* 38: 626–631, 641–648.

Wujek, D.E. (1976) Ultrastructure of flagellated chrysophytes. II. *Uroglena* and *Uroglenopsis. Cytologia,* 41: 665–670.

16 A decade of euglenoid molecular phylogenetics

Richard E. Triemer and Mark A. Farmer

CONTENTS

ABSTRACT

Variously claimed by both protozoologists and phycologists, the euglenoid flagellates have long fascinated biologists with their unique ability to carry out photosynthesis as well as move with "animal-like" fluidity. As such, they have remained an enigma among the algae and are unique in being the only photosynthetic members of a large clade of protists. Morphological studies in the 20th century, as well as the recognition that plastids can be acquired through secondary symbiosis, changed our thinking about their phylogenetic affinities and raised interesting questions regarding the evolutionary history of euglenoids. Today, molecular systematics has allowed for careful comparisons both among and between the photosynthetic members of the group and colourless relatives, the result being a robust phylogeny that more accurately reflects the events leading up to the acquisition of the plastid in green euglenoids and their subsequent divergence and diversification.

HISTORICAL BACKGROUND

Perhaps the oldest of human traits is the desire to describe and understand the world around us. Long before Aristotle tried to order the distinctions between the plant and the animal worlds with his *Scala Naturae*, humans recognized the fundamental differences between these two major groups of living things. To a large extent we continue to view the natural world in this simplistic manner, beginning the game "Twenty Questions" with the query "Animal, vegetable, or mineral?" For most of mankind's history we have been completely unaware of the diverse world of microorganisms that dominates our planet. That perception changed dramatically in 1674 when the Dutch cloth merchant Antonie van Leeuwenhoek wrote to a friend about his discovery of a tiny wriggling creature that was "green in the middle and at either end white." Today, we know this organism to be *Euglena viridis*, the first described microscopic protist. Yet in the 300+ years since its discovery, *Euglena* and its relatives remained largely unknown. Some may know of *Euglena*'s existence, few have ever actually seen it, fewer still would have any hope of distinguishing *Euglena viridis* from

315

its closest relatives. For a group of organisms that may have predated all animals and land plants by as much as a billion years, this seems unwarranted. Euglenoids remain obscure largely because they are tiny. In short, size does matter.

Euglenoid systematics began when C.G. Ehrenberg first recognized and described several of the major photosynthetic taxa: *Euglena* (Ehrenberg, 1830a, 1830b, 1838), *Trachelomonas* (Ehrenberg, 1833), *Cryptoglena* (Ehrenberg, 1831), and *Colacium* (Ehrenberg, 1833). With the addition of *Lepocinclis* (Perty, 1849) and *Phacus* (Dujardin, 1841), most of the major photosynthetic genera had been described by the mid-1800s. The earliest attempt to organize the systematics of the euglenoids was presented by Stein (1878) who recognized four major groups that he distinguished from one another based largely on the presence or absence of chloroplasts, and on the mode of nutrition. The euglenoids were first recognized as a single, distinct group, equal in status to other protistan groups, by Otto Bütschli in 1884. Contemporary schemes by Klebs (1883), Blochmann (1895), Senn (1900), and Lemmermann (1913) largely followed the "green vs. colourless" demarcations of Stein (1878) and Bütschli (1884). Reichenow (1928) was the first to suggest that colourless (Astasiidae) and pigmented forms (Euglenidae) be united into a single group that was distinct from the phagotrophic forms (Peranemidae), but this view was widely opposed by leading protozoologists of the time (Hall and Jahn, 1929). The most extreme example of reliance on the "green vs. colourless" scheme (Calkins, 1933) placed the photosynthetic euglenoids in the kingdom Plantae while all of the colourless forms were considered to be animals.

The taxonomic system in common use throughout the 20th century had its roots in that proposed by Hollande (1942). He created a scheme that was based largely on body shape (radial symmetry vs. asymmetry), nutritional mode (phototrophic, osmotrophic, or phagotrophic), flagellar structure, and degree of metaboly (rigid vs. metabolic, the ability to wriggle). This resulted in three major groupings, the Euglenoidinees, Peranemoidees, and Petalomonadinees, with the Euglenoidinees being subdivided into five families, each of which included both pigmented and colourless genera. The presence or absence of chloroplasts was now considered to be of minor importance. Ten years later, Hollande (1952) revised this scheme to include only four families. Leedale (1967) incorporated new physiological and electron microscopic information into the Hollande scheme and created a modified scheme that recognized six separate orders. Orders were based on a combination of many factors (mode of nutrition, flagellar structure, pellicle construction) and similarly, grouped photosynthetic and colourless forms together. A decade later, Leedale (1978) modified his scheme to account for the likely event that pigmented euglenoids are derived from colourless, phagotrophic forms that acquired chloroplasts through endosymbiosis. This was supported by the work of Gibbs (1981) and others who proposed that the chloroplasts of euglenoids are the result of a secondary endosymbiosis of a photosynthetic eukaryote (i.e., a green alga). Building on the growing body of ultrastructural data, we proposed that the hypothetical colourless, phagotrophic ancestor proposed by Leedale was closer in form to some of the kinetoplastid flagellates than to taxa such as *Euglena* (Farmer, 1988). As discussed later in this chapter, this hypothesis subsequently was supported by molecular data.

THE ADVENT OF EUGLENOID MOLECULAR SYSTEMATICS

When sequence data first entered the realm of systematics, the system of classification used for euglenoids was that proposed by Leedale (1967). Leedale's classification separated the euglenoids into the division Euglenophyta (Euglenida) on the basis of (1) flagellar insertion, number, and morphology; (2) the presence of an eyespot independent of the chloroplast; (3) the helical symmetry of the cell; (4) a cell surface composed of interlocking strips (a pellicle); (5) the presence of a crystalline β 1-3 glucan storage product, paramylon; and (6) a closed mitotic spindle with a persistent nucleolus. The taxa were divided into six orders (Figure 16.1). The Euglenales, Eutreptiales, and Euglenamorphales contained all of the photosynthetic genera and some colourless genera. Nearly all of the colourless genera included in these orders have photosynthetic counterparts sharing many

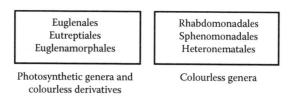

Euglenales Eutreptiales Euglenamorphales	Rhabdomonadales Sphenomonadales Heteronematales
Photosynthetic genera and colourless derivatives	Colourless genera

FIGURE 16.1 Leedale's (1967) taxonomic scheme.

similar morphological features with the exception of chloroplasts, and sometimes eyespots. The remaining orders housed only colourless genera. The Rhabdomonadales contained osmotrophic genera. Members of the Sphenomonadales were osmotrophic or phagotrophic, and the Heteronematales contained all of the phagotrophic euglenoids that possessed a distinct feeding apparatus.

It is in this morphological context that molecular phylogenetics entered. Molecular phylogenetics of euglenoids began in 1986, when Sogin et al. (1986) determined the SSU rRNA sequences from *Euglena gracilis* and *Trypanosoma brucei*. They found that *Euglena* and *Trypanosoma* represented deep, but independent, evolutionary divergences on the eukaryotic tree which "far antedates a period of massive evolutionary radiation that gave rise to the plants, animals, fungi and certain groups of protists" (Sogin et al., 1986). These early studies paved the way for euglenoid molecular phylogenetics, but it was nearly a decade before the first molecular phylogenies incorporating more than a single taxon (*Euglena gracilis*) were performed. The first gene to be explored in a phylogenetic context was the chloroplast *rbc*L gene (Thompson et al., 1995); however, this study focused on the evolution of introns in the genus *Euglena* and not on phylogenetic relationships among genera. Two years later, the first phylogenies based on the SSU rDNA sequences from four representative euglenoid taxa (one photosynthetic, one osmotrophic, and two phagotrophic), provided the framework that modern phylogenies still support (Montegut-Felkner and Triemer, 1997). These earliest phylogenies, using parsimony (Figure 16.2), distance, and likelihood methods, suggested that (1) euglenoids and kinetoplastids arose from a common ancestor; (2) phagotrophy arose early in the euglenoid lineage prior to the advent of photosynthesis; (3) the earliest phagotrophs had a cell surface complex (the pellicle) composed of longitudinally arranged strips and were rigid and not capable of the wriggling motion known as metaboly (see Triemer et al., 2006, for definition of metaboly); and (4) the phagotrophic ancestor of the phototrophic lineage had two emergent flagella, had a pellicle with helically arranged strips, and underwent metaboly. Although

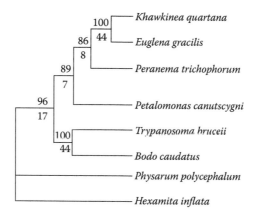

FIGURE 16.2 Most parsimonious tree obtained from 500 bootstrap replicates (percentages shown above branches). Numbers below branches indicate decay indices. (Reprinted with permission from the Phycological Society of America; Montegut-Felkner, A.E. and Triemer, R.E., *Journal of Phycology*, 33, 512–519, 1997.)

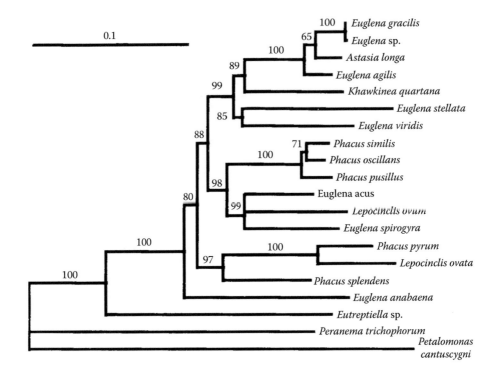

FIGURE 16.3 Distance tree generated for euglenoid species using the Kimura two-parameter model with PAUP*. Numbers above the branches represent the percentage of 2000 bootstrap replications in which the node was found. Scale bar = 10% divergence. (Reprinted with permission from the Phycological Society of America; Linton, E., Nudelman, A., Conforti, V., and Triemer, R.E., *Journal of Phycology*, 36, 740–746, 2000.)

genera have been erected, combined, and resurrected based upon recent molecular analyses, these inferences still hold in most modern phylogenies.

Over the next few years phylogenies remained focused on SSU rDNA and as taxa were added it became clear that several of the commonly accepted euglenoid genera were paraphyletic or polyphyletic (Linton et al., 1999, 2000). For example, *Astasia longa* and *Khawkinea quartana*, both osmotrophic colourless taxa, were positioned in a clade among several species of *Euglena* (Figure 16.3). This suggested that both taxa were closely related to members of the genus *Euglena* and had independently lost the capacity for photosynthesis. This hypothesis was supported by data that demonstrated a reduced plastid genome in *Astasia* (Seimeister and Hachtel, 1989, 1990; Gockel et al., 1994a, 1994b) and the presence of an eyespot in the otherwise colourless, *Khawkinea* (Schuster and Hershenov, 1974; Leedale, 1982). These phylogenies (Linton et al., 1999) also inferred that the first photosynthetic euglenoids (represented by *Eutreptiella* sp.) possessed two emergent flagella and helically arranged pellicle strips making the presence of a single emergent flagellum in members of the Euglenales a derived state. As more species were added to the phylogenies (Linton et al., 2000), it became apparent that even the well-known genus *Euglena* was polyphyletic, with *E. anabaena* arising outside of a large clade containing the other Euglenales, i.e. *Phacus*, *Lepocinclis*, and *Euglena* (Figure 16.3) and *E. acus* and *E. spirogyra* positioned among species of *Phacus* and *Lepocinclis*. These observations set the stage for the major taxonomic changes within the Euglenales that were to come.

While several studies focused on photosynthetic taxa, others investigated relationships among the phagotrophic and osmotrophic genera and provided several new insights (Preisfeld et al., 2000, 2001; Müllner et al., 2001; Busse and Preisfeld, 2003a). While the photosynthetic taxa formed a monophyletic assemblage, osmotrophic taxa were scattered over the tree. As in earlier studies (Figure 16.3) *Astasia*

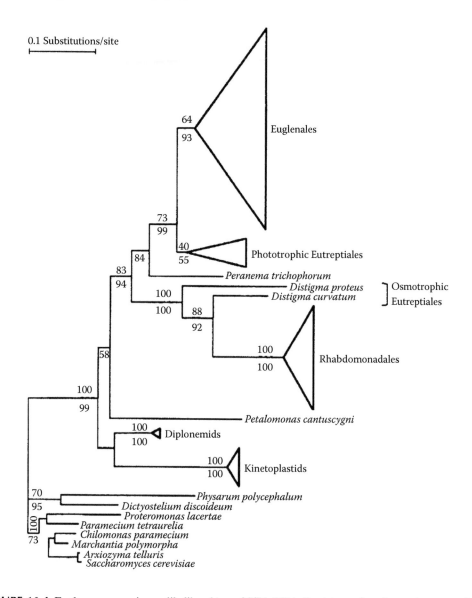

FIGURE 16.4 Euglenozoan maximum-likelihood tree of SSU rDNA. Bootstrap values for maximum-parsimony analysis above; for neighbor joining (Kimura, 1980) below branches. Consensus of 1000 bootstrap replicates. General tree topology was independent of evolutionary model used. (Reprinted with permission from International Journal of Systematic and Evolutionary Microbiology; Preisfeld, A., Busse, I., Klingberg, M., Talke, S., and Ruppel, H.G., *International Journal of Systematic and Evolutionary Microbiology*, 51, 751–758, 2001.)

longa and *Khawkinea quartana* were placed among members of the genus *Euglena*. The five osmotrophic genera (*Gyropaigne, Rhabdomonas, Rhabdospira, Menoidium,* and *Parmidium*) assigned to the Rhabdomonadales (Figure 16.4) all formed a strongly supported clade (although *Rhabdomonas* was not monophyletic and species appeared in three subclades), while the two osmotrophic species assigned to the Eutreptiales (*Distigma proteus* and *D. curvatum*) diverged prior to the Rhabdomonadales. This demonstrated that osmotrophy arose at least three times within the euglenoid lineage, in *Astasia longa*, in *Khawkinea quartana*, and in the *Distigma*/Rhabdomonadalean lineage. The taxa within the *Distigma*/Rhabdomonadalean lineage were considered to be primary osmotrophs (Müllner et al., 2001; Preisfeld et al., 2000, 2001), while *Astasia longa* and *Khawkinea quartana* became osmotrophic secondarily following the loss of photosynthesis. As was soon to be discovered, not all

Astasia species were secondary osmotrophs. When *Astasia curvata, A. comma,* and *A. torta* were added to the data set, they joined *Distigma*/Rhabdomonadalean lineage near the base of the tree (Müllner et al., 2001; Busse and Preisfeld, 2003a). Furthermore, suprageneric issues arose. The photosynthetic genera in the Eutreptiales (*Eutreptia* and *Eutreptiella*) did not clade with the osmotrophic Eutreptiales, but were positioned at the base of the photosynthetic lineage (Figure 16.4). This indicated that the Eutreptiales were not monophyletic and that the morphological characters used to define this taxon (i.e., presence of two emergent flagella and a metabolic cell body) were not congruent with the results of the molecular analyses. Within 5 years, molecular systematic studies had shown that (1) *Euglena, Phacus, Lepocinclis, Astasia,* and *Menoidium* were not monophyletic; (2) the photosynthetic taxa and osmotrophic taxa within the Eutreptiales (*sensu* Leedale, 1967) were highly divergent; and (3) osmotrophy had arisen multiple times within the euglenoid lineage.

RECENT ADVANCES—MORE TAXA, MORE GENES

Much of the attention in recent work has gone to revising the taxonomy to correspond with monophyletic groupings. The number of SSU rDNA sequences available for euglenoid taxa has risen above 100. Several photosynthetic genera (*Colacium, Cryptoglena, Trachelomonas, Strombomonas*) have been added to the data matrix, genera and orders have been emended and re-emended, and new genera have been created (*Discoplastis*, Triemer et al., 2006) or resurrected from the older literature (*Monomorphina*, Mereschkovsky 1877; Marin et al., 2003).

The year 2003 saw a flurry of molecular phylogenies. In February, Nudelman et al. (2003) published a manuscript with 42 taxa representing, what were then, 10 genera of photosynthetic euglenoids. In addition to including more representatives from several genera, the loricate taxa *Trachelomonas* and *Strombomonas* and the colourless osmotroph *Hyalophacus* were included in the phylogenies. Similar to the case for *Khawkinea* and *Astasia* which were positioned among the *Euglena* species (Linton et al., 2000), *Hyalophacus* appeared among the photosynthetic *Phacus* species and represented another instance of secondary loss of photosynthesis. Nudelman et al. also noted that *Astasia longa, Khawkinea quartana,* and *Hyalophacus ocellatus* would need to be renamed to make these clades monophyletic, but did not make the formal changes. The introduction of the loricate taxa introduced a new series of challenges. For example, *Trachelomonas conspersa* and *T. volvocinopsis* were firmly embedded in a clade with *Strombomonas* species. A further examination of these cells using light and electron microscopy demonstrated that *T. conspersa* was not a *Trachelomonas* at all, but conformed to *Strombomonas*. Similarly, *T. volvocinopsis* had also been misidentified in the culture collection. The addition of more taxa also revealed new instances of polyphyly. Taxa assigned to the genus *Phacus* gave rise to two divergent clades, as suspected in some of the previous studies (Müllner et al., 2001). One clade contained small *Phacus* species with an overall teardrop-shaped body and one or two chloroplasts. The other clade contained larger *Phacus* species with numerous small discoid chloroplasts. Here too, generic changes were necessary to create monophyletic genera. The wait for these changes was not far off.

In March 2003, Busse and Preisfeld (2003a) began the process of emending the taxonomy for the osmotrophic taxa. The Eutreptiales (ICBN; Eutreptiina-ICZN) was emended such that it contained only the phototropic genera with two or more emergent flagella and a flexible pellicle (e.g., *Eutreptia, Eutreptiella*). A new superorder, the Rhabdomonadidae (ICBN; Rhabdomonadidia—ICZN), was created as well as a new subgenus, *Parvonema* Busse et Preisfeld, to reconcile conflicts among the primary osmotrophic taxa.

This was followed in April 2003 by a major taxonomic reorganization by Marin et al. (2003) in which most of the photosynthetic euglenoid genera were emended using morphological characters and molecular signatures for diagnosing taxa. Several individual taxa were transferred into other genera based on their position on the phylogenetic tree (e.g., *Astasia longa* and *Khawkinea quartana* were transferred into *Euglena*), but larger changes were proposed as well. Marin et al. (2003) "recognized each major clade of the Euglenales as a separate genus" (Figure 16.5). First, the small

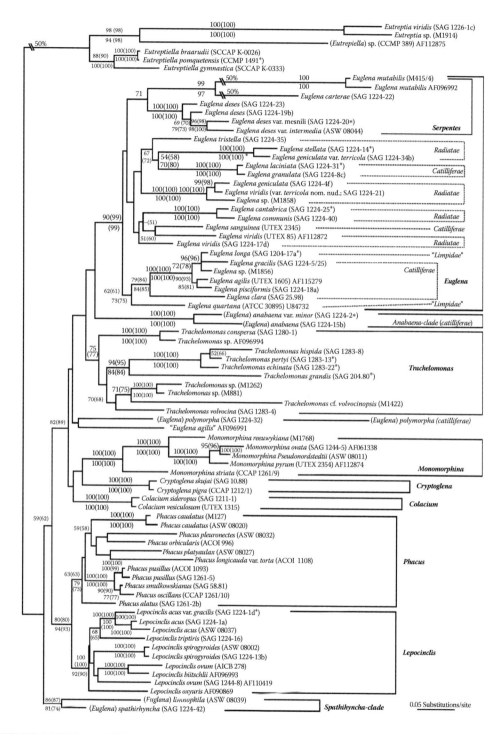

FIGURE 16.5 Unrooted SSU rDNA maximum-likelihood phylogeny of the Euglenales. Numbers above the nodes represent bootstrap values for neighbor-joining analysis using the same evolutionary model as for the maximum likelihood (ML) tree with (and without) *Euglena mutabilis*, a highly divergent taxon on a long branch. Numbers in boldface below the nodes represent maximum parsimony (MP) bootstrap values with (and without) *Euglena mutabilis*. (Reprinted from *Protist*, Vol. 154, Marin, B., Palm, A., Klingberg, M., and Melkonian, M., Phylogeny and taxonomic revision of plastid-containing Euglenophytes based on SSU rDNA sequence comparisons and synapomorphic signatures in the SSU rRNA secondary structure, pp. 99–145, Copyright 2003, with permission from Elsevier.)

Phacus species with an overall teardrop-shaped body and one or two chloroplasts were transferred into the little-known genus, *Monomorphina* Mereschkowsky. This genus had been created by Mereschkowsky in 1877 when he renamed *Lepocinclis pyrum* Perty as *Monomorphina pyrum*. When this taxon was later renamed as *Phacus pyrum* (Archer and Barker, 1871; Stein, 1878), the genus *Monomorphina* was abandoned and forgotten. It has now been rightfully restored. Second, the genus *Phacus* was emended to exclude those taxa now in *Monomorphina* making both genera monophyletic. Third, the genus *Trachelomonas* was emended such that it now included all of the taxa formerly in *Strombomonas*. Fourth, *Lepocinclis* was emended to accommodate those taxa with small discoid chloroplasts lacking pyrenoids that were formerly included in the genus *Euglena* (e.g., *E. acus, E. tripteris, E. spirogyra, E. oxyuris*). This made *Lepocinclis* monophyletic, but the polyphyly in *Euglena* remained. *Euglena anabaena* and *E. polymorpha* both diverged separately prior to the main *Euglena* clade and *E. spathirhyncha* and *E. limnophila* formed a clade sister to the remainder of the Euglenales (Figure 16.5).

NEW GENES AND MULTIPLE GENES

Marin et al. (2003) made major strides into creating a euglenoid taxonomy that was consistent with the available morphological and molecular data. Yet issues such as the polyphyly of *Euglena* remained unresolved and the lack of support or weak support for several of the deeper nodes on the trees (Figure 16.5) raised new questions. Could the SSU rDNA alone provide sufficient signal to resolve these relationships? Would additional data in the form of new genes need to be added to the database? Would sequences from different genes provide phylogenies congruent with those from the SSU rDNA gene?

Four months after the publication of the Marin et al. paper (2003), 31 partial LSU rDNA sequences were added to the data set and combined with SSU rDNA sequences (Brosnan et al., 2003). The combined data set produced a robust phylogeny with seven strongly supported clades corresponding to the major genera in previous studies (Marin et al., 2003); however, the relationships among these clades varied depending upon the method of analysis.

Other investigators moved from the nucleus to the chloroplast. Early studies employing the chloroplast SSU rDNA (Milanowski et al., 2001) again found polyphyly in the genus *Euglena*, but bootstrap support was weak or lacking for most nodes probably due to the limited taxon sampling (five genera were represented by a single species). Indeed, in recent studies using the chloroplast SSU rDNA gene and including more representative species, *Lepocinclis, Trachelomonas, Strombomonas, Phacus*, and *Monomorphina*, all appear as monophyletic (Milanowski et al., 2006). When the chloroplast SSU rDNA data were combined with the SSU rDNA data some relationships among genera shifted, but the individual clades for these genera continued to receive strong bootstrap support. The polyphyly of *Euglena* was still present whether using single or combined data sets.

Most euglenoid phylogenies have focused on ribosomal genes, whether from the chloroplast or nucleus, but one study has utilized structural genes that encode proteins found in the flagellum. The euglenoid flagellum has a typical eukaryotic axoneme, but also contains a proteinaceous rod, the paraxonemal rod, which is composed of two major proteins. The genes that encode these proteins, *par*1 and *par*2, have been used to generate phylogenies (Talke and Preisfeld, 2002). Although these studies were limited by taxon sampling (only six euglenoid taxa were represented), the careful choice of taxa resulted in phylogenies consistent with those of the SSU rDNA trees. The primary osmotrophs grouped together and the phototrophs and secondary osmotrophs formed a separate clade. It was encouraging that nuclear and chloroplast RNA genes, as well as protein-encoding genes, recovered similar clades and supported an overall framework that was largely congruent.

The most recent phylogenetic treatment of photosynthetic euglenoids (Triemer et al., 2006) combined SSU rDNA and partial LSU rDNA for 84 taxa representing 11 genera (Figure 16.6). The Bayesian phylogeny generated from the combined SSU/LSU rDNA data set was congruent

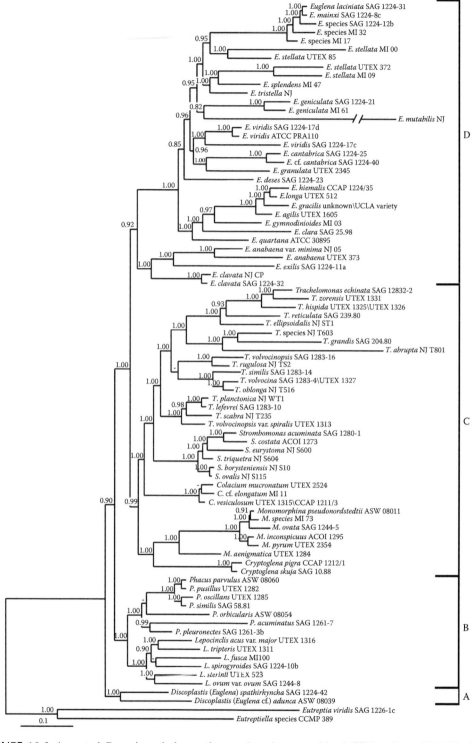

FIGURE 16.6 A rooted Bayesian phylogenetic tree based on combined SSU and partial-LSU rDNA sequences. Numbers on the internodes represent the posterior probability (pp) for the node. Genus names appearing within brackets indicate the former genus of the taxon. The tree was outgroup rooted posteriorly with *Eutreptia viridis* and *Eutreptiella* species. Scale bar represents number of substitutions/site. (Redrawn with permission from the Phycological Society of America; Triemer, R.E., Linton, E., Shin, W., Nudelman, A., Monfils, A., Bennett, M., and Brosnan, S., *Journal of Phycology*, 42, 731–740, 2006.)

with the major clades of photosynthetic genera found in previous studies using the SSU rDNA and also resolved relationships among these taxa. Two of the outlying *Euglena* species, *E. spathirhyncha* and *E.* cf. *adunca*, were transferred to a new genus, *Discoplastis* Triemer (Figure 16.6, clade A), reducing the polyphyly found within *Euglena*. The sister taxon relationship of *Phacus* and *Lepocinclis* in clade B was strongly supported (pp = 1.00); however, relationship of these taxa as sister to those in clades C and D was supported with a posterior probability (pp) of only 0.90.

Clade C contained taxa from five genera, each forming monophyletic groups. *Monomorphina* and *Cryptoglena* group together and were a sister clade to *Trachelomonas*, *Strombomonas*, and *Colacium* (Figure 16.6). Marin et al. (2003) transferred all *Strombomonas* species into *Trachelomonas*. A more recent phylogeny (Figure 16.6), with increased taxon sampling for both genera, showed these taxa forming two strongly supported independent clades and should therefore be retained as separate genera (Triemer et al., 2006). This was further supported on the basis of morphological evidence that showed differing developmental patterns of lorica formation between *Trachelomonas* and *Strombomonas* species (Brosnan et al., 2005).

All included *Euglena* species (Figure 16.6), now monophyletic, are grouped in clade D. Note, however, that there are two major subclades within the genus. *Euglena anabaena*, *E. exilis*, and *E. clavata* form a clade sister to the remaining *Euglena* species, and the posterior probability on the node joining these two clades is only 0.92. If this node collapses with the addition of more taxa, then *Euglena* will no longer be monophyletic. Furthermore, some problematic (fast-evolving?) taxa that arise as independent divergences on long branches, such as *Euglena proxima* (personal observation) and *Euglena polymorpha* (Marin et al., 2003), were not included in this study. In summary, further revisions of the genus *Euglena* are still possible.

THE FUTURE OF EUGLENOID SYSTEMATICS

With most of the major genera now in place on the phylogenetic tree, the future will focus on subgeneric problems ranging from misidentified species to cryptic species. Much of this work has already begun. When several strains conforming to the morphological description of *Euglena viridis*, the type for the genus *Euglena*, were sequenced (Shin and Triemer, 2004), they formed two separate clades. The same was true for several strains of *E. stellata*. Epitypes were designated to help stabilize taxa, but taxa in the genus *Euglena* have a high degree of sequence divergence and cryptic speciation has appeared more than once (Shin and Triemer, 2004). High levels of genetic divergence also have been found within a species (e.g., *Euglena agilis*; Zakrys et al., 2004) further complicating the process of defining species. Some species, such as *Euglena geniculata* and *E. myxocylindracea*, which had been recognized as separate species based on morphological criteria, were sequenced only to find that these were the same taxon and that the morphological differences were due to environmental conditions (Zakrys et al., 2002). These issues recur in other euglenoid genera as well. Kosmala et al. (2005) sequenced multiple strains of several taxa within *Lepocinclis* and had to redefine species and designate epitypes.

Although major strides have been made in euglenoid systematics based upon molecular phylogenetics, the work is far from over. Relationships between and among taxa still need to be clarified, and on a larger scale, the position of the euglenoids among other Euglenozoans is still in dispute.

RELATIONSHIP OF EUGLENOIDS TO OTHER EUGLENOZOANS

While photosynthetic taxa such as *Euglena gracilis* are the best-known euglenoids, they are in fact a highly derived group. The green euglenoids are descended from a colourless phagotrophic ancestor that ingested, but did not digest the chloroplast of a green alga (Gibbs, 1981). As a group, the euglenoids contain phagotrophic, osmotrophic, autotrophic, and perhaps even parasitic taxa. The euglenoids are sister to two other well-defined groups of protists; the diplonemids and the kinetoplastids. Collectively, the euglenoids, diplonemids, and kinetoplastids form a larger taxonomic group, the Euglenozoa (Cavalier-Smith, 1981; Simpson, 1997). Early studies demonstrated a number of ultrastructural features that suggested a close evolutionary relationship between the three groups. These included similarities in flagellar apparatus, feeding apparatus, mitotic structure, and cytoskeleton (Triemer and Farmer, 1991b). More recently, studies based on molecular characters confirmed the monophyletic nature of the Euglenozoa based on SSU rDNA sequences (von der Heyden et al., 2004) and other protein-coding gene sequences (Baldauf et al., 2000; Simpson et al., 2006b). The members of the Euglenozoa share several structural apomorphies that define the group. These apomorphies include a paraxonemal rod comprised of protein homologs of *par*1 and *par*2 (Talke and Preisfeld, 2002) and a flagellar apparatus built on the basic pattern of two basal bodies (often joined by a fibrous connective fiber) and three asymmetrically distributed microtubular roots (Farmer and Triemer, 1988). Most Euglenozoan taxa also have "discoidal" mitochondrial cristae with a shape reminiscent of a table tennis paddle.

The diplonemids are a poorly studied group of marine flagellates that are primarily free living and have sometimes been associated with die-offs of marine grasses (Porter, 1973). They possess a complex feeding apparatus used to ingest bacteria (Montegut-Felkner and Triemer, 1996) and despite lacking a cytoskeleton arranged in pellicular strips, have been reported to exhibit cell shape changes similar to euglenoid metaboly.

Kinetoplastids are the other major group of Euglenozoans and are typically divided into two groups, the biflagellate bodonids (free-living, commensal, and parasitic taxa) and the uniflagellate and obligately parasitic trypanosomes. Most recent phylogenetic studies now recognize the bodonids to be paraphyletic, while the trypanosomes, which are best known from the disease-causing genera *Trypanosoma* and *Leishmania*, are believed to be a monophyletic group (Simpson et al., 2006a). As the causative agents of several human (Chagas Disease, African Sleeping Sickness, "Kala-azar" or Leishmaniasis) and plant diseases (*Phytomonas*), trypanosomes have been extensively studied.

While the monophyly of the Euglenozoa is well established, there is still considerable debate regarding the precise sister relationships among the euglenoids, diplonemids, and kinetoplastids. All three possible groupings have been proposed (Figure 16.7) including

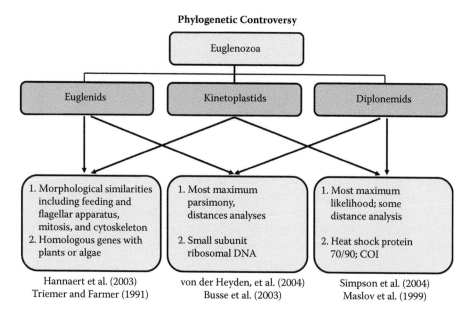

FIGURE 16.7 Three possible phylogenetic relationships among Euglenozoan groups showing sister taxa relationships between euglenoids/kinetoplastids (morphology and plastid-like genes), euglenoids/diplonemids (SSU rDNA via maximum parsimony), and kinetoplastids/diplonemids (SSU rDNA via maximum likelihood and protein-coding genes).

scenarios in which euglenoids and kinetoplastids are most closely related (Hannaert et al., 2003), euglenoids and diplonemids are sister taxa (Busse and Preisfeld, 2003a; von der Heyden et al., 2004) or diplonemids, and kinetoplastids form a clade within the Euglenozoa (Maslov et al., 1999; Simpson and Roger, 2004a). The resolution of the relationships between euglenoids, kinetoplastids, and diplonemids will require additional sequence data from both protein-coding and rDNA sequences. It will also require greater taxon sampling, particularly from organisms that appear to occupy basal positions in Euglenozoan phylogenies, e.g. the enigmatic marine flagellate *Petalomonas cantuscygni*, which has morphological features that are apomorphic for euglenoids (e.g., pellicle composed of alternating groups of cytoskeletal microtubules) as well as for kinetoplastids (e.g., fibrillar mitochondrial inclusion that resembles the kinetoplast) (Leander et al., 2001).

The most controversial idea, that kinetoplastids are descended from a photosynthetic euglenoid ancestor, is based on the sequencing of genes for fructose 1,6-bisphosphatase (FBPase) and sedoheptulose-1,7-bisphosphatase (SBPase) from the kinetoplastid *Trypanosomoa brucei* and the genes for glucose-6-phosphate dehydrogenase (G6PDH) and 6-phosphogluconolactonase (6PGL) from *Leishmania mexicana* (Hannaert et al., 2003). Sequences for these enzymes group most closely with those from plants and other photosynthetic organisms. This led some to speculate that plastids have subsequently been lost from kinetoplastids (Hannaert et al., 2003; Martin and Borst, 2003), an idea that is strongly refuted by morphological evidence that supports the idea that photosynthetic euglenoids are descended from a heterotrophic ancestor that shared a common ancestor with the kinetoplastids (Triemer and Farmer, 1991a; Leander et al., 2001; Leander, 2004). The heterotrophic ancestry of euglenoids, diplonemids, and kinetoplastids is also well supported by a variety of studies that examined molecular gene sequences from a number of Euglenozoan taxa (Preisfeld et al., 2001; Busse and Preisfeld, 2003b; von der Heyden et al., 2004; Maslov et al., 1999; Simpson and Roger, 2004a). Curiously, the ability of euglenoids to detect and orient to light may have been present in the heterotrophic ancestor and may have arisen as an adaptation to phagotrophy of algal prey and not as a by-product of photosynthesis itself (Leander et al., 2001; Saranak and Foster, 2005).

Least resolved is the question of how the Euglenozoa are related to other eukaryotes. Cavalier-Smith (1998) proposed that together with the Percolozoa (a group of protists that includes the Heteroloboseans), the Euglenozoa form a larger clade known as the Discicristata. The apomorphy for the Discicristata is mitochondrial cristae that are discoidal in shape. This feature distinguishes the Euglenozoa from other algal groups that have flattened or tubular mitochondrial cristae. Most recently, the Euglenozoa have been considered to be part of an even larger clade, the Excavata (Simpson, 2003; Adl et al., 2005), which groups them with the parasitic protists *Giardia* and *Trichomonas* that lack mitochondria, as well as free-living organisms such as *Reclinomonas*, whose mitochondrial genome has retained a greater complement of bacterial genes than any other eukaryote (Lang et al., 1997). While molecular studies of some genes weakly support the monophyly of the Excavata (Simpson et al., 2006b), the members of the Euglenozoa lack all of the morphological apomorphies that characterize most of the other members of the group (Simpson, 2003; Simpson and Roger, 2004b). Thus the Euglenozoa remain a distinct group of organisms with uncertain affinities to other protists. Nonetheless, by virtue of their unique cell morphology, gene composition, and ecological diversity, the Euglenozoa remain as one of the most interesting and fascinating groups of protists.

ACKNOWLEDGMENTS

The authors wish to wish to acknowledge financial support provided by the National Science Foundation PEET (Partnership for Enhanced Expertise in Taxonomy) program, grant no. DEB-0329799.

REFERENCES

Adl, S.M., Simpson, A.G.B., Farmer, M.A., Andersen, R.A., Anderson, O.R., Barta, J.R., Bowser, S.S., Brugerolle, G., Fensome, R.A., Fredericq, S., James, T.Y., Karpov, S., Kugrens, P., Krug, J., Lane, C.E., Lewis, L.A., Lodge, J., Lynn, D.H., Mann, D.G., McCourt, R.M., Mendoza, L., Moestrup, Ø., Mozley-Standridge, S.E., Nerad, T.A., Shearer, C.A., Smirnov, A.V., Spiegel, F.W., and Taylor, M.F.J.R. (2005) The new higher level classification of eukaryotes with emphasis on the taxonomy of protists. *Journal of Eukaryotic Microbiology,* 52: 399–451.

Archer, W. and Barker, J. (1871) Proceedings of Societies: Dublin Microscopical Club, 21 July 1870 and 18 August 1870 [remarks on *Phacus* and *Euglena*]. *Quarterly Journal of Microscopical Science,* 11: 98–99.

Baldauf, S.L., Roger, A.J., Wenk-Siefert, I., and Doolittle, W.F. (2000) A kingdom-level phylogeny of eukaryotes based on combined protein data. *Science,* 290: 972–977.

Blochmann, F. (1895) *Die microscopische Pflanzen-und Thierwelt des Süsswassers. Theil II. Abteilung I: Protozoa.* pp. 50–52.

Brosnan, S., Brown, P.J.P., Farmer, M.A., and Triemer, R.E. (2005) Morphological separation of the euglenoid genera *Trachelomonas* and *Strombomonas* (Euglenophyta) based on lorica development and posterior strip reduction. *Journal of Phycology,* 41: 590–605.

Brosnan, S., Shin, W., Kjer, K.M., and Triemer, R.E. (2003) Phylogeny of the photosynthetic euglenophytes inferred from the nuclear SSU and partial LSU rDNA. *International Journal of Systematic and Evolutionary Microbiology,* 53: 1175–1186.

Busse, I. and Preisfeld, A. (2003a) Systematics of primary osmotrophic euglenids: a molecular approach to the phylogeny of *Distigma* and *Astasia* (Euglenozoa). *International Journal of Systematic and Evolutionary Microbiology,* 53: 617–624.

Busse, I. and Preisfeld, A. (2003b) Application of spectral analysis to examine phylogenetic signal among euglenid SSU rDNA data sets (Euglenozoa). *Organisms Diversity and Evolution,* 3: 1–12.

Bütschli, O. (1884) *Mastigophora.* Dr H.G. Bronn's Klassen und Ordnungen des Thier-Reichs: Vol. 1. *Protozoa*: No. 2. Leipzig: C.F. Winter. pp. 617–1097.

Calkins, G. (1933) *The Biology of the Protozoa.* Lea and Febinger, Philadelphia.

Cavalier-Smith, T. (1981) Eukaryote kingdoms: 7 or 9? *BioSystems,* 14: 461–481.

Cavalier-Smith, T. (1998) A revised six-kingdom system of life. *Biological Reviews,* 73: 203–266.

Dujardin, F. (1841) *Histoire Naturelle des Zoophytes Infusoires,* Roret, Paris.

Ehrenberg, C.G. (1830a) Neue Beobachtungen über blutartige Ersheinungen in Aegypten, Arabien and Siberien, nebst einer Uebersicht une Kritik der früher bekannten. *Annalen der Physik Chemie,* 18: 477–514.

Ehrenberg, C.G. (1830b) Beiträge zur Erkenntnis der Infusorien und ihrer geographischen Verbreitung, besonders in Siberien. *Abhandlungen der Königlichen Akademie der Wissenschaften Berlin,* 1–88.

Ehrenberg, C.G. (1831) Über die Entwickelung und Lebensdauer der Infusionsthiere; nebst ferneren Beiträgen zu einer Vergleichung ihrer organischen Systeme. *Abhandlungen der Königlichen Akademie der Wissenschaften Berlin* (1832), 1–154.

Ehrenberg, C.G. (1833) Dritter Beitrag zur Erkenntnis grosser Organisation in der Richtung des Kleinsten Raumes. *Abhandlungen der Königlichen Akademie der Wissenschaften Berlin,* 145–336.

Ehrenberg, C.G. (1838) *Die Infusionsthierchen als vollkommene Organismen. Ein Blick in das tiefere organische Leben der Natur. Nebst einem Atlas von vierundsechszig colorirten Kupfertafeln, gezeichnet vom Verfasser.* Leopold Voss, Leipzig.

Farmer, M.A. (1988) A Re-evaluation of the Taxonomy of the Euglenophyceae Based on Ultrastructural Characteristics. Ph.D. thesis, Rutgers University, Piscataway, NJ.

Farmer, M.A. and Triemer, R.E. (1988) Flagellar systems in the euglenoid flagellates. *BioSystems,* 21: 283–292.

Gibbs, S.P. (1981) The chloroplasts of some algal groups may have evolved from symbiotic green algae. *Annals of the New York Academy of Sciences,* 361: 193–208.

Gockel, G., Baier, S., and Hachtel, W. (1994a) Plastid ribosomal protein genes from the nonphotosynthetic flagellate *Astasia longa. Plant Physiology,* 105: 1443–1444.

Gockel, G., Hachtel, W., Baier, S., Fliss, C., and Henke, M. (1994b) Genes for components of the chloroplast translational apparatus are conserved in the reduced 73-kb plastid DNA of the nonphotosynthetic euglenoid flagellate *Astasia longa. Current Genetics,* 26: 256–262.

Hall, R.P. and Jahn, T.L. (1929) On the comparative cytology of certain euglenoid flagellates and the systematic position of the families Euglenidae Stein and Astasiidae Bütschli. *Transactions of the American Microscopical Society,* 48: 388–405.

Hannaert, V., Saavedra, E., Duffieux, F., Szikora, J.P., Rigden, D.J., Michels, P.A.M., and Opperdoes, F.R. (2003) Plant-like traits associated with metabolism of Trypanosoma parasites. *Proceedings of the National Academy of Science*, 100: 1067–1071.

Hollande, A. (1942) Étude cytologique et biologique de quelques flagellés libres. *Archives de Zoologie Expérimentale et Générale*, 83: 1–268.

Hollande, A. (1952) Classe des Eugléniens (Euglenoidina Bütschli, 1884). *Traité de Zoologie*, 1: 238–284.

Kimura, M. (1980) A simple method for estimating evolutionary rates of base substitutions through comparative studies of nucleotide sequences. *Journal of Molecular Evolution*, 16: 111–120.

Klebs, G. (1883) Über die Organization einiger Flagellatengruppen und ihre Beziehungen zu Algen und Infusorien. *Untersuchungen aus dem botanischen Institut zu Tübingen*, 1: 233–362.

Kosmala, S., Karnkowska, A., Milanowski, R., Kwiatowski, J., and Zakrys, B. (2005) Phylogenetic and taxonomic position of *Lepocinclis* fusca comb. Nov (= Euglena fusca) (Euglenaceae): Morphological and molecular justification. *Journal of Phycology*, 41: 1258–1267.

Lang, B.F., Burger, G., O'Kelly, C.J., Cedergren, R., Golding, G.B., Lemieux, C., Sankoff, D., Turmel, M., and Gray, M.W. (1997) An ancestral mitochondrial DNA resembling a eubacterial genome in miniature. *Nature*, 387: 493–497.

Leander, B.S. (2004) Did trypanosomatid parasites have photosynthetic ancestors? *Trends in Microbiology*, 12: 251–258.

Leander, B.S., Triemer, R.E., and Farmer, M.A. (2001) Character evolution in heterotrophic euglenids. *European Journal of Protistology*, 37: 337–356.

Leedale, G.F. (1967) *Euglenoid Flagellates*. Prentice Hall, Englewood Cliffs, NJ.

Leedale, G.F. (1978) Phylogenetic criteria in euglenoid flagellates. *Biosystems*, 10: 183–187.

Leedale, G.F. (1982) Ultrastructure. In *The Biology of Euglena*, Vol. 3 (ed. D.E. Buetow), Academic Press, New York, pp. 1–27.

Lemmermann, E. (1913) Euglenineae. In *Die Süsswasserflora Deutschlands, Österreichs, und der Schweiz*, Jena.

Linton, E., Hittner, D., Lewandowski, C.L., Auld, T., and Triemer, R.E. (1999) A molecular study of euglenoid phylogeny using small subunit rDNA. *Journal of Eukaryotic Microbiology*, 46: 217–223.

Linton, E., Nudelman, A., Conforti, V., and Triemer, R.E. (2000) A molecular analyses of the euglenophytes using small subunit rDNA. *Journal of Phycology*, 36: 740–746.

Marin, B., Palm, A., Klingberg, M., and Melkonian, M. (2003) Phylogeny and taxonomic revision of plastid-containing Euglenophytes based on SSU rDNA sequence comparisons and synapomorphic signatures in the SSU rRNA secondary structure. *Protist*, 154: 99–145.

Martin, W. and Borst, P. (2003) Secondary loss of chloroplasts in trypanosomes. *Proceedings of the National Academy of Sciences*, 100: 765–767.

Maslov, D.A., Yasuhira, S., and Simpson, L. (1999) Phylogenetic affinities of *Diplonema* within the euglenozoa as inferred from the SSU rRNA gene and partial COI protein sequences. *Protist*, 150: 33–42.

Mereschkowsky, K.S. (1877) Etiudy nad prostieishchimi sievera Rossii. *Trudy Sankt-Peterburgskago Obshchestva estestvoispytatelei* 8 (pt. 2): 203–205, 275–278, 288–299.

Milanowski, R., Kosmala, S., Zakry, B., and Kwiatowski, J. (2006) Phylogeny of photosynthetic euglenophytes based on combined chloroplast and cytoplasmic SSU rDNA sequence analysis. *Journal of Phycology*, 42: 721–730.

Milanowski, R., Zakry, B., and Kwiatowski, J. (2001) Phylogenetic analysis of chloroplast small-subunit genes of the genus *Euglena* Ehrenberg. *International Journal of Systematic and Evolutionary Microbiology*, 51: 773–781.

Montegut-Felkner, A.E. and Triemer, R.E. (1996) Phylogeny of *Diplonema ambulator* (Larsen and Patterson). 2. Homologies of the feeding apparatus. *European Journal of Protistology*, 32: 64–76.

Montegut-Felkner, A.E. and Triemer, R.E. (1997) Phylogenetic relationships of select euglenoid genera based on morphological and molecular data. *Journal of Phycology*, 33: 512–519.

Müllner, A.N., Angeler, D.G., Samuel, R., Linton, E.W., and Triemer, R.E. (2001) Phylogenetic analysis of phagotrophic, phototrophic and osmotrophic euglenoids by using the nuclear 18S rDNA sequence. *International Journal of Systematic and Evolutionary Microbiology*, 51: 783–791.

Nudelman, M.A., Rossi, M.S., Conforti, V., and Triemer, R.E. (2003) Phylogeny of Euglenophyceae based on small subunit rDNA sequences: taxonomic implications. *Journal of Phycology*, 39: 226–235.

Perty, M. (1849) Über verticale Verbreitung mikroskopischer Lebensformen. *Mitteilungen der Naturforschenden Gesellschaft in Bern* (15 Feb 1849): 17–45.

Porter, D. (1973) *Isonema papillatum* sp. n., a new colourless marine flagellate: a light and electron microscopic study. *Journal of Protozoology,* 20: 351–356.

Preisfeld, A., Berger, S., Busse, I., Liller, S., and Ruppel, H.G. (2000) Phylogenetic analyses of various euglenoid taxa (Euglenozoa) based on 18S rDNA sequence data. *Journal of Phycology,* 36: 220–226.

Preisfeld, A., Busse, I., Klingberg, M., Talke, S., and Ruppel, H.G. (2001) Phylogenetic position and inter-relationships of the osmotrophic euglenids based on SSU rDNA data, with emphasis on the Rhab-domonadales (Euglenozoa). *International Journal of Systematic and Evolutionary Microbiology,* 51: 751–758.

Reichenow, E. (1928) Euglenoidina. In *Lehrbuch der Protozoenkunde* (ed F. Doflein), G. Fischer, Jena, pp. 436–864.

Saranak, J. and Foster, K.W. (2005) Photoreceptor for curling behavior in *Peranema trichophorum* and evolution of eukaryotic rhodopsins. *Eukaryotic Cell,* 4: 1605–1612.

Schuster, F.L. and Hershenov, B. (1974) *Khawkinea quartana*, a colorless euglenoid flagellate. I. Ultrastructure. *Journal of Protozoology,* 21: 33–39.

Seimeister, G. and Hachtel, W. (1989) A circular 73 kb DNA from the colourless flagellate *Astasia longa* that resembles the chloroplast DNA of *Euglena*: restriction and gene map. *Current Genetics,* 15: 435–441.

Seimeister, G. and Hachtel, W. (1990) Organization and nucleotide composition of ribosomal RNA genes on a circular 73 kbp DNA from the colourless flagellate *Astasia longa. Current Genetics,* 17: 433–438.

Senn, C. (1900) Euglenineae. In *Die naturlichen Pflanzenfamilien* (eds. A. Engler and K. Prantl), 1: 173–185.

Shin, W. and Triemer, R.E. (2004) Phylogenetic analysis of the genus *Euglena* (Euglenophyceae) with particular reference to the type species *Euglena viridis. Journal of Phycology,* 40: 226–235.

Simpson, A.G.B. (1997) The identity and composition of the Euglenozoa. *Archiv für Protistenkunde,* 148: 318–328.

Simpson, A.G.B. (2003) Cytoskeletal organization, phylogenetic affinities and systematics in the contentious taxon Excavata (Eukaryota). *International Journal of Systematic and Evolutionary Microbiology,* 53: 1759–1777.

Simpson, A.G.B., Inagaki, Y., and Roger, A.J. (2006b) Comprehensive multigene phylogenies of excavate protists reveal the evolutionary positions of "primitive" eukaryotes. *Molecular Biology and Evolution,* 23: 615–625.

Simpson, A.G.B. and Roger, A.J. (2004a) Protein phylogenies robustly resolve the deep-level relationships within Euglenozoa. *Molecular Phylogenetics and Evolution,* 30: 201–212.

Simpson, A.G.B. and Roger, A.J. (2004b) Excavata and the origin of amitochondriate eukaryotes. *In Organelles, Genomes, and Eukaroyote Phylogeny: An Evolutionary Synthesis in the Age of Genomics* (eds *R.P. Hirt and D.S. Horner),* CRC Press, Boca Raton, FL, pp. 27–53.

Simpson, A.G.B., Stevens, J.R., and Lukes, J. (2006a) The evolution and diversity of kinetoplastid flagellates. *Trends in Parasitology,* 22: 168–174.

Sogin, M.L., Hille, E.J., and Gunderson, J.H. (1986) Evolutionary diversity of eukaryotic small-subunit rRNA genes. *Proceedings of the National Academy of Science,* 83: 1383–1387.

Stein, F.R. (1878) *Der Organismus Infusionsthiere. III. Abt. Der Organismus Flagellaten,* William Engelman, Leipzig.

Talke, S. and Preisfeld, A. (2002) Molecular evolution of euglenozoan paraxonemal rod genes par1 and par2 coincides with phylogenetic reconstruction based on small subunit rDNA data. *Journal of Phycology,* 38: 995–1003.

Thompson, M.D., Copertino, D.W., Thompson, E., Favreau, M., and Hallick, R.B. (1995) Evidence for the late origin of introns in chloroplast genes from an evolutionary analysis of the genus *Euglena. Nucleic Acid Research,* 23: 4745–4752.

Triemer, R.E. and Farmer, M.A. (1991a) Ultrastructural organization of the heterotrophic euglenids and its evolutionary implications. In *The Biology of Free-Living Heterotrophic Flagellates* (eds. D.J. Patterson and J. Larsen), Oxford University Press, pp. 185–204.

Triemer, R.E. and Farmer, M.A. (1991b) An ultrastructural comparison of the mitotic apparatus, feeding apparatus, flagellar apparatus and cytoskeleton in euglenoids and kinetoplastids. *Protoplasma,* 164: 91–104.

Triemer, R.E., Linton, E., Shin, W., Nudelman, A., Monfils, A., Bennett, M., and Brosnan, S. (2006) Phylogeny of the Euglenales based upon combined SSU and LSU rDNA sequence comparisons and description of *Discoplastis* gen. nov. (Euglenophyta). *Journal of Phycology,* 42: 731–740.

von der Heyden, S., Chao, E.E., Vickerman, K., and Cavalier-Smith, T. (2004) Ribosomal RNA phylogeny of bodonid and diplonemid flagellates and the evolution of Euglenozoa. *Journal of Eukaryotic Microbiology*, 51: 402–416.

Zakrys, B., Empel, J., Milanowski, R., Gromadka, R., Borsuk, P., Kedzior, M., and Kwiatowski, J. (2004) Genetic variability of *Euglena agilis* (Euglenophyceae). *Acta Societatis Botanicorum Poloniae*, 73: 305–309.

Zakrys, B., Milanowski, R., Empel, J., Borsuk, P., Gromadka, R., and Kwiatowski, J. (2002) Two different species of *Euglena, E. geniculata* and *E. myxocylindracea* (Euglenophyceae), are virtually genetically and morphologically identical. *Journal of Phycology*, 38: 1190–1199.

17 The contribution of genomics to the understanding of algal evolution

Chris Bowler and Andrew E. Allen

CONTENTS

ABSTRACT

The genomes of six eukaryotic algae have now been completely sequenced. These include representatives of the green (*Chlamydomonas reinhardtii*, *Ostreococcus tauri*, and *Ostreococcus lucimarinus*), red (*Cyanidioschyzon merolae*), and brown (*Thalassiosira pseudonana*, *Phaeodactylum tricornutum*) algal lineages. The genomes have provided a wealth of new information about the evolution of these different groups of photosynthetic eukaryotes with respect to each other and with respect to other eukaryotes. For example, the green algal genomes are of particular interest for understanding plant gene evolution, while the genomes from brown algae are revealing insights into the origin and evolution of chromalveolates. In addition, the analysis of algal gene repertoires is providing valuable new information about specialized and unexpected metabolisms, and about ecological adaptations to specialized lifestyles. A notable example is the discovery of an apparently fully functional urea cycle in marine diatoms. This pathway is typically associated with metazoans, and so the way that it is integrated into the metabolism of a photosynthetic cell will be of major interest and may ultimately provide explanations for the profound ecological success of this group of algae in the contemporary ocean.

INTRODUCTION

The science of genomics can be defined as the large-scale analysis of genome architecture, aimed at the discovery and study of many genes simultaneously. It can be divided arbitrarily into functional, environmental, and comparative genomics. Comparative genomics consists principally of examining whole genome structure and gene arrangement and rearrangement, and aims to delineate the evolution of gene families, the molecular basis of adaptation (including the identification of genes potentially involved in niche adaptation), as well as the evolutionary relationships at various taxonomic levels in the tree of life (Baldauf, 2003). Functional genomics comprises the development of genome-wide or gene-related experimental approaches to assess the biochemical, structural, or regulatory roles of gene products with unknown functions and provides knowledge about their interactions. Environmental genomics deals with the understanding of the functional significance of genomic variation in natural biological communities (DeLong and Karl, 2005). This includes the use of various genotyping approaches to delineate the structure of inter- and intra-specific biodiversity, as well as the metagenomics approach, which treats entire ecosystems as a single living organism and involves the cloning and functional analysis of very large DNA fragments from uncultured organisms.

In recent years, the feasibility and utility of whole genome sequencing programmes has been amply demonstrated, beginning in 1995 with the sequencing of the genome of *Haemophilus influenzae* (a bacterium of the human respiratory tract that can cause meningitis and other diseases) and reaching a climax with the announcement of the completion of the human genome just a few years later (see www.ornl.gov/sci/techresources/Human_Genome/home.shtml for an overview). Besides human, the first eukaryotic genomes to be completed were from model organisms for which considerable research on gene function had already been carried out, such as the yeast *Saccharomyces cerevisiae*, the nematode *Caenorhabditis elegans*, the fruitfly *Drosophila melanogaster*, the flowering plant *Arabidopsis thaliana*, and the zebrafish *Danio rerio*. These days it is almost impossible to keep track of all the organisms whose genomes are being sequenced, and close to 1000 whole genome sequences are available, as of 2006. Because prokaryotes generally have smaller genomes and fewer repeating sequences than eukaryotes, the majority of these genomes are prokaryotic. For eukaryotes, completed genomes from representatives of all the crown groups are available, except for the Rhizaria (Figure 17.1). Although several algal species are represented (see below), the sequences are animal and unikont centric, and also include sequences from organisms of relevance to human health (e.g., *Plasmodium*, *Anopheles*, *Trypanosoma*, and *Leishmania*) and agriculture (e.g., rice, *Phytophthora*, and several fungal pathogens).

There are currently only a handful of whole genome sequences from eukaryotic algae, even though more than 30,000 species have already been described and other entirely new groups potentially await discovery. The definition of algae is in fact a rather ill-defined assemblage that includes most aquatic photosynthetic eukaryotes and has little phylogenetic significance, in that three of the five main lineages of the eukaryotic crown represented in Figure 17.1 contain photosynthetic representatives. Photosynthesis in eukaryotes is believed to have first appeared more than 1.5 billion years ago following a primary endosymbiotic event in which a heterotrophic eukaryote engulfed (or was invaded by) a photosynthetic cyanobacterium (Keeling, 2005). This event ultimately generated eukaryotic cells containing a chloroplast with a highly reduced cyanobacterial-derived genome that was surrounded by two membranes. These primary endosymbionts form the Plantae, which subsequently diverged into three main lineages, the Rhodophyta (red algae), the Glaucophyta, and the Viridiplantae (green algae and Streptophyta). Other photosynthetic eukaryotes obtained their plastids through secondary or tertiary endosymbioses, perhaps beginning around 1 billion years ago. Secondary endosymbiotic events, in which heterotrophic eukaryotes engulfed either red or green algae, are thought to have occurred at least three different times. In one case, the engulfment of a red alga is thought to have given rise to the chromalveolates, which possibly includes all of the brown algae: members of the haptophytes, cryptomonads, stramenopiles, and alveolates (Bachvaroff et al., 2005; Yoon et al., 2002). Consequently, chromalveolate plastids are typically surrounded by four membranes. This single origin of the chromalveolates (known as the chromalveolate hypothesis) is supported by the relic nucleus (known as a nucleomorph) that is found between the outer two and inner two plastid membranes in cryptophyte algae (Douglas et al., 2001). It also implies that the non-photosynthetic organisms in these groups have lost their photosynthetic capacity over evolutionary time. Dinoflagellates are also members of the alveolates, and are thought to derive from tertiary endosymbioses (Yoon et al., 2005), which significantly complicates our understanding of them.

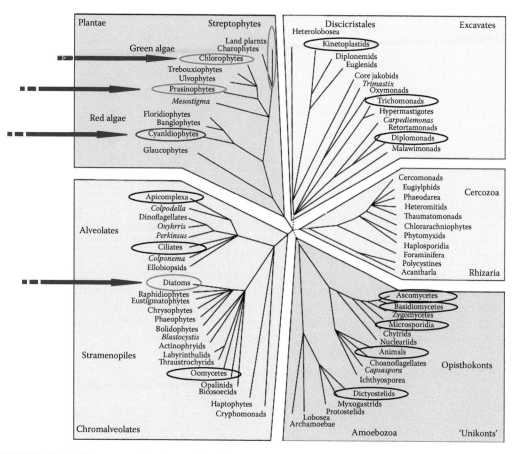

FIGURE 17.1 (Please see colour insert following page 76) Taxa for which complete genome sequences are available are circled. Photosynthetic eukaryotes whose genome sequences have been completed are found within the Plantae and Chromoalevolate groups and are indicated with an arrow. (The eukaryotic phylogentic tree is reproduced with permission from Keeling, P.J., Burger, G., Durnford, D.G., Lang, B.F., Lee, R.W., Pearlman, R.E., Roger, A.J., and Gray, M.W., *Trends in Ecology and Evolution*, 20, 670–676, 2005.)

The photosynthetic eukaryotes whose genome sequences have been completed are found within the Plantae and Chromalveolate groups (Figure 17.1). Within the Plantae, sequences are currently available from *Chlamydomonas reinhardtii* (Chlorophyta), *Ostreococcus tauri* and *O. lucimarinus* (Prasinophyta), and *Cyanidioschyzon merolae* (Cyanidiophyta), and within the Chromalveolates there are sequences available from *Thalassiosira pseudonana* and *Phaeodactylum tricornutum* (Diatomea) (Figure 17.1 and Table 17.1). Additional algal genome sequences that are nearing completion include *Volvox carteri*, a deep-living strain of *O. tauri*, and two strains of *Micromonas pusilla* from the Plantae, and *Emiliania huxleyi*, *Ectocarpus siliculosus*, *Pseudo-Nitzschia multiseries*, and *Fragilariopsis cylindrus* from the Chromalveolates. Although the number of algal genome projects remains rather low, our understanding of their evolution can also benefit from the analysis of genomes from organisms with related phylogeny (e.g., the land plants, Apicomplexans, Ciliates, and Oomycetes). In addition, a large number of ESTs (expressed sequence tags) are becoming available from a range of other algae.

To what extent can the handful of genomes currently available from algae help to address issues of algal evolution? In general, understanding the red and green algae that derived from the primary endosymbiotic event can be aided by comparison of their genomes with those of higher plants and cyanobacteria, whereas the algae derived from secondary endosymbioses present more

TABLE 17.1
Complete Genome Sequences from Algae*

	Chlamydomonas reinhardtii	Ostreococcus tauri	Ostreococcus lucimarinus	Cyanidioschyzon merolae	Thalassiosira pseudonana	Phaeodactylum tricornutum
Genome size (Mbp)	120	12.6	13.2	16	34	26.5
GC (%)	62	59	58	55	47	49
Completeness (%)	81	99	100	99.98	100	100
Number of predicted proteins	19,832	8166	7651	5331	11,242	10,681

* For further details of each genome and of the analytical methods used, readers are advised to consult the original articles.

complex issues. Although they typically contain one nuclear genome, one plastid genome, and one mitochondrial genome, at the initial stage of such endosymbioses, the complex cell likely contained at least five diffferent genomes, the nuclear and mitochondrial genomes of the exosymbiont and the nuclear, mitochondrial, and plastid genomes of the algal endosymbiont. Considerable mixing of genes is therefore likely to have occurred throughout their evolutionary history, and consequently, the only two available nuclear genomes from secondary endosymbionts (from the two diatoms) can only provide the first hints about these events. Notwithstanding these limitations, together with the genomes from related organisms such as Oomycetes, Apicomplexans, and Cilitaes, they are aiding our understanding of several key aspects (see below).

GENERAL FEATURES OF ALGAL GENOMES

The general characteristics of the six available nuclear genomes are summarized in Table 17.1. The *Chlamydomonas* sequence is currently only around 80% complete, due to its extraordinarily high guanine + cytosine (GC) content and repetitive nature, whereas the other sequences are essentially complete. Genome size varies almost tenfold and does not correlate closely with predicted gene content. It should be noted that the *O. tauri* genome is the smallest of any free-living eukaryote known to date, although all these organisms have rather small genomes compared with the multicellular eukaryotes (except *C. reinhardtii*). Such genome sizes appear to be typical for other simple eukaryotes such as the fungi and Oomycetes.

The first complete algal genome was reported in 2004 from a red alga, the extremophile *C. merolae* (Matsuzaki et al., 2004). It was characterized as 16.5 Mb and was predicted to encode 5331 genes on 20 chromosomes. The most unusual features revealed by the genome were the lack of introns in all but 26 putative genes, and the highly reduced nature of the ribosomal DNA unit, being found in only a single copy in three separate locations in the genome. Although the sequence provided support for the common origin of the green and red algal lineages, its highly reduced nature and the adaptation of the species to high temperatures and acid make it unlikely to reveal much information of general relevance to other red algae.

Other complete algal sequences from the Plantae are represented by *Ostreococcus* and *Chlamydomonas*, both green algae. The *Ostreococcus* nuclear genomes are around 13 Mb and are predicted to encode around 8000 genes. The sequences are derived from species isolated from a shallow lagoon in the Mediteranean (*O. tauri*) and from coastal waters off California in the Pacific (*O. lucimarinus*). In spite of their different areas of isolation, they are both derived from surface waters and their genomes appear to be rather similar, and information concerning specific ecological adaptations, e.g., to different light intensities, will have to wait for the genome sequencing of a deep strain of *O. tauri*, scheduled for 2007. A publication describing the *O. tauri* genome sequence is available (Derelle et al., 2006), and it is significant because it is the first sequence from a Prasinophyte, an ancient group that diverged at the base of the green lineage. The genome is highly compact due to the extensive reduction of intergenic regions, although it is structurally complex because it exhibits a level of heterogeneity that has never been observed previously in a eukaryote. In particular, two of the 20 chromosomes have a significantly biased GC content, and the genes on these chromosomes have an altered codon bias, different splicing, and contain the majority of transposons, i.e. genetic elements involved in genomic instability. It has been speculated that one of these chromosomes may represent a sex chromosome (Derelle et al., 2006), because such chromosomes sometimes display distinctive features to avoid recombination, whereas the other encodes many surface membrane proteins that may have bacterial origin, and may therefore be alien in nature. This may be a remnant of the ancient algal endosymbiont or may be derived from a more recent horizontal gene transfer event from bacteria. Although a publication describing the *C. reinhardtii* genome has not yet appeared, it is clear that it does not display this same level of genome heterogeneity. In general, the fact that this genome is an order of magnitude larger than those of *Ostreococcus* makes this genome much more complex. Consequently, introns are larger, intergenic regions are bigger, and gene families contain many more copies, reflected in the total gene count (Table 17.1).

The first diatom genome sequence, from *T. pseudonana*, was published in 2004 (Armbrust et al., 2004). The nuclear genome is 34 Mb and contains 11,242 putative genes arranged on 24 chromosomes, similar to the comparably sized *Neurospora crassa* genome (Galagan et al., 2003). The sequence was of major interest for addressing the secondary endosymbiosis hypothesis, which predicts different possible origins for nuclear-encoded diatom genes: nuclear or mitochondrial genomes of the exosymbiont; or nuclear, plastid, or mitochondrial genomes of the endosymbiont. Indeed, close to half of the encoded diatom proteins show similar alignment scores to their closest homologs in plant, red algal, and animal genomes, underscoring the evolutionarily ancient divergence of Plantae, Opisthokonta (animals/fungi), and the unknown exosymbiont that is proposed to have given rise to the chromalveolate lineage. Interestingly, around 1000 putative diatom proteins aligned with metazoan proteins but not green plant or red algal proteins, perhaps indicating that they were derived from the secondary exosymbiont host. A similar comparison that included cyanobacteria revealed the large number of proteins shared by all groups of photosynthetic eukaryotes, the majority of which are likely involved in chloroplast functions. Analysis of the genome sequence from this first diatom therefore supported the secondary endosymbiosis theory and revealed the extraordinary combination of genes that had been maintained from the exosymbiont and endosymbiont, with the potential to encode innovative metabolisms in these photosynthetic organisms (see below).

A second diatom genome has recently been completed, from *Phaeodactylum tricornutum*. This species belongs to the raphid pennate diatoms, which are considered to have arisen after the centric diatoms, to which *T. pseudonana* belongs. The complexity of diatom phylogeny is such that many more sequences will be required before we can have a full overview of diatom evolution, but these two diatom sequences nonetheless provide an important starting point. Furthermore, *T. pseudonana* is a planktonic diatom, whereas *P. tricornutum* can have a benthic lifestyle, and the latter is motile, whereas *T. pseudonana* is not. The genome size of *P. tricornutum* is somewhat smaller than that of *T. pseudonana*, although gene content appears similar, and like the first diatom genome the *P. tricornutum* genome appears to be a composite derived from the exosymbiont and endosymbiont ancestors.

INSIGHTS INTO ALGAL CELL BIOLOGY

The availability of whole genome sequences from algae allows prediction of the encoded proteome, thus permitting an analysis of the potential metabolic capacities of the cell, as well as its general structure. Although such *in silico*-based methods cannot uncover previously unknown functions, they can provide useful comparative information based on knowledge from better-studied organisms. For algal groups that are important components of marine phytoplankton communities worldwide, such findings can have significant implications for understanding the biogeochemical cycling of key nutrients such as carbon, nitrogen, and silicon.

The *Ostreococcus* and diatom genomes have revealed interesting differences in the carbon-concentrating mechanisms (CCMs), which are used to bring carbon dioxide to the carbon-fixing Rubisco enzyme within the plastid. Two mechanisms are generally considered: one based on the action of carbonic anhydrases, the other based on the delivery of CO_2 to the plastid by the decarboxylation of a C4 compound such as oxaloacetic acid by a plastid-localized malic enzyme (Edwards et al., 2004). The *O. tauri* genome appears to encode only one carbonic anhydrase, likely to be located in the plastid, and also appears to encode a suite of enzymes capable of C4 fixation (Derelle et al., 2006). On the other hand, *T. pseudonana* has several carbonic anhydrases, although none of them appear to be located in the plastid, and a plastid-localized malic enzyme also appears to be missing (Armbrust et al., 2004), in spite of some biochemical evidence for a C4 pathway in related diatoms (Reinfelder et al., 2004). By contrast, *P. tricornutum* has two plastid-localized carbonic anhydrases and does show potential for C4 photosynthesis (Montsant et al., 2005).

Other interesting differences can be predicted concerning nitrogen metabolism. While both *Ostreococcus* and the diatoms can utilize a range of nitrogen sources, such as nitrate, ammonium, and urea, *Ostreococcus* appears to be particularly well equipped for ammonium utilization due to the presence of four ammonium transporter genes in the genome (Derelle et al., 2006), whereas the diatom genomes encode the components of a urea cycle (Armbrust et al., 2004). The identification of enzymes necessary for a complete urea cycle was a surprise because this pathway is a typically metazoan feature and has never before been described in a eukaryotic photoautotroph. Unlike other organisms with a urea cycle, diatoms are unlikely to excrete urea because they possess an active urease and can grow on urea as a sole nitrogen source. In *T. pseudonana*, the enzyme that catalyzes the first step of the urea cycle (carbamoyl phosphate synthase, CPS III) appears to be targeted to mitochondria, as in other organisms with a urea cycle. Green algae and higher plants, on the other hand, possess a plastid-localized version of CPS that uses glutamine rather than NH_4^+ and is required for the first step of pyrimidine biosynthesis. Diatoms also have this form of CPS, but without any organellar targeting sequence, suggesting that pyrimidine biosynthesis occurs in the cytoplasm as in heterotrophs, rather than in plastids as in higher plants/green algae. This observation raises the intriguing question of how pyrimidines are transported across the four plastid membranes.

The urea cycle appears to be fully integrated into diatom metabolism in ways not previously suspected (Allen et al., 2006). Two intermediates of the urea cycle, arginine and ornithine, feed into other pathways present in the diatom. Ornithine is used to make spermine and spermidine, as well as long-chain polyamines required for silica precipitation during frustule formation (Kroger et al., 2000). Ornithine can also be converted directly to proline by ornithine cyclodeaminase. Arginine is used for the synthesis of the signalling molecule nitric oxide via nitric oxide synthase, which in higher plants plays a role in pathogen defense (Crawford and Guo, 2005), as may also be the case in diatoms (Vardi et al., 2006). Both diatoms also appear to be capable of generating the energy storage molecule creatine phosphate, via a urea cycle branch pathway originating from arginine that is again absent in *A. thaliana*, *C. reinhardtii*, and *C. merolae*.

Another interesting feature of diatom metabolism that was revealed from the genome sequences was their ability to break down fatty acids in both the mitochondria and peroxisomes, like animals, and to thereby generate ATP as well as heat from fatty acid oxidation (Armbrust et al., 2004). Plants, on the other hand, possess only the peroxisomal pathway, which cannot generate ATP. In the diatoms, peroxisomal pathway products of fatty acid oxidation presumably feed into the glyoxylate cycle and ultimately into gluconeogenesis for carbohydrate production, whereas their breakdown in the mitochondria could be used for the generation of ATP, which may explain how diatoms can withstand long periods of darkness and begin growing rapidly upon a return to the light.

These examples illustrate how the novel combinations of genes that have been endowed in diatoms from their endo- and exo-symbiont ancestors have been combined into successful metabolisms that are likely to be important factors underlying their ecological success.

Other features of algal genomes indicate that different genes have been exploited to carry out the same functions. A notable example concerns transcription factors, the workhorses of transcriptional control in all organisms. More than 50 different classes of transcription factor are now known (Derelle et al., 2006), and they are particularly common in higher plants, with *Arabidopsis* containing more than 1200 genes encoding a range of different classes (Riechmann et al., 2000). Both *Ostreococcus* and the diatoms contain a similar number of genes encoding transcription factors, around 120 each, however whereas half of these encode the heat shock transcription factor class in diatoms, a similar number encodes the homeodomain class in *Ostreococcus* (Montsant et al., 2005; Derelle et al., 2006). Why these different classes have been targeted for gene expansion in these different organisms is an important evolutionary question to address in future studies.

The function of these different classes of transcription factors in *Ostreococcus* and diatoms must await experimental investigation, as must the large number of diatom-specific and *Ostreococcus*-specific genes with unknown function. However, the novel features that have already been revealed indicate that future experimental work is going to reveal a wealth of novel features previously unknown in animals and plants. It will be particularly interesting to identify the specific features responsible for adaptation and ecological success in each organismal group, e.g., to define CCM and its interplay with photosynthesis and the primary metabolic pathways, to understand the basis of cell size and cell cycle control, to understand sex in an environmental context, to dissect the molecular and biochemical basis for silica nanofabrication in diatoms, and to understand diatom motility.

STRATEGIES OF ADAPTIVE EVOLUTION REVEALED BY GENOMICS APPROACHES

The above examples illustrate how the analysis of whole genome sequences can reveal specialized adaptations in different organismal groups. But from where does this variation derive and how is it maintained? Evolution comprises innovations that proceed through modifications of the gene content, e.g., the birth and decay of key genes or key members of gene families, changes in regulatory elements, and creation of novel interactions between gene products. In addition, specific features are seen to appear at crucial evolutionary transitions (e.g., secondary endosymbiosis). These genome-level events can be studied in an evolutionary context by comparing corresponding genomic sequences in different species. Regions that have been highly conserved during evolution can be considered as coding for essential universal functions, whereas discrepancies should reflect lineage-specific adaptations. For each organismal group, two genome spaces could therefore be envisioned: a core genome, which reflects phylogenetic history, and a pan genome, in which specialized genic adaptations lead to environmental niche adaptation, which ultimately drives speciation. The information available from eukaryotic algae is much too sparse currently to evaluate this objectively, but studies of prokaryotic genomes and community metagenomics approaches, which consider the ecosystem as one organism with one gene pool, are revealing a wealth of relevant information (De Long, 2005; Tringe et al., 2005). In addition, the potential for transfer of useful genes by cyanophages has been recently brought to light (Coleman et al., 2006), which is further facilitated by the clustering of genes encoding related functions in bacterial genomes (Mussmann et al., 2005). Hence, in contrast to terrestrial environments, barriers to genetic exchange may be few and far between in aquatic and ocean environments, and there must be significant potential for the propagation of aquired properties in single-celled organisms compared with multicellular organisms. Darwinís barriers to speciation derived from his study of finches may therefore be of only limited significance for the algae.

Ultimately, a major aim is to know how genetic variation is shaped by the environment, how changes in environmental gene expression reflect ecosystem function, how an organism's lifestyle influences genomic plasticity, and how intraspecific variation in gene expression affects adaptation, population divergence, and eventual speciation. Environmental genomics is challenging, precisely because one must link ecology, evolution, and ultimately the organism, with the ecosystem. Genome-wide surveys of transcription using high-throughput genotyping and microarrays in ecology are rapidly changing the way scientists think by forcing them to change their entire perspective on processes and interactions between populations and the environment (Ranz and Machado, 2006), and to think about how polygenic processes are controlled. Use of neutral markers and more conventional methods will continue, but instead of working with just a handful of loci, the new generation of population genetics will involve hundreds (or even thousands) of loci. This will allow the analysis of the balance between genetic drift (neutral or stochastic processes) and selection (deterministic processes) using many loci simultaneously. These types of data are crucial for understanding fluctuations in population size, historical demography, and gene flow, all of which are central for understanding adaptation over evolutionary time scales and for understanding population dynamics.

PERSPECTIVES

Whole genome sequences have already provided a wealth of new and unexpected information about the ecology, evolution, and metabolism of green, red, and brown algae. Notwithstanding, our information is only fragmentary due to the sparsity of whole genome sequences. Fortunately, several other algal species have already been slated for sequencing, and many more seem likely to follow. The community clearly needs a genome sequence from a representative red algal species, and picoeukaryotes need to be explored in much more detail. It is also important to cover diatom phylogeny in more detail, and in addition to the genome sequences from pennate and bi- and multi-polar centrics (represented by *P. tricornutum* and *T. pseudonana*, respectively), it will be important to identify araphid pennate and radial centric species (Medlin and Kaczmarska, 2004) that should be targeted for whole genome sequencing.

With more genomes from reference organisms available, it will be possible to explore the dynamics of speciation in an environmental context, as is currently being done for the prokaryotes (DeLong and Karl, 2005). However, the identification of brown algal viruses that appear to carry host-derived regulatory genes (Delaroque et al., 2003), and the fact that genes with related functions are often clustered in the algal genomes sequenced to date may indeed indicate that lateral gene transfer may be significant in these organisms, as it appears to be in the marine prokaryotes. Species concepts may therefore require significant revision.

ACKNOWLEDGMENTS

Research in the Bowler laboratories is partially supported from the EU-funded FP5 MarGenes project (QLRT-2001-01226), the EU-FP6 Diatomics project (LSHG-CT-2004-512035), the EU-FP6 Marine Genomics Network of Excellence (GOCE-CT-2004-505403), an ATIP "Blanche" grant (Centre National de la Recherche Scientifique [CNRS]), and the Agence Nationale de la Recherche (France).

REFERENCES

Allen, A.E., Vardi, A., and Bowler, C. (2006) An ecological and evolutionary context for integrated nitrogen metabolism and related signaling pathways in marine diatoms. *Current Opinion in Plant Biology*, 9: 264–273.

Armbrust, E.V., Berges, J.A., Bowler, C., Green, B.R., Martinez, D., Putnam, N.H., Zhou, S.G., Allen, A.E., Apt, K.E., Bechner, M., Brzezinski, M.A., Chaal, B.K., Chiovitti, A., Davis, A.K., Demarest, M.S., Detter, J.C., Glavina, T., Goodstein, D., Hadi, M.Z., Hellsten, U., Hildebrand, M., Jenkins, B.D., Jurka, J., Kapitonov, V.V., Kroger, N., Lau, W.W.Y., Lane, T.W., Larimer, F.W., Lippmeier, J.C., Lucas,

S., Medina, M., Montsant, A., Obornik, M., Parker, M.S., Palenik, B., Pazour, G.J., Richardson, P.M., Rynearson, T.A., Saito, M.A., Schwartz, D.C., Thamatrakoln, K., Valentin, K., Vardi, A., Wilkerson, F.P., and Rokhsar, D.S. (2004) The genome of the diatom *Thalassiosira pseudonana*: ecology, evolution, and metabolism. *Science*, 306: 79–86.

Bachvaroff, T.R., Puerta, M.V.S., and Delwiche, C.F. (2005) Chlorophyll c-containing plastid relationships based on analyses of a multigene data set with all four chromalveolate lineages. *Molecular Biology and Evolution*, 22: 1772–1782.

Baldauf, S.L. (2003) The deep roots of eukaryotes. *Science*, 300: 1703–1706.

Coleman, M.L., Sullivan, M.B., Martiny, A.C., Steglich, C., Barry, K., Delong, E.F., and Chisholm, S.W. (2006) Genomic islands and the ecology and evolution of *Prochlorococcus*. *Science*, 311: 1768–1770.

Crawford, N.M. and Guo, F.Q. (2005) New insights into nitric oxide metabolism and regulatory functions. *Trends in Plant Sciences*, 10: 195–200.

Delaroque, N., Boland, W., Muller, D.G., and Knippers, R. (2003) Comparisons of two large phaeoviral genomes and evolutionary implications. *Journal of Molecular Evolution*, 57: 613–622.

DeLong, E.F. (2005) Microbial community genomics in the ocean. *Nature Reviews Microbiology*, 3: 459–469.

DeLong, E.F. and Karl, D.M. (2005) Genomic perspectives in microbial oceanography. *Nature*, 437: 336–342.

Derelle, E., Ferraz, C., Rombauts, S., Rouzé, P., Worden, A.Z., Robbens, S., Partensky, F., Degroeve, S., Echeynié, S., Cooke, R., Saeys, Y., Wuyts, J., Jabbari, K., Bowler, C., Panaud, O., Benoit, Piégu, Ball, S.G., Ral, J.-P., Bouget, F.-Y., Piganeau, G., Baets, B.D., Picard, A., Delseny, M., Demaille, J., Peer, Y.V.D., and Moreau, H. (2006) Genome analysis of the smallest free-living eukaryote Osterococcus tauri unveils many unique features. *Proceedings of the National Academy of Sciences USA*, 103: 11647–11652.

Douglas, S., Zauner, S., Fraunholz, M., Beaton, M., Penny, S., Deng, L.T., Wu, X.N., Reith, M., Cavalier-Smith, T., and Maier, U.G. (2001) The highly reduced genome of an enslaved algal nucleus. *Nature*, 410: 1091–1096.

Edwards, G.E., Franceschi, V.R., and Voznesenskaya, E.V. (2004) Single-cell C4 photosynthesis versus the dual-cell (kranz) paradigm. *Annual Reviews of Plant Biology*, 55: 173–196.

Galagan, J.E., Calvo, S.E., Borkovich, K.A., Selker, E.U., Read, N.D., Jaffe, D., Fitzhugh, W., Ma, L.J., Smirnov, S., Purcell, S., Rehman, B., Elkins, T., Engels, R., Wang, S.G., Nielsen, C.B., Butler, J., Endrizzi, M., Qui, D.Y., Ianakiev, P., Pedersen, D.B., Nelson, M.A., Werner-Washburne, M., Selitrennikoff, C.P., Kinsey, J.A., Braun, E.L., Zelter, A., Schulte, U., Kothe, G.O., Jedd, G., Mewes, W., Staben, C., Marcotte, E., Greenberg, D., Roy, A., Foley, K., Naylor, J., Stabge-Thomann, N., Barrett, R., Gnerre, S., Kamal, M., Kamvysselis, M., Mauceli, E., Bielke, C., Rudd, S., Frishman, D., Krystofova, S., Rasmussen, C., Metzenberg, R.L., Perkins, D.D., Kroken, S., Cogoni, C., Macino, G., Catcheside, D., Li, W.X., Pratt, R.J., Osmani, S.A., Desouza, C.P.C., Glass, L., Orbach, M.J., Berglund, J.A., Voelker, R., Yarden, O., Plamann, M., Seller, S., Dunlap, J., Radford, A., Aramayo, R., Natvig, D.O., Alex, L.A., Mannhaupt, G., Ebbole, D.J., Freitag, M., Paulsen, I., Sachs, M.S., Lander, E.S., Nusbaum, C., and Birren, B. (2003) The genome sequence of the filamentous fungus Neurospora crassa. *Nature*, 422: 859–868.

Keeling, P.J., Burger, G., Durnford, D.G., Lang, B.F., Lee, R.W., Pearlman, R.E., Roger, A.J., and Gray, M.W. (2005) The tree of eukaryotes. *Trends in Ecology and Evolution*, 20: 670–676.

Kroger, N., Deutzmann, R., Bergsdorf, C., and Sumper, M. (2000) Species-specific polyamines from diatoms control silica morphology. *Proceedings of the National Academy of Sciences USA*, 97: 14133–14138.

Matsuzaki, M., Misumi, O., Shin-I, T., Maruyama, S., Takahara, M., Miyagishima, S.Y., Mori, T., Nishida, K., Yagisawa, F., Yoshida, Y., Nishimura, Y., Nakao, S., Kobayashi, T., Momoyama, Y., Higashiyama, T., Minoda, A., Sano, M., Nomoto, H., Oishi, K., Hayashi, H., Ohta, F., Nishizaka, S., Haga, S., Miura, S., Morishita, T., Kabeya, Y., Terasawa, K., Suzuki, Y., Ishii, Y., Asakawa, S., Takano, H., Ohta, N., Kuroiwa, H., Tanaka, K., Shimizu, N., Sugano, S., Sato, N., Nozaki, H., Ogasawara, N., Kohara, Y., and Kuroiwa, T. (2004) Genome sequence of the ultrasmall unicellular red alga *Cyanidioschyzon merolae* 10D. *Nature*, 428: 653–657.

Medlin, L.K. and Kaczmarska, I. (2004) Evolution of the diatoms: V. Morphological and cytological support for the major clades and a taxonomic revision. *Phycologia*, 43: 245–270.

Montsant, A., Jabbari, K., Maheswari, U., and Bowler, C. (2005) Comparative genomics of the pennate diatom Phaeodactylum tricornutum. *Plant Physiology*, 137: 500–513.

Mussmann, M., Richter, M., Lombardot, T., Meyerdierks, A., Kuever, J., Kube, M., Glockner, F.O., and Amann, R. (2005) Clustered genes related to sulfate respiration in uncultured prokaryotes support the theory of their concomitant horizontal transfer. *Journal of Bacteriology*, 187: 7126–7137.

Ranz, J.M. and Machado, C.A. (2006) Uncovering evolutionary patterns of gene expression using microarrays. *Trends in Ecology and Evolution,* 21: 29–37.

Riechmann, J.L., Heard, J., Martin, G., Reuber, L., Jiang, C.Z., Keddie, J., Adam, L., Pineda, O., Ratcliffe, O.J., Samaha, R.R., Creelman, R., Pilgrim, M., Broun, P., Zhang, J.Z., Ghandehari, D., Sherman, B.K., and Yu, C.L. (2000) Arabidopsis transcription factors: genome-wide comparative analysis among eukaryotes. *Science,* 290: 2105–2110.

Reinfelder, J.R., Milligan, A.J., and Morel, F.M.M. (2004) The role of the C-4 pathway in carbon accumulation and fixation in a marine diatom. *Plant Physiology,* 135: 2106–2111.

Tringe, S.G., Von Mering, C., Kobayashi, A., Salamov, A.A., Chen, K., Chang, H.W., Podar, M., Short, J.M., Mathur, E.J., Detter, J.C., Bork, P., Hugenholtz, P., and Rubin, E.M. (2005) Comparative metagenomics of microbial communities. *Science,* 308: 554–557.

Vardi, A., Formiggini, F., Casotti, R., De Martino, A., Ribalet, F., Miralto, A., and Bowler, C. (2006) A stress surveillance system based on calcium and nitric oxide in marine diatoms. *PloS Biology,* 4: 411–419.

Yoon, H.S., Hackett, J.D., Pinto, G., and Bhattacharya, D. (2002) The single, ancient origin of chromist plastids. *Proceedings of the National Academy of Sciences, USA,* 99: 15507–15512.

Yoon, H.S., Hackett, J.D., Van Dolah, F.M., Nosenko, T., Lidie, L., and Bhattacharya, D. (2005) Tertiary endosymbiosis driven genome evolution in dinoflagellate algae. *Molecular Biology and Evolution,* 22: 1299–1308.

18 Algal molecular systematics: a review of the past and prospects for the future

Linda K. Medlin, Katja Metfies, Uwe John and Jeanine L. Olsen

CONTENTS

ABSTRACT

The advent of the polymerase chain reaction (PCR) and DNA sequencing nearly 20 years ago revolutionized our fundamental understanding of algal evolution. Today, we sit at the beginning of a new era of comparative phylo-genomics, an area in which algae will play a central role in elucidating the network of life. In this review we tour the original "morphology vs. molecules" debate, the renewed role of ultrastructure, the discovery of cryptic species and the many paraphyletic taxa that pervade virtually all lineages of algae, evaluate molecular clocks, and discuss what to make of the new DNA barcoding. Full integration of taxonomy, phylogenetics, classification, and nomenclature under the systematics umbrella is more cogent than ever as losses of biodiversity continue to increase and the role of herbarium collections and databases becomes ever more important in assessing the relationship between biodiversity and ecosystem function over all scales. We round out our review with a look into the new field of genomics that will take the revolution to the deepest and shortest internodes, to the paraphyletic tips of species and populations. Use of many concatenated genes and resequencing technologies are already here, thus opening new possibilities to understand the very process of speciation through adaptive population divergence. The 21st century systematist will be cross-trained as a molecular naturalist including knowledge of bioinformatics.

INTRODUCTION

The polymerase chain reaction (PCR) was first published in 1988 (Saiko et al., 1988), and within the same year, the first universal primers for ribosomal genes appeared (Medlin et al., 1988). These primers were for the small sub-unit rRNA gene (SSU rRNA), which had already been recognized as a gene suitable for deep phylogenetic analysis (Woese, 1987). Shortly thereafter, universal primers for the rRNA internal transcribed spacers, ITS1 and ITS2, were introduced for species-level comparisons (White et al. 1990), thus sweeping the full taxonomic range of diversity. Prior to PCR, most genes had to be cloned from a genomic library, which could take up to a year or more for each gene. Ribosomal genes, however, could be accessed by reverse transcriptase sequencing of the coding strand (Lane et al., 1985). These early, pre-PCR phylogenetic trees were an important milestone, but were criticized because they were based on a single strand and partial sequences of only a few hundred nucleotides.

Phycologists were quick to realize that PCR could provide a vehicle for studying the evolution and systematics of the algae (Olsen, 1990). Within a decade the first broad-brush phylogenetic surveys appeared for the green algae (Buchheim and Chapman, 1991; Zechman et al., 1990; Wilcox et al., 1993), the red algae (Ragan et al., 1994; Freshwater et al., 1994; Saunders and Kraft, 1997), the brown algae (Saunders and Druehl, 1992; Tan and Druehl, 1993; Draisma et al., 2001; Rousseau et al., 2001), the diatoms (Medlin et al., 1993, 1996), the dinoflagellates (Fensome et al., 1999), and the haptophytes (Medlin et al., 1994). Four new algal classes were also discovered—first with molecular data and then supported by morphological data (i.e., the bolidophytes, the pelagophytes, the prasinophytes, and the pinguinophytes) (Moestrup, 1991; Andersen et al., 1993; Guillou et al., 1999; Kawachi et al., 2002). A flurry of class, ordinal, family, and generic-level surveys soon followed within all of the major algal divisions. Other genes and spacer sequences besides the ribosomal operon were explored (e.g., actin, *rbc*L, RUBISCO spacer [rhodophytes and heterokonts], *psb*a, *tuf*A, and other chloroplast genes), but expanding the repertoire of suitable genes and spacers, particularly those with higher resolving power at the species-level studies, proved to be surprisingly difficult. Even today, only a handful of different gene sequences regularly appear in the literature.

THE MORPHOLOGY VS. MOLECULES DEBATE

The 1980s witnessed a change from largely typological and phenetic approaches to classification based on comparative morphology and ultrastructure, to cladistic methods of classification based on homology and the strict recognition of monophyletic groups (Hillis et al., 1996; Felsenstein, 2004). The impact of the new phylogenetic methodologies was profound and acrimonious (Hull, 1988), in part because many cherished principles of morphological organization proved to be grades rather than clades. Convergence and parallel evolution have also proven to be extraordinarily common in the algae.

Initial distrust of molecular sequence data led many systematists to believe that molecular and morphological trees had to agree. A typical argument of the time was that congruence of the trees proved that molecular data were unnecessary, whereas incongruence proved that the molecular data were wrong. There was also a short-lived notion that molecular data were "purer" and not subject to homoplasy in the way morphological data were because of the underlying model assumption of neutral molecular evolution. Today we recognize that all data types are subject to homoplasy and that disagreements between morphological and molecular-based trees are instructive rather than divisive. Morphological characters used in identification need not *per se* be phylogenetically informative, i.e., synapomorphic. Autapomorphic characters and even some sympleiosomorphic characters have proven to be useful. Classification, on the other hand, needs to reflect evolutionary history of identity by descent through homology. When convergence of morphological characters is strong and the number of synapomorphic characters few, conflicts between traditional phenetic and phylogenetic classifications will inevitably arise.

ULTRASTRUCTURE—AN EARLIER REVOLUTION

Prior to the molecular revolution was the ultrastructural revolution, which revealed the traces of early endosymbiotic events reflected in chloroplasts, flagella, mitosis, cytokinesis, cell wall composition, and ornamentation (see classic works, such as Pickett-Heaps, 1975; Round and Crawford, 1981). Today, some 30 years later, ultrastructural data are being re-evaluated in their role to define deeper branches in our trees. For example, the dinoflagellate genus, *Gymnodium*, has been divided into three new genera based on the depth and shape of the apical groove in the epicone (Daugjberg et al., 2000). In the diatoms, the distinction between centrics and pennates has been abandoned in favour of three groups that can be defined by the ultrastructure of the specialised diatom zygote, the auxospore (Medlin and Kaczmarska, 2004). In the red algae, phylogenetic information based on ultrastructure of the pit-plug connections (Pueschel, 1980) is in agreement with molecular analyses (Saunders and Bailey, 1997), although additional work is needed to establish homology of plug ultrastructure across a broader range of rhodophytes.

CRYPTIC SPECIES, PARAPHYLY, AND NOMENCLATURE

One of the major benefits of molecular trees is that we can easily recognize cryptic species and paraphyletic lineages. At the same time, "what to do" about these from a nomenclatural point of view remains problematic and controversial. What is the utility of naming species that cannot be recognized in the absence of molecular data; and if we demand strict monophyly for a taxon, is it useful to apply the rules of nomenclature to such an extent that commonly recognized taxa are lost under new names? Is Phylocode (www.ohiou.edu/phylocode) a better alternative than Linneaus' 300-year-old system? The response to these questions varies.

Early examples of cryptic species and paraphyly came from unicellular green algae and diatoms (Figure 18.1) (Huss and Sogin, 1990; Medlin et al., 1991). In some cases where molecular data

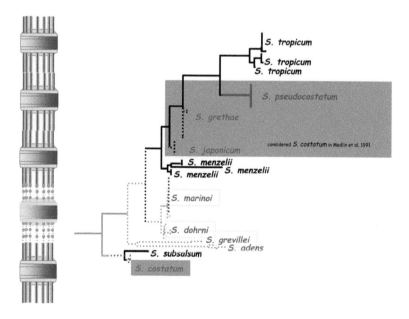

FIGURE 18.1 Changing concepts of cryptic diversity in the diatom *Skeletonema costatum*. The part of the tree shown in the grey box represents the diversity found by Medlin et al. (1991) in which *S. grethae* and *S. japonicum* were regarded as *Skeletonema costatum*. A recently expanded analysis of the genus (Sarno et al., 2005) indicates that the real *Skeletonema costatum* is sister species to *S. subsalsum* and that six new species have been discovered (dotted lines).

have revealed cryptic species and paraphyly, taxonomic changes have been made (Daugjberg et al., 2000; Edvardsen et al., 2000; Medlin and Kaczmarska, 2004), but in most cases, the paraphyletic lineages have remained, with investigators either willing to live with non-monophyletic taxa, unable to find new characters to define the new monophyletic groups, or unwilling to go against conventional wisdom that would lead to the demise of long-standing taxa. A couple of dramatic examples from seaweeds include *Ulva* and the kelps. *Ulva* and *Enteromorpha* contain many cryptic species and paraphyly is rampant. Hayden et al. (2003) entitled their paper, "Linnaeus was right all along: *Ulva* and *Enteromorpha* are not distinct" and proceeded to make the necessary nomenclatural changes. In the case of the kelps, this has not occurred. Members of the genus *Laminaria* are paraphyltic yet we live with the duality.

There is no question that the Linnaean system is typological and antiquated; the problem is finding a practical alternative. One hybrid approach is to adopt a partial non-hierarchical system (Soares, 2004; Cantino and de Queiroz 2005), such as Phylocode (Figure 18.2). In this system, the clades in the tree are given names and relationships among taxa are empirically derived. Clades and species are then named using the traditional rank-based system. In this manner, phylogenetic principles can flexibly equate taxa with clades and species without losing the trail of earlier circumscriptions or inherent instabilities of tree topology associated with changes in taxon sampling. One advantage of this system is that as new species are added to the phylogenetic tree, only changes in clade representation are noted, whereas ranks of taxa do not change.

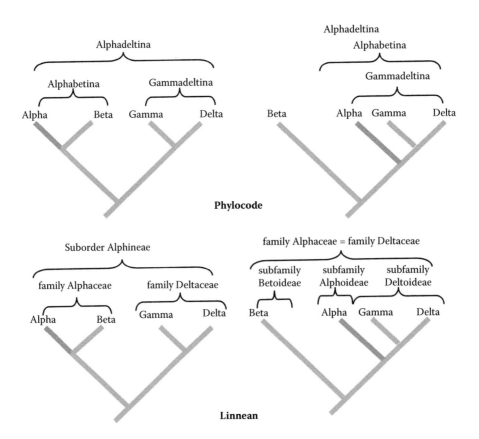

FIGURE 18.2 A comparison of nomenclatural rules between Phylocode and the Linnaean system. If a genus is found to belong to another family clade (right), the Linnaean rules could result in the loss of a family name, whereas the Phylcode rules would prevent this. (Redrawn from www.ohiou.edu/phylocode.)

BARCODING—TAXONOMIC IDENTIFICATION INDEPENDENT OF PHYLOGENY

Taxonomic identification can be based on any set of diagnostic characters. Although this has traditionally been morphological, with vastly accumulating sequence data, barcoding (Moritz and Cicero, 2004) of the algae is not an unreasonable goal. The challenge is in finding an appropriate gene sequence that contains sufficient resolving power to discriminate species. Barcoding based on the mitochondrial *cox1* gene works well for single metazoan groups such as fish or copepods, but the algae are represented by many different and very deep lineages, not all of which evolved at the same time or at the same rate. Given that algae are protists, mitochondrial DNA may prove useful in some groups, although present candidates under consideration include rDNA-ITS, *rbc*L, and other chloroplast genes. Although barcoding is appealing, it is essentially a modern-day typological method. For naturally paraphyletic taxa (e.g., nascent species, hybrids, and so on), this poses a problem. More insidiously, given the efforts to place biodiversity in an evolutionary framework of classification, it is important to remember that the fact that the barcode is a DNA fingerprint does not imply that it is saying something about the evolutionary history of the species (Moritz and Cicero, 2004) unless the barcode sequence is also used in a formal phylogenetic analysis.

INCONGRUENCE AMONG PHYLOGENIES

One of the most troubling difficulties in molecular phylogenetics is the occurrence of incongruence between trees generated from different genes, i.e., the gene-tree, species-tree problem. Although many two- and three-gene phylogenies agree, many do not. Incongruencies have been identified at all taxonomic levels. The reasons for this are both analytical and biological. The former includes taxon sampling, choice of gene(s), amount of data, choice of optimality criteria, and assumptions of the model of DNA evolution. The latter includes the role of genetic drift and natural selection, neither of which are typically considered in species- and higher-level phylogenetic reconstructions. Populations are assumed to be infinitely large and in equilibrium; and the genes used are assumed to be evolving under neutrality.

As more and more genes become available through genome-sequencing projects, the ability to use many genes in so-called concatenated gene sequences is growing steadily. New phylogenies based on clusters of genes or on total genomes (Phillipe et al., 2005) will be more common in the future and are expected to resolve most of the problems encountered by single- or few-gene phylogenies. Not only gene sequences but also gene order or even gene loss (Nozaski et al., 2003) can be examined with this new phylo-genomic approach. Most critically, genome-wide coverage can be achieved which will allow for better control over the relative effects of selection and constraint. In a recent review of genome-scale approaches to resolving incongruence in molecular phylogenies, Rokas et al. (2003) determined that a set of 15 to 20 genes are necessary for 95% confidence level for a species-level tree. Whereas this number of genes seems high at the moment, developments in high-throughput sequencing technology and lower costs will make this approach tractable for most labs within the near future. Certainly for the heterokonts and chlorophytes, where several full genomes are already available, chip-based resequencing approaches are rapidly becoming available. Nevertheless, before genes can be concatenated, there are several aspects of the individual genes that must be considered. These include length (>200 to 250 amino acids), coverage across taxa, alignability, no evidence for gene transfer, conserved function, absence of paralogy, and clear ability to identify orthologs if in a gene family. Programs are available that evaluate whether or not data sets (genes) can be concatenated (see Cunningham, 1997).

MOLECULAR CLOCKS

Molecular clocks continue to intrigue and to frustrate (Arbogast et al., 2002; Shields, 2004). The basic principle is to take a measure of genetic distance between extant species and convert it to divergence time from a common ancestor—usually in millions of years or in number of generations.

Calibration requires fossil or other palaeogeological or climatic evidence, the quality of which varies enormously. The result is almost always a wide confidence interval in the regression of divergence time (Knowlton and Weigt, 1998). Ribosomal and mitochondrial clocks are reasonably robust but even with these, lineage-specific differences can and do cause problems. Although PAUP (Phylogenetic Analysis Using Parsimony; Swofford, 2002) permits only a strict molecular clock to be applied to a data set, there are other programs (e.g., r8s; Sanderson, 2006) that allow for more relaxed molecular clocks and even for different rates of evolution to be applied to different parts of the tree. Molecular clocks have been calibrated from fossil data for several algal groups (e.g., Kooistra and Medlin, 1996, for the diatoms; Olsen et al., 1994, for the Dasycladales; Hillis et al., 1998, for *Halimeda*). They have also been used to infer divergence times of groups without a fossil record (Bakker et al., 1995; Lange et al., 2002), to confirm the appearance of cryptic species (Sáez et al., 2003), to infer the biogeographic history of taxa (John et al., 2003), and to infer the number of endosymbiotic events in the mega evolution of the algae (Yoon et al., 2004).

With the development of Bayesian maximum likelihood and coalescence methods Felsenstein (2004) applied to haplotypes scattered across the genome, possibilities to statistically infer population-level histories (Brumfield et al., 2003; Drummond et al., 2005) backwards in time based on generations is becoming increasingly tractable. This will be especially strong in timeframes of less than one million years.

PHYLOGENETICS AT THE SPECIES AND POPULATION LEVELS

With most groups having a broad-based phylogeny now in place, analyses are beginning to shift toward assessment of genetic diversity below the species level. Most of the focus has so far been on the identification, distribution, and diversity of biogeographic populations. Population genetic studies focusing on, e.g., hybridization, speciation, life history traits, and selection are still relatively limited.

Population-level surveys of putatively single species have uncovered many cryptic species—especially bacterio-plankton, where environmental clone libraries have revealed massive amounts of hidden diversity (Venter et al., 2000). Earlier notions about globally distributed, clonal species and associated marine panmixia (Palumbi, 1994) have fallen by the way. Among both unicellular and multicellular algal lineages, novel biogeographic clades are common (Coyer et al., 2006; John et al., 2003; Medlin et al., 2006). Data from population-level surveys may come from sequences (especially rDNA-ITS, RUBISCO spacer), haplotypes frequencies (mainly from mitochondrial or chloroplast DNA sequences), or allele frequency data (mainly microsatellite loci)—each analyzed in a data-specific manner.

It is here that differences in population size between pelagic and benthic algae become important. In extremely large populations—typical of oceanic phytoplankton—even small selection pressures may have large effects as compared with the stochastic process of genetic drift, which are dampened by the large size. Conversely, smaller (though still large) benthic algal populations are likely to be more affected by genetic drift, requiring a stronger selection pressure to effect change (Peijnenberg et al., 2005). Although still speculative, the influence of ultra-large populations could provide at least one explanation for the enormously high diversity detected in marine plankton.

Among the phytoplankton, *Emiliania huxleyi* has many genetically distinct populations and has the ability to form and maintain blooms ranging from sub-polar to tropical latitudes. Low gene flow has been reported between areas geographically close to each other, such as the North Atlantic and the Norwegian fjords (Igelias-Rodriques et al., 2006) proving that both coastal and open-ocean compartmentalization occur. Another interesting finding is that populations of *Pseudo-nitzschia pungens* are genetically diverse but appear to belong to a single panmictic global population (Evans et al., 2005) supported by breeding studies among global strains (Chepernov et al., 2005). Conversely, the identification of discrete populations of *Cyclotella menegheniana* in rivers only 40 km apart (Beszteri et al., 2005, 2007) is a testament to the statement that, "not everything is everywhere"

(Finlay et al., 2002). Whereas these examples support the role of reproductive isolation as a criterion for species recognition, other studies equally illustrate that reproductive isolation can be weak or absent as in *Fucus* species (Strate et al., 2002; Coyer et al., 2002a, 2002b, 2006; Billard et al., 2005; Engel et al., 2004, 2005; Coyer et al., 2006).

In studies of benthic algae, latitudinal and habitat variation in genetic diversity is well documented, along with isolation-by-distance over a range of scales from meters to hundreds of kilometres. The role of hybridization and speciation in *Fucus* (Coyer et al., 2002a, 2002b; Billard et al., 2005) illustrates the rapid radiation of this taxon over the past 20,000 yrs. In these cases, the influence of effective population size (N_e), crucial to the interpretation of population expansions, has played a strong role which, in turn, has affected the topology of population-level phylogenies including molecular clocks. At this stage, phylogeographic studies have had little impact on algal systematics other than to emphasize the fact that the paraphyly associated with nascent speciation of ecotypes is extremely common. In the coming decade we expect this to change as species concepts become increasingly dynamic and, from a systematics perspective, move away from typological definitions that are so strongly enforced by the Linnaean system.

COMPARATIVE GENOMICS

Phylo-genomics in algae is an exciting prospect as more and more algal genomes are sequenced. At present there are 11 algal genomes in or out of the pipeline (www.jgi.doe.gov). These include *Thalassiosira pseudonana, Phaeodactylum tricornutum, Ectocarpus siliculosus, Chondrus crispus, Porphyra yezoensis, Cyanidioschyzon merolae, Chlamydomonas reinhardtii, Micromonas pupilla, Ostreococcus tauri, Volvox carteri,* and *Emiliania huxleyi.* Because of the expected enormous size of many eukaryotic algal genomes, algal genome projects will, in the near future, remain restricted to particular subsets. Lineages such as the dinoflagellates may never be completely sequenced given an approximate genome size of more than a 100 times that of the human genome, which is itself 3000 Mb. Therefore, the "poor man's" genome approach (Rudd, 2003) (i.e., to sequence and identify the expressed genes) will play a strong role in the coming 5 years or so. Expressed sequence tags (ESTs) remain the best strategy for genomic characterisation of expressed genes under various conditions (e.g., stress or development). A by-product of the numerous EST sequencing projects underway is the development of literally thousands of new markers for phylogenetics at all levels.

The phylogenomic challenges, in our opinion, are in understanding short inter-nodal distances in the deep roots of divisions and classes, and unravelling paraphyly in the tips where speciation is underway. The opportunity to utilize multi-gene phylogenies will, in any case, refine and stabilize already existing phylogenies.

ECOLOGICAL GENOMICS

At the level of ecological genomics, which focuses on identification and differential gene expression in response to environmental variation through time and space, we see two important areas with implications for systematics.

IDENTIFICATION—WHO IS WHERE AND WHEN USING MICROARRAY FINGERPRINTING

New techniques, such as microarray fingerprints or phylochips (a more elaborate form of DNA barcoding) (Metfies and Medlin, 2004) will help to resolve many of our problems with identification of pro- and eukaryotic microalgae, propagules, and gametes, and with some easy-to-use markers, e.g. AFLPs (amplified fragment length polymorphisms) being too variable for analysis between populations (John et al., 2004b). Current generation "phylochips" mostly utilize rRNA genes and are already proving their worth in studies of microbial biodiversity, because they make it possible to identify taxa from very broadly circumscribed groups in a systematic fashion without the need of a prior cultivation step. Phylochips have been applied for the analysis of the diversity in the picoplankton-fraction of samples taken during the autumn blooms (Medlin et al., 2006). Newer phylochips are experimenting with thousands of random oligonucleotide sequences that can be analyzed informatically to assess relatively fine-scale diversity differences. Examples of this general approach can be found in Kingsley et al. (2002).

Genetic identification and variation of taxa can also be assessed using a new generation of markers called single nucleotide polymorphisms (SNPs) (Laviolette et al., 2000; Wang et al., 1998). The analysis of SNPs with DNA microarray technology provides a high-throughput method for the analysis of genetic variation (Lindblad-Toh et al., 2000; Raitio et al., 2001). Unlike the phylochips, SNP development depends on *a priori* comparative knowledge of a genomic region among a panel of individuals. At present this is mostly restricted to model organisms (e.g., *Chlamydomonas reinhardtii*; Vysotskaia et al., 2001) but is expected to play an important role in the near future.

SELECTION AS A DRIVER FOR ECOTYPIC DIFFERENTIATION AND EVENTUAL SPECIATION

Differential gene expression will help us to understand the ecological niche of the species and how the species responds to environmental change. It may also help to elucidate the process of population differentiation through ecotypic differentiation leading to speciation—a process that necessarily generates paraphyly.

These approaches also utilize EST libraries and microarray technologies (Gibson, 2002) to compare suites of genes that become up- and down-regulated in response to particular conditions, e.g., temperature, salinity, nutrient cues, and so on. For example, sudden crashes of blooms of *E. huxleyi* have been observed and linked to the occurrence of viruses. Different strains of *E. huxleyi* display different susceptibility to viral infection. Understanding the differences in the genome or transcriptome of the different strains will contribute to the understanding of the viral resistance of *E. huxleyi* and how the species is able to survive viral attack. Preliminary work suggests that the virus shuts down the photosynthetic system of *E. huxleyi* (Kegel et al., unpublished). In another example, ESTs from the sea ice diatom *Fragilariopsis cylindrus* showed that genes for photosynthesis increased their expression when both temperature and light decreased during ambient winter conditions in the Antarctic (Mock and Valentin 2004). In seaweeds, EST libraries for *Fucus vesiculosus* and *F. serratus* are being developed to assess ecotypic differentiation to desiccation stress over an intertidal gradient (Hoarau et al., unpublished). Similar studies are being pursued in other seaweeds such as *Laminaria, Chondrus, Graciliaria*, and *Caulerpa* (unpublished data). In the field of toxic algae, the EST approach has been applied for several species e.g., dinoflagellates, a group of species with a genome estimated at more than 100× the human genome (John et al., 2004a; Lidie et al., 2005). For such large genomes, the limited genomic approach of generating an EST library and the design and production of microarrays will probably be in the near future the only way to gain more understanding in phenomena, such as the genetic regulation growth and toxin production (Synder et al., 2005). At present, most is known about the cyanophyceae (reviewed by Börner and Dittmann, 2005) although progress is being made for eukaryotic microalgae. For many species, such as *Chrysochromulina polylepis, Prymnesium parvum, Alexandrium ostenfeldii, Prorocentrum lima*, and *Karenia brevis*, the predicted toxins are known to be polyether compounds.

21ST CENTURY SYSTEMATIST

Rounding out this review and perspective, we conclude with a description of what a 21st century systematist will need to have in house. First and foremost, there is no substitute for empirical knowledge of the group of interest. This includes field knowledge and knowledge of the literature. Excellent Web-based databases, such as AlgaeBase, CoML (Census of Marine Life), and the online

Index Nomen Algarum, are outstanding examples of integration of herbarium data, field keys, and relevant professional literature. Nevertheless, it is essential that the new generation know how to do more than access the relevant literature. With the decline in taxonomists, it is all the more urgent that every phycologist makes it his or her duty to train at least one or two students in their special systematic knowledge. Second, students need to develop a theoretical background with respect to issues and differences between taxonomic identification, classification, and the formal rules governing the application of nomenclature (both Linnaean and Phylocode). This includes herbarium training as well as theoretical training in the principles of phylogenetic systematics, and to some extent, philosophy and history of the field. Third, students need training in at least the basics of molecular evolution and population genetics. Both of these fields have become more accessible over the past decade with the development of user-friendly software and tutorials. Nevertheless, becoming an intelligent user of software remains crucial. Fourth, students need experience with the basics of PCR, cloning, sequencing, genotyping, and data mining.

In summary, the major challenges for algal systematists in the coming decades will be to:

1. Elucidate the very deep and short internodes of the tree of life.
2. Elucidate speciation and population differentiation at the tips of the tree of life.
3. Develop genome-wide sets of genes for definitive phylogenies at all levels.
4. Develop phylochips to detect microscopic cell diversity.
5. Develop new classes of neutral and selective markers with genome-wide representation enabling us to utilize new-generation phylogenetic and coalescence methods for exploring populations and species' histories.
6. Ensure training of the next generation of algal systematists to include classical training as well as molecular training.

ACKNOWLEDGMENTS

This work was partially sponsored by the EU Network of Excellence Marine-Genomics-Europe MGE-GOCE-CT-2004-505403.

REFERENCES

Andersen, R.A., Saunders, G.W., and Paskind, M.P. (1993) Ultrastructure and 18S rRNA gene sequence for *Pelagomonas calceolata* gen. and sp. nov. and the description of a new algal class, the Pelagophyceae *classis nov. Journal of Phycology*, 29: 701–715.

Arbogast, B.S., Edwards, S.V., Wakeley, J., Beerli, P., and Slowinski, J.B. (2002) Estimating divergence times from molecular data and phylogenetic and population genetic timescales. *Annual Review of Ecology and Systematics*, 33: 707–740.

Bakker, F.T., Olsen, J.L., and Stam, W.T. (1995) Evolution of nuclear rDNA ITS sequences in the *Cladophora albida/sericea* clade (Chlorophyta). *Journal of Molecular Evolution*, 40: 640–651.

Beszteri, B., Acs, E., and Medlin, L.K. (2005) Ribosomal DNA sequence variation among sympatric strains of the *Cyclotella meneghiniana* complex (Bacillariophyceae) reveals cryptic diversity. *Protist*, 156: 317–333.

Beszteri, B., Uwe J., and Medlin, L.K. (2007) Congruent variation at a nuclear and a plastid locus suggests that the diatom *Cyclotella meneghiniana* is a species complex. *European Journal of Phycology*, 42: 47–60.

Billard, E., Serrão, E.A., Pearson, G.A., Engel, C.R., Destombe, C., and Valero, M. (2005) Analysis of sexual phenotype and prezygotic fertility in natural populations of *Fucus spiralis, F. vesiculosus*, (Fucaceae, Phaeophyceae) and their putative hybrids. *European Journal of Phycology*, 40: 397–407.

Billot, C., Engel, C.R., Rousvoal, S., Kloareg, B., and Valero, M. (2003) Current patterns, habitat discontinuities and population genetic structure: the case of the kelp *Laminaria digitata* in the English Channel. *Marine Ecology Progress Series*, 253: 111–121.

Börner, T. and Dittmann, E. (2005) Molecular biology of cyanobacterial toxins. In *Harmful Cyanobacteria* (eds J. Huisman, H.C.P. Matthijs, and P.M. Visser), Springer Verlag, Dordrecht, Netherlands, pp. 25–40.

Brumfield, R., Beerlie, P., Nickerson, A., and Edwards, S.V. (2003) The utility of single nucleotide polymorphisms in inferences of population history. *Trends in Ecology and Evolution*, 18: 249–255.

Buchheim, M.A. and Chapman, R.L. (1991) Phylogeny of the colonial green flagellates: a study of 18S and 26 S rRNA sequence data. *BioSystems*, 25: 85–100.

Cantino, P.D. and de Queiroz, K. (2005) Phylocode: a phylogenetic code of biological nomenclature (www.ohiou.edu/phylocode).

Chepurenov, V.A., Mann, D.G., Sabbe, K., Vannerum, K., Casteleyn, G., Verleyen, E., Peperzak, L., and Vyverman, W. (2005) Sexual reproduction, mating system, chloroplast dynamics, and abrupt size reduction in *Pseudo-nitzschia pungens* from the North Sea. *European Journal of Phycology*, 40: 379–395.

Coyer, J.A., Hoarau, G., Oudot-Le Secq, M.-P., Stam, W.T., and Olsen, J.L. (2006) Rapid radiation within the brown algal genus *Fucus* (Heterokontophyta: Phaeophyta) as determined by mtDNA sequences. *Molecular Phylogenetics Evolution*, 39: 209 222.

Coyer, J.A., Peters, A.F., Hoarau, G., Stam, W.T., and Olsen, J.L. (2002a) Hybridisation of the marine *seaweeds Fucus serratus* and *F. evanescens* (Heterokontophyta: Phaeaophyceae) in a century-old zone of secondary contact. *Proceedings of the Royal Society B*, 269: 1829–1834.

Coyer, J.A., Peters, A.F., Hoarau, G., Stam, W.T., and Olsen, J.L. (2002b) Inheritance patterns of ITS1, chloroplasts, and mitochondria in artificial hybrids of the seaweeds *Fucus serratus* and *F. evanescens* (Heterokontophyta: Fucales). *European Journal of Phycology*, 37: 173–178.

Coyer, J.A., Peters, A.F., Stam, W.T., and Olsen, J.L. (2003) Post-ice age recolonization and differentiation of *Fucus serratus* L. (Fucaceae: Phaeophyta) populations in Northern Europe. *Molecular Ecology*, 12: 1817–1829.

Cunningham, C.W. (1997) Can incongruence tests predict when data should be combined? *Molecular Biology and Evolution*, 14: 733–740.

Daugjberg, N., Hansen, G., Larsen, J., and Moestrup, Ø. (2000) Phylogeny of some of the major genera of dinoflagellates based on ultrastructure and partial LSU rDNA sequence data, including the erection of three new genera of unarmoured dinoflagellates. *Phycologia*, 39: 302–317.

Draisma, S.G.A., Prud'homme van Reine, W.F., Stam, W.T., and Olsen, J.L. (2001) A reassessment of phylogenetic relationships within the Phaeophyceae based on Rubisco large subunit and ribosomal DNA sequences. *Journal of Phycology*, 37: 586–603.

Drummond, A.J., Rambant, A., Shapiro, B., and Pybus, O.G. (2005) Bayesian coalescent inference of past population dynamics from molecular sequences. *Molecular Biology and Evolution*, 22: 1185–1192.

Edvardsen, B., Eikrem, W., Green, J.C., Andersen, R.A., Moon-Van Der Staay, S.Y., and Medlin, L.K. (2000) Phylogenetic reconstructions of the Haptophyta inferred from rRNA sequences and available morphological data. *Phycologia*, 39: 19–35.

Engel, C.R., Daguin, C., and Serrão, E.A. (2005) Genetic entities and mating system in hermaphroditic *Fucus spiralis* and its close dioecious relative *F. vesiculosus* (Fucaceae, Phaeophyceae). *Molecular Ecology*, 24: 2033–2046.

Engel, C.R., Destombe, C., and Valero, M. (2004) Mating system and gene flow in the red seaweed *Gracilaria gracilis*: effect of haploid-diploid life history and intertidal rocky shore landscape on fine-scale genetic structure. *Heredity*, 92: 289–298.

Evans, K.M., Kühn, S.F., and Hayes, P.K. (2005) High levels of genetic diversity and low levels of genetic differentiation in North Sea *Pseudo-nitzschia pungens* (Bacillariophyceae) populations. *Journal of Phycology*, 41: 506–514.

Felsenstein, J. (2004) *Inferring Phylogenies*. Sinauer Associates, Sunderland, MA.

Fensome, R.A., Saldarriage, J.F., and Taylor, F.J.R. (1999) Dinoflagellate phylogeny revisited: reconciling morphological and molecular based phylogenies. *Grana*, 38: 66–80.

Finlay, B.J., Monaghan, E.B., and Maberly, S.C. (2002) Hypothesis: the rate and scale of dispersal of freshwater diatom species is a function of their global abundance. *Protist*, 153: 261–273.

Freshwater, D.W., Fredericq, S., Butler, B.S., Hommersand, M.H., and Chase, M.W. (1994) A gene phylogeny of the red algae (Rhodophyta) based on the plastid *rbc*L. *Proceedings of the National Academy of Sciences USA*, 91: 7281–7285.

Gibson, G. (2002) Microarrays in ecology and evolution: a preview. *Molecular Ecology*, 11: 17–24.

Guillou, L., Chretiennot-Dinet, M.-J., Medlin, L.K., Claustre, H., Loiseaux-de Goer, S., and Vaulot, D. (1999) *Bolidomonas:* a new genus with two species belonging to a new algal class, the Bolidophyceae (Heterokonta). *Journal of Phycology*, 35: 368–381.

Hayden, H.S., Blomster, J., Maggs, C.A., Silva, P.C., Stanhope, M.J., and Waaland, J.R. (2003) Linnaeus was right all along: *Ulva* and *Enteromorpha* are not distinct genera. *European Journal of Phycology*, 38: 277–294.

Hillis, D.M., Moritz, C., and Mable, B.K. (1996) *Molecular Systematics* (2nd ed.), Sinauer Associates, Sunderland, MA.

Hillis, L.W., Engman, J.A., and Kooistra, W.H.C.F. (1998) Morphological and molecular phylogenies of *Halimeda* (Chlorophyta, Bryopsidales) identify three evolutionary lineages. *Journal of Phycology*, 34: 669–671.

Hull, D.L. (1988) *Science as a Process*. University of Chicago Press, Chicago, IL.

Huss, V.A.R. and Sogin, M.L. (1990) Phylogenetic position of some *Chlorella* species with the Chlorococcales based upon complete small-subunit ribosomal RNA sequences. *Journal of Molecular Evolution*, 31: 432–442.

Iglesias-Rodriguez, M.D., Schofield, O.M., Batley, P.J., Medlin, L.K, and Hayes P.K. (2006) Extensive intraspecific genetic diversity in the marine coccolithophore *Emiliania huxleyi*: the use of microsatellite analysis in marine phytoplankton population studies. *Journal of Phycology*, 42: 526–536.

John, U., Fensome, R.A., and Medlin, L.K. (2003) The application of a molecular clock to explain the biogeographic distribution of the dinoflagellate genus *Alexandrium*. *Molecular Biology and Evolution*, 20: 1015–1027.

John, U., Groben, R., Beszteri, B., and Medlin, L.K. (2004b) Utility of amplified fragment length polymorphisms (AFLPs) to analyse genetic structures within the *Alexandrium tamarense* species complex. *Protist*, 155: 169–179.

John, U., Mock, T., Valentin, K., Cembella, A.D., and Medlin, L.K. (2004a) Dinoflagellates come from outer space but haptophytes and diatoms do not. In *Harmful Algae (2002) Proceedings of the Xth International Conference on Harmful Algae* (eds K.A. Steidinger, J.H. Landsberg, C.R. Tomas, and G.A. Vargo), Florida Fish and Wildlife Conservation Commission and Intergovernmental Oceanographic Commission of UNESCO, pp. 428–431.

Kawachi, M., Inouye, I., Honda, D., et al. (2002) The Pinguiophyceae classis nova, a new class of chromophyte algae whose members produce large amounts of omega-3 fatty acids. *Phycological Research*, 50: 31–47.

Kingsley, M.T., Straub, T.M., Call, D.R., Daly, D.S., Wunschel, S.C., and Chandler, D.P. (2002) Fingerprinting closely related *Xanthomonas* pathovars with random nonamer oligonucleotide microarrays. *Applied and Environmental Microbiology*, 68: 6361–6370.

Knowlton, N. and Weigt, L.A. (1998) New dates and new rates for divergence across the Isthmus of Panama. *Proceedings of the Royal Society*, 265: 2257–2263.

Kooistra, W.H.C.F. and Medlin, L.K. (1996) Evolutionary of the diatoms (Bacillariophyta): IV. A reconstruction of their age from small subunit rRNA coding regions and the fossil record. *Molecular Phylogenetics and Evolution*, 6: 391–407.

Lane, D.J., Pace, B., Olsen, G.J., Stahl, D.A., Sogin, M.L., and Pace, N.R. (1985) Rapid determination of 16S ribosomal RNA sequences for phylogenetic analyses. *Proceedings of the National Academy of Sciences USA*, 82: 6955–6959.

Lange, M., Chen, Y.-Q., and Medlin, L.K. (2002) Molecular genetic delineation of *Phaeocystis* species (Prymnesiophyceae) using coding and non-coding regions of nuclear and plastid genomes. *European Journal of Phycology*, 37: 77–92.

Laviolette, J.P., Ardlie, K., Reich, D.E., Robinson, E., Sklar, P., Shah, N., Thomas, D., Fan, J.B., Gingeras, T., Warrington, J., Patil, N., Hudson, T.J., and Lander, E.S. (2000) Large-scale discovery and genotyping of single-nucleotide polymorphisms in the mouse. *Nature Genetics*, 24: 381–386.

Lidie, K.B., Ryan, J.C., Barbier, M., and Van Dolah, F.M. (2005) Gene expression in Florida red tide dinoflagellate *Karenia brevis*: analysis of an expressed sequence tag library and development of DNA microarray. *Marine Biotechnology*, 7: 481–493.

Lindblad-Toh, K., Winchester, E., Daly, M.J., Wang, D.G., Hirschhorn, J.N., Laviolette, J.P., Ardlie, K., Reich, D.E., Robinson, E., Sklar, P., Shah, N., Thomas, D., Fan, J.B., Gingeras, T., Warrington, J., Patil, N., Hudson, T.J., and Lander, E.S. (2000) Large-scale discovery and genotyping of single-nucleotide polymorphisms in the mouse. *Nature Genetics*, 24, 381–386.

Medlin, L.K., Barker, G.L.A., Baumann, M., Hayes, P.K., and Lange, M. (1994) Molecular biology and the systematics of the Prymnesiophyta. In *The Haptophyte Algae* (eds. J.C. Green and B.S.C. Leadbeatter), Systematics Association Special Vol. 51, Oxford University Press, Clarendon, pp. 393–412.

Medlin, L., Elwood, H.J., Stickel, S., and Sogin, M.L. (1988) The characterization of enzymatically amplified eukaryotic 16S-like rRNA coding regions. *Gene*, 71: 491–499.

Medlin, L.K., Elwood, H.J., Stickel, S., and Sogin, M.L. (1991) Genetic and morphological variation within the diatom *Skeletonema costatum* (Bacillariophyta): evidence for a new species, *Skeletonema pseudocostatum*. *Journal of Phycology*, 27: 514–524.

Medlin, L.K., Gersonde, R., Kooistra, W.H.C.F., and Wellbrock, U. (1996) Evolution of the diatoms (Bacillariophyta): II. Nuclear-encoded small-subunit rRNA sequences comparisons confirm a paraphyletic origin for the centric diatoms. *Molecular Biology and Evolution*, 13: 67–75.

Medlin, L.K. and Kaczmarska, I. (2004) Evolution of the diatoms: V. Morphological and cytological support for the major clades and a taxonomic revision. *Phycologia*, 43: 245–270.

Medlin, L.K., Metfies, K., Mehl, H., Wiltshire, K., and Valentin, K., (2006) Picoplankton diversity at the Helgoland Time Series site as assessed by three molecular methods. *Molecular Ecology*, 167: 1432–1451.

Medlin, L.K., Williams, D.M., and Sims, P.A. (1993) The evolution of the diatoms (Bacillariophyta): I. Origin of the group and assessment of the monophyly of its major divisions. *European Journal of Phycology*, 28: 261–275.

Metfies, K. and Medlin, L.K. (2004) Microarrays—the fluorescent wave of the future. *Nova Hedwigia*, 79: 321–327.

Mock, T. and Valentin, K. (2004) Photosynthesis and cold acclimation: molecular evidence from a polar diatom. *Journal of Phycology*, 40: 732–741.

Moestrup, Ø. (1991) Further studies of presumedly primitive green algae, including the description of Pedinophyceae class. nov. and *Resultor* gen. nov. *Journal of Phycology*, 27: 119–133.

Moritz, C. and Cicero, C. (2004) DNA barcoding: promises and pitfalls. *Public Library of Science Biology*, 2: 1529–1531.

Nozaski, H., Matsusaki, M., Takahara, M., Misumi, O., Kuroiwa, H., Hasegawa, I.T., Shin, Y., Kohara, N., Ogagsawara, N., and Kuroiwwa, T. (2003) The phylogenetic position of red algae revealed by multiple nuclear genes from mitochondria-containing eukaryotes and an alternative hypothesis on the origin of plastids. *Journal of Molecular Evolution*, 56: 485–497.

Olsen, J.L. (1990) Nucleic acids in algal systematics. *Journal of Phycology*, 26: 209–214.

Olsen, J.L., Stam, W.T., Berger, S., and Menzel, D. (1994) 18S rDNA and evolution in the Dasycladales (Chlorophyta): modern living fossils. *Journal of Phycology*, 30: 729–744.

Palumbi, S.R. (1994) Genetic divergence, reproductive isolation, and marine speciation. *Annual Review of Ecology and Systematics*, 25: 547–572.

Peijnenburg, K.T.C.A., Fauvelot, C., Breeuwer, J.A.J., and Menkine, S.B.J. (2005) Population structure at different spatial scales of the planktonic *Sagitta setosa* (Chaetognatha) in European seas, as revealed by mitochondrial and nuclear DNA. In *Structure at Open Sea: Genetics of Zooplankton Populations*. PhD thesis of Peijnenburg, University of Amsterdam, 2005.

Phillippe, H., Delsuc, F., Brinkmann, H., and Lartillot, N. (2005) Phylogneomics. *Annual Review of Ecology and Systematics*, 36: 541–562.

Pickett-Heaps, J.D. (1975) *The Green Algae: Structure, Reproduction and Evolution of Selected Genera*. Sinauer Associates, Sunderland, MA.

Pueschel, C.M. (1980) A reappraisal of the cytochemical properties of rhodophycean pit plugs. *Phycologia*, 19: 210–217.

Ragan, M.A., Bird, C.J., Ricc, E.L., Gutell, R.R., Murphy, C.A., and Singh, R.K. (1994) A molecular phylogeny of the marine red algae (Rhodophyta) based on the nuclear small-subunit rRNA gene. *Proceedings of the National Academy of Sciences USA*, 91: 7276–7280.

Raitio, M., Lindroos, K., Laukkanen, M., Pastinen, T., Sistonen, P., Sajantila, A., and Syvanen, A.C. (2001) Y-chromosomal SNPs in Finno-Ugric-speaking populations analyzed by minisequencing on microarrays. *Genome Research*, 11: 471–482.

Rokas, A., Williams, B.L., King, N., and Carroll, S.B. (2003) Genome-scale approaches to resolving incongruence in molecular phylogenies. *Nature*, 425: 798–804.

Round, F.E. and Crawford, R.M. (1981) The lines of evolution of the Bacillariophyta. I. Origin. *Proceedings of the Royal Society B*, 211: 237–260.

Rousseau, F., Burrowes, R., Peters, A.F., Kuhlenkamp, R., and De Reviers, B. (2001) A comprehensive phylogeny of the Phaeophyceae based on nrDNA sequences resolves the earliest divergences. *Compte Rendus des Academies Sciences Paris, Sciences de la Vie*, 324: 1–15.

Rudd, S. (2003) Expressed sequence tags: alternative or complement to whole genome sequences? *Trends in Plant Sciences*, 8: 321–329.

Sáez, A.G., Probert, I., Geisen, M., Quinn, P., Young, J.R., and Medlin, L.K. (2003) Pseudo-cryptic speciation in Coccolithophores. *Proceedings of the National Academy of Sciences USA*, 100: 6893–7418.

Saiko, R., Gelfand, D.H., Stoffel, A., Scharf, S.J., Higuchi, R., Horn, G.T., Mullis, K.B., and Erlich, H.A. (1988) Primer directed enzymatic amplification of DNA with a termostable DNA polymerase. *Science*, 239: 487–491.

Saunders, G.W. and Bailey, J.C. (1997) Phylogenies of pig-plug–associated feature in the Rhodophyta: inferences from molecular systematic data. *Canadian Journal of Botany*, 75: 1436–1447.

Saunders, G.W. and Druehl, L.D. (1992) Nucleotide sequences of the small-subunit ribosomal RNA genes from selected Laminariales (Phaeophyta): implications for kelp evolution. *Journal of Phycology*, 28: 544–549.

Saunders, G.W. and Kraft, G.T. (1997) A molecular perspective on red algal evolution: focus on the Florideophycidae *Plant Systematics and Evolution*, 11: 115–138.

Saunderson, M. (2006) r8s version 1.71. *Estimating rates of molecular evolution.* (http://ginger.ucdavis.edu/r8s).

Shields, R. (2004) Pushing the envelope on molecular dating. *Trends Genetics*, 20: 221–222.

Snyder, V.R., Guerrero, M.A., Sinigalliano, C.D., Winshell, J., Perez, R., Lopez, J.V., and Rein, K.S. (2005) Localization of polyketide synthase encoding genes to the toxic dinoflagellate *Karenia brevis*. *Phytochemistry*, 66: 1767–1780.

Soares, C. (2004) What's in a name? *Scientific American*, 291: 36–38.

Strate, H.J. vand der, Zande, L. van de, Stam, W.T., and Olsen, J.L. (2002) The contribution of haploids, diploids and clones to fine-scale population structure in the seaweed *Cladophoropsis membranacea* (Chlorophyta). *Molecular Ecology*, 11: 329–345.

Swofford, D.L. (2002) *PAUP*, Phylogenetic Analysis Using Parsimony,* Version 4.0 Beta version 8, program and documentation. Illinois Natural History Survey, University of Illinois, Champaign, IL.

Tan, I. and Druehl, L.D. (1993) Phylogeny of the north east Pacific brown algal (Phaeophycean) orders as inferred from 18S rRNA gene sequences. *Hydrobiologia*, 260/261: 699–704.

Venter, J.C., Remington, K., Heidelberg, J.F., Halpern, A.L., Rusch, D., Eisen, J.A., Dongying Wu, D., Paulsen, I., Nelson, K.E., Nelson, W., Fouts, D.E., Levy, S., Knap, A.H., Lomas, M.W., Nealson, K., White, O., Peterson, J., Hoffman, J., Parsons, R., Baden-Tillson, H., Pfannkoch, C., Rogers, Y., and Smith, H.O. (2004) Environmental genome shotgun sequencing of the Sargasso Sea. *Science*, 304: 66–74.

Vysotskaia, V.S., Curtis, D.E., Voinov, A.V., Kathir, P., Silflow, C.D., and Lefebvre, P.A. (2001) Development and characterization of genome-wide single nucleotide polymorphism markers in the green alga *Chlamydomonas reinhardtii*. *Plant Physiology*, 127: 386–389.

Wang, D.G., Fan, J.B., Siao, C.J., Berno, A., Young, P., Sapolsky, R., Ghandour, G., Perkins, N., Winchester, E., Spencer, J., Kruglyak, L., Stein, L., Hsie, L., Topaloglou, T., Hubbell, E., Robinson, E., Mittmann, M., Morris, M.S., Shen, N., Kilburn, D., Rioux, J., Nusbaum, C., Rozen, S., Hudson, T.J., Lipshutz, R., Chee, M., and Lander, E.S. (1998) Large-scale identification, mapping, and genotyping of single-nucleotide polymorphisms in the human genome. *Science*, 280: 1077–1082.

White, T.J., Bruns, T.D., Lee, S.B., and Taylor, J.W. (1990) Analysis of phylogenetic relationships by amplification and direct sequencing of ribosomal genes. In *PCR Protocols* (eds. M.A. Innis, D.H. Glefand, J.J. Sninsky, and T.J. While), Academic Press, New York, pp. 315–322.

Wilcox, L.W., Fuerst, P.A., and Floyd, G.L. (1993) Phylogenetic relationships of four charophycean green algae inferred from complete nuclear-encoded small subunit rRNA gene sequences. *American Journal of Botany*, 80: 1028–1033.

Woese, C.R. (1987) Bacterial evolution. *Microbiological Reviews*, 5: 221–271.

Yoon, H.S., Hackett, J.D., Cingula, C., Pinto, G., and Bhattacharya, D. (2004) A timeline for the origin of photosynthetic eukaryotes. *Molecular Biology and Evolution*, 21: 809–818.

Zechman, F.W., Theriot, E.C., Zimmer, E.A., and Chapman, R.L. (1990) Phylogeny of the Ulvophyceae (Chlrophyta): cladistic analysis of nuclear-encoded rRNA sequence data. *Journal of Phycology*, 26: 700–710.

Glossary

Akinete   Non-motile asexual spore, with a wall incorporated in that of the parent cell. In cyanobacteria a thick walled resting spore.

Allopatric   Two or more species evolving in different places and geographically isolated from each other.

Allozyme   Form of enzyme which has differing amino acid sequence to others of the same enzyme, encoded by one allele at a single locus.

Amoeboid   Movement of cells, often in naked form, where the cell shape changes frequently and dramatically.

Amphiesma   Outer wall of cell (of dinoflagellates).

Anisogamous   Having gametes of unequal sizes.

Anisokont   Having flagella of different lengths or structure (e.g. one smooth, one with scales).

Antheridia   Organs producing male gametes.

Aplastidic   Without plastids.

Apomorphic   A new feature arising through evolution.

Araphid   Lacking a raphe.

Athecate   Without a wall element (for example, in dinoflagellates the lack of thecal plates).

Autapomorphic/y   Feature unique to clade.

Autolysins   Enzymes that hydrolyse cells in which they are produced.

Autospores   Spore without motility with the same shape as parent cell.

Auxiliary cell   In sexual reproduction in the Rhodophyceae the auxiliary cell receives the diploid nucleus from fertilisation.

Auxospores   Diatom cell resulting from gametic fusion or autogamy (i.e. zygote), or (more rarely) one formed by vegetative processes, which swells to maximum size.

Auxosporulation   The formation of an auxospore by gametic fusion or autogamy.

Axenic   Contains only one species of microorganism.

Axoneme   Microtubule arrangement of flagella consisting of two central with nine surrounding doublets.

Axopodium (pl. axopodia)   A particular type of pseudopodial structure of amoeboid protists where the cell extension (pseudopod) is supported by microtubules.

Bayesian analysis   Analysis where evidence is used to assess a hypothesis.

Bikont   Eukaryotic cell possessing two flagella.

Biological species concept   A concept whereby species are groups of interbreeding natural populations that are reproductively isolated from other such groups.

Cabozoan theory   Theory that the Rhizara and Excavata are sister groups and their common ancestor enslaved a single green algal symbiont, which was ancestral to both chlorarachnean and euglenoid pastids.

Carpogonium (pl. carpogonia)   Female sexual cell, consisting of a basal portion, the contents of which function as a gamete, and an elongated apical receptive portion, the trichogyne.

Carpospore	The diploid spore produced in the Rhodophyceae.
Carposporophyte	The diploid generation in the Rhodophyceae (derived from the zygote) which produces carpospores.
CBC clade	Group of organisms with no compensating two-sided base pair changes in ITS and are thus sexually compatible.
Cenancestral	Pertaining to last common ancestor.
Chimaera	Organism with material of two genetic types, the result of fusion of two different organisms.
Chloroplast	Membrane bound organelle, known also as a plastid, which contains chlorophyll.
Chromalveolate hypothesis	Postulate that the chlorophyll *c* containing algae and protists share a common red algal secondary endosymbiont.
Cilium (pl. cilia)	A form of flagellum composed of a central bundle of microtubules surrounded by cytoplasm and bounded by the outer surface of the cell membrane.
Cladistics	Determination of evolutionary relationships through analysis of their shared characteristics.
Cladogenesis	Mode of evolution where each branch of the tree is a clade.
Coccoid	Spherical in shape.
Coccoliths	Calcareous scale of Prymnesiophyceae.
***Codiolum* stage**	A stage in a type of life history that is frequent in green algae where the unicellular zygote resembles species of *Codiolum*.
Concatenated(ation)	Joining of items.
Conceptacles	Cavity in algae where the structures forming gametes (gametangia) are produced.
Conchocelis	A morphological phase in the life history consisting of endolithic filaments; confined to the Bangiophycidae in red algae.
Conjugation	Fusion of two cells to exchange genetic information.
Convergent evolution	Independent development of similar morphological features in differing species.
Cortical alveoli	Flattened vesicle-like structures, located underneath the plasma membrane in the alveolates.
Cortication	Having a specialised outer wall layer or cortex.
cox1	Cyclooxygenase-1, protein that acts as an enzyme to speed up the production of certain chemical messengers.
Cristae	Chambers formed by folding of inner membrane of mitochondria.
Crown groups	Group that includes all living members.
Cryptic species	Species that cannot be distinguished using morphological criteria but are nevertheless distinguishable using other criteria (notably sequence analysis).
Cryptist	Of Cryptophyceae.
c-type chlorophyll	A form of the green pigment chlorophyll.
Cyanelle	Endosymbiotic cyanobacterium.
Cyanophage	Virus affecting cyanobacteria.
Cystocarp	The carposporophyte and surrounding reproductive tissue in the Rhodophyceae.
Cytokinesis	Division of cell into two cells.
Cytology	The study of the structure and function of cells.
Cytosolic	Of the cytoplasm (cytosol) of a cell.

Deme	A locally interbreeding population.
Desmotubule	Cylindrical membrane that runs through the centre of a plasmodesma connecting adjacent cells.
Diadinoxanthin	Xanthophyll pigment found in several algal groups.
Diatomics	Study of diatoms.
Diatoxanthin	Xanthophyll pigment found in a range of algae including the diatoms.
Diazotrophic	Nitrogen-fixing organism using nitrogenase to reduce N_2 to ammonia.
Dimorphism	Having two shapes.
Dinosporin	Type of sporopollenin produced in the wall of dinoflagellate cysts.
Dioecious	Male and female gametangia on separate plants/thalli.
Diploid	With two complete sets of chromosomes (genetic material).
Diplontic	The diploid phase of the life cycle.
DNA barcoding	Use of short diagnostic sequences of DNA to distinguish between species.
Ecotypes	Group of individuals within a species adapted genetically to a particular habitat but able to cross freely with other ecotypes of the same species.
Ejectosomes	Cells able to discharge contents.
Embryophytes	Multicellular mostly photosynthetic eukaryotes possessing specialist reproductive organs (trees, flowers, ferns and mosses).
Endocytobiosis	Symbiosis where symbiont dwells within host cell.
Endophytic	Living within a plant but not parasitic.
Endosymbiosis	Where one organism lives within another.
Eobacteria	One of three families proposed for all living things by Woese, Eobacteria encompasses normal bacteria (as opposed to Archaea or Eukarya).
Epicone	The upper part of a cell. Above the girdle in the dinoflagellates and diatoms.
Epilithic	Living on the surface of rocks or stones.
Epipelic	Living on mud.
Epiphytic	Living on a plant (alga) but attached to the surface only.
Epitype	Specimen or illustration selected to serve as an interpretative type when the holotype, lectotype or previously designated neotype, or all original material associated with a validly published name, is demonstrably ambiguous and cannot be critically identified for purposes of the precise application of the name of a taxon.
Exosymbiont	Symbiont that lives outside the host tissue; for example in internal cavities.
Fibula	A bridge of silica subtending the raphe system; restricted to some diatom genera.
Filopodium	Protoplasmic thread emerging from cell surface.
Flagellum	Whip-like organelle of many single celled organisms. Composed of a central bundle of microtubules surrounded by cytoplasm and bounded by the outer surface of the cell membrane.
Form genera	Artificial taxon for morphologically similar organisms where true relationships are obscure.
Fucoxanthin	Xanthophyll pigment found in a variety of algae, notably the Phaeophyceae.

Fultoportulae (**strutted process**) A tube penetrating the silica wall of a diatom, supported internally by two or more buttresses; the external opening may or may not project above the valve surface, but is usually simple.

Gametophyte Morphological thallus, either haploid or diploid, which bears the reproductive organs or gametangia.

Gene tree Diagrammatic depiction of inferred historical relationships of genes or groups of genes.

Genomics Study of genes and their function.

Glycobacteria Informal (paraphyletic) group of bacteria including Cyanobacteria, Eurybacteria and Gracillicutes.

Glyoxysomes Microbodies or small organelles, bounded by a single membrane, (0.5–1.0 mm in diameter) containing enzymes of the glyoxylate cycle.

Gonimoblasts In the Rhodophyceae the cells that form the carposporangia.

Gyrogonites Sporollenin-encrusted spores in stoneworts.

Haploid Cells with one set of chromosomes.

Haplostichous Single row of cells to each branchlet or bract cell of *Chara*.

Haplotype Haploid genotype of a chromosome or genome.

Haptonema Specialised appendage located between the flagella in prymnesiophytes.

Haptophyte-type **chloroplast** Plastid derived from a tertiary endosymbiosis.

Helotism Where one species works for another (dominant) species.

Hennigian Theory Theory of cladistics.

Heterococcolithophorid Bearing heterococcoliths—coccoliths composed of differing elements—calcium carbonate is deposited in a uniform way but continues to grow.

Heterocyst Thick walled enlarged cell in cyanobacteria.

Heterokont With different kinds of flagella.

Heteromorphic Refers to a life history consisting of reproductive phases with different morphologies.

Heterothallic With different male and female gametes produced on different plants.

Heterotrophic Organisms that use organic compounds as sources of nutrition.

Heterozygous With at least two versions of at least one gene.

Holococcolithophorid Bearing holococcoliths—coccoliths composed of identical elements—calcium carbonate is deposited as uniform crystals.

Holophyletic/y Where all the descendents of a common ancestor are linked in one lineage.

Homologous/y Sharing the same origin.

Homoplasy Phenomena leading to similarity in character states for reasons other than inheritance from a common ancestor.

Homothallic Producing male and female gametes on the same thallus. Able to self fertilize.

Homozygous With two identical genes.

Hormogonia Pieces of trichome that detach from parent filament in cyanobacteria and develop into new filaments.

Hypnospore Spore with a greatly thickened wall.

Hypnozygote	A thick walled zygote produced as a result of fusion of two motile gametes (2N).
Incongruence	Phylogenetic trees produced using differing genes that do not match.
Isogamous/y	With male and female gametes of the same size.
Isomorphic	Refers to a life history consisting of reproductive phases with similar morphologies.
Isozyme	Enzymes that differ with different amino acid sequence but catalyse the same reaction encoded by different genetic loci.
Kleptidoplasty	Process where plastids from consumed algae are retained by host organism, they are maintained and continue to fix carbon in this state.
Laminarin	Storage polysaccharide in Phaeophyceae.
Lentic	Living in still freshwater systems (e.g. lakes).
Leucoplasts	Plastids without pigments.
Long Branch Attraction	Where rapidly evolving species are interpreted (in a phylogenetic tree) as being closely related despite their true evolutionary relationship.
Lorica (pl. loricae)	Outer shell of organism not attached to protoplast, notably in Chrysophytes.
Lotic	Living in flowing freshwater (e.g. streams).
Mastigonemes	Hair on the surface of flagella.
Maximum Likelihood Analysis	Probabilistic statistical technique used to infer parameters of distribution from a data set.
Meiosporangia	Structures where the spores have been produced by meiosis.
Meiospores	Spore formed through meiosis.
Mesospore	Middle layer of zygospore.
Meta-algae	Eukaryote-eukaryote chimaera.
Metaboly	Motion of cell produced through shape-changing contraction as found in some euglenids.
Microsatellite	Short (2–6 nucleotides) DNA sequence, tandemly repeated; found at many loci in the genome.
Microspecies	Where species reproduce asexually with modified meiosis giving rise to groups which differ only slightly to each other.
Mixotrophic	Photosynthetic organism also capable of taking up organic material.
Minicircles	Circular DNA found within dinoflagellate chloroplasts each encoding a single gene.
Monadoid species TP	Species of unicellular habit.
Monoecious	With male and female gametangia on the same plant (alga).
Monophyletic/y	Derived from a common ancestor; taxa derived from and including a single founder species.
Monosporangium	Spore producing structure that forms monospores in Rhodophyceae.
Monospore	Asexual spore in Rhodophyceae.
Monotypic	Of one type.
Morphological species concept	Species that appear identical based on morphological criteria.
Morphospecies	Species determined by morphological characteristics.
Mutualism	Where two species coexist and neither is harmed but both derive benefit.

Myzocytosis   Feeding by piercing cell wall of another organism and withdrawing contents.

Nanofabrication   In algae—the construction of calcite or silicate elements through particle assembly.

Nanoplankton   Plankton passing through a 20 μm filter but retained on a 2 μm filter.

Negibacteria   One of two subkingdoms of eubacteria.

Neighbour Joining Analysis   Method of constructing phylogenetic trees based on iterative linking of least distant nodes in a phylogenetic cluster along with their common ancestor.

Nucleomorph   Relict of the nucleus of a eukaryotic symbiont.

Ontogeny   Development of a zygote to adult.

Oogamous/y   With a small male gamete and large female gamete.

Oogonia   Organs producing female gametes.

Orthologs   Genes derived from same ancestral locus.

Osmotrophic   Feeding by absorption across cell wall.

Palaeoecology   Study of past ecology through the fossil record.

Pallial feeding   Feeding behaviour where a pallium (structure resembling a cloak) is used to engulf and digest prey.

Palmelloid   Mucilaginous matrix containing single non-motile cells.

Panmictic   Population that mates randomly.

Parallel evolution   Independent evolution of similar character states.

Paralogs   Genes that have been duplicated and then diverge from the ancestral condition.

Paralogy   Relationship between genes that have been derived through duplication.

Paraphyletic/y   Groups derived from a single ancestral species but which do not contain all the descendants of that ancestor.

Paraphysis (pl. paraphyses)   A sterile filament or cell usually associated with reproductive organs (sporangia or gametangia).

Parasitism   Two organisms living together but where one causes harm to the other.

Parenchymatous   With cells that divide in more than one direction.

Parietal chloroplasts Chloroplasts that are positioned peripheral to the cell, lying along the wall.

Parsimony Analysis   Analysis that derives the simplest way to explain relationships.

Pellicle   Outer wall of Euglenophyceae consisting of proteinaceous material.

Peridinin-type chloroplast   Plastids found in dinoflagellates containing peridinin pigments.

Periplast   Cell wall.

Perizonium   The wall of the auxospore, which usually comprises delicate siliceous scales (in radially symmetrical centric diatoms), or closed or split siliceous bands (in pennate diatoms).

Peroxisome   Organelle bound by a single membrane which contains enzymes that catalyse the production and breakdown of hydrogen peroxide.

Phagocytosis   Ingestion of particles in food vesicle or phagocyte.

Phagotrophy/ic   Mode of nutrition by ingestion of solid food particles as above.

Phenetic   Observable traits of a species.

Phenotypic	Net result of genotypic and environmental variance.
Phototrophic	Obtaining nutrition through light.
Phragmoplast	Wall formed by joining of golgi vesicles between spindle microtubules, found in the Chlorophyceae and land plants.
Phycoplast	Microtubular structure formed perpendicular to the spindle where the division of the cells takes place.
Phycobilisomes	Found on thylakoid membranes, protein complexes that are the light harvesting mechanism for photosystem II.
Phylocode	Set of rules (under development) governing phylogenetic nomenclature. Principally concerned with regulating the naming of clades.
Phylogenomic	Phylogentic analysis of protein-encoding genes that are functionally linked.
Physodes	Vesicle in the Phaeophyceae containing tannins.
Picoeukaryotes	Eukaryotes of less than 2μm in size.
Picoplankton	Plankton of very small size, generally defined as passing through a 2μm filter but being retained on a 0.2μm filter.
Pit connection	A small pore between two adjacent cells occluded by a pit plug; confined to the Rhodophyceae.
Pit plug	Occlusion of aperture between two adjacent cells in the Rhodophyceae—can have complex structure.
Placoderm desmids	True desmids composed of two halves separated by a suture or extended constriction (isthmus).
Plasmodesmata	Singular plasmodesma. Cytoplasmic threads that connect adjacent cells through openings in cell walls.
Plastid	Organelle containing photosynthetic structures.
Plesiomorphic	Ancestral character.
Plurilocular	Containing many loculi (the individual compartments of plurilocular sporangium) said of reproductive organs.
Polymorphism	Ability to change shape.
Polyphasic	With more than one stage in the life cycle or to use multiple approaches to classify an organism.
Polyphyletic/y	Group having origin in several different lines of descent.
Polystichous	Truly parenchymatous structures formed cell divisions in various planes.
Polytomies	Nodes of a tree with more than two descendants.
Posibacteria	One of the two subkingdoms of eubacteria.
Properizonium	Band-like elements of expanding auxospores in diatoms.
Protandric oogamy	Having male organs early in life history and female later.
Protist	An organism belonging to the protista, includes bacteria, protozoans, single celled algae and fungi.
Protoplast	The material of a cell without its wall.
Pseudo-cryptic species	Species that were morphologically indistinguishable (cryptic) but shown to be genetically separated, for which subsequent further investigation has lead to morphological separation.
Pseudopod/ia	Temporary projections of typically amoeboid cells, can be used for locomotion.
Pseudo-raphe	Area on the frustule of a diatom which is free of ornament where the raphe would have occurred had it been present.

Pyrenoid	Area of a chloroplast where material is stored.
Raphe	Slit in the frustule of some diatoms.
Raphid	With a raphe.
Retronemes	Rigid, thrust-reversing, tubular, ciliary hairs.
Rhizostyle	Flagellar root that may be contractile.
Rimoportula (labiate process)	A tube penetrating the silica wall of a diatom, opening to the inside by a slot, and to the outside by a simple aperture.
Rubisco	Ribulose 1,5-bisposphate carboxylase oxygenase; the enzyme which catalyzes the reaction that fixes CO_2 into carbohydrates and is located in the stromal space of the plastid; encoded by *rbc*L and *rbc*S genes.
Sarcinoid	Small numbers of cells grouped together.
Semicryptic	Morphologically similar demes that can only be distinguished using geographic origin.
Sibling species	Two or more species that are morphologically similar but not able to interbreed.
Siphonocladous algae	With thalli composed of multinucleate cells.
Somatonemes	Tubula hair-like appendages found on the surface of some protists.
Sori	Surface organs where sporangia are produced.
Speciation	Formation of new species by genetic separation.
Sporangium (pl. sporangia)	Organ where spores are produced by meiosis.
Sporangium autolysin	Enzyme that dissolves the cell wall of sporangia.
Sporophyte	The diploid generation bearing spores.
Sporopollenin	Highly resistant carotenoid based material, forming the walls of pollen grains, fungal spores and a variety of algal resting stages.
Sporozoa	Phylum of unicellular heterotrophic protists without flagellate stages (with the exception of the male gametes).
ß-tubulin	Subunit of the dimeric protein polymer that microtubules are composed of.
Sternum	Thickened rib orientated apically on a pennate diatom, may bear the raphe or not when it might be termed a pseudoraphe.
Stoma	A pore.
Stramenopiles	Very diverse group of protists linked by having tubular cristae in their mitochondria.
Subaeriel algae	Algae that grow in the open air living on the surface of the substrate.
Symbiogenesis	Merging of two different organisms to form a single new organism.
Symbiont	Organism living together with another organism of a different species in a mutually beneficial association.
Symbiosis	Condition where two different organisms live together.
Sympatric	Two or more species evolving in the same place.
Symplesiomorphic/y	A trait shared in two or more taxa also shared with other taxa with an earlier common ancestor.
Synapomorphic/y	A derived character state shared by two or more taxa, inherited from their most recent ancestor.
Syngamy	Reproduction through the fusion of gametes.
Systematics	Study of the classification of organisms.
Taxon	A group defined through taxonomy.

Tetrasporangia	Structure producing tetraspores.
Tetraspores	Asexual spores produced in fours in tetrasporangia.
Thallus	Plant (alga) of comparatively simple organisation.
Thecate (ae)	Possesses a theca or rigid cell covering (in diatoms a silica frustule, in armoured dinoflagellates cellulose plates).
Thylakoid	Membrane-bound phospholipid bi-layer found within chloroplasts.
Transcriptome	Expressed genome.
Transposons	Sequences of DNA that can move around within the genome of a single cell, also referred to as 'jumping genes'.
Trichocysts	Organelles containing threads that can be ejected by exocytosis.
Trichomes	Hair on the surface of a plant (alga).
Type species	The first species to be described from a particular genus.
Unikont	Eukaryote possessing a single flagellum.
Unilocular	Containing a single loculus or cell, said of reproductive organs. See also entry for plurilocular.
Unilocular sporangia	Spore producing structures with a single cavity.
Vestibulum	Depression or cavity from which flagella emerges in cryptophytes.
Z clade	Group of organisms producing viable zygotes.
Zoidangia	Zoid producing structure.
Zoosporangia	Zoospore producing structure.
Zoospores	Flagellated spore.
Zygote	Diploid cell produced from the fusion of two haploid gametes.

Index

A

Achnanthes, raphe origination and, 60–61
Actinocyclus, uniflagellate sperm of, 233
Agardh, Carl Adolph, 61, 62, 63
Aglaothamnion, rhodobindin gamete recognition proteins in, 116
Akashiwo sanguinea, 8
Akkesiphycaceae, creation of, 275
Algae
  breakthroughs in study of, 1
  cell membrane organization, 21
  chromophyte, 21
  definition of, 22
  description of, 1
  difficulty of defining, 2
  drier, 9
  eukaryotic, genome sequencing of, 331, 332
  first, 22
  as functional group of organisms, 49
  historical context, 22, 24–25
  importance of, 2
  major groups with secondary photosynthetic plastids, 13
  monophyly versus polyphyly in, 2
  origins of, 2
  phylo-genomics in, 347
  relationships of, *see* Evolution and relationships of algae
  secondary, 172
Algal molecular systematics, 341–353
  barcoding, 345
  comparative genomics, 347
  cryptic species, paraphyly, and nomenclature, 343–344
  ecological genomics, 347
  incongruence among phylogenies, 345
  microarray fingerprinting, 347–348
  molecular clocks, 345–346
  morphology vs. molecules debate, 342
  phylogenetics at the species and population levels, 346–347
  selection, 348
  21st century systematist, 348–349
  ultrastructure, 343
Algal photosynthetic unity, 48
Algal systematics, progress in, 4
Alveolates, photosynthetic diversity of, 42
Amnesic shellfish poisoning, 236
Amplified fragment length polymorphism, 124, 139, 164
Anabaena
  phenotypic characterisation of, 99
  taxonomic problems within, 98
  taxonomy of, 97–98
  toxins accumulated by, 96
Apical furrow system, 222, 223–224

B

Apicomplexa, non-photosynthetic plastids of, 14
Apicomplexans, photosynthetic, free-living, 44
Arabidopsis genome, 16
Astasia longa, osmotrophy in, 318
Aulacoseira, molecular systematic study of, 239, 241
Autolysin concept, 143

Bacillariales
  clade of raphid diatoms, 248
  molecular systematic study of, 239, 240, 241
Bacillariophytina, 76
Bacteria, enslaved, protein-coding genes of, 23
Bangiophytes
  ordinal classification of, 105
  pit-plug ultrastructure, 106
  Schmitz–Kylin classification, 105
Barcode of Life Data System, 116
Barcoding, 345
Benthic diatoms, diversification of, 256
Bergey's Manual of Systematic Bacteriology, 95
Biddulphia, 234
Biddulphiophycidae, 80, 81
Bigelowiella
  life cycle vegetative stage, 179
  longifila
    amoeboid cell in, 178
    life cycle of, schematic drawing of, 177
    mucilaginous colonies of, 173
  natans
    cell size, 173
    life cycle of, 177
    nucleomorph genome sequence, 175
Bigyra, 37
Bikont(s), 22
  diversification, 31
  pattern of ciliary transformation, 31
Blastodiniales, 223
Blastodinium, position in SSU trees, 223
Blue green algae, 7
Bonnemaisonia hamifera Hariot, life histories, 105
Bostrychia, cryptic diversity in, 112
Brown algae, *see also* Phaeophyceae, classification of
  accessory pigments, 269
  classes of, 267, 270
  classification, 270, 277–278
  definition of, 269
  DNA-sequence data of, 268
  evolution, reconstruction of, 272
  freshwater, 268
  heterokont flagellated cells of, 267
  tropical waters, 268
Buchnera, amino acids provided by, 23

Systematics Association
Publications

[*] Published by Oxford University Press for the Systematics Association

[†] Published by the Association (out of print)

SYSTEMATICS ASSOCIATION SPECIAL VOLUMES

[*] Published by Academic Press for the Systematics Association
[†] Published by the Palaeontological Association in conjunction with Systematics Association
[‡] Published by the Oxford University Press for the Systematics Association
[**] Published by Chapman & Hall for the Systematics Association